Advances in Food Processing and Preservation

Advances in Food Processing and Preservation

Editor: Sarah Scott

R CALLISTO REFERENCE

www.callistoreference.com

Callisto Reference,
118-35 Queens Blvd., Suite 400,
Forest Hills, NY 11375, USA

Visit us on the World Wide Web at:
www.callistoreference.com

ISBN: 978-1-64116-629-4 (Hardback)

Cataloging-in-Publication Data

Advances in food processing and preservation / edited by Sarah Scott.
      p. cm.
Includes bibliographical references and index.
ISBN 978-1-64116-629-4
1. Food industry and trade. 2. Food--Preservation. 3. Food preservatives. 4. Food processing plants.
5. Food. I. Scott, Sarah.
TP370 .A38 2022
664--dc23

Table of Contents

Permissions

List of Contributors

Index

Preface

Food processing is a process that transforms agricultural products into food or one kind of food to other forms. Grinding grain to make raw flour, home cooking to complex industrial methods that are used to make convenience food are some of the forms that are involved in food processing. There are three types of food processing methods - primary, secondary and tertiary. The aim of primary food processing is to make food edible, while secondary food processing deals with the conversion of the ingredients to familiar food. Food preservation is a method that prevents the growth of microorganisms. It also stops oxidation of fats that cause rancidity in food. It involves preventing processes which results in visual deterioration such as enzymatic browning in fruits and vegetables after being cut. Some of the methods for food preservation are cooling, freezing, pickling, boiling, pasteurization, vacuum packing, irradiation, etc. These methods help in maintaining or creating nutritional value, texture and flavor of the food. The topics covered in this extensive book deal with the core aspects of food processing and preservation. It discusses the fundamentals as well as modern methods of food processing and preservation. This book will provide comprehensive knowledge to the readers.

After months of intensive research and writing, this book is the end result of all who devoted their time and efforts in the initiation and progress of this book. It will surely be a source of reference in enhancing the required knowledge of the new developments in the area. During the course of developing this book, certain measures such as accuracy, authenticity and research focused analytical studies were given preference in order to produce a comprehensive book in the area of study.

This book would not have been possible without the efforts of the authors and the publisher. I extend my sincere thanks to them. Secondly, I express my gratitude to my family and well-wishers. And most importantly, I thank my students for constantly expressing their willingness and curiosity in enhancing their knowledge in the field, which encourages me to take up further research projects for the advancement of the area.

Editor

Effect of Freezing on the Shelf Life of Salmon

Paul Dawson ⓘ, Wesam Al-Jeddawi, and Nanne Remington

Department of Food, Nutrition and Packaging Sciences, Clemson University, Clemson, SC 29634, USA

Correspondence should be addressed to Paul Dawson; pdawson@clemson.edu

Academic Editor: Salam A. Ibrahim

Food shelf-life extension is important not only to food manufacturers, but also to home refrigeration/freezing appliance companies, whose products affect food quality and food waste. While freezing and refrigerating both extend the shelf life of foods, food quality deterioration continues regardless of the preservation method. This review article discusses the global fish market, the composition of fish meat, and the effects of freezing and thawing on salmon quality.

1. Introduction

Consumer preference is vital to sustaining any food commodity including the fish industry. Since 1961 international fish intake has increased as fast at 3.6% per year [1]. Whether locally caught and consumed fresh or distributed frozen, the health benefits of adding fish to the daily diet has impacted consumption [1]. Atlantic salmon (*Salmo Salar*) shows cardiovascular, cancer inhibiting, and joint health benefits mostly due to the presence of omega-3 long-chain fatty acids, eicosapentaenoic acid (EPA), and docosahexaenoic acid (DHA), some of which are essential and important nutrients for human body [2]. It is also a highly oily fish known to be low in mercury, like tuna, catfish, and cod [2]. Therefore, the quality of Atlantic salmon is important for palatability especially for frozen fish [3]. While freezing will slow the biological, chemical, and physical deterioration of food, degradation of food quality such as color, texture, enzymatic activity, lipid oxidation, and ice crystal structural damage still occur. Many researchers have reported that fast freezing results in rapid ice nucleation within the intracellular areas of food products creating smaller and more uniform ice crystals causing less structural damage to the product [4]. The denaturation of the protein is one problem caused by slow freezing with protein denaturation-dependent upon temperature [5]. Freezing at low temperatures provided small ice crystals which increased light scattering and absorption across all wavelengths in the visible region [6]. Several

researchers showed that freezing even in the short term changed physical properties such as weight loss, color, and texture of the Atlantic salmon and other types of fish [7]. Long-term frozen storage leads to slow deterioration in the quality of salmon which can differ due to the storage temperature. Biological and chemical reactions such as enzymatic activity and lipid oxidation have a significant impact on fish quality during long-term frozen storage [8, 9]. Lipid oxidation decreases the sensory quality of fish and fish products and is influenced by handling and processing of fish which can also impact nutritional quality, texture, and color [10, 11].

Freezing preserves quality allowing an expanded distribution range for raw fish; thus research into different freezing methods and their effect on quality is important to the seafood industry. Quality measures affected by freezing include changes in color, texture, water holding capacity, and intracellular/extracellular ice crystal growth effects on structure. Faster freezing rates maintained structural quality and lowered chemical activity of Atlantic sea bass, Atlantic salmon, hake, chicken, and other types of meat [12–18]. However, there was a threshold-freezing rate after which an increase in freezing rate did not impact overall product quality during storage [19]. Therefore, more research is needed to find a balance between maintaining food quality, while maximizing energy efficiency of household and commercial freezers.

2. Global Fish Consumption

Atlantic salmon *(Salmo salar)* is a migratory fish found widely in the northern Atlantic Ocean and adjacent freshwater. During the few past decades, it has become an important marine fish species in food markets (Figure 1).

In 2012, the total harvest of Atlantic salmon was 1.78 million tones. As the largest species of edible salmonids, Atlantic salmon can be prepared in many ways such as smoking, grilling, and sushi. Atlantic salmon processing industry requires high quality salmon, especially for sushi consumption. Inland farming of Atlantic salmon occurs in Norway, Chile, UK, North America, and New Zealand/Tasmania (Figure 2). Recent data shows North America has become the second largest Atlantic salmon market in the world having only 37 present of entire demand fulfilled by the harvest of its own.

Salmon is harvested and frozen in Norway and South America and then transported to North America by cargo ship. Generally, freezing of Atlantic salmon is necessary according to both industrial procedures and federal legislation. However, processing and retail establishments have a high demand for fresh Atlantic salmon and the shelf-life of the fresh salmon is an important factor influencing the salmon industry.

Worldwide fish consumption per capita in 1960 and 2015 was 9.9 kg and 20.4 kg, respectively [20]. According to FAO, fish consumption will continue to increase in the coming years and is predicted to increase to 21.8 kg in 2025 which is equivalent to another 28 million tons of seafood [20]. The greatest increase in fish consumption is expected to be primarily in Asia and developing countries leading to an increase in world trade [23]. Fish consumption in developing countries is predicted to increase by 57% from 62.7 million tons in 1997 to 98.6 million tons in 2020. While in comparison, developed country fish consumption is estimated to increase only 4% from 28.1 million tons in 1997 to 29.2 million tons in 2020. The consumption of salmon in the United States varies by species, product, origin (domestic and imported), and type (wild and farmed) [24]. The average annual total US salmon consumption from 2000 to 2004 was 284,000 metric tons and was comprised of 105,000 metric tons (37%) of Pacific salmon (chinook, sockeye, coho, pink, and chum) and 180,000 metric tons (63%) of Atlantic salmon. US consumption of fresh salmon over the same period was 63%, while canned salmon was 16% and frozen salmon 21%. In addition, 68% of US salmon consumption was imported and 65% was farmed [24]. Total US salmon consumption increased dramatically from 130,000 metric tons in 1989 to more than 300,000 metric tons in 2004 [24]. Part of the reason for the increased consumption is due to perceived health benefits from regular consumption of fish which include protection against human pathogens [25], reduction of heart diseases, and risk of developing dementia, including Alzheimer's disease [26].

3. Proximate Composition of Fish

The proximate composition of fish is in the range of 16-21% protein, 0.2-5% fat, 1.2-1.5% mineral, 0-0.5% carbohydrate,

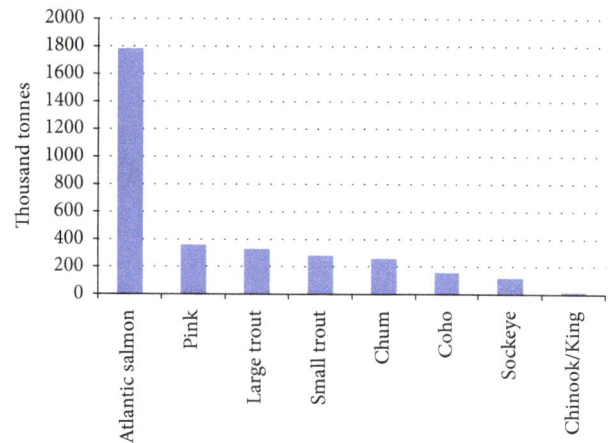

FIGURE 1: 2012 Worldwide Salmonids Harvest (species).

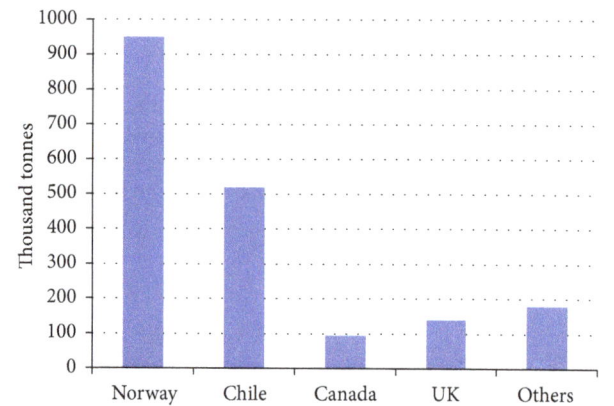

FIGURE 2: 2011 Worldwide Salmonids Harvest (place of production) (created from [20]).

and 66-81% moisture [27, 28]. However, fish of various species do not provide the same nutrient profile to the consumer [29] and the nutritive value of a fish varies with season [30]. The major component in fish is moisture which is a determinant of the value of the products, sensory attributes, and shelf-life in fish [31]. To reduce moisture or drip loss during frozen storage and thawing, the commercial fish production has improved the retention and addition of water to fish during harvest, processing, and storage. However, water addition to make up for moisture losses and excessive extraneous water addition for economic gain can negatively affect quality and is fraudulent [31]. Protein is the most predominant component in fish other than moisture and most finfish muscle tissue such as salmon consists of 18–22% crude protein [32]. Fish protein has a complete amino acid profile and high degree of digestibility [33]. The third and fourth major components in fish are lipid and ash. The total lipid and ash content of fish vary with the size of the fish and may also vary with the season and habitat from which the fish is harvested [34]. Fat content may vary widely due to fish species, muscle, and how a fillet is cut [35]. For example, the fat content of Norwegian salmon fillets ranged from 11% to 19% [36]. Fish oil contains a high amount of polyunsaturated fatty acids (omega-3 fatty acids)

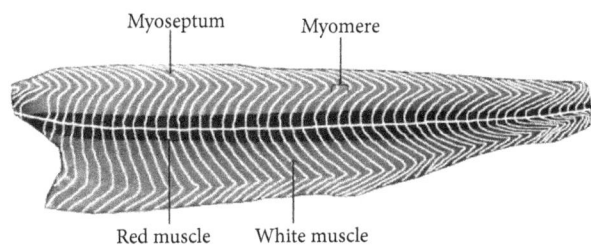

FIGURE 3: Salmon fish fillet in longitudinal section, beneath the skin, to present the W-shape of myomere and the two muscle types (Figure 3 is reproduced from [21].

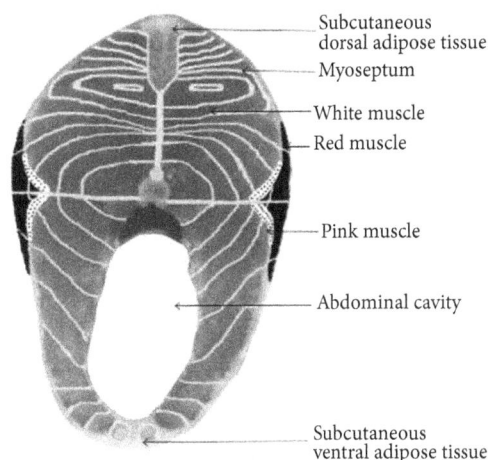

FIGURE 4: Organization and distribution of muscle mass on a trout cutlet (Figure 4 is reproduced from [21].

which can decrease the serum cholesterol and prevent heart diseases [37, 38]. Moreover, fish provide a great source of vitamins and minerals such as vitamins A and D, phosphorus, magnesium, selenium, and iodine [20].

4. Fish Muscle Structure

Fish have a distinctive muscle structure that rapidly degrades postmortem. Fish structure can vary based upon species, seasonality, maturity, and living environment but have a similar red and white muscle structure (Figures 3 and 4). Fish muscle can be structurally divided into myotomes, which are separated by connective tissue called myocomma or myoseptum [21]. There are red, white, and mosaic skeletal muscle groups, all having different functions. Red muscle represents around 30% of fish skeletal muscle and is the primary muscle type involved in highly aerobic activities, such as swimming. Red muscles contain high levels of lipids and are subsequentially subject to lipid oxidation, especially in fatty fish, such as Atlantic salmon. Trimethylamine oxides are also found within the red muscle that can be enzymatically or nonenzymatically degraded, resulting in products such as dimethylamine (DA) and formaldehyde (FA) [39, 40]. White muscle makes up around 70% of fish muscle structure and is associated with anaerobic activity. White muscle more readily reduces glycogen to lactic acid than other fiber types [41]. Finally, mosaic muscles are muscle locations in fish where mixtures of red and white muscle coexist. These areas in fish muscle are seen more frequently in fatty fish, such as Atlantic salmon. Both the lipid oxidative capabilities of fish muscle and chemical side reactions between lipids and proteins causing protein aggregation and denaturation are linked to muscle structure activity within Atlantic salmon contributing to quality change during short- and long-term storage [42].

5. Quality Measurements Affected by Frozen Storage

Major physical and chemical attributes that change during freezing and frozen storage in fish are color, texture, enzymatic activity, lipid oxidation, and ice crystal structural damage. Most studies on Atlantic salmon and other types of fish show physical changes during freezing (short-term effects),

such as weight loss, color change, and structural/texture changes because of ice crystal nucleation and growth [6, 7]. During longer term storage, physical attributes continue to slowly deteriorate; however, chemical characteristics such as enzymatic activity, lipid oxidation, and microbial growth become increasingly important factors affecting meat quality [8, 43]. Because of the difference in quality that short- and long-term freezing effects can have on products, studying both phases of freezing is necessary in determining the effects on Atlantic salmon.

5.1. Weight Loss. Weight loss by ice sublimation has been widely studied [4, 17, 44–47]. Freeze and thaw weight loss for salmon decreased with faster freezing rates [4, 17]. Weight loss within frozen and thawed salmon occurred because of damaged induced by ice crystal growth during the freezing process [17]. There has not been extensive research on freeze loss specifically as a quality measure in weight loss analysis, as drip loss (thaw/cook loss) is more common. However, with ice crystal pore morphology studies, the freezing process transfers freestanding water molecules into a uniform crystal lattice unit [13]. Atlantic salmon or other fish frozen at higher freezing rates freeze more quickly and retain structural integrity in the intracellular muscle structure since more and smaller ice crystals are formed resulting in fewer freestanding or thermodynamically unstable water/ice molecules. Salmon placed in higher temperatures freezers freeze more slowly and as a result accrue larger, less uniform ice crystals. As ice crystals form in extracellular and intracellular areas around fish muscle structure, cell membrane damage causes less water to be bound within the muscle structure [13]. Thawing can further damage meat structure and the period during which the damage from slow freezing manifests itself. During thawing ice crystals melt, and if formed intracellularly or around muscle tissue, moisture would remain within the fish. However, samples with cell membrane damage cannot retain unbound water [4]. Air blast freezing caused greater drip loss in Atlantic salmon compared to faster freezing methods such as pressure shift freezing [17]. Muscle fiber shrinkage will also cause higher drip loss during freezing, which occurred in salmon samples frozen more slowly [48]. When water is frozen slowly and unbound from muscle structure, muscle fibers retract due to dehydration of concentrated proteins and minerals [49]. Unlike with pork as reported by [50], salmon freeze and thaw loss were statistically different between treatment samples during freezing/thawing.

5.2. Color.

Color or appearance is a physical attribute that can change during freezing resulting from deterioration at the food surface, although changes in pigment appearance can be due to both chemical and biological actions. Color can affect product perception without affecting nutrition or flavor [42]. Different forms of color analysis of food products are available to predict consumer acceptance and track color changes in frozen products [51]. Fish appearance and color originate from meat-water binding properties and pigmentation within the skin or meat surface. Depending on the fish species, pigmentation can be oxidized resulting in darkening or fading. Salmon meat has a pink pigmentation in its natural state and with freezing; the pink color tends to fade [42]. Fading or increase in lightness is related to ice crystal formation during freezing [6]. Higher freezing rates form small, more numerous ice crystals within salmon, which then reflect light more intensely. Slower freezing rates form larger and fewer ice crystals in salmon, resulting in light refraction and a darkening effect of the meat surface. To analyze color changes in products the International Commission of Illumination (CIE) proposed a universal method in 1931 to be used in analyzing color. This method distinguishes color into three different tristimulus values. More recently the Munsell system simplifies quantifying color even further through a multidimensional method. L*, a*, b*, h, and C readings can be quantified and compared. L represents the overall lightness of a sample. a* value denotes redness or greenness in a sample, while b* value denotes yellowness and blueness. Hue (h) and Chroma (C) values are derived from a* and b* values with hue expressed in radians or degrees of the angle within the color space and chroma as a measure of intensity as distance from the achromatic center of the color space. Color analysis can indicate surface degradation of product and color difference value (ΔE) can be helpful in distinguishing the difference between storage treatments.

The number generated from this equation can be used as a comparative value against a control sample and utilizes L*, a*, and b* values to express sample color difference. Studies on freezing Atlantic salmon and other varieties of fish have shown similar results with a relationship between freezing rate and color change [7, 17]. Lightness values (L) are seen to increase with freezing rate, while a* and b* values tended to vary [7, 17]. Zhu et al. [7] studied how color difference (ΔE) was affected by freezing rate more so than the freezing, thawing, or cooking methods.

$$\Delta E^*_{ab} = \left[\left(\Delta L^* \right)^2 + \left(\Delta a^* \right)^2 + \left(\Delta b^* \right)^2 \right]^{1/2} \qquad (1)$$

5.3. Salmon Preservation.

Freezing is an important method of preservation of Atlantic salmon. Modern refrigeration is often employed in the Atlantic salmon industry and research has revealed that freezing can maintain Atlantic salmon quality at a high level. Indergård et al. [52] tested biochemical, structural, sensory, and microbiological factors of Atlantic salmon after long-term frozen storage. -25°C frozen storage maintained the salmon at an acceptable level after 12 months. -60°C frozen storage reduced drip loss but had no other significant quality improvements than the –25°C frozen. Duun et al. [53] tested the parameters of superchilled Atlantic salmon with vacuum package at -1.4 and -3.6°C. They reported that the shelf-life of vacuum packed salmon fillets was doubled with the utilization of superchilled storage at -1.4°C and -3.6°C. Some processes require thawed Atlantic salmon for preparation to retail markets. Repeated freezing and thawing will inevitably damage salmon fillet tissue. Einen et al. [48] compared how freezing and thawing of fresh or frozen Atlantic salmon fillets affected drip loss, gaping, texture, color, and rigor contraction. They concluded that freezing decreased color quality and firmness and unfrozen fillets had less fillet gaping, higher color score, lower drip loss, and firmer texture. Moreover, as an indispensable step in Atlantic salmon industry, thawing is often implemented more slowly than freezing and usually causes damage to salmon tissue. Although many new freezing and thawing techniques have been utilized in food industry, particular characteristics of Atlantic salmon result in many difficulties in implementing these freezing and thawing techniques. For example, high-pressure thawing (HPT) has been applied in many food marketing aspects. However, Zhu et al. [7] reported that HPT of Atlantic salmon resulted in significant drip loss and structural cracking compared to other thawing methods. Even though HPT significantly accelerated the thawing process, it is not a favorable method for Atlantic salmon thawing because of the quality loss.

There are several challenges and opportunities in the preservation of Atlantic salmon. Damage during repeated freezing and thawing has led to studies to protect stored salmon. Ice-chilling, vacuum, and other fundamental preservation methods are often employed to preserve Atlantic salmon in an attempt to satisfy the rising demand for salmon. Many researchers are studying more efficient ways to preserve Atlantic salmon. Gallart-Jornet et al. [54] showed that superchilling was more effective compared to freezing and ice-packing in preserving raw Atlantic salmon.

5.4. Ice Crystallization.

Another effect of freezing foods is ice crystal nucleation and morphology. Ice crystallization and recrystallization affect food structure and food texture. There are three main steps to ice crystallization: (1) nucleation of the crystalline lattice; (2) crystal growth and continuation of ice crystal nucleation; and (3) recrystallization [55]. Freezing rate affects crystal formation and uniformity. Slow freezing leads to larger crystal formation in the extracellular areas of food products, especially within different species of meat and fish. Large crystals as well as small thermodynamically unstable crystals form during the nucleation process and any fluctuation in temperature after freezing causes ice to melt and refreeze, a process also known as recrystallization. Recrystallization during slow freezing causes larger, irregularly spherulite, ice spear-like ice crystals to form. The irregular and extracellular nature of crystals formed from recrystallization damages muscle structure, especially in meat and fish, as connective tissue surfaces deteriorate. Rapid freezing results in ice nucleation within the intracellular areas of food products. These ice crystals are smaller and more uniform, therefore creating a more structurally stable product [4]. For example, different freezing rates affected cell wall structure in salmon (Figure 5). Images "bottom right" and "bottom left" were both subjected to faster freezing methods as compared to images "top right" and "top left", which were frozen raw and more slowly. Because of different freezing rates, especially shown in image "b", there is less cell wall damage [4]. Temperature fluctuations and ice recrystallization damage occurs readily in Atlantic salmon muscle structure during four weeks of frozen storage (Figure 6) [13]. Image "a" shows small ice crystals formed uniformly throughout the fish tissue directly after freezing. Immediately after freezing, a sample would not be subjected to recrystallization due to temperature fluctuations, therefore resulting in smaller, more uniform ice crystal pores. Images "b" and "c" represent samples exposed to an increase in their core temperature close to a glassy state and the onset temperature of ice crystals melting. Image "b" has larger ice crystal pores that are still uniform as compared to image "a". This shows that samples are more structurally stable at a -27°C(-17°F) glassy state. Image "c" shows the ice crystal pores growing larger and varying in uniformity, at a temperature range between -27°C(-17°F) and -17°C(1°F), which allows for ice crystals to melt and recrystallize. Finally, image "d" shows a sample when the temperature surpassed the onset temperature of ice crystal melting with the largest, most irregular ice crystal pores from the whole study. Temperatures above -17°C (1°F) did not provide the most ideal conditions for Atlantic salmon during long-term storage. This study showed the physical damage that can occur to samples frozen to different core temperatures, an area where increased research is needed in the frozen products industry [13]. Thawing allows intracellular water to become permanently extracellular; thus, incurring greater cell wall or muscle damage during freezing will increase water relocation during thawing [56]. Measuring ice crystal pore damage is a good indicator of structural damage imposed by freezing foods. After samples are frozen they can be subjected to freeze-drying, freeze concentration, or free substitution to isolate the pores left behind by ice crystal growth [14].

Freeze-drying has proven to be an effective method to determine ice crystal pore size although it is slower and more expensive than other methods, mainly because ice crystals are vaporized leaving behind a physical pore that emulates the ice crystal's shape [4]. After the pore has been stabilized different techniques can be used to examine pore morphology. Arnaud [57] used optical microscopy to study pore structure and size in ice cream. Fractal, environmental scanning electron microscopy, CT X-ray, and cold stage scanning electron microscopic analysis techniques have been used on different types of fish including Atlantic salmon and sea bass [4, 13, 14]. With these techniques micrograph images can be used to quantify and qualify ice crystal frequency and size. These studies support the objective of the current research, as increasingly more evidence needs to be collected about different freezing rates and how they affect the physical integrity of Atlantic salmon.

5.5. *Texture.* Because of ice crystal growth and recrystallization, another quality parameter that is compromised in fish is texture. Freezing storage temperature affected texture quality in Atlantic salmon fillets more than thawing techniques [17]. Control of pressure and temperature affected ice crystal growth and distribution [17]. Toughness is related to protein and fat content and properties within fish meat. Texture analysis of fish meat differs from other meat types because of variability in muscle structure [21]. There are also added variables such a species, composition, and seasonality that make texture a more challenging attribute to measure. However, trends such as higher resistance and toughness in fish that are in frozen storage for extended periods of time at higher temperatures have been established [16]. Texture changes during frozen storage have also been directly linked to protein denaturation within fish [58]. Salt-water fish, such as Atlantic salmon, may contain higher levels of trimethylamine oxides (TMAO) within red muscle as compared to fresh water fish [59]. TMAO degrades in the presence of TMAase, an enzyme located within fat tissue. The products dimethylamine and formaldehyde are then susceptible to form both intra- and intermolecular crosslinks with protein side chains. The aggregation of these cross linkages causes toughness in fish [17]. Slow freezing could also cause for a tougher sample, as larger ice crystals tend to break down protein structure within fish [42]. Different texture analyses methods include the Kramer Shear Cell method, Warner-Bratzler shear cell method, puncture test, and texture profile analysis (TPA). Each of these methods uses a blade or probe to measure a maximum force for food samples. The Kramer Shear and puncture method can test multiple locations per sample unlike the Warner-Bratzler and TPA methods, which only allow for one or two maximum force readings per sample [16].

5.6. *Pore Size due to Freezing.* The size and number of pores formed during freezing (and refreezing during freezer cycling) affect structural stability within salmon since larger, less uniform pores cause cell wall damage and dehydration in fish muscle fiber structure [60]. Therefore, ice crystal damage may be attributed to the number of nucleated ice crystals first and then to the specific average size of the ice crystals formed [4]. Just as in fish, ice cream storage time and temperature affect ice crystal morphology. Ice crystal formation starts during the churning process and ice crystal size and shape change during storage. A creamy and smooth final ice cream product is associated with a high number of small ice crystals [4, 61]. The lack of statistical difference between average surface pore size in some treatments could be related to the fact that samples frozen more slowly produce

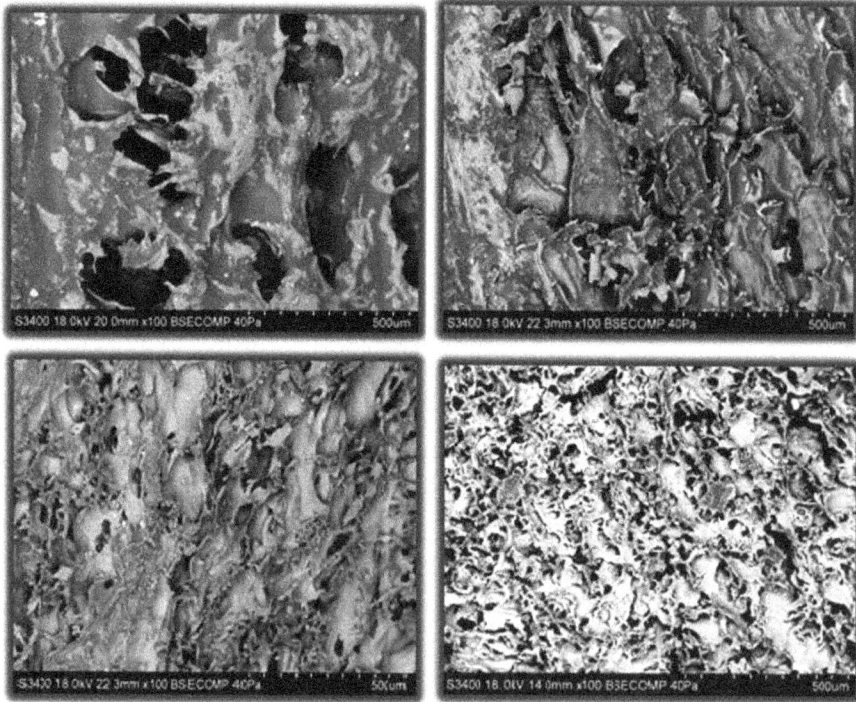

FIGURE 5: Effect of freezing rate on the microstructure of salmon. Photo micrographs of surface pore. Top left: -7°C(-7°C) corresponds to a treatment frozen at in the -7°C unit to a core temperature of -7°C. Top right: -18°C(-7°C) corresponds to a treatment frozen in the -18°C unit to a core temperature of -7°C. Middle left: -29°C(-7°C) corresponds to a treatment frozen in the -29°C unit to a core temperature of -7°C. Middle right: -106°C(-7°C) corresponds to a treatment frozen in the -77°C to a core temperature of -7°C. (Figure 5 is reproduced from [22].

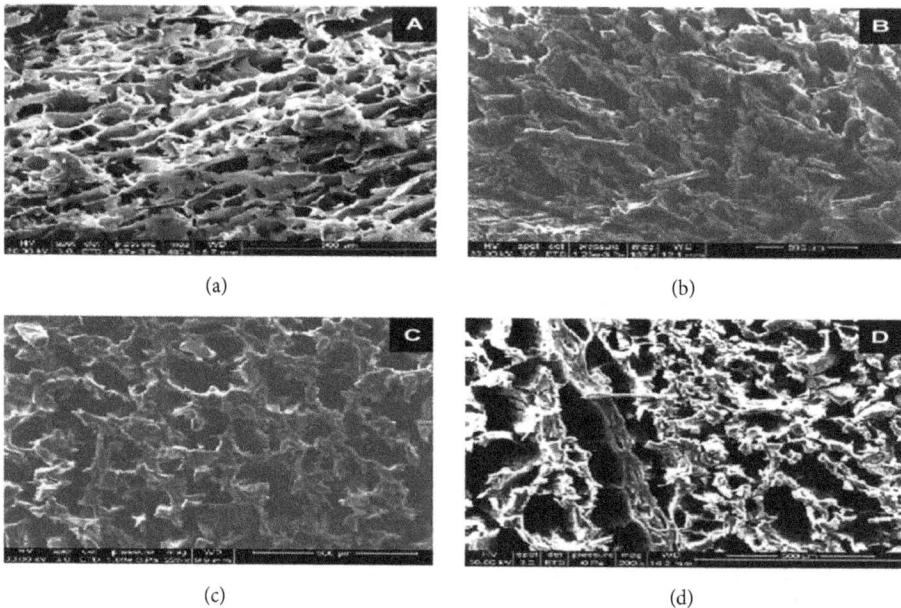

FIGURE 6: Environmental scanning electron microscopy of freeze-dried salmon. Before freeze-drying, frozen salmon was subjected to state transitions during 4 weeks' storage. (a) Immediately after freezing, (b) T<-27°C(-17°F); (c) -27°C(-17°F) < T < 17°C(1°F); (d) T > 17°C(1°F) (Figure 6 is reproduced from [22].

large nucleated ice crystals with large variation in size, as well as small thermodynamically unstable ice crystals [4]. Since the food surface freezes more quickly than the center, core ice crystal nucleation and morphology changes occur more slowly than those on the surface. Studies have shown that the differences in ice crystal characteristics on the food surface reflect those seen in the center of food products due to freezing rate [14] but pore number and average size at the center of cylindrical gelatin gels decreased with increasing diameter when frozen at the same freezing rate [14].

5.7. Water Holding Capacity (WHC). Water holding capacity of food products is an important quality parameter due to weight loss during transport and storage, the drip loss during thawing, and the juiciness and tenderness of the meat [62–66]. Water holding capacity is commercially very important due to loss of profit and for consumer acceptance of appearance and texture. Water holding capacity is closely related to textural properties, and a low of WHC is often related to postmortem structural changes in the muscle. Changes include myofilament lattice degradation, denaturation of myosin, and increase of extracellular space [67]. Myofibrils are long rod-like organelles found in cardiac and skeletal muscle comprising nearly 80% of the muscle cell volume [64, 68]. Moreover, almost 85% of the water in a muscle cell is held in the myofibrils [63, 64, 68]. There are three types of water in food which are free, immobilized, and bound. Approximately 90% of water in tissues of fish are immobilized (held by capillary action), mainly in intracellular locations [69, 70]. Free water is easily lost from the tissue and freezing can change immobilized water into free water. Freshly frozen then thawed samples tend to have good water holding, but the ability of the structure to retain immobilized water upon thawing diminishes as length of storage increases due to continued structural changes [71]. Changes associated with water holding capacity have negative outcome on fish texture [72]. NMR relaxation measurements have been used for characterizing changes in water location and water holding capacity during frozen storage of fish [73]. How water molecules react to the imposition and then release (relaxation) of an electric magnetic field are indicative of how tightly they are bond and their location within the meat. The relaxation of an NMR signal in the study of water in meat is reported in terms of two parameters. One parameter, the T1 relaxation, is responsible for the loss of signal intensity while the other parameter, the T2 relaxation, is responsible for the broadening of the signal. The spin-spin (T2) relaxation time changes during frozen storage, from a single peak to a broader multiexponential peak which can be related to texture deterioration [71]. In thawed fish, some of the water that has been isolated from muscle fibers by the formation of ice crystals is not retained or regained by the fibers. The T2 relaxation then shows a long T2 signal peak having a relaxation response similar to that of water [71]. Salmon fillets that were shell/partially frozen using an impingement freezer maintained optimal ice crystal features such as size, distribution, and shape provided that the super chilling is carried rapidly [74]. In addition to physical damage to fibers

due to ice crystal formation, an decrease in WHC may result from proteolytic activity in the muscle during storage [75] which causes a loss of water described as a "leaking out" effect [76].

5.8. Protein Stability. Protein denaturation correlates well with textural characteristics. Several researchers have proposed that frozen storage decreases protein quality in fish due to denaturation [77]. Furthermore, protein denaturation during frozen storage causes a decrease in protein solubility due to loss of intermolecular hydrogen or hydrophobic bonds, as well as disulfide bonds and ionic interactions [78–80]. Protein denaturation causes textural changes, especially as a result of changes in the myofibrillar proteins. Protein denaturation is affected by ultimate freezing temperature, storage duration, storage temperature, rigor state at freezing, process of thawing, and temperature fluctuation during storage [4, 81–87]. Protein denaturation reduces the amount of soluble proteins [53, 60]. However, salt-soluble proteins have also been reported to increase during the frozen storage of salmon due to the combined effects of denaturation and proteases to increase extractability [53]. An increase in salt concentration in unfrozen water phase due to the removal of pure water from this phase by freezing leads to myofibril denaturation [88]. Kaale and Eikevik [89] reported no significant difference in water and salt-soluble proteins between surface and center parts of the superchilled salmon samples indicating that rapid freezing minimized the concentration of salts in the unfrozen water phase. However other researchers found that SH groups and the formation of S-S bonds were lost during frozen storage [90–92]. For lizardfish, a consistent decrease in sulfhydryl groups was found throughout storage which was presumed to be caused by formaldehyde formation in this species induced by the oxidation of sulfhydryl groups [93]. Moreover, the masking of sulfhydryl groups by protein aggregates was also presumed to lead to the decrease in free sulfhydryl groups [93]. The S-S formation has been a causative factor for protein aggregation during frozen storage [79, 94, 95]. Changes in proteins could also influence antioxidant activity in the muscle resulting in an increased rate of rancidity during long-term frozen storage. Protein aggregation has been suggested to inhibit the interaction of exogenous hemoglobin with membranes [96]. Therefore, protein aggregation and denaturation occurring through frozen storage may alter the environment of the membranes and could make it more difficult for the antioxidants to interact with oxidative sensitive sites such as phospholipid membranes, thus reducing the efficiency of antioxidants.

5.9. Fat Stability and Lipid Oxidation. Storage time and temperature are major factors affecting quality loss and the shelf life of fish [97] with the lipid fraction subject to mainly autoxidative and hydrolytic changes during frozen storage [98]. Several researchers reported that the fat content of fish decreased during frozen storage [99–101]. Arannilewa et al. [101] found that the total lipid content of Tilapia decreased from 9.72% to 7.20% during frozen storage for 60 days [101] primarily due to oxidation resulting in losses in the triglyceride fraction [102, 103]. Similarly, storage at - 18°C for 6 months resulted in oxidative changes and decrease in unsaturated fatty acids level in goat meat fat [104]. Peroxidation affects mainly phospholipids, which are in the cell membrane and are the most exposed to attack by free radicals [105]. Other researchers found that the fatty acid of (C16:1) decreased in meat fat during frozen storage [98, 106], while there was no decrease in polyunsaturated fatty acids (PUFA) in frozen meat.

Lipid oxidation is a major factor in the shelf life of fish because it adversely affects the flavor and nutritional value [107]. Fish lipids are rich in long-chain PUFA known for positive health effects; however, they are highly susceptible to oxidation [108]; thus oxidation reduces fish nutritional, texture, and color quality. Many studies showed that the PUFA, especially arachidonic acid (C20:4n-6), eicosanoic acid (EPA), and docosahexaenoic acid (DHA; C22:6n3), decreased as the storage time increased with freezing and refrigeration time [109–111].

Frozen storage reduced the PUFA and increased the saturated fatty acids (SFA) which indicated a substantial loss of nutritional value in the fillets of rainbow trout [110]. The decrease in PUFA might be due to its susceptibility to oxidation; therefore, free PUFA may undergo oxidation as a greater extent than SFA. The first stage of oxidation is the reaction of oxygen with the unsaturated fatty acid molecules creating hydroperoxides which are a primary indicator product of oxidation. Peroxide value (PV) is an early indicator of oxidation (hydroperoxide formation). Peroxide value was lower in cobia frozen at -40 and -80°C compared to -20°C after 6 months of storage [107]. Thiobarbituric acid value which primarily quantifies malondialdehyde is another chemical analysis used as an indicator of lipid oxidation and was found to increase in numerous studies during the storage of fish [107, 112–114]. As expected, a low freezing temperature - 80°C resulted in slower oxidation rate which significantly increased during the storage of anchovies compared to -20°C and -40°C [108]. Lipid oxidation of salmon results in the formation of volatile products such as aldehydes and ketones which are detected by humans as rancid flavors and odors [115]. The rancid off flavor of frozen salmon is primarily because of increase in three aldehydes, (E, Z)-2,6-nonadienal with a cucumber odor, (Z)-3-hexenal with a green odor, and (Z, Z)-3,6-nonadienal with a fatty odor [115, 116], which are formed from the oxidation of n-3 unsaturated fatty acids.

In summary, effects of freezing salmon on quality can be divided into the effects that primarily occur during freezing which are mostly physical effects and effects that occur during frozen storage, which are both physical and chemical/biological. Faster freezing rates form smaller and more numerous ice crystals within salmon, which cause less structural damage and more evenly reflect light. Slower freezing rates form larger and fewer ice crystals, resulting in more structural damage and greater light refraction and a darkening effect of the meat surface. Freshly frozen then thawed samples tend to have good water holding, but the ability of the meat to retain good water holding capacity upon on thawing diminishes the longer the frozen meat is held. Frozen storage factors include physical deterioration

due to cycling of temperature which causes water to thaw and refreeze within the meat tissue resulting in the formation of larger ice crystals. Thus, a lower and more consistent holding temperature will slow the formation of large ice crystals and slow the damage to meat tissue. Although lower temperatures slow enzymatic and chemical deterioration, these reactions continue during frozen storage. Browning and autooxidation reactions will proceed during frozen storage to affect flavor, appearance, and nutritional quality of salmon. Salmon is a high quality and highly desirable food and freezing can be a sound method to preserve eating quality in the ever-expanding worldwide market.

Conflicts of Interest

The authors declare that they have no conflicts of interest.

Acknowledgments

The authors acknowledge the Electrolux Corporation for equipment support.

References

[1] USDA and HHS, *Dietary Guidelines for Americans*, U.S. Department of Agriculture and U.S. Department of Health and Human Services, Washington, Wash, USA, 2010.

[2] J. A. Foran, D. H. Good, D. O. Carpenter, M. C. Hamilton, B. A. Knuth, and S. J. Schwager, "Quantitative analysis of the benefits and risks of consuming farmed and wild salmon," *Journal of Nutrition*, vol. 135, no. 11, pp. 2639–2643, 2005.

[3] J. G. Bell, R. J. Henderson, D. R. Tocher, and J. R. Sargent, "Replacement of dietary fish oil with increasing levels of linseed oil: Modification of flesh fatty acid compositions in atlantic salmon (Salmo salar) using a fish oil finishing diet," *Lipids*, vol. 39, no. 3, pp. 223–232, 2004.

[4] G. Petzold and J. M. Aguilera, "Ice morphology: Fundamentals and technological applications in foods," *Food Biophysics*, vol. 4, no. 4, pp. 378–396, 2009.

[5] W. A. Johnston, F. J. Nicholson, A. Roge, and G. D. Stroud, "Freezing and refrigerated storage in fisheries," *FAO Fisheries Technical Paper*, no. 340, p. 143, 1994.

[6] S. Ottestad, G. Enersen, and J. P. Wold, "Effect of freezing temperature on the color of frozen salmon," *Journal of Food Science*, vol. 76, no. 7, pp. S423–S427, 2011.

[7] S. Zhu, H. S. Ramaswamy, and B. K. Simpson, "Effect of high-pressure versus conventional thawing on color, drip loss and texture of Atlantic salmon frozen by different methods," *LWT-Food Science and Technology*, vol. 37, no. 3, pp. 291–299, 2004.

[8] G. Strasburg, Y. L. Xiong, and W. Chiang, "Physiology and chemistry of edible muscle tissues," in *Food Chemistry*, pp. 958-959, CRC Press, Boca Raton, Fla, USA, 4th edition, 2008.

[9] K. S. Bahçeci, A. Serpen, V. Gökmen, and J. Acar, "Study of lipoxygenase and peroxidase as indicator enzymes in green beans: Change of enzyme activity, ascorbic acid and chlorophylls during frozen storage," *Journal of Food Engineering*, vol. 66, no. 2, pp. 187–192, 2005.

[10] S. Eymard, E. Carcouët, M.-J. Rochet, J. Dumay, C. Chopin, and C. Genot, "Development of lipid oxidation during manufacturing of horse mackerel surimi," *Journal of the Science of Food and Agriculture*, vol. 85, no. 10, pp. 1750–1756, 2005.

[11] C. Nasopoulou, E. Psani, E. Sioriki, C. A. Demopoulos, and I. Zabetakis, "Evaluation of sensory and in vitro cardio protective properties of sardine (Sardina pilchardus): The effect of grilling and brining," *Journal of Food and Nutrition Sciences*, vol. 4, no. 9, pp. 940–949, 2013.

[12] Z. M. Abdel-Kader, "Lipid oxidation in chicken as affected by cooking and frozen storage," *Nahrung-Food*, vol. 40, no. 1, pp. 21–24, 1996.

[13] R. M. Syamaladevi, K. N. Manahiloh, B. Muhunthan, and S. S. Sablani, "Understanding the influence of state/phase transitions on ice recrystallization in atlantic salmon (salmo salar) during frozen storage," *Food Biophysics*, vol. 7, no. 1, pp. 57–71, 2012.

[14] D. Chevalier, A. Le Bail, and M. Ghoul, "Freezing and ice crystals formed in a cylindrical food model: Part I. Freezing at atmospheric pressure," *Journal of Food Engineering*, vol. 46, no. 4, pp. 277–285, 2000.

[15] B. Woinet, J. Andrieu, M. Laurent, and S. G. Min, "Experimental and theoretical study of model food freezing. Part II. Characterization and modelling of the ice crystal size," *Journal of Food Engineering*, vol. 35, no. 4, pp. 395–407, 1998.

[16] M. Barroso, M. Careche, and A. J. Borderías, "Quality control of frozen fish using rheological techniques," *Trends in Food Science & Technology*, vol. 9, no. 6, pp. 223–229, 1998.

[17] E. Alizadeh, N. Chapleau, M. De Lamballerie, and A. LeBail, "Effects of freezing and thawing processes on the quality of Atlantic salmon (Salmo salar) fillets," *Journal of Food Science*, vol. 72, no. 5, pp. E279–E284, 2007.

[18] M. D. Ayala, O. López Albors, A. Blanco et al., "Structural and ultrastructural changes on muscle tissue of sea bass, Dicentrarchus labrax L., after cooking and freezing," *Aquaculture*, vol. 250, no. 1-2, pp. 215–231, 2005.

[19] M. M. Farouk, K. J. Wieliczko, and I. Merts, "Ultra-fast freezing and low storage temperatures are not necessary to maintain the functional properties of manufacturing beef," *Meat Science*, vol. 66, no. 1, pp. 171–179, 2004.

[20] "Salmon Farming Industry Handbook, Marine Harvest," 2016, http://marineharvest.com/product/.

[21] A. Listrat, B. Lebret, I. Louveau et al., "How muscle structure and composition influence meat and flesh quality," *The Scientific World Journal*, vol. 2016, Article ID 3182746, 14 pages, 2016.

[22] M. Remington, *The effect of freezing and refrigeration on food quality. Masters Thesis [M.S., thesis]*, Clemson University, Clemson, SC, USA, 2017.

[23] "World bank report number 83177-GLB," 2013, http://documents.worldbank.org/curated/en/458631468152376668/pdf/831770WP0P11260ES003000Fish0to02030.pdf.

[24] "The great salmon run: competition between wild and farmed salmon," 2007, http://www.iser.uaa.alaska.edu/people/knapp/personal/pubs/TRAFFIC/The_Great_Salmon_Run.pdf.

[25] S. Ravichandran, K. Kumaravel, G. Rameshkumar, and T. T. Ajithkumar, "Antimicrobial peptides from the marine fishes," *Research Journal of Immunology*, vol. 3, no. 2, pp. 146–156, 2010.

[26] W. B. Grant, "Dietary links to Alzheimers disease," *Alzheimer's Disease Review*, vol. 2, pp. 42–55, 1997.

[27] W. Steffens, "Freshwater fish-wholesome foodstuffs," *Bulgarian Journal of Agricultural Science*, vol. 12, pp. 320–328, 2006.

[28] R. M. Love, *The Chemical Biology of Fishes: With a Key to the Chemical Literature*, Academic Press, London, UK, 1970.

[29] K. Takama, T. Suzuki, K. Yoshida, H. Arai, and T. Mitsui, "Phosphatidylcholine levels and their fatty acid compositions in teleost tissues and squid muscle," *Comparative Biochemistry and Physiology—Part B: Biochemistry & Molecular Biology*, vol. 124, no. 1, pp. 109–116, 1999.

[30] J. Varljen, S. Šulić, J. Brmalj, L. Batičić, V. Obersnel, and M. Kapović, "Lipid classes and fatty acid composition of Diplodus vulgaris and Conger conger originating from the Adriatic Sea," *ood Technology and Biotechnology*, vol. 41, no. 2, pp. 149–156, 2003.

[31] S. van Ruth, E. Brouwer, A. Koot, and M. Wijtten, "Seafood and water management," *Foods*, vol. 3, no. 4, pp. 622–631, 2014.

[32] UK Association of Frozen Food Producers, British Frozen Food Federation, British Retail Consortium, British Hospitality Association, Sea Fish Industry Authority; LACOTS, and Association of Public Analysist, "Code of practice on the declaration of fish content in fish products," http://www.seafish.org/media/Publications/Fish_Content_CoP.pdf.

[33] N. Louka, F. Juhel, V. Fazilleau, and P. Loonis, "A novel colorimetry analysis used to compare different drying fish processes," *Food Control*, vol. 15, no. 5, pp. 327–334, 2004.

[34] M. Hassan, *Influence of pond fertilization with broiler dropping on the growth performance and meat quality of major carps [Ph.D. thesis]*, University of Agriculture, Faisalabad, Faisalabad, Pakistan, 1996.

[35] L. H. Stien, A. Kiessling, and F. Manne, "Rapid estimation of fat content in salmon fillets by colour image analysis," *Journal of Food Composition and Analysis*, vol. 20, no. 2, pp. 73–79, 2007.

[36] K. Fjellanger, A. Obach, and G. Rosenlund, "Proximate analysis of fish with special emphasis on fat," in *Farmed Fish Quality, Fishing News Books*, S. C. Kestin and P. D. Warris, Eds., pp. 307–317, Blackwell Science, Oxford, UK, 2000.

[37] A. Nordøy, R. Marchioli, H. Arnesen, and J. Videbæk, "n-3 polyunsaturated fatty acids and cardiovascular diseases," *Lipids*, vol. 36, pp. S127–S129, 2001.

[38] A. Türkmen, M. Türkmen, Y. Tepe, and I. Akyurt, "Heavy metals in three commercially valuable fish species from İskenderun Bay, Northern East Mediterranean Sea, Turkey," *Food Chemistry*, vol. 91, no. 1, pp. 167–172, 2005.

[39] G. J. Benoit Jr. and E. R. Norris, "Studies of Trimethylamine Oxide II. The Origin of Trimethylamine Oxide in Young Salmon," *The Journal of Biological Chemistry*, vol. 158, pp. 439–442, 1944.

[40] J. W. Jebsen and M. Riaz, "Breakdown products of trimethylamineoxide in air dried stock fish. Means of enhancing the formation of the formaldehyde and dimethylamine," *Fiskeridirektoratets Skrifter Serie Erireing*, vol. 1, no. 4, pp. 145–153, 1978.

[41] I. A. Johnston, "A comparative study of glycolysis in red and white muscles of the trout (Salmo gairdneri) and mirror carp (Cyprinus carpio)," *Journal of Fish Biology*, vol. 11, no. 6, pp. 575–588, 1977.

[42] E. E. M. Santos-Yap, *Freezing Effects on Food Quality*, L. E. Jeremiah, Ed., Marcel Dekker, New York, NY, USA, 1996.

[43] D. R. Heldman and R. W. Hartel, "Freezing and Frozen-Food Storage," in *Principles of Food Processing*, Heldman. and D. R, Eds., pp. 131-132, Aspen Publishers, Gaithersburg, Md, USA, 1998.

[44] L. A. Campañone, L. A. Roche, V. O. Salvadori, and R. H. Mascheroni, "Monitoring of Weight Losses in Meat Products

during Freezing and Frozen Storage," *Food Science and Technology International*, vol. 8, no. 4, pp. 229–238, 2002.

[45] L. A. Espinoza Rodezno, S. Sundararajan, K. M. Solval et al., "Cryogenic and air blast freezing techniques and their effect on the quality of catfish fillets," *LWT- Food Science and Technology*, vol. 54, no. 2, pp. 377–382, 2013.

[46] D. P. Crane, C. C. Killourhy, and M. D. Clapsadl, "Effects of three frozen storage methods on wet weight of fish," *Fisheries Research*, vol. 175, pp. 142–147, 2016.

[47] H. Turan, Y. Kaya, and I. Erkoyuncu, "Effects of glazing, packaging and phosphate treatments on drip loss in rainbow trout (Oncorhynchus mykiss) during frozen storage," *Turkish Journal of Fisheries and Aquatic Sciences*, vol. 3, pp. 105–109, 2003.

[48] O. Einen, T. Guerin, S. O. Fjæra, and P. O. Skjervold, "Freezing of pre-rigor fillets of Atlantic salmon," *Aquaculture*, vol. 212, no. 1-4, pp. 129–140, 2002.

[49] Z. Sikorski, J. Olley, S. Kostuch, and H. S. Olcott, "Protein Changes in Frozen Fish," *C R C Critical Reviews in Food Science and Nutrition*, vol. 8, no. 1, pp. 97–129, 1976.

[50] T. M. Ngapo, I. H. Babare, J. Reynolds, and R. F. Mawson, "Freezing and thawing rate effects on drip loss from samples of pork," *Meat Science*, vol. 53, no. 3, pp. 149–158, 1999.

[51] Y. Pomeranz and C. E. Meloan, "Measurement of color," in *Food Analysis Theory and Practice*, Food. Analysis, Ed., pp. 85–96, Van Nostrand Reinhold Company, New York, NY, USA, 2nd edition, 1987.

[52] E. Indergård, I. Tolstorebrov, H. Larsen, and T. M. Eikevik, "The influence of long-term storage, temperature and type of packaging materials on the quality characteristics of frozen farmed Atlantic Salmon (Salmo Salar)," *International Journal of Refrigeration*, vol. 41, pp. 27–36, 2014.

[53] A. S. Duun and T. Rustad, "Quality of superchilled vacuum packed Atlantic salmon (Salmo salar) fillets stored at -1.4 and -3.6°C," *Food Chemistry*, vol. 106, no. 1, pp. 122–131, 2008.

[54] L. Gallart-Jornet, T. Rustad, J. M. Barat, P. Fito, and I. Escriche, "Effect of superchilled storage on the freshness and salting behaviour of Atlantic salmon (Salmo salar) fillets," *Food Chemistry*, vol. 103, no. 4, pp. 1268–1281, 2007.

[55] R. W. Hartel, "Crystallization in foods," in *Handbook of Industrial Crystallization*, pp. 287–304, Elsevier, 2002.

[56] R. P. Sign and D. R. Heldman, *Introduction of Food Engineering*, 3rd edition, 2001.

[57] L. Arnaud, M. Gay, J. Barnola, and P. Duval, "Imaging of firn and bubbly ice in coaxial reflected light: a new technique for the characterization of these porous media," *Journal of Glaciology*, vol. 44, no. 147, pp. 326–332, 1998.

[58] S. Y. K. Shenouda, "Theories of protein denaturation during frozen storage of fish flesh," *Advances in Food and Nutrition Research*, vol. 26, no. C, pp. 275–311, 1980.

[59] K. Yamada, "Occurrence and origin of TMAO in fishes and marine invertebrates," *Bulletin of Japanese Society for the Science of Fish*, vol. 33, no. 6, pp. 591–603, 1967.

[60] R. M. Syamaladevi, S. S. Sablani, J. Tang, J. Powers, and B. G. Swanson, "Stability of anthocyanins in frozen and freeze-dried raspberries during long-term storage: in relation to glass transition," *Journal of Food Science*, vol. 76, no. 6, pp. E414–E421, 2011.

[61] R. W. Hartel, "Ice crystallization during the manufacture of ice cream," *Trends in Food Science & Technology*, vol. 7, no. 10, pp. 315–321, 1996.

[62] A. S. Duun and T. Rustad, "Quality changes during superchilled storage of cod (Gadus morhua) fillets," *Food Chemistry*, vol. 105, no. 3, pp. 1067–1075, 2007.

[63] M. J. A. den Hertog-Meischke, R. J. L. M. van Laack, and F. J. M. Smulders, "The water-holding capacity of fresh meat," *Veterinary Quarterly*, vol. 19, no. 4, pp. 175–181, 1997.

[64] E. Huff-Lonergan, in *Water-holding capacity of fresh. Meat American Meat Science Association*, 2002, https://www.pork.org/research/influence-of-early-postmortem-factors-on-water-holding-capacity-and- tenderness-of-fresh-pork/.

[65] M. Irie, A. Izumo, and S. Mohri, "Rapid method for determining water-holding capacity in meat using video image analysis and simple formulae," *Meat Science*, vol. 42, no. 1, pp. 95–102, 1996.

[66] G. R. Shaviklo, G. Thorkelsson, and S. Arason, "The influence of additives and frozen storage on functional properties and flow behaviour of fish protein isolated from haddock (melanogrammus aeglefinus)," *Turkish Journal of Fisheries and Aquatic Sciences*, vol. 10, no. 3, pp. 333–340, 2010.

[67] A. S. Duun, *Superchilling of muscle food storage stability and quality aspects of salmon (Salmo salar), cod (Gadus morhua) and pork (Doctoral theses) [Ph.D. thesis]*, Dep. Biotechnology, NTNU, Trondheim, Norway, 2008.

[68] E. Huff-Lonergan and S. M. Lonergan, "Mechanisms of water-holding capacity of meat: the role of postmortem biochemical and structural changes," *Meat Science*, vol. 71, no. 1, pp. 194–204, 2005.

[69] A. Jonsson, S. Sigurgisladottir, H. Hafsteinsson, and K. Kristbergsson, "Textural properties of raw Atlantic salmon (Salmo salar) fillets measured by different methods in comparison to expressible moisture," *Aquaculture Nutrition*, vol. 7, no. 2, pp. 81–89, 2001.

[70] G. Offer and J. Trinick, "On the mechanism of water holding in meat: the swelling and shrinking of myofibrils," *Meat Science*, vol. 8, no. 4, pp. 245–281, 1983.

[71] I. Medina, M. J. González, J. Iglesias, and N. D. Hedges, "Effect of hydroxycinnamic acids on lipid oxidation and protein changes as well as water holding capacity in frozen minced horse mackerel white muscle," *Food Chemistry*, vol. 114, no. 3, pp. 881–888, 2009.

[72] R. Lakshmanan, J. A. Parkinson, and J. R. Piggott, "High-pressure processing and water-holding capacity of fresh and cold-smoked salmon (Salmo salar)," *LWT- Food Science and Technology*, vol. 40, no. 3, pp. 544–551, 2007.

[73] K. R. Brownstein and C. E. Tarr, "Importance of classical diffusion in NMR studies of water in biological cells," *Physical Review A: Atomic, Molecular and Optical Physics*, vol. 19, no. 6, pp. 2446–2453, 1979.

[74] L. D. Kaale, T. M. Eikevik, T. Bardal, E. Kjorsvik, and T. S. Nordtvedt, "The effect of cooling rates on the ice crystal growth in air-packed salmon fillets during superchilling and superchilled storage," *International Journal of Refrigeration*, vol. 36, no. 1, pp. 110–119, 2013.

[75] L. D. Kaale, T. M. Eikevik, T. Rustad, and T. S. Nordtvedt, "Changes in water holding capacity and drip loss of Atlantic salmon (Salmo salar) muscle during superchilled storage," *LWT- Food Science and Technology*, vol. 55, no. 2, pp. 528–535, 2014.

[76] G. B. Olsson, R. Ofstad, J. B. Lødemel, and R. L. Olsen, "Changes in water-holding capacity of halibut muscle during cold storage," *LWT—Food Science and Technology*, vol. 36, no. 8, pp. 771–778, 2003.

[77] A. MILLS, "Measuring changes that occur during frozen storage of fish: a review," *International Journal of Food Science & Technology*, vol. 10, no. 5, pp. 483–496, 1975.

[78] J. J. Matsumoto, "Chemical deterioration of muscle proteins during frozen storage," in *Chemical Deterioration of Proteins*, J. R. Whitaker and M. Fujimoto, Eds., vol. 123 of *ACS Symposium Series*, pp. 95–124, American Chemical Society, Washington, Wash, USA, 1980.

[79] T. Akahane, *Freeze denaturation of fish muscle proteins [Ph.D. thesis]*, Sophia University, Tokyo, Japan, 1982.

[80] F. Badii and N. K. Howell, "A comparison of biochemical changes in cod (Gadus morhua) and haddock (Melanogrammus aeglefinus) fillets during frozen storage," *Journal of the Science of Food and Agriculture*, vol. 82, no. 1, pp. 87–97, 2002.

[81] S. Benjakul and W. Visessanguan, "Impacts of freezing and frozen storage on quality changes of seafoods," in *Physicochemical Aspects of Food Engineering and Processing*, S. Devahastin, Ed., vol. 20100931 of *Contemporary Food Engineering*, pp. 283–306, CRC Press, New York, NY, USA, 2010.

[82] G. Blond and M. L. Meste, "Principes of frozen storage," in *Handbook of Frozen Foods*, K. D. Murell, Y. H. Hui, W.-K. Nip, M. H. Lim, I. G. Lugarreta, and P. Cornillon, Eds., CRC Marcel Dekker, 2004.

[83] D. Chevalier, A. Sequeira-Munoz, A. L. Bail, B. K. Simpson, and M. Ghoul, "Effect of freezing conditions and storage on ice crystal and drip volume in turbot (Scophthalmus maximus) evaluation of pressure shift freezing vs. air-blast freezing," *Innovative Food Science and Emerging Technologies*, vol. 1, no. 3, pp. 193–201, 2000.

[84] T. Hagiwara, H. Wang, T. Suzuki, and R. Takai, "Fractal analysis of ice crystals in frozen food," *Journal of Agricultural and Food Chemistry*, vol. 50, no. 11, pp. 3085–3089, 2002.

[85] S.-T. Jiang and T.-C. Lee, "Changes in Free Amino Acids and Protein Denaturation of Fish Muscle during Frozen Storage," *Journal of Agricultural and Food Chemistry*, vol. 33, no. 5, pp. 839–844, 1985.

[86] H. Kiani and D.-W. Sun, "Water crystallization and its importance to freezing of foods: A review," *Trends in Food Science & Technology*, vol. 22, no. 8, pp. 407–426, 2011.

[87] G. S. Mittal and M. W. Griffiths, *Pulsed Electric Field Processing of Liquid Foods and Beverage*, D.-W. Sun, Ed., Emerging technologies for food processing, Food Science and Technology, International Series, Elsevier, 2005.

[88] K. Takahashi, N. Inoue, and H. Shinano, "Effect of Storage Temperature on Freeze Denaturation of Carp Myofibrils with KCl or NaCl," *Nippon Suisan Gakkaishi*, vol. 59, no. 3, pp. 519–527, 1993.

[89] L. D. Kaale and T. M. Eikevik, "Changes of proteins during superchilled storage of Atlantic salmon muscle (Salmo salar)," *Journal of Food Science and Technology*, vol. 53, no. 1, pp. 441–450, 2016.

[90] H. K. LIM and N. E. HAARD, "Protein insolubilization in frozen greenland halibut (Reinhardtius hippoglossoides)," *Journal of Food Biochemistry*, vol. 8, no. 3, pp. 163–187, 1984.

[91] E. L. LeBlanc and R. J. LeBlanc, "Determination of hydrophobicity and reactive groups in proteins of cod (Gadus morhua) muscle during frozen storage," *Food Chemistry*, vol. 43, no. 1, pp. 3–11, 1992.

[92] Y. Sultanbawa and E. C. Y. Li-Chan, "Structural changes in natural actomyosin and surimi from ling cod (Ophiodon elongatus) during frozen storage in the absence or presence of cryoprotectants," *Journal of Agricultural and Food Chemistry*, vol. 49, no. 10, pp. 4716–4725, 2001.

[93] S. Benjakul, W. Visessanguan, C. Thongkaew, and M. Tanaka, "Comparative study on physicochemical changes of muscle proteins from some tropical fish during frozen storage," *Food Research International*, vol. 36, no. 8, pp. 787–795, 2003.

[94] H. Buttkus, "Accelerated denaturation of myosin in frozen solution," *Journal of Food Science*, vol. 35, no. 5, pp. 558–562, 1970.

[95] E. L. LeBLANC and R. J. LeBLANC, "Separation of Cod (Gadus morhua) Fillet Proteins by Electrophoresis and HPLC after Various Frozen Storage Treatments," *Journal of Food Science*, vol. 54, no. 4, pp. 827–834, 1989.

[96] M. Pazos, I. Medina, and H. O. Hultin, "Effect of pH on hemoglobin-catalyzed lipid oxidation in cod muscle membranes in vitro and in situ," *Journal of Agricultural and Food Chemistry*, vol. 53, no. 9, pp. 3605–3612, 2005.

[97] K. J. Whittle, "Sea food from producer to consumer, Integrated approach to quality," in *Sea Food from Producer to Consumer, Integrated Approach to Quality. Proceedings of The International Seafood Conference on The 25Th Anniversary of WEFTA*, J. B. Luten, T. Borrosen, and J. Oehlenschlager, Eds., Elseveer, Netherlands, Amsterdam, 1995.

[98] M. Zymon, J. Strzetelski, H. Pustkowiak, and E. Sosin, "Effect of freezing and frozen storage on fatty acid profile of calves' meat," *Polish Journal of Food and Nutrition Sciences*, vol. 57, no. 4(C), pp. 647–650, 2007.

[99] J. S. Omotosho and O. O. Olu, "The effect of food and frozen storage on the nutrient composition of some African fishes.," *Revista de Biología Tropical*, vol. 43, no. 1-3, pp. 289–295, 1995.

[100] M. Kamal, M. N. Islam, M. A. Mansur, M. A. Hossain, and M. A. I. Bhuiyan, "Biochemical and sensory evaluation of hilsa fish (Hilsa ilisha) during frozen storage," *Indian Journal of Marine Sciences*, vol. 25, no. 4, pp. 320–323, 1996.

[101] S. T. Arannilewa, S. O. Salawu, A. A. Sorungbe, and B. B. Ola-Salawu, "Effect of frozen period on the chemical, microbiological and sensory quality of frozen tilapia fish (Sarotherodun galiaenus)," *African Journal of Biotechnology*, vol. 4, no. 8, pp. 852–855, 2005.

[102] J. A. Dudakov, A. M. Hanash, R. R. Jenq et al., "Interleukin-22 drives endogenous thymic regeneration in mice," *Science*, vol. 336, no. 6077, pp. 91–95, 2012.

[103] R. Gandotra, "Change In Proximate Composition And Microbial Count By Low Temperaturepreservation In Fish Muscle Of Labeo Rohita(HamBuch)," *IOSR Journal of Pharmacy and Biological Sciences*, vol. 2, no. 1, pp. 13–17, 2012.

[104] J. M. Santos-Filho, S. M. Morais, D. Rondina, F. Beserra, J. N. M. Neiva, and E. F. Magalhães, "Effect of cashew nut supplemented diet, castration, and time of storage on fatty acid composition and cholesterol content of goat meat," *Small Ruminant Research*, vol. 57, no. 1, pp. 51–56, 2005.

[105] H. O. Hultin, "Oxidation of lipids in Seafoods," in *Chemistry, Processing Technology and Quality*, p. 49, Chapman and Hall, London, UK, 1994.

[106] E. De Pedro, M. Murillo, J. Salas, and F. Peña, "Effect of storage time on fatty acid composition of subcutaneous fat," Unpublished work, supported by CEE (Project n.800-ct90-0013), 1999.

[107] S. Taheri, A. A. Motallebi, A. Fazlara, Y. Aftabsavar, and S. P. Aubourg, "Influence of vacuum packaging and long term storage on some quality parameters of cobia (Rachycentron

canadum) fillets during frozen storage," *American-Eurasian Journal of Agricultural and Environmental Sciences*, vol. 12, no. 4, pp. 541–547, 2012.

[108] I. Aydin and N. Gokoglu, "Effects of temperature and time of freezing on lipid oxidation in anchovy (Engraulis encrasicholus) during frozen storage," *European Journal of Lipid Science and Technology*, vol. 116, no. 8, pp. 996–1001, 2014.

[109] M. Chaijan, S. Benjakul, W. Visessanguan, and C. Faustman, "Changes of lipids in sardine (Sardinella gibbosa) muscle during iced storage," *Food Chemistry*, vol. 99, no. 1, pp. 83–91, 2006.

[110] C. Chávez-Mendoza, J. A. García-Macías, A. D. Alarcón-Rojo, J. Á. Ortega-Gutiérrez, C. Holguín-Licón, and G. Corral-Flores, "Comparison of fatty acid content of fresh and frozen fillets of rainbow trout (Oncorhynchus mykiss) Walbaum," *Brazilian Archives of Biology and Technology*, vol. 57, no. 1, pp. 103–109, 2014.

[111] N. Tenyanga, H. M. Womenib, B. Tiencheub, P. Villeneuved, and M. Lindere, "Effect of refrigeration time on the lipid oxidation and fatty acid profiles of catfish (Arius maculatus) commercialized in Cameroon," *Grasas Y Aceite*, vol. 68, no. 1, 2017.

[112] S. P. Aubourg, A. Rodríguez, and J. M. Gallardo, "Rancidity development during frozen storage of mackerel (Scomber scombrus): Effect of catching season and commercial presentation," *European Journal of Lipid Science and Technology*, vol. 107, no. 5, pp. 316–323, 2005.

[113] S. Simeonidou, A. Govaris, and K. Vareltzis, "Effect of frozen storage on the quality of whole fish and fillets of horse mackerel (Trachurus trachurus) and mediterranean hake (Merluccius mediterraneus)," *European Food Research and Technology*, vol. 204, no. 6, pp. 405–410, 1997.

[114] S. Sathivel, Q. Liu, J. Huang, and W. Prinyawiwatkul, "The influence of chitosan glazing on the quality of skinless pink salmon (Oncorhynchus gorbuscha) fillets during frozen storage," *Journal of Food Engineering*, vol. 83, no. 3, pp. 366–373, 2007.

[115] C. Milo and W. Grosch, "Changes in the Odorants of Boiled Salmon and Cod As Affected by the Storage of the Raw Material," *Journal of Agricultural and Food Chemistry*, vol. 44, no. 8, pp. 2366–2371, 1996.

[116] W. Grosch, "Reactions of hydroperoxides - Products of low molecular weight," in *Autoxidation of Unsaturated Lipids*, pp. 95–139, Academic Press, London, UK, 1987.

2

The use of Partial Least Square Regression and Spectral Data in UV-Visible Region for Quantification of Adulteration in Indonesian Palm Civet Coffee

Diding Suhandy[1] and Meinilwita Yulia[2]

[1]Laboratory of Bioprocess and Postharvest Engineering, Department of Agricultural Engineering, The University of Lampung, Jl. Soemantri Brojonegoro No. 1, Gedong Meneng, Bandar Lampung, Lampung 35145, Indonesia
[2]Department of Agricultural Technology, Lampung State Polytechnic, Jl. Soekarno Hatta No. 10, Rajabasa, Bandar Lampung, Lampung, Indonesia

Correspondence should be addressed to Diding Suhandy; diding.sughandy@fp.unila.ac.id

Academic Editor: Thierry Thomas-Danguin

Asian palm civet coffee or kopi luwak (Indonesian words for coffee and palm civet) is well known as the world's priciest and rarest coffee. To protect the authenticity of luwak coffee and protect consumer from luwak coffee adulteration, it is very important to develop a robust and simple method for determining the adulteration of luwak coffee. In this research, the use of UV-Visible spectra combined with PLSR was evaluated to establish rapid and simple methods for quantification of adulteration in luwak-arabica coffee blend. Several preprocessing methods were tested and the results show that most of the preprocessing spectra were effective in improving the quality of calibration models with the best PLS calibration model selected for Savitzky-Golay smoothing spectra which had the lowest RMSECV (0.039) and highest RPD_{cal} value (4.64). Using this PLS model, a prediction for quantification of luwak content was calculated and resulted in satisfactory prediction performance with high both RPD_p and RER values.

1. Introduction

Coffee is one of the most important food commodities worldwide. Among all commodity traded in the world, coffee is number two after crude oil [1]. There are two important species of coffee which has economic significance in the global coffee trade, species *arabica* (*Coffea arabica*) and *robusta* (*Coffea canephora*). Another important type of coffee is luwak coffee or Asian palm civet coffee or kopi luwak (Indonesian words for coffee and palm civet) which is well known as the world's priciest and rarest coffee [2].

Luwak coffee is any coffee bean (arabica or robusta) which has been eaten and passed through the digestive tract of Asian palm civet (*Paradoxurus hermaphroditus*), which uses its keen senses to select only the best and ripest berries. As a result, its rarity as well as the coffee's exotic and unique production process ultimately accounts for its high selling price, approximately a hundred times higher than regular coffee (International Coffee Organization, http://www.ico.org/prices/pr-prices.pdf).

As one of the most profitable trading products, luwak coffee has been a target for fraud trading by mixing luwak coffee with other cheaper coffee. In order to protect the authenticity of luwak coffee and protect consumer from luwak coffee adulteration, it is very important to develop a robust and easy method for adulteration detection and quantification in luwak coffee. Recently, food authentication is a major challenge that has become increasingly important due to the drive to guarantee the actual origin of a product and for determining whether it has been adulterated with contaminants or filled out with cheaper ingredients [3].

At present, there is no internationally accepted method of verifying whether a roasted bean is luwak coffee or non-luwak coffee. Traditionally, coffee aroma has been used to characterize coffee quality. Sensory panel evaluation is commonly used

TABLE 1: Descriptive statistic of luwak content in coffee samples used for developing calibration and prediction in luwak-arabica coffee blend.

Item	Calibration and validation sample set	Prediction sample set
Number of samples	58	40
Range	1.0~0.5	1.0~0.5
Mean	0.828	0.888
Standard deviation	0.181	0.171
Unit	Gram	Gram

to assess the aroma profile of coffee. However, this technique has some limitations. For example, it is quite difficult to train the panel effectively in order to limit subjectivity of human response to odors and the variability between individuals [4]. Jumhawan et al. [5] used gas chromatography coupled with quadruple mass-spectrometry (GC-Q/MS) to discriminate luwak and regular coffee which resulted in high coefficient of determination (R^2) = 0.965. However, this method is quite expensive analysis with chemical waste included. Indonesia as one of the most important players in luwak coffees production is now just starting to develop an advanced technology for coffee processing. It is including a search for a novel inspection system for luwak coffees characterization. This technology is very important for coffee industry to protect high expensive luwak coffees from any adulteration.

In the previous study, Souto et al. [6] reported the use of UV-Visible spectroscopy as a simple analytical method for the identification of adulterations in ground roasted coffees (due to the presence of husks and sticks). This UV-Vis based analytical method is one of the most common and inexpensive techniques used in routine analysis and it will be compatible with situation in Indonesia for further technology development. Therefore, in this research, we attempt to use UV-Visible spectra combined with chemometrics methods (PLSR/partial least squares regression) to establish a rapid and simple method for quantification of adulteration in luwak-arabica coffee blend.

2. Materials and Methods

2.1. Sample Preparation. An amount of 1 kg ground roasted luwak robusta coffee (Indonesian wild palm civet coffee) was collected directly from coffee farmers at Liwa, Lampung, Indonesia (Hasti coffee Lampung). Another 1 kg ground roasted arabica coffee was also provided for making luwak coffee adulteration. All coffees were roasted in a home coffee roaster (Feike Roaster, W3000) at temperature of 210°C for 15 minutes (medium roasting). All coffees were grinded using home coffee grinder. Since particle sizes in coffee powder have significant influence on spectral analysis, it is important to use the same particle size in coffee powder samples [7]. In this research we use particle size of 420 μm by sieving through a nest of US standard sieves (mesh number of 40) on a Meinzer II sieve shaker (CSC Scientific Company, Inc., USA) for 10 minutes. The experiments were performed at room temperature (around 27–29°C). In this research, we prepared 98 samples of coffee samples which consist of two types of samples, unadulterated (49 samples) and adulterated

samples (49 samples). Unadulterated samples consist of 100% luwak coffee only and adulterated samples consist of luwak coffee with adulteration (adulterated with arabica coffee in the level of adulteration 10% (10 samples), 20% (10 samples), 30% (10 samples), 40% (10 samples), and 50% (9 samples)).

For developing and evaluating calibration model, the samples were divided into two groups: calibration and prediction sample set, respectively. Calibration sample set has 58 samples (24 unadulterated and 34 adulterated samples) and it is going to be used for developing calibration model with full cross-validation method. Prediction sample set has 40 samples (25 unadulterated and 15 adulterated samples) and this set is going to be used for evaluating the performance of developed calibration model. Table 1 shows the detailed information on the samples used in this study.

An aqueous extraction procedure of the coffee samples was performed as described by Souto et al. [6] and Yulia and Suhandy [8]. First, 1.0 g of each sample was weighed and placed in a glass beaker. Then, 10 mL of distilled water was added at 90–98°C and then mixed with magnetic stirring (Cimarec™ Stirrers, model S130810-33, Barnstead International, USA) at 350 rpm for 5 min. Then the samples were filtered using a 25 mm pore-sized quantitative filter paper coupled with an Erlenmeyer. After cooling process to room temperature (for 20 min), all extracts were then diluted in the proportion of 1 : 20 with distilled water. UV-Vis-NIR spectra from the aqueous extracts were acquired using a UV-Vis spectrometer (GENESYS™ 10S UV-Vis, Thermo Scientific, USA).

2.2. Instrumentation and Spectra Data Acquisition. UV-Vis spectra in the range of 190–700 nm were acquired by using a UV-Visible spectrometer (GENESYS 10S UV-Vis, Thermo Scientific, USA) equipped with a quartz cell with optical path of 10 mm and spectral resolution of 1 nm at 27–29°C. Before the measurement step, blank (the same distilled water used in extraction process) was placed inside of the sample cell to adjust the 100% transmittance signal.

2.3. Spectral Data Analysis. All recorded spectra data were transferred to computer via USB flash disk and then converted the spectra data from *.csv* extension into an excel data (*.xls*). Spectral preprocessing is required to remove physical phenomena in the spectra and to remove any irrelevant information such as noise and scattering effect. Recently many preprocessing methods are available in the commercial chemometric analysis tools. Some preprocessing methods were applied, including smoothing (moving average, median filter, and Savitzky-Golay smoothing), multiplicative scatter correction (MSC), and standard normal variate (SNV). The averaging technique is used to reduce the number of wavelengths or to smooth the spectrum of coffee solutions. It is also used to optimize the signal-to-noise ratio [9]. The MSC and SNV are designed to reduce the (physical) variability between samples due to scatter and adjust for baseline shifts between samples [10]. The MSC and SNV have the capability to remove both additive and multiplicative effects in the spectra [11].

Principal component analysis (PCA) was performed before developing the calibration model to determine any relevant and interpretable structure in the data and to detect outliers through the analysis of the Hotelling's T^2 and squared residuals statistics [12]. PCA searches for directions of maximum variability in sample grouping and uses them as new axes called principal components (PC) that can be used as new variables, instead of the original data, in further calculations [13]. PCA results showed that there were no outliers detected in calibration and prediction data sets.

Partial least squares (PLS) regression was used to develop the calibration model for original and preprocessing spectra. PLS finds the directions of greatest variability by considering not only spectral data but also luwak content data, with new axes, called PLS factors (F) or latent variables [13]. The best number of latent variables (LVs) is then chosen according to a commitment between the lowest root mean square error of cross-validation (RMSECV) and the lowest number of latent variables [14, 15]. The quality of the calibration model was evaluated using the following statistical parameters: coefficient of determination between predicted and measured luwak content in luwak-arabica blend (R^2), root mean square error of calibration (RMSEC), root mean square error of cross-validation (RMSECV), bias between actual and predicted luwak content, and ratio prediction to deviation (RPD) value ($RPD_{cal} = SD_{validation\ set}/RMSECV$) [16]. A value of R^2 indicates the percentage of the variance in the Y variable (luwak content in luwak-arabica blend) that is accounted for by the X variable (spectral data). As mentioned by Saeys et al. [17], a calibration model with R^2 value greater than 0.91 is considered to be an excellent calibration, while an R^2 value between 0.82 and 0.90 results in good prediction [18, 19]. A small difference between RMSEC and RMSECV value was also important to avoid "overfitting" in the calibration model [20]. The calibration model should have as high as possible RPD value. The RPD value is desired to be larger than 3 for an acceptable calibration [21]. Calibrations with RPD values between 1.4 and 2 indicate a satisfactory performance of the model which can be useful for rapid screening of samples and may be improved using different sampling strategies or modelling methods and <1.4 indicated an unacceptable model [22].

Spectra preprocessing, PCA, and PLS regression were performed using The Unscrambler® version 9.8 (CAMO, Oslo, Norway), a statistical software for multivariate analysis. A student's paired t-test was performed using Statistical Package for the Social Science (SPSS) version 11.0 for Windows in order to evaluate the significance level of the developed model.

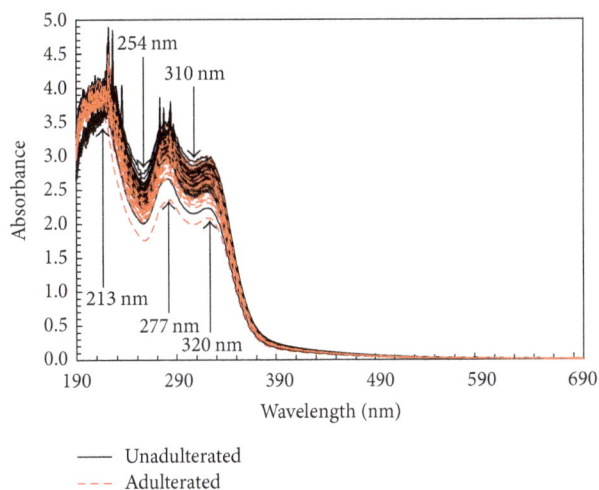

FIGURE 1: Original spectra of unadulterated and adulterated coffee samples in the UV-Vis region.

3. Results and Discussions

3.1. UV-Visible Spectra of Coffee Solution Samples in Range 190–700 nm. Figure 1 shows the original spectra of 98 coffee solution samples in range 190–700 nm. Several peaks can be observed at 213, 277, and 320 nm. It can be seen that all the spectra have similarity in spectral shape and absorbance. The spectra of unadulterated (solid line with black color) and adulterated samples (dashed line with red color) overlap, and it is difficult to detect obvious division between them. The high noise was also observed. Thus, it is necessary to apply appropriate multivariate analysis methods to extract useful information from the spectra, minimize the noise, and build calibration models for quantification of luwak content in luwak-arabica coffee blend.

3.2. Developing PLS Calibration Model for Quantification Luwak Content. Using the PLS regression method the calibration and validation were performed for original and preprocessing spectra (Table 2). The calibration model with the original spectra resulted in a high coefficient of determination ($R_{cal}^2 = 0.97$). In terms of RMSECV, all the preprocessing of spectra (except for mean centering) was effective in improving the quality of calibration model. For smoothing spectra, the calibration model was improved by moving average, media filter, and Savitzky-Golay (SG). Using MSC and SNV spectra, the PLS calibration model was significantly improved as the RMSECVs were decreased. The best PLS calibration model was selected for SG smoothing spectra with window width of 13 points (6-1-6) and polynomial order = 2 which had the lowest RMSECV (0.039) and highest RPD_{cal} value (4.64). This calibration model has 7 optimal numbers of LVs as indicated in Figure 2.

This PLS calibration model was comparable to that reported by Wang et al. [23] for Kona coffee content determination in several brands of commercial Kona coffee blend with $R^2 = 0.996$ for ground Kona coffee blends and R^2 value of 0.999 for brewed Kona coffee. Using metabolomics

TABLE 2: Calibration and validation results for determination of luwak content in luwak-arabica blend using original and preprocessing spectra in the range 200–450 nm.

Type of spectra	F	R_{cal}^2	RMSEC	RMSECV	Bias	RPD_{cal}
Original	7	0.97	0.029	0.062	−0.003	2.92
Moving average smoothing with 5 segments	7	0.97	0.029	0.042	−0.002	4.31
Savitzky-Golay smoothing with window width of 13 points and polynomial order 2	*7*	*0.98*	*0.028*	*0.039*	*−0.002*	*4.64*
Median filter smoothing with 3 segments	7	0.98	0.027	0.050	−0.003	3.62
MSC	7	0.98	0.026	0.055	−0.001	3.29
SNV	7	0.98	0.025	0.054	−0.001	3.35
Baseline offset	7	0.98	0.027	0.060	−0.002	3.02
Mean centering	7	0.97	0.029	0.062	−0.003	2.92

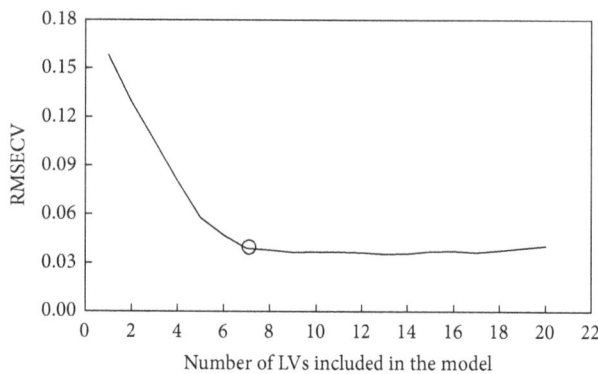

FIGURE 2: Number of LVs versus RMSECV for PLS calibration model for determination of luwak content.

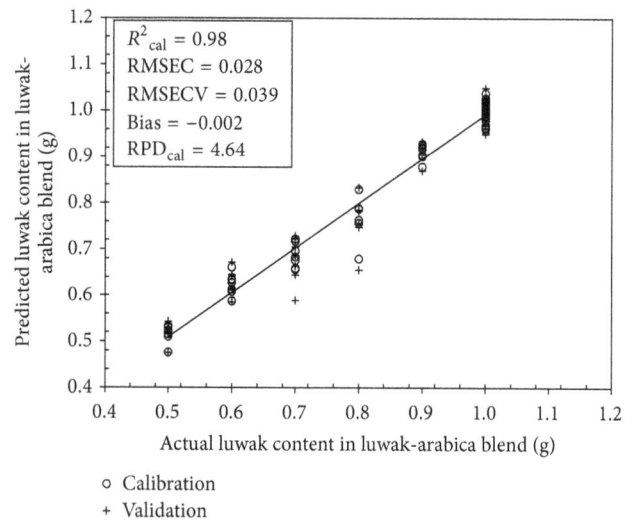

FIGURE 3: PLS calibration and validation model for luwak content determination using SG smoothing spectra in the range 190–700 nm.

approach and orthogonal projection to latent structures (OPLS) prediction technique, Jumhawan et al. [24] developed two prediction OPL models to quantify the degree of coffee adulteration for certified and commercial luwak coffee with $R^2 = 0.975$ and $R^2 = 0.987$, respectively. The scatter plot between actual and predicted luwak content in the best PLS calibration model using SG smoothing spectra is presented in Figure 3.

3.3. Prediction Result Using the Best PLS Calibration Model. To evaluate the performance of the best PLS calibration model, the independent prediction sample set (the sample used for prediction is different with sample used for developing calibration model) is projected onto the best PLS calibration model yielding the prediction set results. From this projection, the root mean square error of prediction (RMSEP), the coefficient of prediction (R^2_p), the range error ratio (RER) (RER = (maximum − minimum)$_{reference\ value}$/RMSEP) [25], and the RPD for prediction ($RPD_p = SD_{prediction\ set}$/RMSEP) were obtained. Both RPD_p and RER are good indicators of evaluating model performance [26, 27]. As for guidance, when RPD_p is greater than 3 and RER is greater than 10 the calibration model is considered to be successful [28–30].

Figure 4 shows the results for luwak content determination based on the best PLS calibration model with SG

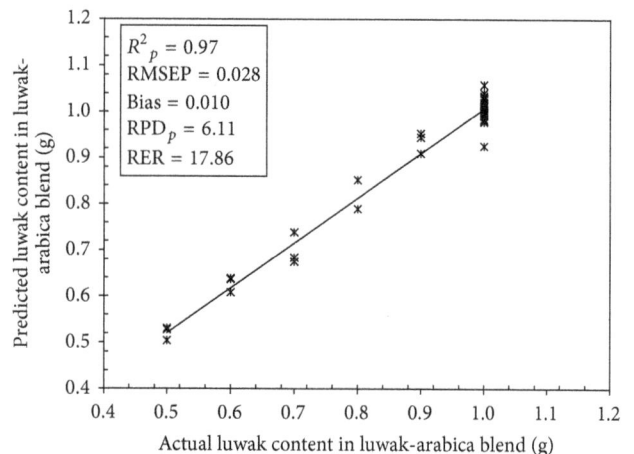

FIGURE 4: Scatter plot of actual versus predicted luwak content calculated using the best PLS calibration model.

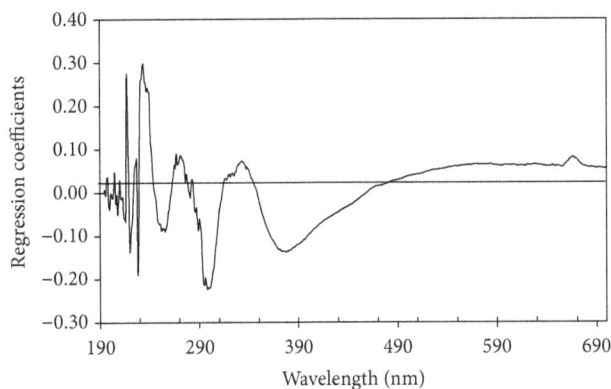

FIGURE 5: Regression coefficients versus wavelength of coffee samples.

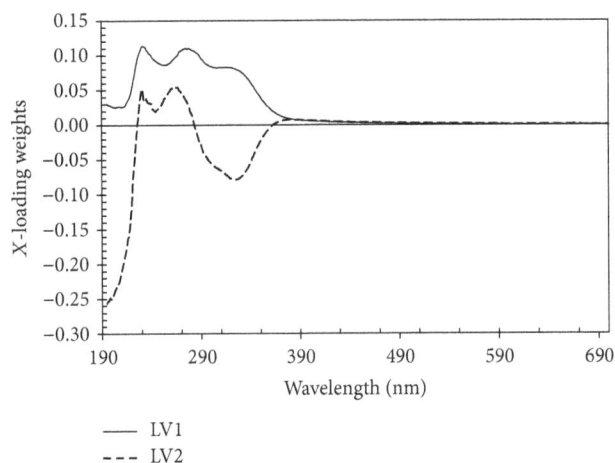

FIGURE 6: X-loading weights versus wavelength of the top two latent variables (LV1 and LV2) of coffee samples.

smoothing spectra. It has high R^2_p = 0.97 with low RMSEP = 0.028. From the RPD_p value, it can be seen that the RMSEP was much lower than the standard deviation (SD = 0.171) of reference data which resulted in high RPD_p value. The obtained RER is also quite good (17.86). By a 95% confidence paired t-test there were no significant differences between actual and predicted luwak content. This indicates that an accurate calibration model can be developed for the determination of luwak content in luwak-arabica coffee blends using UV-Vis spectroscopy and PLS regression.

3.4. Selection of Important Wavelengths. In order to understand the complexity of developed PLS model, regression coefficients and x-loading weights of the best PLS model were presented in Figures 5 and 6, respectively. The x-loadings show how well the x-variable (wavelengths) is taken into account by the model components. It can be used to understand how much each x-variable (wavelengths) contributes to the meaningful variety variation in the data and to interpret variable relationships. It is also useful to interpret the meaning of each model component. The loading weights

show how much each wavelength (x-variables) contributes to explaining the response variation (degree of adulteration) along each model component. The loading weights are normalized, so that their lengths could be interpreted as well as their directions. Wavelengths (x-variables) with large loading weight values are important for the determination of luwak content in luwak-arabica blend. With a similar function, regression coefficients are primarily used to check the effects of different wavelengths (x-variables) in determination of luwak content in luwak-arabica blend. Large absolute values indicate the importance and significance of the effects of the wavelengths. According to Figure 5 we can see several peaks and valleys at certain wavelengths which were considered to be more important for determination of luwak content in luwak-arabica blend, such as 228, 256, 274, 299, 332, and 376 nm. In Figure 6 we can notice several wavelengths with high contribution to the developed PLS model at 228, 246, 274, and 320 nm. We can see that all important wavelengths are in the ultraviolet spectral region. In the visible region we could not find any important wavelengths indicated with low x-loading and regression coefficients in the visible region. It is shown that the determination of luwak content in luwak-arabica blend is mainly characterized in the ultraviolet region. The several observed important wavelengths in this study are closely related to the absorbance of several important chemical compositions in roasting coffee. For example, the wavelength at 276 nm can be found both in x-loading and in regression coefficient plot and this wavelength is related to the absorbance of caffeine while wavelength at 320 nm is related to absorbance of caffeic acid [6]. The wavelengths at 246, 299, and 320 nm are closely related to the absorbance of chlorogenic acids (CGA) [31]. The wavelength at 256 nm is closely related to the absorbance of vanillic acid.

4. Conclusion

In this research, the determination of luwak content in luwak-arabica coffee blends was achieved by using UV-Visible spectroscopy and PLS regression. The best PLS calibration model with Savitzky-Golay smoothing spectra resulted in satisfactory prediction with excellent value both for RPD and for RER. Several wavelengths with high contribution to the luwak content determination were confirmed including 276 nm which is related to the absorbance of caffeine while wavelength at 320 nm is related to absorbance of caffeic acid. This research shows the possibility of developing a simple, rapid, and economic method for determining luwak content in luwak-arabica coffee blends using UV-Visible spectroscopy and multivariate analysis.

Conflicts of Interest

The authors declare that they have no conflicts of interest.

Acknowledgments

The authors gratefully acknowledge support of the Indonesian Ministry of Research, Technology and Higher Education

(KEMENRISTEKDIKTI) via Penelitian Strategis Nasional (STRANAS) 2016 (no. 419/UN26/8/LPPM/2016). The authors also would like to acknowledge the University of Lampung, Indonesia, for providing the laboratory facilities and Hasti Coffee Lampung for providing them with the coffee samples. They also thank Professor Garry John Piller (Graduate School of Agriculture, Kyoto University, Japan) for his help and useful discussions.

References

[1] P. Esquivel and V. M. Jiménez, "Functional properties of coffee and coffee by-products," *Food Research International*, vol. 46, no. 2, pp. 488–495, 2012.

[2] M. F. Marcone, "Composition and properties of Indonesian palm civet coffee (Kopi Luwak) and Ethiopian civet coffee," *Food Research International*, vol. 37, no. 9, pp. 901–912, 2004.

[3] G. P. Danezis, A. P. Tsagkaris, F. Camin, V. Brusic, and C. A. Georgiou, "Food authentication: techniques, trends & emerging approaches," *TrAC Trends in Analytical Chemistry*, vol. 85, pp. 123–132, 2016.

[4] N. F. Shilbayeh and M. Z. Iskandarani, "Quality control of coffee using an electronic nose system," *American Journal of Applied Sciences*, vol. 1, pp. 129–135, 2004.

[5] U. Jumhawan, S. P. Putri, Yusianto, E. Marwani, T. Bamba, and E. Fukusaki, "Selection of discriminant markers for authentication of asian palm civet coffee (Kopi Luwak): a metabolomics approach," *Journal of Agricultural and Food Chemistry*, vol. 61, no. 33, pp. 7994–8001, 2013.

[6] U. T. C. P. Souto, M. F. Barbosa, H. V. Dantas et al., "Identification of adulteration in ground roasted coffees using UV–Vis spectroscopy and SPA-LDA," *LWT—Food Science and Technology*, vol. 63, no. 2, pp. 1037–1041, 2015.

[7] D. Suhandy, S. Waluyo, C. Sugianti et al., "The use of UV-Vis-NIR spectroscopy and chemometrics for identification of adulteration in ground roasted arabica coffees -Investigation on the influence of particle size on spectral analysis," in *Proceeding of Seminar Nasional Tempe*, Bandar Lampung, Indonesia, May 2016.

[8] M. Yulia and D. Suhandy, "Indonesian palm civet coffee discrimination using UV-visible spectroscopy and several chemometrics methods," *Journal of Physics: Conference Series*, vol. 835, no. 1, pp. 1–6.

[9] H. Cen and Y. He, "Theory and application of near infrared reflectance spectroscopy in determination of food quality," *Trends in Food Science and Technology*, vol. 18, no. 2, pp. 72–83, 2007.

[10] A. Rinnan, F. van den Berg, and S. B. Engelsen, "Review of the most common pre-processing techniques for near-infrared spectra," *Trends in Analytical Chemistry*, vol. 28, no. 10, pp. 1201–1222, 2009.

[11] Y. Bi, K. Yuan, W. Xiao et al., "A local pre-processing method for near-infrared spectra, combined with spectral segmentation and standard normal variate transformation," *Analytica Chimica Acta*, vol. 909, pp. 30–40, 2016.

[12] T. Naes, T. Isaksson, T. Fearn, and T. Davies, *A User-Friendly Guide to Multivariate Calibration and Classification*, NIR Publications, Chichester, UK, 2002.

[13] M. Blanco and I. Villarroya, "NIR spectroscopy: a rapid-response analytical tool," *Trends in Analytical Chemistry*, vol. 21, no. 4, pp. 240–250, 2002.

[14] D. Suhandy, M. Yulia, Y. Ogawa, and N. Kondo, "Prediction of L-ascorbic acid using FTIR-ATR terahertz spectroscopy combined with interval partial least squares (iPLS) regression," *Engineering in Agriculture, Environment and Food*, vol. 6, no. 3, pp. 111–117, 2013.

[15] M. Yulia, D. Suhandy, Y. Ogawa, and N. Kondo, "Investigation on the influence of temperature in l-ascorbic acid determination using FTIR-ATR terahertz spectroscopy: calibration model with temperature compensation," *Engineering in Agriculture, Environment and Food*, vol. 7, no. 4, pp. 148–154, 2014.

[16] R. G. Brereton, "Introduction to multivariate calibration in analytical chemistry," *Analyst*, vol. 125, no. 11, pp. 2125–2154, 2000.

[17] W. Saeys, A. M. Mouazen, and H. Ramon, "Potential for onsite and online analysis of pig manure using visible and near infrared reflectance spectroscopy," *Biosystems Engineering*, vol. 91, no. 4, pp. 393–402, 2005.

[18] D. Suhandy, T. Suzuki, Y. Ogawa et al., "A quantitative study for determination of glucose concentration using attenuated total reflectance terahertz (ATR-THz) spectroscopy," *Engineering in Agriculture, Environment and Food*, vol. 5, no. 3, pp. 90–95, 2012.

[19] P. Williams, *Near-infrared Technology-Getting the Best Out of Light*, Nanaimo, Canada, 2003.

[20] A. H. Gómez, Y. He, and A. G. Pereira, "Non-destructive measurement of acidity, soluble solids and firmness of Satsuma mandarin using Vis/NIR-spectroscopy techniques," *Journal of Food Engineering*, vol. 77, no. 2, pp. 313–319, 2006.

[21] P. Williams, *Grains and seeds. In Near-Infrared Spectroscopy in Food Science and Technology*, vol. 7, John Wiley and Sons, Hoboken, N.J, 2007.

[22] C. W. Chang, D. A. Laird, M. J. Mausbach, and C. R. Hurburgh, "Near-infrared reflectance spectroscopy-principal components regression analyses of soil properties," *Soil Science Society of America Journal*, vol. 65, pp. 480–490, 2001.

[23] J. Wang, S. Jun, H. C. Bittenbender, L. Gautz, and Q. X. Li, "Fourier transform infrared spectroscopy for kona coffee authentication," *Journal of Food Science*, vol. 76, no. 5, pp. 385–391, 2009.

[24] U. Jumhawan, S. P. Putri, Yusianto, T. Bamba, and E. Fukusaki, "Quantification of coffee blends for authentication of Asian palm civet coffee (Kopi Luwak) via metabolomics: a proof of concept," *Journal of Bioscience and Bioengineering*, vol. 122, no. 1, pp. 79–84, 2016.

[25] L. M. Magalhães, S. Machado, M. A. Segundo, J. A. Lopes, and R. N. M. J. Páscoa, "Rapid assessment of bioactive phenolics and methylxanthines in spent coffee grounds by FT-NIR spectroscopy," *Talanta*, vol. 147, pp. 460–467, 2016.

[26] A. D. Girolamo, V. Lippolis, E. Nordkvist, and A. Visconti, "Rapid and non-invasive analysis of deoxynivalenol in durum and common wheat by fourier-transform near infrared (FT-NIR) spectroscopy," *Food Additives and Contaminants. Part A Chemistry, Analysis, Control, Exposure and Risk Assessment*, vol. 26, no. 6, pp. 907–917, 2009.

[27] K. H. Esbensen, D. Guyot, F. Westad, and L. P. Houmoller, "Multivariate data analysis – in practice: an introduction to multivariate data analysis and experimental design," *Journal of Chemometrics*, 2004.

[28] P. Williams and D. Sobering, "How do we do it: a brief summary of the methods we use in developing near infrared calibrations," in *Near Infrared Spectroscopy: The Future Waves*, A. M. C. Davies and P. Williams, Eds., NIR Publications, Chichester, UK, 1996.

[29] D. F. Malley, L. Yesmin, and R. G. Eilers, "Rapid analysis of hog manure and manure-amended soils using near-infrared spectroscopy," *Soil Science Society of America Journal*, vol. 66, no. 5, pp. 1677–1686, 2002.

[30] C. Lorenzo, T. Garde-Cerdán, M. A. Pedroza, G. L. Alonso, and M. R. Salinas, "Determination of fermentative volatile compounds in aged red wines by near infrared spectroscopy," *Food Research International*, vol. 42, no. 9, pp. 1281–1286, 2009.

[31] A. Belay and A. V. Gholap, "Characterization and determination of chlorogenic acids (CGA) in coffee beans by UV-Vis spectroscopy," *African Journal of Pure and Applied Chemistry*, vol. 3, pp. 234–240, 2009.

3

Good Manufacturing Practices and Microbial Contamination Sources in Orange Fleshed Sweet Potato Puree Processing Plant in Kenya

Derick Nyabera Malavi ⓘ,[1,2] Tawanda Muzhingi ⓘ,[2] and George Ooko Abong'[1]

[1]Department of Food Science, Nutrition and Technology, University of Nairobi, P.O. Box 29053, Nairobi 00625, Kenya
[2]International Potato Centre (CIP), Sub-Saharan Africa (SSA) Regional Office, Old Naivasha Road, P.O. Box 25171, Nairobi 00603, Kenya

Correspondence should be addressed to Derick Nyabera Malavi; nyaberad26@gmail.com

Academic Editor: Marie Walsh

Limited information exists on the status of hygiene and probable sources of microbial contamination in Orange Fleshed Sweet Potato (OFSP) puree processing. The current study is aimed at determining the level of compliance to Good Manufacturing Practices (GMPs), hygiene, and microbial quality in OFSP puree processing plant in Kenya. Intensive observation and interviews using a structured GMPs checklist, environmental sampling, and microbial analysis by standard microbiological methods were used in data collection. The results indicated low level of compliance to GMPs with an overall compliance score of 58%. Microbial counts on food equipment surfaces, installations, and personnel hands and in packaged OFSP puree were above the recommended microbial safety and quality legal limits. Steaming significantly ($P < 0.05$) reduced microbial load in OFSP cooked roots but the counts significantly ($P < 0.05$) increased in the puree due to postprocessing contamination. Total counts, yeasts and molds, Enterobacteriaceae, total coliforms, and *E. coli* and *S. aureus* counts in OFSP puree were 8.0, 4.0, 6.6, 5.8, 4.8, and 5.9 $\log_{10}$ cfu/g, respectively. In conclusion, equipment surfaces, personnel hands, and processing water were major sources of contamination in OFSP puree processing and handling. Plant hygiene inspection, environmental monitoring, and food safety trainings are recommended to improve hygiene, microbial quality, and safety of OFSP puree.

1. Introduction

Sweet potato is one of the most important food crops in Kenya. According to FAOSTAT [1], sweet potato production in Kenya stood at 697,364 tonnes in the year 2016. Kenya is the sixth largest producer of sweet potato in Africa with an average yield of 8.2 tonnes/ha [2]. In a review by Abong' et al. [3], major sweet potato producing counties in Kenya as per the year 2014 were Homa Bay, Busia, Migori, and Bungoma counties. Many cultivars of sweet potatoes differentiated by color and shape are cultivated in Kenya. The flesh color ranges from white, cream, purple, yellow, and orange [4]. Breeding and utilization of biofortified Orange Fleshed Sweet Potato (OFSP) variety is being promoted in Kenya and other sub-Saharan (SSA) countries by research organizations

due its high beta-carotene (provitamin A) content [5–8]. OFSP is an important food crop for income generation, addressing vitamin A deficiency, and food insecurity in SSA [6, 9]. Depending on the region, OFSP cultivars grown in Kenya include Kabode, Vitaa, SPK 004, and Ejumula [10]. Sweet potato is often processed into purees that are subsequently incorporated as a food ingredient in baby foods, puddings, doughnuts, buns, breads, cakes, cookies, soups, and beverages [6, 11]. Since the year 2015, the International Potato Centre (CIP) has been working with a privately owned OFSP puree processing company operated on a small-scale basis and one of the largest retail chain stores in Kenya to promote utilization of OFSP puree in bakery applications and enhance intake of vitamin A among the urban population [12].

Most studies are currently focusing on nutritional and socioeconomic benefits of OFSP but little effort has been directed towards enhancing food safety along the OFSP puree value chain that has gained prominence in Kenya. Food safety problems are more pronounced in developing countries where food production is frequently done under unsanitary conditions [13]. Microbial quality and safety of foods can constantly be achieved by implementation and adherence to Good Hygiene and Good Manufacturing Practices (GMPs) in processing. There is a potential of contamination in sweet potato puree along the process line that could result from poor hygiene practices and contamination from the processing environment [11]. Microbial food contamination is associated with incidences of foodborne illness and food spoilage [14]. Food contamination emanates from the use of contaminated raw materials and ingredients in processing, poor personal hygiene, ineffective cleaning and sanitation of food contact surfaces, and contamination from food processing environment [13, 15–17].

Several microorganisms ranging from spoilage and pathogenic and indicator microorganisms are important in assessing safety, hygiene, and sanitary quality of foods and processing environments. These classes of microorganisms are comprised of total viable counts (TVC), yeasts and molds, *Staphylococcus aureus*, Enterobacteriaceae, coliforms, and *Escherichia coli*. TVC, yeasts, and molds are indicators of hygiene, sanitation, and microbial quality of both raw and processed foods. High total aerobic counts in foods are often associated with accelerated spoilage, hence deterioration in food quality [18]. High counts of *S. aureus* in foods and processing environment depict extensive handling and poor hand washing hygiene practices by food handlers. Consequentially, *S. aureus* counts above 10^5 CFU/g produce heat-stable toxins responsible for staphylococcal food poisoning [19]. Coliforms, *E. coli*, and Enterobacteriaceae are useful indicators of water quality, personnel hygiene, and efficacy of cleaning and sanitation programs in food processing plants [20–22]. Control of contamination from persistent microorganisms in food processing environments can be achieved by application of quality assurance approaches such as GMPs and Environmental Monitoring Programs (EMPs). EMPs identify harborage niches for pathogens, spoilage, and indicator microorganisms that may act as a source of contamination and verifies adherence and implementation of GMPs in food processing environments [23].

Despite the economic and nutritional benefits accrued from OFSP puree processing, upholding food safety regulations is still a challenge that needs to be addressed. Like many small-scale food processing industries, OFSP puree processing is prone to microbial contamination attributed to low level of food safety knowledge and practices from food handlers [24] and poor hygiene within the processing environment [25]. Relatively high pH and high water activity in sweet potato puree further provide an excellent environment for growth of a wide array of both spoilage and pathogenic microorganisms [11]. There is lack of information on the level of compliance to GMPs and levels and sources of microbial contamination in OFSP puree processing in Kenya. There is need to generate data for identifying food safety risk areas and

provide recommendations for improving hygiene, microbial quality, and safety of OFSP puree. The objective of the current study was therefore to determine the level of compliance to GMPs, sources, and levels of microbial contamination in OFSP puree processing plant in Homa Bay County, Kenya.

2. Material and Methods

2.1. Study Setting and Design. The study was conducted at a privately owned OFSP puree processing plant in Homa Bay County, Kenya. A cross-sectional analytical study design was used for data collection. Intensive observation and interviews guided by a structured GMPs checklist was used to assess the level of compliance to GMPs at the processing plant [25, 26]. The facility and its operations were evaluated for compliance to GMPs on aspects of suitability of buildings and sanitary facilities for food production, personal hygiene, food contact surfaces and equipment, pest control, and process control. The findings were categorized as either compliant or noncompliant, totaled, and converted into a percentage.

Samples for microbiological analysis were randomly collected from OFSP puree processing environment as described by Barros et al. [16]. A total of 62 samples comprising environmental samples, processing water, and OFSP samples were collected for microbial analysis. Swab samples from equipment, walls, floors, drains, and personnel hands were collected using 3M buffered swab sponges [27]. Sterile papers were used to outline areas of 30 cm² and 60 cm² on surfaces for swabbing. Samples from surfaces of equipment and installations were collected after cleaning. Samples from personnel hands were collected during working hours. OFSP samples were collected from three different batches at different processing stages: after washing, steaming, cooling, and cutting and packaging. All the samples were transported in a cold box filled with ice packs to the Department of Food Science, Nutrition and Technology, Upper Kabete Campus, University of Nairobi, and immediately analyzed for total aerobic counts, yeasts and molds, Enterobacteriaceae, coliforms, *Escherichia coli*, and *Staphylococcus aureus*.

2.2. Sample Preparation, Microbial Analysis, and Enumeration. All swab sample sponges (each presoaked in 10 ml buffer) were diluted with 90 ml sterile saline water (0.85% NaCl). The swab sponges were squeezed to release microbes from the surface before making successive serial dilutions. Twenty-five grams of process water and OFSP samples was each diluted with 225 ml of 0.85% NaCl before making subsequent serial dilutions as described by Gungor and Gokoglu [27].

2.2.1. Determination of Total Viable Counts (TVC). Total viable counts (TVC) were determined by spread plate method on Plate Count Agar (PCA, LAB, UK). The plates were incubated at 35°C for 48 hours as described by Pérez-Díaz et al. [11].

2.2.2. Determination of Yeasts and Molds. Yeasts and molds were enumerated by plating 0.1 mL of each sample on Petri dishes with solidified Potato Dextrose Agar (PDA) (Oxoid,

TABLE 1: Assessment of Good Manufacturing Practices for equipment used in OFSP puree processing.

Item	Yes	No	Status
(1) Are all cleaned food equipment surfaces sanitized as necessary to prevent contamination of the product?		√	Noncompliant
(2) Is the equipment designed or otherwise suitable for use in OFSP puree processing?	√		Compliant
(3) Is there a build-up of food or other material on the equipment?		√	Compliant
(4) Is there any build-up or seepage of cleaning solvents or lubricants on your equipment which can contaminate food?		√	Compliant
(5) Is the equipment hard to dissemble for clean-up and inspection?		√	Compliant
(6) Is there a lot of dead space in and around the machinery where the food and other debris can collect and act as nest for insects and bacteria?		√	Compliant
(7) Can the surface of the equipment be sanitized?	√		Compliant

Hampshire). The plates were incubated at 25°C for 5 days as previously described by Gungor and Gokoglu [27].

2.2.3. Determination of Enterobacteriaceae.

Enterobacteriaceae group of microorganisms were determined by spread plating 0.1 mL of each sample on Violet Red Bile Glucose (VRBG) Agar (Oxoid, Hampshire, England). The plates were incubated at 37°C for 24 hours as described in a method by Hervert et al. [28].

2.2.4. Detection of Coliforms and Escherichia coli.

The presence of coliforms and *E. coli* was examined by plating 0.1 mL of each sample on Brilliance *E. coli*/coliform agar (Oxoid, Hampshire, England) according to the method by Sylvia et al. [29]. The plates were incubated at 37°C for 24 hours. Dark blue colonies were enumerated as *E. coli* while pink colonies were recorded as total coliforms.

2.2.5. Detection of Staphylococcus aureus.

Staphylococcus aureus was determined as per the method previously used by Gungor and Gokoglu [27]. Plating of 0.1 mL of each sample was done on Baird Parker agar (Oxoid, Hampshire, England). The plates were incubated at 37°C for 48 hours. Typical *S. aureus* colonies were enumerated and streaked in Brain Heart Infusion (BHI) broth (Oxoid) and further incubated at 37°C for 24 h. Typical *S. aureus* colonies were confirmed by coagulase test [30]. Test for coagulation was done by aseptically adding 0.1 mL of BHI culture to 0.3 mL of rabbit plasma in sterile hemolysis tubes. The tubes were incubated at 37°C and observed for coagulation after 6 hours.

2.2.6. Enumeration of Microbial Colonies.

Enumeration was done for plates with 30–300 colonies. All microbial counts were expressed as $\log_{10}$ CFU/g for OFSP samples, $\log_{10}$ CFU/ml for water sample, and $\log_{10}$ CFU/cm^2 for swab samples.

2.3. Statistical Data Analysis.

Compliance to GMPs was presented in tables. TVC, yeasts and molds, *S. aureus*, Enterobacteriaceae, coliforms, and *E. coli* counts were converted to $\log_{10}$ CFU units in Microsoft Excel (MS Office 365), exported to SPSS (IBM SPSS Version 20) for analysis before being tabulated as means and standard deviations. Analysis of variance and Tukey's test were used to determine statistical differences in the level of microbial contamination in the samples with a preset P value of 0.05.

3. Results and Discussion

3.1. Suitability and State of Hygiene of Equipment Used for OFSP Puree Processing.

The suitability, design, cleaning, and sanitation of equipment for OFSP puree processing are shown in Table 1. The equipment surfaces were well designed for use in food processing but lack of sanitation procedures at the plant resulted in high microbial load of the surfaces as shown in Table 2. Lowest total aerobic counts were detected in packaging bags (6.6 ± 0.3 log cfu·cm^{-2}) and the highest level of contamination was detected from weighing spoons (9.5 ± 0.0 cfu·cm^{-2}). The pureeing machine was least contaminated with yeasts and molds (4.3 ± 1.0 log cfu·cm^{-2}). Low Enterobacteriaceae counts were detected in packaging bags and cooling trays with mean counts of 5.8 ± 0.6 and 5.8 ± 1.2 log cfu·cm^{-2}, respectively. Highest counts of *S. aureus*, Enterobacteriaceae, and coliform counts with means 6.5 ± 0.0, 7.0 ± 0.0, and 6.7 ± 0.0 log cfu·cm^{-2}, respectively, were detected on the inside cabin surface of the truck used in transportation of OFSP roots and OFSP puree. Highest yeasts and molds and *E. coli* counts with means 6.8 ± 0.5 and 5.3 ± 0.8 log cfu·cm^{-2}, respectively, were obtained from buckets used for washing OFSP roots. The mean counts among different food equipment surfaces were statistically different ($P < 0.05$).

High counts above 10^5 cfu·cm^{-2} for aerobic plate counts, *S. aureus*, yeasts and molds, coliforms, and Enterobacteriaceae were detected on more than 90% of all equipment surfaces indicating inadequate cleaning and sanitation procedures. This is similar to findings by Schlegelová et al. [22] that reported relatively high counts for total counts, enterococci, *E. coli*, and *Staphylococci* spp. on food equipment surfaces in dairy and meat processing plants. The high contamination level from equipment is attributed to lack of adherence to documented cleaning procedures by food handling personnel and lack of sanitation program at the establishment. Inefficient cleaning and sanitation of equipment surfaces lead to formation of biofilms, a potential source of food contamination [14, 22]. High counts on knives,

TABLE 2: Microbial Counts on equipment surfaces in OFSP puree processing plant[1] (log Mean CFU·cm^{-2}).

Sample	TVC	Yeast-molds	S. aureus	Enterobacteriaceae	Coliforms	E. coli
OFSP buckets	7.71 ± 0.24^{bcde}	6.79 ± 0.45^{ef}	5.08 ± 0.32^{bcd}	7.33 ± 0.09^{bcd}	6.51 ± 0.10^{efgh}	5.29 ± 0.75^{b}
OFSP scrub brushes	7.83 ± 0.02^{bcde}	6.32 ± 0.01^{def}	4.52 ± 0.21^{bcd}	7.28 ± 0.03^{bcd}	6.44 ± 0.13^{efgh}	4.34 ± 0.15^{b}
Knives	6.57 ± 0.32^{bc}	4.35 ± 0.29^{bcde}	5.06 ± 0.13^{bcd}	6.24 ± 0.02^{bcd}	5.51 ± 0.15^{def}	4.79 ± 0.33^{b}
Tables	8.12 ± 1.24^{bcde}	5.73 ± 1.13^{cdef}	4.44 ± 0.84^{bcd}	6.71 ± 0.07^{bcd}	6.35 ± 0.54^{defgh}	2.16 ± 3.74^{a}
Cooling trays	7.45 ± 1.37^{bcde}	4.74 ± 0.27^{cdef}	4.68 ± 0.55^{bcd}	5.83 ± 1.20^{bcd}	5.45 ± 1.16^{cde}	2.68 ± 2.10^{a}
Puree machine	7.63 ± 0.81^{bcde}	4.29 ± 1.03^{bcd}	4.43 ± 0.63^{bcd}	6.34 ± 0.45^{bcd}	5.58 ± 0.20^{defg}	2.04 ± 2.23^{a}
Weighing spoons	9.47 ± 0.03^{e}	6.45 ± 0.08^{def}	5.41 ± 0.14^{cd}	6.43 ± 0.10^{bc}	6.17 ± 0.12^{defgh}	1.81 ± 2.57^{a}
Packaging bags	6.28 ± 0.09^{bc}	4.83 ± 1.35^{cdef}	3.16 ± 0.43^{bc}	5.83 ± 0.64^{bc}	5.65 ± 0.73^{defgh}	nd*
Packaging machine	9.10 ± 0.74^{de}	6.15 ± 0.51^{cde}	5.15 ± 0.04^{bcde}	6.77 ± 0.05^{cde}	6.45 ± 0.02^{defg}	3.58 ± 0.08^{b}
Freezers	7.46 ± 0.13^{bcde}	5.95 ± 0.83^{cdef}	4.89 ± 1.53^{bcd}	6.64 ± 0.11^{bcd}	6.32 ± 0.40^{defgh}	1.21 ± 2.11^{b}
Cold boxes	7.97 ± 0.13^{bcde}	5.48 ± 0.55^{cdef}	5.65 ± 0.11^{cd}	6.27 ± 0.17^{cd}	5.58 ± 0.24^{defg}	4.36 ± 0.76^{b}
Shipment vehicle	8.33 ± 0.01^{bcde}	5.28 ± 0.00^{cde}	6.45 ± 0.02^{e}	7.43 ± 0.03^{e}	6.68 ± 0.01^{fgh}	4.70 ± 0.96^{b}

[1] All values reflect mean counts and standard deviation. Values bearing different superscript letters in each column are significantly different ($P < 0.05$); TVC: total viable counts; nd*: microbial parameter not detected.

TABLE 3: Assessment of GMPs for buildings, grounds, and structures for OFSP puree processing.

Item	Yes	No	Status
(1) Is the OFSP puree processing plant located in an environment free of dust?	√		Compliant
(2) Is the area around the plant clear of litter, weeds, grass and brush?	√		Compliant
(3) Is there any standing water on the ground which might also attract pests?		√	Compliant
(4) Are floors, walls and drains properly cleaned?		√	Noncompliant
(5) Are floors made of alkali and acid resistant material?	√		Compliant
(6) Do floors have sufficient slope to avoid water stagnation?		√	Compliant
(7) Do production area doors and windows to the outside have fine mesh screens to keep out insects? If not are they tightly sealed?	√		Compliant
(8) Have all holes and cracks been filled so as not provide hiding places or entry points for pests?	√		Compliant
(9) Are there any presence of domestic animals such as cats and dogs?		√	Compliant
(10) Are the hand washing facilities furnishedwith soap and paper towels?		√	Noncompliant
(11) Are there any leaks in the roof, sky lights, windows or overhead piping?		√	Compliant
(12) Are drains adequately designed to handle the volume of waste water?		√	Noncompliant
(13) Are all drains fitted with screens and waste traps to prevent pest entry into the processing areas?	√		Compliant
(14) Are the overhead lights covered with shields to prevent contamination of products by broken glass in case the lamps burst?		√	Noncompliant

cooling trays, tables, and pureeing machine were identified as primary sources of contamination in OFSP puree. Efficient cleaning and sanitation following documented procedures should always be done at the beginning of each work day, after every batch processing, and at the end of the day after processing to prevent formation of biofilms on equipment surfaces and contamination in OFSP puree processing.

3.2. Quality of Water Used in Orange Fleshed Sweet Potato Puree Processing. The level of microbial contamination in water for processing OFSP puree is shown in Figure 1. It was highly contaminated ($>10^4$ CFU/ml) with Enterobacteriaceae, coliforms, and *Escherichia coli* due to nonexistence of water treatment (disinfection) program at the facility. As stipulated by Environmental Protection Agency [31], total coliforms and *E. coli* should be absent in 100 ml of water sample to be deemed suitable for drinking and food processing. It further recommends total counts not to exceed 500 cfu/ml. Use of untreated water from unknown sources contaminates equipment and foods prepared on these surfaces [32]. High Enterobacteriaceae, coliforms, and *E. coli* counts in water for OFSP puree processing indicated contamination with fecal matter, deterioration in water quality, and likelihood presence of enteric pathogens [32–34]. The water at the facility was therefore not suitable for use in processing. It was a possible source of contamination on equipment and personnel hands and consequently in OFSP puree. There was an urgent need to establish water disinfection program involving chlorination at the facility for preventing water to puree contamination.

3.3. Suitability of Buildings and State of Hygiene of Installations for OFSP Puree Processing. Prerequisite programs with respect to design of buildings, sanitary facilities, and pest control program in OFSP puree processing are shown in

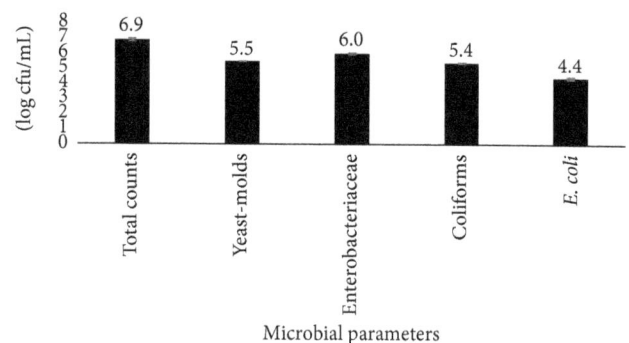

FIGURE 1: Microbial load in water for processing Orange Fleshed Sweet Potato puree; the values above the bars represent the mean ± SEM (standard error of the mean).

Tables 3, 4, and 5, respectively. Cleaning of walls, floors, and drains was not done efficiently as evidenced from our observation. This resulted in high level of contamination on installations as shown in Table 6. Total aerobic counts from floors, walls, and drains were identical with lowest counts (8.0 ± 0.9 log cfu·cm^{-2}) recorded from floors and highest counts (8.7 ± 0.7 log cfu·cm^{-2}) from drains. Yeasts and molds counts were above 10^5 CFU·cm^{-2} but insignificant ($P > 0.05$) among the three installation points. Drains had the lowest contamination level for yeasts and molds (5.5 ± 1.3 log cfu·cm^{-2}) while the walls had the highest counts (6.2 ± 0.6 log cfu·cm^{-2}). The level of *Staphylococcus aureus* was low on floors (5.3 ± 0.4 log cfu·cm^{-2}) and almost identical but high on walls (5.5 ± 0.0 log cfu·cm^{-2}). Low and high Enterobacteriaceae counts were obtained from floors and drains with mean counts of 7.0 ± 0.4 and 7.2 ± 0.1 log cfu·cm^{-2}, respectively. The level of contamination with coliforms was

TABLE 4: Good Manufacturing Practices assessment for sanitary facilities in puree processing.

Item	Yes	No	Status
(1) Is trash, debris and clutter picked up, both inside and outside, so as not to provide hiding places for pests?	√		Compliant
(2) Are OFSP puree handlers provided with designated areas for eating, drinking and using tobacco products?		√	Noncompliant
(3) Is food spilled cleaned up quickly so as not to attract pests or breed bacteria?	√		Compliant
(4) Is garbage quickly removed and dumped in appropriate bins?	√		Compliant
(5) Is the garbage kept covered?		√	Noncompliant
(6) Is the water used in the plant treated?		√	Noncompliant

TABLE 5: Assessment of GMPs on pest control programs for OFSP puree processing.

Item	Yes	No	Comment
(1) Do you have professional pest control services?		√	Noncompliant
(2) Do you have documentation on what chemicals are being used?		√	Noncompliant
(3) Are mites, weevils or other insects apparent in the plant?	√		Noncompliant
(4) Do you have enough bait stations?	√		Compliant
(5) Are safety rules observed during fumigation?	√		Compliant
(6) Are the pest control logs and documentation readily available?		√	Noncompliant
(7) Are pesticides or application equipment readily available?		√	Noncompliant

TABLE 6: Microbial counts of installations in OFSP puree processing plant[1] (log Mean CFU·cm^{-2}).

Sample	TVC	Yeast-Molds	S. aureus	Enterobacteriaceae	Coliforms	E. coli
Floors	8.03 ± 0.87^{bcde}	5.71 ± 0.94^{cdef}	5.31 ± 0.40^{cd}	6.98 ± 0.35^{cd}	6.66 ± 0.22^{gh}	2.75 ± 3.20^{a}
Walls	8.69 ± 0.65^{bcde}	6.17 ± 0.56^{cdef}	5.54 ± 0.03^{cd}	7.00 ± 0.26^{cd}	6.43 ± 0.54^{defgh}	1.45 ± 2.51^{a}
Drains	8.71 ± 0.73^{bcde}	5.46 ± 1.29^{cdef}	5.33 ± 0.90^{cd}	7.20 ± 0.06^{cd}	6.80 ± 0.36^{h}	3.50 ± 3.14^{b}

[1] All values reflect mean counts and standard deviation. Values bearing different superscript letters in each column are significantly different ($P < 0.05$); TVC: total viable counts.

not significantly ($P > 0.05$) different. The lowest coliform counts were recorded from walls ($6.4 \pm 0.5 \log$ cfu·cm^{-2}) while highest counts were from drain surfaces ($6.8 \pm 0.4 \log$ cfu·cm^{-2}). E. coli counts from drains were significantly ($P < 0.05$) high as compared to floors and walls.

Floors, walls, and drains are high risk areas for bacterial growth and contamination in food processing plants [16]. Floors transfer contamination to food handlers' shoes who consequently circulate and disseminate the microorganisms within the establishment. Drains and floors offer a favorable environment for microbial growth if cleaning and sanitation are not done appropriately. High total counts, coliforms, and Enterobacteriaceae counts (>$10^5 \log$ CFU·cm^{-2}) from walls, floors, and drains in OFSP puree processing facility were attributed to inefficient cleaning of these areas. Similar results have been reported from meat processing environments in studies by Barros et al. and Ali et al. [16, 18]. Other than transferring contamination to trolleys and food handler's shoes, contaminated floors and walls in the facility can recontaminate personnel hands and equipment such as buckets, pallets, brushes, and cold boxes stored on the floor. Routine inspections, supervision of cleaning, and maintenance of installations and sanitary facilities can help in preventing proliferation and spread of microbial contamination within the OFSP puree facility.

3.4. Level of Personal Hygiene and Level of Microbial Load on OFSP Puree Handlers' Hands. Personal hygiene practices by OFSP puree handlers are shown in Table 7. Only 69% of the assessed practices on personal hygiene were considered compliant to food safety regulations for OFSP puree processing. The level of contamination on personnel's hands in OFSP puree processing in a decreasing order was 8.3 ± 0.6; 6.9 ± 0.4; 6.6 ± 0.2; 6.0 ± 1.0; 5.1 ± 0.9; and $2.7 \pm 0.4 \log$ cfu·cm^{-2} for total counts, Enterobacteriaceae, coliforms, yeasts and molds, Staphylococcus aureus, and E. coli, respectively (Figure 2).

High total counts and presence of potential pathogens (S. aureus and E. coli) on OFSP puree handlers' hands indicate low compliance to good hand washing hygiene practices by OFSP puree handlers and thus a potential source of contamination during processing of OFSP puree. Adherence to good personal hygiene by food handlers is important in preventing cross-contamination in food processing. The contact between food handlers and contaminated surfaces of equipment, phones, and walls classifies them as a potential source of contamination [35]. It is estimated that 10–20% of

TABLE 7: Assessment of personal hygiene practices in OFSP puree processing.

Item	Yes	No	Status
(1) Are there instructions on how to be suitably dressed to enter production areas?	√		Compliant
(2) Do food handlers wash their hands in clean water before handling and preparation of food?	√		Compliant
(3) Do operators wash their hands each time after visiting the toilet?	√		Compliant
(4) Are employees provided paper towels and hand sanitizers?		√	Noncompliant
(5) Are operator's clothes clean and presentable?	√		Compliant
(6) Are the food handlers observed to have any illnesses, infections, or injuries (boils, cuts) that can contaminate food in the production area?		√	Compliant
(7) Do food handlers use protective clothing when handling and preparing food?	√		Compliant
(8) Do food handlers handle food with bare hands?	√		Noncompliant
(9) Are disposable or reusable gloves provided?		√	Noncompliant
(10) Do operators have clean short nails?	√		Compliant
(11) Is their hair covered when handling and processing food?	√		Compliant
(12) Do food handlers use mobile phones while handling and preparing food?	√		Noncomplaint
(13) Do food handlers wear jewelry, rings, watches, fingernail polish or bandages in the processing establishment?		√	Compliant
(14) Do food handlers smoke/chew while handling and preparing food?		√	Compliant
(15) Do operators use the same equipment and surfaces in preparing raw and processed food?		√	Compliant
(16) Do food handlers blow their nose/cough while handling and preparing food?		√	Compliant

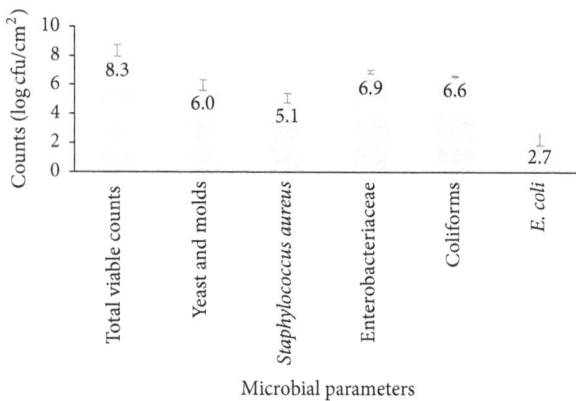

FIGURE 2: Microbial counts on Orange Fleshed Sweet Potato puree handlers' hands; the values above the bars represent the mean ± SEM (standard error of the mean).

foodborne disease outbreaks occur as a result of contamination by food handlers [36]. All counts from personnel hands in this study were $>10^5$ log CFU·cm^{-2} except for *E. coli*. *Staphylococcus aureus* was detected in all hand swab samples (100%) from OFSP puree handlers. This is contrary to findings by Al-Bahry et al. [37] that reported positive results for *S. aureus* in only 34% of asymptomatic food handlers in their study. From our observation, handling of OFSP in all the stages of processing was done with bare hands and more than 50% of all OFSP puree handlers were not regularly using soap during hand washing despite soap and hand washing instructions being supplied at every hand washing station. More than 50% of OFSP puree handlers failed to wash their hands after using their mobile phones during processing.

Mobile phones have been reported as a source of bacterial and fungal contamination in food handling [35].

Provision of necessary food safety resources in a food production facility enhances food safety [38]. Gloves for handling OFSP puree and paper towels for hand drying were not provided at the OFSP puree processing facility. Gloves are crucial in preventing personnel's' hands to food contamination [39, 40]. It is documented that the transmission of bacteria is more likely to occur from wet hands than from dry hands [41]. This makes proper hand drying an essential component for effective hand hygiene in food processing facilities. Another study by Choi et al. [42] argues that provision of appropriate hand washing resources is not enough in enforcing proper hand hygiene. Results from their study indicated that 50% of food handlers in retail establishments failed to practice good hand washing hygiene despite being provided with necessary resources for hand hygiene. Training on personal hygiene, management support, and provision of food safety supplies such as gloves and paper towels should be considered in efforts to improve food safety in OFSP puree processing.

3.5. Process Control and Changes in Microbial Load in OFSP Puree during Processing. Compliance to quality control process parameters and changes in microbial load during processing of OFSP puree are shown in Tables 8 and 9, respectively. Only 44% of all quality control procedures were considered appropriate for OFSP puree processing. Total viable counts, yeasts and molds, *S. aureus*, Enterobacteriaceae, and coliforms counts were destroyed after steaming OFSP roots. Steaming is the main critical control point (CCP) for enhancing keeping quality and safety of OFSP puree. Several studies have reported cooking as an effective method in eliminating or reducing microorganisms in foods to safe

TABLE 8: Good Manufacturing Practices for process control in OFSP puree processing.

Item	Yes	No	Status
(1) Are OFSP roots and puree stored on a first in, first out basis to reduce the possibility of contamination through spoilage?	√		Compliant
(2) Is a thermometer for recording heating temperature for OFSP roots provided?		√	Noncompliant
(3) Is OFSP puree dated to ensure a proper rotation and for internal tracking purposes		√	Noncompliant
(4) Are raw materials and the puree overstocked?		√	Compliant
(5) Are trucks inspected, cleaned and sanitized?		√	Noncompliant
(6) Are incoming OFSP roots inspected for damage or contamination so that they can be rejected?	√		Compliant
(7) Are freezer temperatures monitored and recorded?		√	Noncompliant
(8) Are food related items stored together with non-food related items?		√	Compliant
(9) Do you have an effective recall procedure set up?		√	Noncompliant

TABLE 9: Microbial counts of OFSP at different stages of processing in OFSP puree processing plant[1] (log Mean CFU/g).

Processing stages	TVC	Yeasts-molds	S. aureus	Enterobacteriaceae	Coliforms	E. coli
Raw OFSP	7.19 ± 0.27^{bcd}	2.51 ± 0.30^{b}	3.01 ± 0.15^{b}	4.57 ± 0.20^{b}	3.74 ± 0.19^{b}	nd*
Steaming	nd*	nd*	nd*	nd*	nd*	nd*
Cooling and Slicing	7.12 ± 0.21^{bcd}	nd*	2.93 ± 2.54^{b}	4.62 ± 0.15^{b}	4.37 ± 0.13^{bc}	nd*
Packaging	7.96 ± 0.57^{bcde}	4.01 ± 0.33^{bc}	5.88 ± 0.53^{cd}	6.55 ± 0.18^{cd}	5.82 ± 0.13^{defgh}	4.77 ± 0.45^{b}

[1] All values reflect mean counts and standard deviation. Values bearing different superscript letters in each column are significantly different ($P < 0.05$); TVC: total viable counts; nd*: not detected.

levels [11, 27]. It is however a challenge destroying heat resistant spores and heat-stable toxins produced by pathogens such as *S. aureus* in a process involving mild heat treatment of foods. *S. aureus* counts in foods above 10^5 CFU/g initiate production of heat-stable enterotoxins [37]. Staphylococcal food poisoning is one of the leading causes of foodborne illnesses worldwide caused by ingestion of food contaminated with preformed *S. aureus* enterotoxins [43]. Routine implementation of appropriate hygiene procedures and process control in OFSP puree processing can be an effective tool in reducing or eliminating contamination by pathogens or their toxins.

Microbial load significantly ($P < 0.05$) increased after cutting, cooling, pureeing, and packaging processes. High total counts and yeasts and mold counts in the puree indicated deterioration in keeping quality and hence accelerated spoilage. Haile et al. [44] reported lower TVC and yeasts and mold counts in porridge prepared from Orange Fleshed Sweet Potato (OFSP) contrary to the current findings. The high counts are attributed to contamination from equipment such as knives, cooling trays, pureeing machine, and packaging bags and contaminated water and poor personal hygiene by OFSP handlers. The presence of Enterobacteriaceae, coliforms, and *E. coli* in OFSP puree indicates fecal contamination and probable presence of enteropathogens due to cross-contamination from equipment, processing water, and food handling personnel. Additionally, high *S. aureus* counts detected in the puree was associated with contaminated equipment and poor hand washing hygiene practices from OFSP puree handlers. Contamination in OFSP puree can be eliminated by use of clean and sanitized equipment;

FIGURE 3: Overall level of compliance to GMPs in OFSP puree processing plant; the values above the bars represent the mean percentage score for compliance/noncompliance to GMPs.

clean and disinfected water; and adherence to good personal hygiene and appropriate handling practices by OFSP puree handlers.

3.6. Overall Level of Compliance to Good Manufacturing Practices in OFSP Puree Processing Plant. The summarized levels of compliance to GMPs in puree processing with respect to buildings, sanitary facilities, personal hygiene, equipment, pest, and process control are shown in Figure 3.

The overall level of compliance to GMPs in puree processing was low with only 58% of all GMPs items in the present study being compliant. There is a need for urgent improvement on low scores for GMP items that covered areas on pest control, process control, sanitary facilities, personal hygiene, and suitability of the puree processing unit. Low compliance to GMPs is an impediment towards achieving food safety and quality standard requirements in OFSP puree processing. Kadariya et al. [43] emphasize strict adherence and implementation of GMPs as one of the best approaches for controlling microbial contamination in food processing. The Orange Fleshed Sweet Potato puree produced at the establishment was of unacceptable microbiological quality if it was to be commercialized as a ready-to-eat (RTE) food. The microbial load on equipment, food contact surfaces, personnel hands, processing water, and the processing environment was also above the recommended food safety legal limits. Even though there is limited data on safe microbial tolerable limits for food ingredients and food contact/preparation surfaces, some standards and regulations have been developed and adopted in some countries based on specifications provided by the International Commission for Microbiological Specification and other research studies. According to Food Standards Australia New Zealand [45], Centre for Food Safety [46], and Idris Ali and Immanuel [47], the levels considered acceptable/satisfactory in RTE foods are $<10^4$ cfu for total viable counts, $<10^2$ cfu for yeasts and molds, <20 cfu for *E. coli* and coliforms, $<10^2$ cfu for Enterobacteriaceae, and $<10^2$ cfu for *S. aureus*. The European Commission [48] previously suggested microbial level ranging from 0 to 10 cfu/cm^2 on food equipment and food preparation surfaces and in processing environments as acceptable. Despite the existence of legal framework on food safety and quality in Kenya, the findings suggest that most small-scale food processors probably operate in disregard of food safety and quality controls [49, 50]. This is attributed to lack of qualified personnel, infrastructure, equipment, and other food safety resources necessary for hygienic processing, handling, storage, and distribution of food products [49, 50].

4. Conclusion

The present study revealed low compliance to GMPs in the only OFSP puree processing plant in Kenya. High microbial contamination levels on equipment, processing environment, and personnel hands revealed poor hygiene practices within the establishment. OFSP puree contamination emanated from equipment, processing water, and violation of food safety practices by puree handlers. It is therefore recommended to integrate Good Manufacturing Practices (GMPs), Good Hygiene Practices (GHPs), environmental monitoring programs, microbial risk assessments, and food safety trainings as quality control tools for enhancing food safety and quality of Orange Fleshed Sweet Potato puree.

Conflicts of Interest

The authors declare that there are no conflicts of interest regarding the publication of this work.

Acknowledgments

The authors would like to acknowledge the International Potato Centre (CIP-SSA) for their financial support and resources that facilitated this work.

References

[1] Food and Agriculture Organization of the United Nations (FAOSTAT), 2016, http://www.fao.org.

[2] M. S. Mukras, J. M. Odondo, and G. Momanyi, "Determinants of demand for sweet potatoes at the farm, retail and wholesale markets in Kenya," *Advances in Economics and Business*, vol. 1, no. 2, pp. 150–158, 2013.

[3] G. O. Abong', V. C. M. Ndanyi, A. Kaaya et al., "A review of production, post-harvest handling and marketing of sweetpotatoes in Kenya and Uganda," *Current Research in Nutrition and Food Science*, vol. 4, no. 3, pp. 162–181, 2016.

[4] B. Vimala, B. Nambisan, and B. Hariprakash, "Retention of carotenoids in orange-fleshed sweet potato during processing," *Journal of Food Science and Technology*, vol. 48, no. 4, pp. 520–524, 2011.

[5] C. M. Donado-Pestana, J. M. Salgado, A. de Oliveira Rios, P. R. dos Santos, and A. Jablonski, "Stability of carotenoids, total phenolics and in vitro antioxidant capacity in the thermal processing of orange-fleshed sweet potato (Ipomoea batatas Lam.) cultivars grown in Brazil," *Plant Foods for Human Nutrition*, vol. 67, no. 3, pp. 262–270, 2012.

[6] J. W. Low and P. J. van Jaarsveld, "The potential contribution of bread buns fortified with β-carotene-rich sweet potato in Central Mozambique," *Food and Nutrition Bulletin*, vol. 29, no. 2, pp. 98–107, 2008.

[7] R. A. Nungo, P. J. Ndolo, R. Kapinga, and S. Agili, "Development and promotion of sweet potato products in Western Kenya," in *Proceedings of the 13th ISTRC Symposium*, pp. 790–794, Kampala, Uganda, 2007.

[8] S. Tumwegamire, R. Kapinga, R. Mwanga, C. Niringiye, B. Lemaga, and J. Nsumba, "Acceptability studies of orange fleshed sweet potato verities in Uganda," in *Proceeding of the 13th ISTRC Symposium*, pp. 807–813, Kampala, Uganda, 2007.

[9] M. Jenkins, C. B. Shanks, and B. Houghtaling, "Orange-fleshed sweet potato: Successes and remaining challenges of the introduction of a nutritionally superior staple crop in Mozambique," *Food and Nutrition Bulletin*, vol. 36, no. 3, pp. 327–353, 2015.

[10] HarvestPlus, "Disseminating orange fleshed sweetpotato, Uganda country report," Tech. Rep., Washington, Wash, USA, 2012.

[11] I. M. Pérez-Díaz, V.-D. Truong, A. Webber, and R. F. McFeeters, "Microbial growth and the effects of mild acidification and preservatives in refrigerated sweet potato puree," *Journal of Food Protection*, vol. 71, no. 3, pp. 639–642, 2008.

[12] I. Tedesco and S. Tanya, "Sweetpotato value chains in Western Kenya: a business opportunity for puree procesing and the role for commercial fresh root storage," NRI report, University of Greenwich, Chatham, UK, February 2015.

[13] C. P. De Sousa, "The impact of food manufacturing practices on food borne diseases," *Brazilian Archives of Biology and Technology*, vol. 51, no. 4, pp. 815–823, 2008.

[14] A. Othman, "Isolation and microbiological identification of bacterial contaminants in food and household surfaces: how to deal safely," *Egyptian Pharmaceutical Journal*, vol. 14, no. 1, pp. 50–55, 2015.

[15] M. Addis and D. Sisay, "A Review on Major Food Borne Bacterial Illnesses," *Journal of Tropical Diseases*, vol. 3, no. 4, pp. 176–183, 2015.

[16] M. D. A. F. Barros, L. A. Nero, A. A. Monteiro, and V. Beloti, "Identification of main contamination points by hygiene indicator microorganisms in beef processing plants," *Ciência e Tecnologia de Alimentos*, vol. 27, no. 4, pp. 856–862, 2007.

[17] Food and Agriculture Organization/World Health Organization, "Assuring food safety and quality: guidelines for strengthening national food control systems," http://www.fao.org/3/a-y8705e.pdf.

[18] N. H. Ali, A. Farooqui, A. Khan, A. Y. Khan, and S. U. Kazmi, "Microbial contamination of raw meat and its environment in retail shops in Karachi, Pakistan," *The Journal of Infection in Developing Countries*, vol. 4, no. 6, pp. 382–388, 2010.

[19] R. L. Buchanan and R. Oni, "Use of microbiological indicators for assessing hygiene controls for the manufacture of powdered infant formula," *Journal of Food Protection*, vol. 75, no. 5, pp. 989–997, 2012.

[20] M. Gwida, H. Hotzel, L. Geue, and H. Tomaso, "Occurrence of *Enterobacteriaceae* in raw meat and in human samples from egyptian retail sellers," *International Scholarly Research Notices*, vol. 2014, Article ID 565671, pp. 1–6, 2014.

[21] H. Tassew, A. Abdissa, G. Beyene, and S. Gebre-Selassie, "Microbial flora and food borne pathogens on minced meat and their susceptability to antimicrobial agents," *Ethiopian Journal of Health Sciences*, vol. 20, no. 3, pp. 137–143, 2010.

[22] J. Schlegelová, V. Babák, M. Holasová et al., "Microbial contamination after sanitation of food contact surfaces in dairy and meat processing plants," *Czech Journal of Food Sciences*, vol. 28, no. No. 5, pp. 450–461, 2018.

[23] R. B. Tompkin, "Environmental sampling-A tool to verify the effectiveness of preventive hygiene measures," *Releases from Food and Hygiene Investigation*, vol. 51, pp. 45–51, 2004.

[24] D. N. Malavi, G. O. Abong', and T. Muzhingi, "Food safety knowledge, attitude and practices of orange-fleshed sweet-potato puree handlers in Kenya," *Food Science and Quality Management*, vol. 67, pp. 54–63, 2017.

[25] C. Mukantwali, H. Laswai, B. Tiisekwa, and S. Wiehler, "Good manufacturing and hygienic practices at small and medium scale pineapple processing enterprises in Rwanda," *Food Science and Quality Management*, vol. 13, pp. 15–31, 2013.

[26] Iowa State University of Science and Technology, "FDA good manufacturing practices checklist for human food," https://www.fshn.hs.iastate.edu.

[27] E. Gungor and N. Gokoglu, "Determination of microbial contamination sources at a Frankfurter sausage processing line," *Turkish Journal of Veterinary & Animal Sciences*, vol. 34, no. 1, pp. 53–59, 2010.

[28] C. J. Hervert, N. H. Martin, K. J. Boor, and M. Wiedmann, "Survival and detection of coliforms, Enterobacteriaceae, and gram-negative bacteria in Greek yogurt," *Journal of Dairy Science*, vol. 100, no. 2, pp. 950–960, 2017.

[29] A. B. Sylvia, M. RoseAnn, and B. K. John, "Hygiene practices and food contamination in managed food service facilities in Uganda," *African Journal of Food Science*, vol. 9, no. 1, pp. 31–42, 2015.

[30] D. El-Hadedy and S. Abu El-Nour, "Identification of Staphylococcus aureus and Escherichia coli isolated from Egyptian food by conventional and molecular methods," *Journal of Genetic Engineering and Biotechnology*, vol. 10, no. 1, pp. 129–135, 2012.

[31] Environmental Protection Agency, "Parameters of water quality, interpretation and standards," https://www.epa.ie.

[32] G. Liguori, I. Cavallotti, A. Arnese, C. Amiranda, D. Anastasi, and I. F. Angelillo, "Microbiological quality of drinking water from dispensers in Italy," *BMC Microbiology*, vol. 10, article 19, 2010.

[33] A. Rompré, P. Servais, J. Baudart, M.-R. De-Roubin, and P. Laurent, "Detection and enumeration of coliforms in drinking water: current methods and emerging approaches," *Journal of Microbiological Methods*, vol. 49, no. 1, pp. 31–54, 2002.

[34] WHO, *Guidelines for Drinking-Water Quality*, vol. 1, World Health Organization, 3rd edition, 2011, http://www.who.int/water_sanitation_health/dwq/fulltext.pdf.

[35] A. Al-Abdalall, "Isolation and identification of microbes associated with mobile phones in Dammam in eastern Saudi Arabia," *Journal of Family and Community Medicine*, vol. 17, no. 1, p. 11, 2010.

[36] M. Anuradha and R. H. Dandekar, "Knowledge, attitude and practice among food handlers on food borne diseases: a hospital based study in tertiary care hospital," *International Journal of Biomedical and Advance Research*, vol. 5, no. 4, 2014.

[37] S. N. Al-Bahry, I. Y. Mahmoud, S. K. Al-Musharafi, and N. Sivakumar, "Staphylococcus aureus contamination during food preparation, processing and handling," *International Journal of Chemical Engineering and Applications*, vol. 5, no. 5, pp. 388–392, 2014.

[38] L. R. Green and C. Selman, "Factors Impacting food workers' and managers' safe food preparation practices: a qualitative study," *Food Protection Trends*, vol. 25, no. 12, pp. 981–990, 2005.

[39] B. Michaels, C. Keller, M. Blevins et al., "Prevention of food worker transmission of foodborne pathogens: risk assessment and evaluation of effective hygiene intervention strategies," *Food Service Technology*, vol. 4, no. 1, pp. 31–49, 2004.

[40] R. Montville, Y. Chen, and D. W. Schaffner, "Glove barriers to bacterial cross-contamination between hands to food," *Journal of Food Protection*, vol. 64, no. 6, pp. 845–849, 2001.

[41] C. Huang, W. Ma, and S. Stack, "The hygienic efficacy of different hand-drying methods: A review of the evidence," *Mayo Clinic Proceedings*, vol. 87, no. 8, pp. 791–798, 2012.

[42] J. Choi, H. Norwood, S. Seo, S. A. Sirsat, and J. Neal, "Evaluation of food safety related behaviors of retail and food service employees while handling fresh and fresh-cut leafy greens," *Food Control*, vol. 67, pp. 199–208, 2016.

[43] J. Kadariya, T. C. Smith, and D. Thapaliya, "*Staphylococcus aureus* and staphylococcal food-borne disease: an ongoing challenge in public health," *BioMed Research International*, vol. 2014, Article ID 827965, pp. 1–9, 2014.

[44] F. Haile, S. Admassu, and A. Fisseha, "Effects of pre-treatments and drying," *American Journal of Food Science and Technology*, vol. 3, no. 3, pp. 82–88, 2015.

[45] Food Standards Australia New Zealand, "Compendium of microbiological criteria for food," https://www.foodstandards.gov.

[46] Centre for Food Safety, "Microbiological guidelines for ready-to-eat food," 2007.

[47] A. Idris Ali and G. Immanuel, "Assessment of hygienic practices and microbiological quality of food in an institutional food service establishment," *Journal of Food Processing & Technology*, vol. 08, no. 08, 2017.

[48] European Commission, 2001, 2001/471/EC: Commission Decision of 8 June 2001 laying down rules for the regular checks on

the general hygiene carried out by the operators in establishments according to Directive 64/433/EEC on health conditions for the production and marketing of fresh meat and Directive 71/118/EEC on health problems affecting the production and placing on the market of fresh poultry meat (Text with EEA relevance) (notified under document number C(2001) 1561).

[49] J. Oloo, "Food safety and quality management in Kenya: an overview of the roles played by various stakeholders," *African Journal of Food, Agriculture, Nutrition and Development*, vol. 10, no. 11, pp. 4379–4397, 2011.

[50] D. N. Malavi, G. O. Abong', and T. Muzhingi, *Food safety knowledge and hygiene practices among orange-fleshed sweetpotato puree handlers: microbial contamination in puree processing company in Kenya and impact of training [M.S. thesis]*, University of Nairobi, 2017, http://erepository.uonbi.ac.ke.

4

Antioxidant Properties of "Natchez" and "Triple Crown" Blackberries using Korean Traditional Winemaking Techniques

Youri Joh,[1] Niels Maness,[2] and William McGlynn[1,2]

[1]Robert M. Kerr Food & Agricultural Products Center, Oklahoma State University, Stillwater, OK 74078, USA
[2]Department of Horticulture and Landscape Architecture, Oklahoma State University, Stillwater, OK 74078, USA

Correspondence should be addressed to Youri Joh; youj@okstate.edu

Academic Editor: Amy Simonne

This research evaluated blackberries grown in Oklahoma and wines produced using a modified traditional Korean technique employing relatively oxygen-permeable earthenware fermentation vessels. The fermentation variables were temperature (21.6°C versus 26.6°C) and yeast inoculation versus wild fermentation. Wild fermented wines had higher total phenolic concentration than yeast fermented wines. Overall, wines had a relatively high concentration of anthocyanin (85–320 mg L^{-1} malvidin-3-monoglucoside) and antioxidant capacity (9776–37845 μmol Trolox equivalent g^{-1}). "Natchez" berries had a higher anthocyanin concentration than "Triple Crown" berries. Higher fermentation temperature at the start of the winemaking process followed by the use of lower fermentation/storage temperature for aging wine samples maximized phenolic compound extraction/retention. The Korean winemaking technique used in this study produced blackberry wines that were excellent sources of polyphenolic compounds as well as being high in antioxidant capacity as measured by the Oxygen Radical Absorbance Capacity (ORAC) test.

1. Introduction

Winemaking and wine consumption are becoming more popular as they are known to provide health beneficial products that are high in antioxidants [1–3]. Blackberry (*Rubus* spp.) wines are good sources of antioxidants because they contain relatively high concentrations of anthocyanins and other phenolic compounds [4–6]. Fermentation processes have been shown to increase the level of antioxidant activity by facilitating the extraction of anthocyanins and other phenolic compounds from the pomace and by forming new polymerized pigments and polyphenols [7]. The Korean traditional wine processing method, which typically employs wild microorganisms for fermentation, may provide different types and levels of health related compounds compared to common grape wine production methods.

Blackberry phenolic composition has been shown to vary on the basis of growing temperature, growing season, geographic location, maturity at harvest, environmental stress, soil type, UV light exposure, hydrophobicity of compounds, genetics, extraction/processing methods, and processing storage conditions [3, 8–15]. Relatively little research has been done on "Natchez" and "Triple Crown" blackberries, cultivars that are suitable for growing in the Midwest section of the United States. The suitability of these blackberries for winemaking of phenolic compounds of wines made from these berries has not previously been studied. The part of the research was presented in ASEV National Conference before, but full information was provided in this research paper [16].

The objectives of this study were to evaluate the winemaking potential of "Natchez" and "Triple Crown" blackberries grown in Oklahoma as well as to examine the chemical properties of phenolic compounds of blackberry wines made using variations on traditional Korean winemaking techniques. The fermentation parameters were fermentation temperature, that is 21.6°C and 26.6°C, and yeast inoculation fermentation versus wild fermentation. The pH, % soluble solids, titratable acidity, and % alcohol of berries and wines were assayed to assess basic quality parameters. Also, the chemical properties of the berries and wines were evaluated by quantifying their total phenolic concentration, anthocyanin concentration, and antioxidant capacity.

2. Materials and Methods

2.1. Blackberry Collection and Storage. Fruit from two blackberry cultivars (*Rubus* spp.), "Natchez" and "Triple Crown," were collected from the Oklahoma Agricultural Experiment Stations, Cimarron Valley Research Station (Perkins, OK, USA). Blackberries were collected after they turned fully purplish black over a period of two years, 2011 and 2012. All blackberries were hand-harvested starting from the third week of May and ending about the third week of July. The ripening time of "Natchez" berries was approximately one month earlier than that of "Triple Crown" berries. During the harvest period, the berries were collected every other day. Blackberries were placed into polyethylene bags and placed into the freezer (−15°C) within one hour of harvest for storage and subsequent experimental use.

2.2. Preparation of Juice Sample. Frozen whole blackberries were placed in a refrigerator at about 4°C for a day and then held at room temperature until they came to temperature equilibrium in about 3 to 4 hours. Fresh juice samples were collected by manually pressing 100 to 150 blackberries against a 2 mm mesh screen. Juice samples of at least 100 mL were collected into 120 mL brown amber bottles for future analysis.

2.3. Korean Traditional Blackberry Winemaking Process

2.3.1. Prefermentation Handling. A modified combination of Korean traditional winemaking techniques was used in this research [17–19]. Figure 1 shows an overview of the Korean traditional winemaking process [20]. Prewashed 12 L traditional Korean earthenware jars (Sin-il Earthenware, Inchon, South Korea) were used as fermentation vessels. Blackberries (≈4.5 kg) thawed as previously described were placed in each Korean earthenware jar and ≈20% raw brown sugar (Cumberland Packing Co., Brooklyn, NY, USA) by blackberry weight was added. Alternating layers (≈5 cm thick) of blackberries and sugar were laid down in each jar such that the jars were ≈2/3 full by volume.

After filling the blackberries and sugar into the Korean earthenware jars, the jars were covered with thin paper (breathable) secured around the neck of the jar with a string. The treatment factors applied were two cultivars ("Natchez" and "Triple Crown"), two fermentation temperatures (26.6°C and 21.6°C), and two fermentation microflora (no added yeast and added yeast). For the yeast, 5 g of Enoferm L2226 (Scott Laboratories Inc., Petaluma, CA, USA) in 50 mL water was added in each jar.

Three Korean earthenware jars were used for each treatment combination; each jar was considered an experimental unit for purposes of statistical analyses. Thus, each treatment combination was replicated three times. An environmental chamber (Ultimate Hot Pack Inc., Lander, WY, USA) was used for temperature control.

2.3.2. Fermentation. A two-part fermentation process was used for all samples [20]. The first fermentation took 1 to 2 weeks. During the first fermentation period, samples

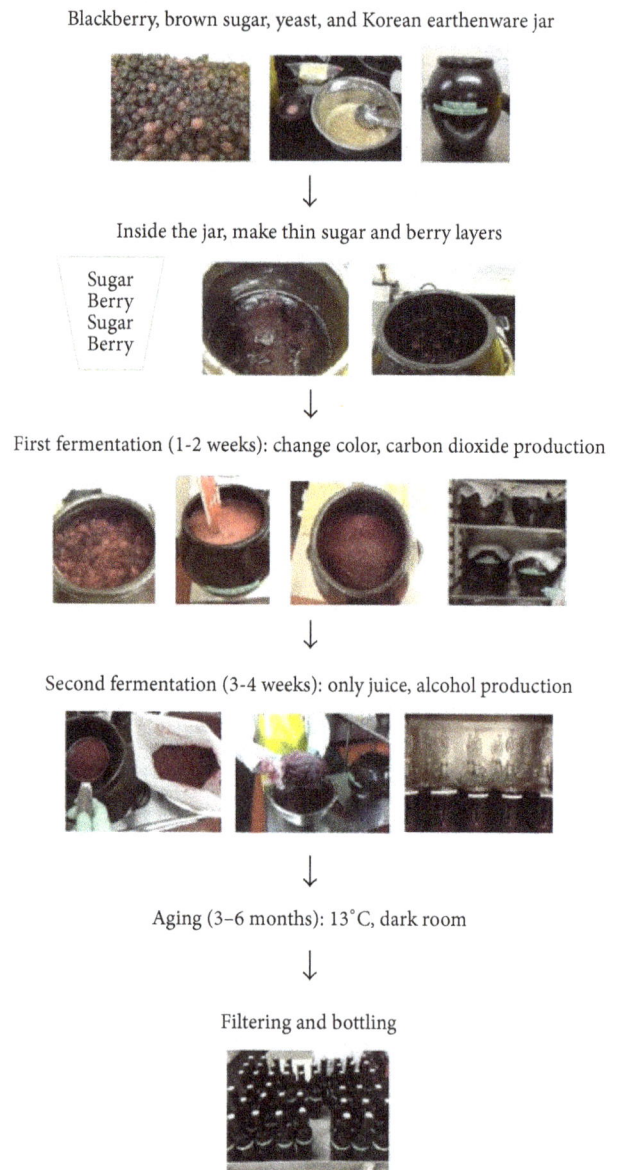

Blackberry, brown sugar, yeast, and Korean earthenware jar

↓

Inside the jar, make thin sugar and berry layers

Sugar
Berry
Sugar
Berry

↓

First fermentation (1-2 weeks): change color, carbon dioxide production

↓

Second fermentation (3-4 weeks): only juice, alcohol production

↓

Aging (3–6 months): 13°C, dark room

↓

Filtering and bottling

FIGURE 1: Overview of Korean traditional winemaking process.

were mixed with a spatula every morning and evening to help insure sufficient aeration. The Korean earthenware jar facilitated this process as the container was breathable with many small pores (1 to 20 μm) that allowed relatively more gas transfer into the samples than plastic or glass jars would have done [21]. While active fermentation was under way, the blackberries changed color from purple to pink and the individual berries lost their structural integrity. The end of the first fermentation stage occurred when CO_2 release rate slowed and the soluble solids concentration dropped below 10° brix.

The second fermentation stage took 3 to 4 weeks. Blackberry skins and seeds were removed using a nylon straining bag (small fine size, 10″ × 23″, LD Carlson, Kent, OH, USA). The strained pomace was collected into polyethylene bags and stored in a −15°C freezer for further analysis. The strained juice was then transferred into a second type of fermentation vessel. During the second stage of fermentation, the goal was to limit oxygen contact. For this reason, glass or plastic fermentation vessels were used and each vessel was filled to within ≈4/5 full [19]. The finishing point of the second fermentation stage occurred when no further production of CO_2 gas was noted via airlock apparatus. Fermentation temperature was controlled at 21.6°C or 26.6°C during both fermentation stages. Triplicate samples were collected into 120 mL brown bottles at the end of 1st and 2nd fermentation and stored in a −15°C freezer for further analysis.

2.3.3. Aging Wine.
After the second fermentation stage was complete, the wine was racked (decanted off the sediment at the bottom of the vessel) and filled into 950 mL brown amber glass bottles to the top and the bottles were tightly sealed with screw caps. The wine was stored at 13°C [19] and 100 mL samples were collected once a month into 120 mL brown bottles for three months for further analyses.

2.4. Quality Analysis for Whole Blackberry, Juice, and Wine

2.4.1. pH.
The pH of the blackberry juice was measured using an Accumet AB 15 pH meter (Buffalo, NY, USA). Duplicate samples were measured and averaged for each replication.

2.4.2. Soluble Solids.
Blackberry juice sugar concentration was estimated as percent soluble solids using a Leica Auto ABBE refractometer (Buffalo, NY, USA). Duplicate samples were measured and averaged for each replication.

2.4.3. Titratable Acidity.
The titratable acidity of blackberry juice and wine samples was measured manually using 0.1 N sodium hydroxide (Arcos Organics, Fair Lawn, NJ, USA) as per the method described in Joh [20]. Two duplicate readings were taken from each blackberry juice and wine sample and then averaged.

2.4.4. Percent Alcohol.
The percent alcohol (w/w) of wine samples was measured using an Alcolyer Wine M (Anton Paar, Ashland, VA, USA). This instrument uses a patented method (US 6,690,015; AT 406711) based on near infrared (NIR) spectroscopy to determine the alcohol content in a highly alcohol-specific wavelength range between 1150 nm and 1200 nm [22]. Samples of the aged wines were collected into 60 mL brown glass bottles. A volume of ≈30 mL of wine per sample was used in the analysis. Two duplicate readings were taken from each wine sample and averaged.

2.5. Antioxidant Activity Analyses

2.5.1. Modified Harbertson-Adams Assay

(1) *Total Phenolic Concentrations.* A volume of 75 μL of blackberry juice or wine and 800 μL resuspension buffer was add to a reduced volume cuvette and then held for 10 minutes at room temperature. Samples were read at 510 nm to generate a value for the iron-reactive phenolics background. In the same cuvette, 125 μL of ferric chloride solution was added and held for another for 10 minutes at room temperature. Samples were read at 510 nm to generate a final value for the iron-reactive phenolics concentration [20].

(2) *Total Anthocyanin Concentrations.* A volume of 400 μL of model wine, 100 μL of blackberry juice or wine sample, and 1 mL of anthocyanin buffer was added to a reduced volume cuvette and then held for 5 minutes at room temperature. Samples were then read at 520 nm [20].

All samples were measured in duplicate. Final value calculations were made using the Skogerson-Boulton Model Assay Input spreadsheet (Boulton Research: Skogerson-Boulton Model Assay Input v.1.3) [20].

2.5.2. Oxygen Radical Absorbance Capacity (ORAC) Assay.
All blackberry juice, wine, and pomace samples were added at a ratio of 1 : 2000 (v/v) to phosphate buffer prior to being tested for antioxidant capacity using a slightly modified version of the Oxygen Radical Absorbance Capacity (ORAC) assay described in Huang and others [23]. The details of the method used may be found in Joh [20]. The final results of the ORAC assay were calculated as μmol Trolox equivalent (TE) per gram of blackberry juice, wine, or pomace. All samples were measured in duplicate.

2.6. Statistical Analysis.
Statistical Analyses were performed using SAS 9.3 (SAS institution, Cary, NC). For all analyses, an analysis of variance (ANOVA) for each set of data was conducted using a three-factor factorial treatment scheme in a completely randomized design with repeated measures. Means were separated using least significant differences (LSD) with a 95% confidence interval ($p < 0.05$).

3. Results and Discussions

3.1. Quality Analysis for Whole Blackberry Juice and Wine

3.1.1. pH.
The mean pH values of blackberry juice samples made from berries harvested in 2011 and 2012 are shown in Table 1. Values ranged from 2.88 to 3.15. The pH values of blackberry wine samples are shown in Table 2. Values ranged from 2.60 to 3.12.

The pH values in blackberry juice and wine matched several previous studies [7, 24–26]. Some researchers found higher pH values, from 3.2 to 4.2, likely due to differences observed among cultivars, growing locations, and/or berry ripeness [27–29].

"Triple Crown" berries showed significant differences between years. It appears that weather condition such as amount of rainfall affected the pH of the berries. The acidity level in blackberries has been observed to decrease under warmer, drier weather conditions [27]. In 2012, the average rainfall of July was 0.2 cm compared to the average rainfall of 1.9 cm in 2011 [30]. The drier conditions in 2012 may have helped to ripen "Triple Crown" berries faster that year and provided less acidic berries.

TABLE 1: Average mean concentration of pH, titratable acidity, and soluble solid values of pressed blackberry juice ($n = 2$).

Cultivars	Natchez		Triple Crown	
Harvest year	2011	2012	2011	2012
pH	2.88	3.01	2.92 a[a]	3.15 b
Titratable acidity (% malic acid)	0.386	0.403	0.391 a	0.422 b
% soluble solids	11.64 a	10.04 b c[b]	11.08	12.06 d

[a]Means with a and b letters indicate significant differences between years ($p < 0.05$). [b]Mean with c and d letters indicate significant differences between cultivars ($p < 0.05$).

Wine samples showed statistically significant differences between fermentation temperatures ($p < 0.05$). Within the cultivar and inoculation treatment, lower fermentation temperature samples had higher pH values than higher fermentation temperature samples. However, comparing cultivars within inoculation treatments, only samples at the higher fermentation temperature had statistically significant differences: "Triple Crown" berries had higher pH values than "Natchez" berries. Also, "Triple Crown" berries showed that yeast-inoculated samples had higher pH values than wild treatment samples.

3.1.2. Titratable Acidity. Titratable acidity of blackberry juice was expressed as % malic acid (MA) and the mean values are shown in Table 1. Observed values ranged from 0.386 to 0.422% MA. The pattern of the results was the same as that seen for pH values. "Triple Crown" berries showed significant difference between years ($p < 0.05$): year 2012 had higher titratable acidity than year 2011.

The mean titratable acidity values of blackberry wines are shown in Table 2. Observed values ranged from 0.350 to 0.420% MA. Our titratable acidity range was similar to the range measured in previous research, from 0.33 to 0.41% MA [8]. Most researchers have recorded somewhat higher titratable acidity values than those observed in the current study [24, 26, 27, 31]. Titratable acidity can be influenced by the weather conditions: lack of sunshine, low temperature, or high rainfall. The relatively low titratable acidity values measured in this study suggest that the berries were less tart than some. However, fermentation likely played a role in the differences observed as well. All samples showed statistically significant differences between fermentation temperatures as shown in Table 2 ($p < 0.05$). Within the cultivars, samples with lower fermentation temperature had higher titratable acidity than samples with higher fermentation temperature. Comparing the two cultivars, "Triple Crown" berries had higher titratable acidity than "Natchez" berries in the higher fermentation temperature samples. In addition, "Triple Crown" berries with yeast inoculation samples showed higher titratable acidity values than wild fermentation.

3.1.3. Soluble Solids. The mean sugar concentrations of blackberry juice samples, expressed as % soluble solids, are shown in Table 1. Values ranged from 10.04 to 12.06%. The soluble solids concentration of blackberry juice observed in this study (Table 1) was about 10 to 12%, which is close to the range seen in other research [32]. However, many previous researchers have reported slightly lower soluble solid concentrations of below 10% [8, 27]. It is well known that environmental condition such as weather and planting location can affect the sugar level of blackberries. Sugar levels in fruits have been shown to be affected by weather conditions leading up to and during harvest [27]. More sunshine and less rain or clouds during berry development could help to increase blackberry sugar concentration as well as the formation of good sugar/acid balance, and this may account for some of the differences in sugar content seen among treatments [1].

Statistically significant differences in soluble solids content were shown between the two years and cultivars ($p < 0.05$). Between the two years, "Natchez" berries in 2011 had higher percent soluble solids values than samples from 2012. "Natchez" berries were harvested in June. According to the Oklahoma Mesonet weather data [30], higher average rainfall was observed in June 2012 (5.5 cm) than in June 2011 (4.3 cm) and this may have led to berries that were less sweet in 2012. Between the cultivars, "Triple Crown" berries in 2012 had higher percent soluble solids than "Natchez" berries in 2012. This may be explained by the fact that "Triple Crown" berries were harvested in July, which had less rainfall (0.2 cm) than June (5.5 cm) in 2012.

3.1.4. Percent Alcohol. The average percent alcohol of blackberry wines is shown in Table 2. The alcohol concentrations measured in the blackberry wines ranged from 13.26 to 15.76%. All treatments except for the "Triple Crown" yeast had 13-14% alcohol concentration, while the "Triple Crown" yeast exceeded 15% alcohol concentration. In general, alcohol percentages were higher than those recorded by some other researchers [28]. Because sugar was added while processing the wine, it is not surprising that a relatively high alcohol concentration was seen in the wines. Other research showed that the alcohol concentration of commercial blackberry wines is about 9 to 15% and the wines were measured at 7 to 24% alcohol [29].

The wines made in this study were well within this range. Alcohol content in wines is a function of the sugar concentration in the starting material, up to the alcohol tolerance of the yeast doing the fermentation, presuming that the fermentation goes to completion. However, other factors may affect fermentation efficiency, such as available fermentable nitrogen and other yeast nutrients.

Within cultivar, inoculation treatment showed statistically significant differences ($p < 0.05$). Yeast-inoculated samples had higher percent alcohol than wild fermentation samples. In this study, "Triple Crown" berries produced more alcohol with yeast inoculation than "Natchez" wines. Since the same ratio of sugar was added based on the total volume of berries used, alcohol should have been produced at a similar rate. In addition, almost all samples showed statistically significant differences between the two cultivars and "Triple Crown" showed higher percent alcohol than "Natchez" berries ($p < 0.05$). The "Triple Crown" berries had somewhat higher starting soluble solids concentration

TABLE 2: Mean values of pH, titratable acidity, and % alcohol in blackberry wines by inoculation type and fermentation temperature ($n = 3$).

Cultivar	Natchez				Triple Crown			
Inoculation treatment	Yeast		Wild		Yeast		Wild	
Fermentation temperature (°C)	21.6	26.6	21.6	26.6	21.6	26.6	21.6	26.6
pH	3.1 a[a]	2.6 b c[b]	3.08 a	2.64 b c	3.12 a e	2.87 b d e	3.05 a f	2.76 b d f
Titratable acidity (% malic acid)	0.42 a	0.35 b c	0.41 a	0.35 b c	0.42 a e	0.39 b d e	0.41 a f	0.37 b d f
% alcohol	13.91 c e[c]	13.57 c	13.46 f	13.26 c	15.76 a d e	15.01 b d e	13.68 f	13.67 d f

[a]Means with a and b letters indicate significant differences between fermentation temperatures within cultivar and inoculation treatment ($p < 0.05$). [b]Means with c and d letters indicate significant differences between cultivars within inoculation treatment and fermentation temperature ($p < 0.05$). [c]Means with e and f letters indicate significant differences between inoculation treatments within cultivar and fermentation temperature ($p < 0.05$).

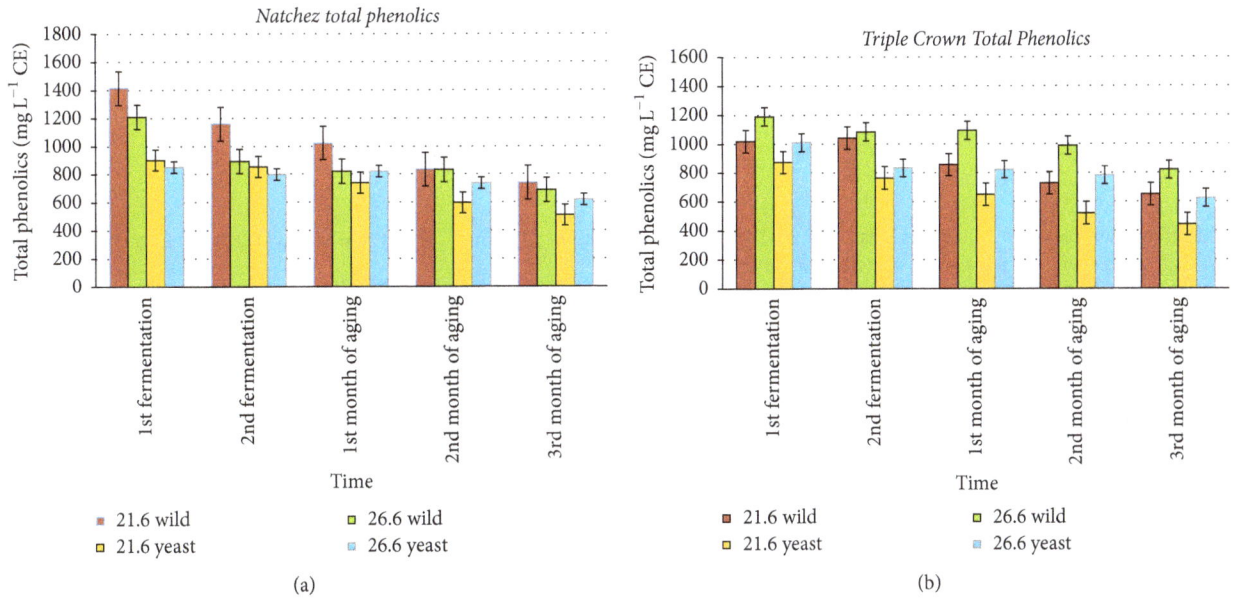

FIGURE 2: Total phenolic concentrations ($N = 120$) of "Natchez" blackberry (a) and "Triple Crown" blackberry (b) fermented juices and wines. Error bars represent ± standard deviation of the mean ($n = 2$).

than "Natchez" berries but not high enough to account for the observed final difference in wine alcohol concentration. This suggests that other fermentation factors affected the final alcohol concentrations seen. The "Natchez" berries did not provide favorable condition for yeast growth as "Triple Crown" berries and this may have limited yeast activity in "Natchez" wines. Although the wild fermentation for "Triple Crown" berries had lower alcohol concentration than the yeast inoculation, it was more consistent in terms of final alcohol concentration than "Natchez," indicating that fermentation conditions may have been better for both yeast and wild microflora in the "Triple Crown" musts.

3.2. Antioxidant Activity Analyses

3.2.1. Modified Harbertson-Adams Assay

(1) Total Phenolic Concentration. The total phenolic concentration in blackberry fermented juice and wine, expressed as mg L^{-1} catechin equivalents (CE), is shown in Figure 2. Total phenolic concentrations of blackberry juices and wines were ranged from 440 to 1420 mg L^{-1} CE.

Our results for total phenolic concentrations (Figure 2) were similar to other research [24], showing the concentrations between 601 and 1624 mg L^{-1} CE. Blackberry wine showed a similar lower end for total phenolics with values ranging from 380 to 520 mg L^{-1} CE [33]. This article examined the effect of storage on total phenolics and, similar to our results (Figure 2), the levels of total phenolics decreased over time. Some blackberry wine research [3, 34] had total phenolic concentrations between 1608 and 2836 mg CE L^{-1} which were higher than our results.

Statistically significant differences were seen between cultivars, fermentation temperatures, and inoculation treatments ($p < 0.05$). Fermentation temperature in combination with cultivar type could have affected total phenolic concentrations this study, particularly early in the winemaking process. Higher fermentation temperatures were generally correlated with higher total phenolics concentrations for "Triple Crown" but not "Natchez" wines; this may reflect different extraction kinetics for phenolics in "Triple Crown" berries.

Wild fermented wines were also generally higher in total phenolic concentration than yeast fermented wines. One possible reason for this result is that wild fermentation

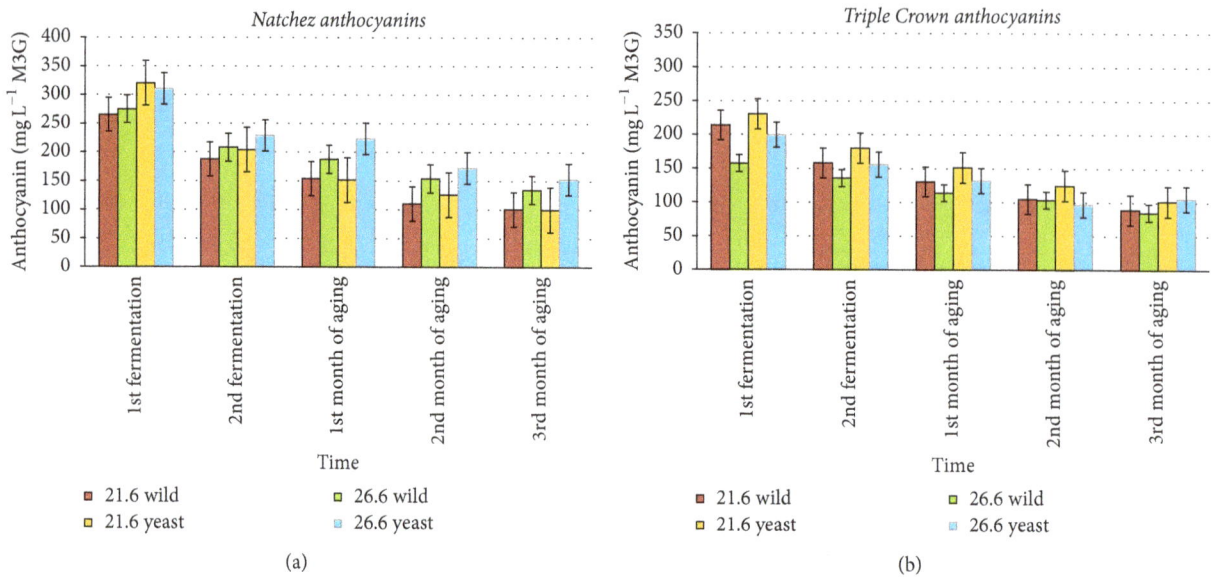

(a)

(b)

FIGURE 3: Total anthocyanin concentrations ($N = 120$) of "Natchez" blackberry (a) and "Triple Crown" blackberry (b) fermented juices and wines. Error bars represent ± standard deviation of the mean ($n = 2$).

microorganisms created a fermentation environment that was more conducive to the preservation of phenolic compounds, perhaps by inhibiting phenolic polymerization and complex formation.

(2) Anthocyanin Concentration. Anthocyanin concentrations of blackberry fermented juices and wines are shown in Figure 3. Anthocyanin concentrations are expressed as $mg\,L^{-1}$ malvidin 3-monoglucoside (M3M). The anthocyanin concentrations of this study ranged from 85 to 320 $mg\,L^{-1}$ M3M.

Our results showed somewhat higher anthocyanin concentration than some other reports (Figure 3). A research [34] showed that the range of anthocyanins was from 12 to 167 $mg\,L^{-1}$ M3M in blackberry wine, which was twofold lower than our result. It is important to note that these wines were prepared using different methods compared to the wines processed in the current study. It is possible that the fermentation method used in the current study affected anthocyanin concentrations of blackberry juice and wine. The blackberry cultivars used were different as well.

Statistically significant differences were found between cultivars, fermentation temperatures, and inoculation levels ($p < 0.05$). In this study, wines made from "Natchez" berries showed higher anthocyanin concentrations in the berries than wines made from "Triple Crown" berries. "Natchez" berries were relatively large and soft-skinned. They may have broken down more easily at the higher fermentation temperature and this may have led to a more complete extraction of anthocyanins. On the other hand, "Triple Crown" berries were relatively small and firm. They may not have broken down as well and thus the longer fermentation time seen at the lower fermentation temperature may have allowed a more complete extraction. For inoculation levels, yeast inoculation samples had higher anthocyanin concentration than wild fermentation. Yeast-inoculated samples may helped to break

down polymeric pigments and provide higher anthocyanin concentration.

Overall, the anthocyanin concentration decreased during wine aging, likely due to polymerization and copigmentation. During storage, anthocyanins gradually disappear as monomeric compounds and are transformed into polymeric forms, resulting in loss of color [35]. Also, newly formed pigments derived from chemical reactions between anthocyanins and various wine compounds, including phenolic compounds, could contribute to the color characteristics of aged wines. It has been suggested that higher temperature extraction could accelerate formation of new, more stable wine pigments [3, 24, 36].

3.2.2. Oxygen Radical Absorbance Capacity (ORAC). The ORAC values for blackberry fermented juice and wines are shown in Figure 4. The range of ORAC values observed in this study was between 9776 and 37845 μmol Trolox equivalent (TE) g^{-1}. Overall, cultivar and fermentation temperature had the greatest influence ORAC values, but the effects were not consistent over time.

The average mean ORAC values of blackberry pomace are shown in Table 3. Values ranged from 16625 to 23200 TE g^{-1}. "Triple Crown" pomace generally showed higher ORAC values than "Natchez" pomace. For pomace samples, blackberry seeds and skin residue had higher antioxidant activity than aged wine samples, indicating that the pomace retains substantial amounts of antioxidant compounds even after wine processing. In general, wild fermentation samples showed higher antioxidant capacity than yeast-inoculated samples. Samples with higher fermentation temperature showed statistically significant differences ($p < 0.05$).

As with the anthocyanin concentrations previously reported, the ORAC values recorded in this study (Figure 4)

(a)

(b)

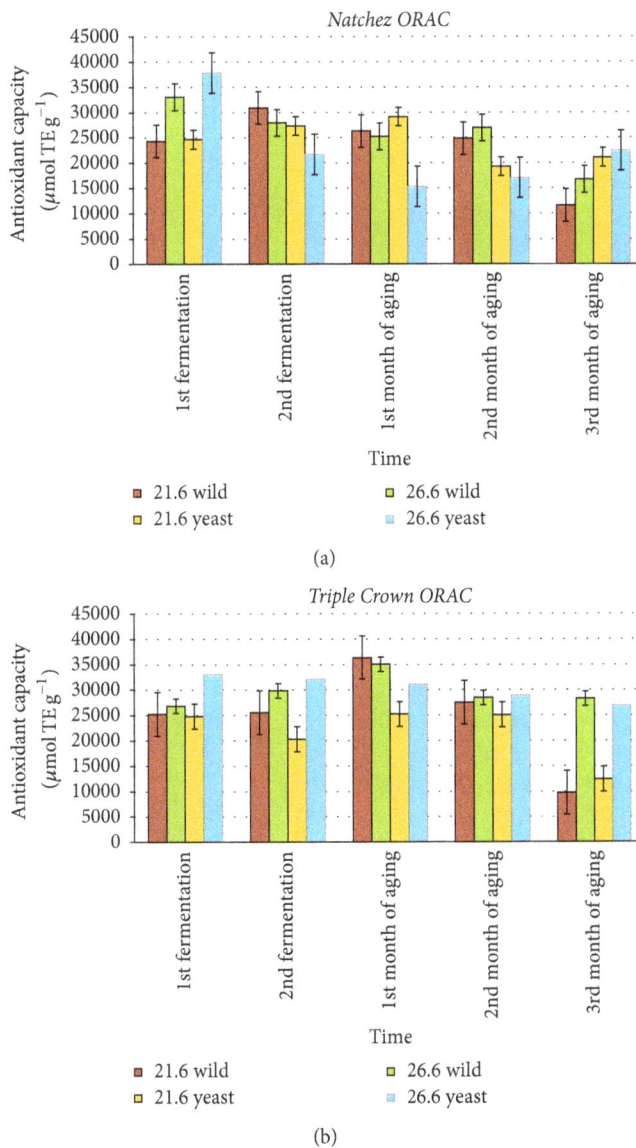

FIGURE 4: Oxygen radical absorbance capacity ($N = 120$) of "Natchez" blackberry (a) "Triple Crown" blackberry (b) fermented juices and wines. Error bars represent ± standard deviation of the mean ($n = 2$).

TABLE 3: Mean ORAC values for blackberry pomace ($n = 3$).

Cultivars	Natchez				Triple Crown			
Inoculation treatment	Yeast		Wild		Yeast		Wild	
Fermentation temp. (°C)	21.6	26.6	21.6	26.6	21.6	26.6	21.6	26.6
ORAC (μmol TE/g)	16920	18628 a[a]	16625	19055 a	17209 c[b]	12481 b d e[c]	19428	23200 b f

$N = 24$. [a]Means with a and b letters indicate significant differences between cultivars within fermentation temperature and inoculation treatment ($p < 0.05$). [b]Means with c and d letters indicate significant differences between fermentation temperatures within cultivar and inoculation treatment ($p < 0.05$). [c]Means with e and f letters indicate significant differences between inoculation treatments within cultivar and fermentation temperature ($p < 0.05$).

were somewhat higher than those reported in other literature. A Spanish researcher [2] showed that the range of ORAC values observed in blackberry juices was 39160 μmol TE g^{-1}. This range was close to our result but slightly higher. Other articles [32, 37] showed that the ORAC values varied a great deal depending on variety, geographic growing location, and extraction method. As fermentation techniques and extraction methods are known to influence ORAC values, these results may indicate that the Korean fermentation style used in this study was more efficient in extracting and/or preserving antioxidant activity in the final wine than some other fermentation methods examined in previous studies.

Fermentation temperature influenced the antioxidant capacity in blackberry juice and wine in this study (Figure 4), but the results were not consistent over time. At times, "Natchez" wines showed higher antioxidant activity at the lower fermentation temperature and "Triple Crown" wines showed higher antioxidant activities associated with the higher fermentation temperature. At other times, the reverse was true. By the end of the three-month aging process, higher antioxidant activity was generally correlated with higher fermentation temperatures for both cultivars and inoculation treatments. Similarly, the influences of cultivar and inoculation type on ORAC values (Figure 4) were not consistent over time. By the end of the three-month aging process, no clear pattern for the effects of cultivar and/or inoculation type on antioxidant activity was discernable.

Overall, there was not clear pattern of correlation between ORAC, total phenolic concentration, and anthocyanin concentrations. However, common patterns in ORAC values and total phenolic concentrations were seen with respect to cultivar and fermentation temperature. One possible cause for the mixed results seen overall for correlations among ORAC values, anthocyanins, and total phenolic concentrations could be variations in cultivar genetics. As noted previously, the two cultivars had notably different berry tissue consistencies and this could have influenced the relative efficiency of phenolic extraction overall and among fermentation temperature treatments.

Also, it is important to note that other compounds may have had a significant influence on the measured ORAC values. Anthocyanins are one class of phenolic compounds, but other types may have had a greater or lesser impact on measured ORAC values over time as the aging process progressed and various polymers and complexes evolved. Further research would be required to elucidate the relationship between all of the compounds present in the wine and the observed antioxidant capacity. In addition, the ORAC test employed in this study provides a relative quantification of the wines' ability to quench peroxyl radicals; other antioxidant tests could be employed in future research to measure antioxidant capacity relative to other oxygen radical species. This could demonstrate additional correlations.

4. Conclusion

This research showed that the blackberry wines made in this study were relatively high in polyphenolic compounds as well as antioxidant capacity. As the Korean wine making method is of relatively low cost and easy to adapt to small-scale production, it may be especially well suited to helping small local growers and/or processors to add value to the blackberry crop such as "Natchez" in the Midwestern United States.

Conflicts of Interest

The authors declare that they have no conflicts of interest.

Acknowledgments

Thanks are due to the Robert M. Kerr Food and Agricultural Product Center for support in funding and laboratory equipment.

References

[1] M. Gündoğdu, T. Kan, and İ. Canan, "Bioactive and antioxidant characteristics of blackberry cultivars from East Anatolia," *Turkish Journal of Agriculture and Forestry*, vol. 40, no. 3, pp. 344–351, 2016.

[2] J. Marhuenda, M. D. Alemán, A. Gironés-Vilaplana et al., "Phenolic composition, antioxidant activity, and in vitro availability of four different berries," *Journal of Chemistry*, vol. 2016, Article ID 5194901, 7 pages, 2016.

[3] M. N. Mitic, M. V. Obradovic, S. S. Mitic, A. N. Pavlovic, J. L. J. Pavlovic, and B. T. Stojanovic, "Free radical scavenging activity and phenolic profile of selected Serbian red fruit wines," *Revista De Chimie*, vol. 64, no. 1, pp. 68–73, 2013.

[4] J. Ivanovic, V. Tadic, S. Dimitrijevic, M. Stamenic, S. Petrovic, and I. Zizovic, "Antioxidant properties of the anthocyanin-containing ultrasonic extract from blackberry cultivar 'Čačanska Bestrna,'" *Industrial Crops and Products*, vol. 53, pp. 274–281, 2014.

[5] A. Kostecka-Gugała, I. Ledwozyw-Smoleń, J. Augustynowicz, G. Wyzgolik, M. Kruczek, and P. Kaszycki, "Antioxidant properties of fruits of raspberry and blackberry grown in central Europe," *Open Chemistry*, vol. 13, no. 1, pp. 1313–1325, 2015.

[6] A. A. F. Zielinski, C. Goltz, M. A. C. Yamato et al., "Blackberry (*Rubus spp.*): influence of ripening and processing on levels of phenolic compounds and antioxidant activity of the 'Brazos' and 'Tupy' varieties grown in Brazil," *Ciencia Rural*, vol. 45, no. 4, pp. 744–749, 2015.

[7] M. H. Johnson, E. G. De Mejia, J. Fan, M. A. Lila, and G. G. Yousef, "Anthocyanins and proanthocyanidins from blueberry-blackberry fermented beverages inhibit markers of inflammation in macrophages and carbohydrate-utilizing enzymes in vitro," *Molecular Nutrition and Food Research*, vol. 57, no. 7, pp. 1182–1197, 2013.

[8] V. M. Celant, G. C. Braga, J. A. Vorpagel, and A. B. Salibe, "Phenolic composition and antioxidant capacity of aqueous and ethanolic extracts of blackberries," *Revista Brasileira de Fruticultura*, vol. 38, no. 2, Article ID e-411, 2016.

[9] J. R. Clark and C. E. Finn, "Blackberry breeding and genetics," *Fruit, Veg and Cereal Sci Biotech*, vol. 5, no. 1, pp. 27–43, 2011.

[10] W.-Y. Huang, H.-C. Zhang, W.-X. Liu, and C.-Y. Li, "Survey of antioxidant capacity and phenolic composition of blueberry, blackberry, and strawberry in Nanjing," *Journal of Zhejiang University: Science B (Biomedicine & Biotechnology)*, vol. 13, no. 2, pp. 94–102, 2012.

[11] L. Kaume, L. R. Howard, and L. Devareddy, "The blackberry fruit: a review on its composition and chemistry, metabolism and bioavailability, and health benefits," *Journal of Agricultural and Food Chemistry*, vol. 60, no. 23, pp. 5716–5727, 2012.

[12] J. Oszmiański, P. Nowicka, M. Teleszko, A. Wojdyło, T. Cebulak, and K. Oklejewicz, "Analysis of phenolic compounds and antioxidant activity in wild blackberry fruits," *International Journal of Molecular Sciences*, vol. 16, no. 7, pp. 14540–14553, 2015.

[13] E. Sariburun, S. Şahin, C. Demir, C. Türkben, and V. Uylaşer, "Phenolic content and antioxidant activity of raspberry and

blackberry cultivars," *Journal of Food Science*, vol. 75, no. 4, pp. C328–C335, 2010.

[14] V. R. De Souza, P. A. P. Pereira, T. L. T. Da Silva, L. C. De Oliveira Lima, R. Pio, and F. Queiroz, "Determination of the bioactive compounds, antioxidant activity and chemical composition of Brazilian blackberry, red raspberry, strawberry, blueberry and sweet cherry fruits," *Food Chemistry*, vol. 156, pp. 362–368, 2014.

[15] L. Zhang, J. Zhou, H. Liu, M. A. Khan, K. Huang, and Z. Gu, "Compositions of anthocyanins in blackberry juice and their thermal degradation in relation to antioxidant activity," *European Food Research and Technology*, vol. 235, no. 4, pp. 637–645, 2012.

[16] Y. Joh, "Antioxidant properties of blackberry wine produced using Korean traditional winemaking techniques," in *Proceedings of the American Society for Enology and Viticulture National Conference*, Portland, USA, 2015, Technical abstracts.

[17] M. K. Park, *Healthy Fermented Drink Making*, Tae-Woong Inc., Seoul, South Korea, 4th edition, 2009.

[18] M. K. Park, *Enzyme and Health: Life Depends on Enzyme Level*, Tae-Woong Inc., Seoul, South Korea, 2nd edition, 2010.

[19] K. S. Yun, *Easy and Simplicity of Winemaking*, Yas Media, Seoul, South Korea, 2007.

[20] Y. Joh, *Antioxidant Capacity, Phenolic and Volatile Compound Composition of Blackberry Wines Produced Using Korean Wine-Making Techniques [DPhill thesis]*, Oklahoma State University, Stillwater, OK, USA, 2014, Available from: shareok advancing Oklahoma scholarship,research and institutional memory.

[21] S. K. Park, "Breathable container, Onggi- their pores and fermentation products," Onggi Folk Museum, 2016, http://onggimuseum.org/bbs/view.php?id=data&page=1&sn1=&div-page=1&sn=off&ss=on&sc=on&select_arrange=headnum&desc=asc&no=29#.

[22] Anton-Parr.com, "Alcohol meter for wine: Alcolyzer wine," 2016, http://www.anton-paar.com/kr-en/products/details/alcolyzer-wine-mme-wine-analysis-system/.

[23] D. Huang, B. Ou, M. Hampsch-Woodill, J. A. Flanagan, and R. L. Prior, "High-throughput assay of oxygen radical absorbance capacity (ORAC) using a multichannel liquid handling system coupled with a microplate fluorescence reader in 96-well format," *Journal of Agricultural and Food Chemistry*, vol. 50, no. 16, pp. 4437–4444, 2002.

[24] Í. Arozarena, J. Ortiz, I. Hermosín-Gutiérrez et al., "Color, ellagitannins, anthocyanins, and antioxidant activity of andean blackberry (Rubus glaucus Benth.) wines," *Journal of Agricultural and Food Chemistry*, vol. 60, no. 30, pp. 7463–7473, 2012.

[25] T. Milosevic, E. Mratinic, N. Milosevic, I. Glisic, and J. Mladenovic, "Segregation of blackberry cultivars based on the fruit physico-chemical attributes," *Tarim Bilimleri Dergisi*, vol. 18, no. 2, pp. 100–109, 2012.

[26] J. K. Rutz, G. B. Voss, and R. C. Zambiazi, "Influence of the degree of maturation on the bioactivecompounds in backberry (*Rubus* spp.) cv. Tupy," *Food and Nutrition Sciences*, vol. 03, no. 10, pp. 1453–1460, 2012.

[27] L. Ali, B. W. Alsanius, A. K. Rosberg, B. Svensson, T. Nielsen, and M. E. Olsson, "Effects of nutrition strategy on the levels of nutrients and bioactive compounds in blackberries," *European Food Research and Technology*, vol. 234, no. 1, pp. 33–44, 2012.

[28] J. Gao, Z. Xi, J. Zhang et al., "Influence of fermentation method on phenolics, antioxidant capacity, and volatiles in blackberry wines," *Analytical Letters*, vol. 45, no. 17, pp. 2603–2622, 2012.

[29] M. H. Johnson and E. Gonzalez de Mejia, "Comparison of chemical composition and antioxidant capacity of commercially available blueberry and blackberry wines in Illinois," *Journal of Food Science*, vol. 77, no. 1, pp. C141–C148, 2012.

[30] University of Oklahoma, "Station monthly summaries," Mesonet, 2016, http://www.mesonet.org/index.php/weather/station_monthly_summaries.

[31] J. Ortiz, M.-R. Marín-Arroyo, M.-J. Noriega-Domínguez, M. Navarro, and I. Arozarena, "Color, phenolics, and antioxidant activity of blackberry (Rubus glaucus Benth.), blueberry (Vaccinium floribundum Kunth.), and apple wines from ecuador," *Journal of Food Science*, vol. 78, no. 7, pp. C985–C993, 2013.

[32] R. H. Thomas, F. M. Woods, W. A. Dozier et al., "Cultivar variation in physicochemical and antioxidant activity of Alabama-grown blackberries," *Small Fruits Review*, vol. 4, no. 2, pp. 57–71, 2005.

[33] M. Kopjar, B. Bilic, and V. Pilizota, "Influence of different extracts addition on total phenols, anthocyanin content and antioxidant activity of blackberry juice during storage," *Croatian Journal of Food Science and Technology*, vol. 3, no. 1, pp. 9–15, 2011.

[34] I. Mudnic, D. Budimir, D. Modun et al., "Antioxidant and vasodilatory effects of blackberry and grape wines," *Journal of Medicinal Food*, vol. 15, no. 3, pp. 315–321, 2012.

[35] S. Yuksel and I. Koka, "Color stability of blackberry nectars during storage," *Journal of Food Technology*, vol. 694, pp. 166–169, 2008.

[36] S. De Pascual-Teresa and M. T. Sanchez-Ballesta, "Anthocyanins: from plant to health," *Phytochemistry Reviews*, vol. 7, no. 2, pp. 281–299, 2008.

[37] P. Denev, M. Ciz, G. Ambrozova, A. Lojek, I. Yanakieva, and M. Kratchanova, "Solid-phase extraction of berries' anthocyanins and evaluation of their antioxidative properties," *Food Chemistry*, vol. 123, no. 4, pp. 1055–1061, 2010.

5

Effect of Starter Culture and Low Concentrations of Sodium Nitrite on Fatty Acids, Color, and *Escherichia coli* Behavior during Salami Processing

Carla María Blanco-Lizarazo,[1] Indira Sotelo-Díaz ⓘ,[2] José Luis Arjona-Roman,[3] Adriana Llorente-Bousquets,[3] and René Miranda-Ruvalcaba ⓘ[4]

[1]*Agroindustrial Process Research Group, University of La Sabana, Colombia*
[2]*Titular Professor, EICEA, Agroindustrial Process Research Group, University of La Sabana, Colombia*
[3]*Engineering and Technology Department, Faculty of Advanced Studies Cuautitlán, National Autonomous University of Mexico (UNAM), Mexico*
[4]*Chemistry Department, Faculty of Advanced Studies Cuautitlan, National Autonomous University of Mexico (UNAM), Mexico*

Correspondence should be addressed to Indira Sotelo-Díaz; indira.sotelo@unisabana.edu.co

Academic Editor: Alejandro Castillo

The reduction of $NaNO_2$ and safety in meat products have been a concern to the meat industry for the last years. This research evaluated the changes in total fatty acids (TFAs) and myoglobin forms by adding starter culture (*Lactobacillus sakei/Staphylococcus carnosus*) and 50 ppm of $NaNO_2$ during salami processing. In the postripening stage, the starter culture influenced the concentration of the palmitic, oleic, vaccenic, and γ-linolenic TFAs, whereas the metmyoglobin concentration was lower (which could be related to the antioxidant effect of the starter culture). In this stage, an increase in enthalpy, specific heat, and onset temperature was found when adding starter culture and $NaNO_2$, which is directly related to polyunsaturated TFA. However, when adding just the starter culture without 50 ppm $NaNO_2$, the *E. coli* population was reduced in 4 log CFU/g. This study proposes the analysis of changes in meat product processing like salami in a holistic form, where the application of starter culture with low nitrite concentrations could be in the meat industry an upward trend for reducing this additive.

1. Introduction

During the processing of salami, physical, physicochemical, biochemical, and microbiological transformations occur. These changes are influenced by factors such as raw material, sodium nitrite, and starter culture. This process has an impact on the total fatty acid profile and it increases the unsaturation level and the susceptibility of oxidation [1, 2]. Furthermore, a concentration of 50 ppm sodium nitrite contributes in delaying lipid oxidation in meat products as a result of its reaction with iron [3]. The addition of $NaNO_2$ influences the formation of the bright red color that corresponds to the nitrosylmyoglobin complex. Moreover, the addition of starter cultures influences changes in the concentration and types of fatty acids and oxidative phenomena of the lipid fraction [4].

Lipid oxidation, including autoxidation and enzyme-catalysed oxidation of fatty acids in fermented and dry-cured meat products, is involved in several aspects of meat products quality (functional, sensory, and nutritional) [5, 6]. Primary products of lipid oxidation generate changes in myoglobin such as the oxidation of oxymyoglobin (OMB) to metmyoglobin (MMB) inducing color changes [1, 7]; these changes result in a decrease in heme redox stability, rather than the oxidation of specific amino acid residues. This reaction generally proceeds in parallel to the lipidic oxidation, where the products of both reactions can mutually accelerate pigmentation and lipid oxidation [2, 8].

In addition, during the processing of salami a_w and pH are reduced; as a result, in these products foodborne pathogens are inhibited. However, deviations in the process parameters (temperature and/or humidity) affect the assurance of the elimination of food-borne pathogens in the final products [9, 10]. As a consequence, it is of great importance for the safety of the product the addition of starter cultures and the validation of the reduction of E. coli during salami processing.

Therefore, the design of the starter culture is critical for salami quality, where the addition of lactic acid bacteria (LAB) is aimed to ensure product safety and coagulase-negative staphylococci (CNS) is related with formation and stabilization of color and prevention of rancidity due to nitrate reductase and catalase activity [11]. In fact, the LAB *Lactobacillus sakei* and the CNS *Staphylococcus carnosus* have been selected as starter culture for their technological properties, where *L. sakei* decrease the pH through the fermentation of carbohydrates [12]. Furthermore, *L. sakei* produces NO and N_2O due to its nitrite reductase and catalase activity heme independent [13]. In addition, *S. carnosus* can inhibit the oxidation of fatty acids due to its high nitrate reductase activity, which generates nitrites (NO_2^-) with antioxidant properties, taking part in nitrosomyoglobin formation, being involved in the development of color in fermented meat products [11].

Differential scanning calorimetry (DSC) is a thermo analytical technique that has been applied for lipid characterization. Thermal properties reported good relation with major triacylglycerols (TAG) and total fatty acids (TFA) and minor components free fatty acids (FFA), and oxidation products [14]. Multiple melting behaviors are explained by the melting of TAG and TFA with different melting peaks and crystal reorganization effects; also, enthalpy transitions are coupled with each other and strongly overlap. Consequently, the objective of this study was to evaluate changes of total fatty acids profile by CG-EM and MDSC, and myoglobin forms, during salami processing led by starter culture and the addition of 50 ppm of $NaNO_2$. Furthermore, the reduction of E. coli was validated in salami with starter culture addition.

2. Materials and Methods

2.1. Salami Preparation. Four treatments were designed for the preparation of salamis (30 samples for each treatment): control (C) without the addition of the starter culture or $NaNO_2$; control (C^+) without the starter culture and 50 ppm of $NaNO_2$; and treatments with the starter culture without $NaNO_2$ (*Ls-Sc*) and with the starter culture and 50 ppm $NaNO_2$ (*Ls-Sc^+*).

Each treatment was evaluated in 3 different batches. Several salami samples units of 25±1 g each were prepared, the experimental samples were designed from mixtures based on pork meat (42% w/w) and pork back fat (23% w/w) under the formulation and procedures of a local industry (Bogotá, Colombia).

A commercial starter culture composed of *Lactobacillus sakei* and *Staphylococcus carnosus* (T-SC-150 Bactoferm™, CHR-Hansen, Hoersholm, Denmark) was added to the *Ls-Sc* and *Ls-Sc^+*. The starter culture was previously activated in a 2% w/v dextrose (Scharlau Chemie SA, Barcelona, Spain) solution in sterile water and incubated at 25±0.1°C for 20 h. For every 10 kg of the mixture, 2.5g was added. The meat and fat were ground through a 10 mm diameter and mixed

with 2.5% m/m NaCl, 0.1% m/m sodium erythorbate and $NaNO_2$ according to the treatment and then stuffed into natural casings of 5 cm diameter.

Salami processing was performed at 80% of relative humidity in 4 stages: (i) the prefermentation stage, the first 12 h at 4±0.1°C; (ii) the postfermentation stage, the following 48 h at 21±0.5°C; (iii) ripening up to 14 days at 17±0.5°C. (iv) The last day of processing was considered post-ripening.

Measurements of pH and a_w were done according to protocols described by Blanco–Lizarazo et al. [15]. The initial pH for all treatments was 5.89±0.14 and in postripening stage was for *Ls-Sc* 4.67±0.01, *Ls-Sc^+* 4.85±0.02, and C 4.43±0.01 and for C^+ 4.52±0.04. The initial a_w was 0.910±0.12 and in postripening stage was 0.810±0.03 for all treatments.

2.2. Determination of Total Fatty Acids. Fatty acids were determined on the lipid extract from the salami samples at different processing times. Samples of 10g of salami were placed within an extraction cartridge and inserted into a reflux flask using petroleum ether under the Soxhlet methodology during 5 h. The obtained extracts were concentrated using a vacuum evaporator at 45°C.

Extracted lipids were methylated and transesterified using sodium methoxide at pH 3, according to the method of Park and Goins [16]. Subsequently, twice the volume of a saturated sodium chloride solution and 3 times the volume of ethyl acetate was added in a separatory funnel. Finally, 5g of anhydrous sodium sulfide was added and decanted. The samples were distilled at 40°C in vacuum.

The fatty acids methyl esters obtained from the TFAs and triglycerides of fatty acids were separated and quantified with a gas chromatographer (Agilent model 62630A, China) coupled to a mass spectrometer model 5975C with the Agilent triple axis detector (USA). The GC-MS system was equipped with a DB-5 column (Agilent, USA, 30 m x 0.25 mm x 0.25 μm). Helium was used as the carrier gas at a flow rate of 1.0 ml/min, and the injection ratio was 1:10. The oven temperature was adjusted to 150°C with a holding time of 2 min, an initial ramp of 5°C/min up to 200°C, followed by a ramp of 3°C/min up to 215°C, and finally a ramp of 10°C/min up to 260°C, which was held for 2 min. The injector and detector temperatures were adjusted to 250°C and 290°C, respectively, and the injection volume was 1 μl. The fatty acids methyl esters were identified by comparison of the respective retention times and analysis of their corresponding mass spectra with a mass spectral database of fatty acid methyl esters, ISBN: 978-1-1181-4394-0, MSD ChemStation (E.02.00.493, Agilent, USA) Software, (November 2011). Quantification was performed through areas under the curve of each peak based on total methyl esters. All analyses were performed by triplicate.

2.3. Modulated Differential Scanning Calorimetry (MDSC) Analysis of Salami Lipids. For the analysis of nonisothermal transitions, 5±0.01 g samples of salami of the treatment *Ls-Sc^+* were collected at 0 and 14 days. To ensure the homogeneity and reproducibility of the sample, the 10 samples for each treatment were subjected to cryomilling with liquid nitrogen for 1 min and were milled by impact for five cycles of 10 s each with a grinder (IKA A11 Basic S1, USA).

A Modulated DSC 2920 (TA Instruments, New Castle, DE, USA) was used. A total of 12±1 mg of sample was placed in a hermetically sealed aluminum cell. The experimental conditions for the samples are based on the back fat pork analysis proposed by Sasaki et al. [17], with an initial temperature of 10°C and a final temperature of 100°C, a rate (ß) of 5°C.min-1, and modulation of ±0.796°C min-1. An empty cell was used as the reference. Indium was used for the temperature calibration, and Sapphire was used for the heat capacity.

2.4. Measurement of the Reflectance for Determinate Concentrations of Myoglobin Forms.

Nine 17 mm thick samples were evaluated with 5 replicates for each salami processing time (0, 1, 2, 3, 7, and 14 days). The reflectance was measured between 400 nm and 700 nm in 10 nm intervals in accordance with protocols proposed by the American Meat Science Association [18] with the ColorquestXE spectro-colorimeter (HunterLab, Reston, USA) (light source A, standard geometric observation at 10° to 1 in of distance, and white standard). The concentrations of the chemical forms of myoglobin were calculated according to the protocols proposed by Tang *et al.* [19]. The reflectance values (503, 557, and 582 nm) were calculated using linear interpolation.

2.5. Microbial Analysis during the Processing of Salami

2.5.1. Bacterial Growth.
For the analysis of bacterial growth, 25±0.1 g of the sample was taken at 0, 3, 6, 9, 12, 16, 20, 24, 48, 72, 240, and 336 h. Then, 250 ml of 0.1% peptone water was added (Merck, Darmstadt, Germany) and homogenised for 120 s using a Stomacher (BA7021, Serward, England). Six 10-fold serial dilutions were performed with 0.1% peptone water and deep cultured in PCA agar plates (Scharlau Chemie S.A., Barcelona, Spain). This step was performed in triplicate and incubated at 37±1°C for 48 h to obtain the total bacterial count. LAB were cultured in Man Rogosa and Sharpe agar (MRS) (Scharlau Chemie S.A., Barcelona, Spain) and CNS in Baird Parker with egg yolk and potassium tellurite (Scharlau Chemie S.A., Barcelona, Spain).

2.5.2. E. coli Reduction.
For the analysis of *E. coli* reduction, salami samples for the treatments *Ls-Sc* and *Ls-Sc*+ were inoculated with 2.5±0.1 ml (4.4±0.19 Log CFU/g) of *E. coli* ATCC 25922™ previously activated with glass beads in 100 mL of BHI medium (Scharlau Chemie SA, Barcelona, Spain) and incubated (Friocell 22 - Comfort, MMM group, Munich, Germany) for 12 h at 37±1°C. The strain was maintained in glass beads with glycerol/Nutrient Broth (Scharlau Chemie SA, Barcelona, Spain) (20/80%) (v/v) and stored at -80°C in liquid nitrogen. The salami samples inoculated with *E. coli* were cultured in Eosin methylene blue agar (EMB) under identical conditions described above for analysis of bacterial growth.

2.5.3. Modelling of Bacterial Behavior.
To analyse the bacterial growth during the processing of salami the model developed by Baranyi and Roberts was applied [20]. This model was selected to determine the maximum growth rate (μ_{max}) and lag phase (λ). The software Matlab R2010a (The MathWorks, USA) was used. For the analysis of *E. coli*

reduction during the processing of salami was applied the model developed by Coroller et al. [21].

2.6. Statistical Analysis.
This experiment was based on a completely randomized design with 3 replicates corresponding to 3 batches of salami for fatty acid analysis by GC-MS. For each TFA from each treatment at the evaluated and time points was compared through independent analysis by Duncan's test.

For the analysis of the concentration of myoglobin forms, 3 samples of 3 different batches of salami were analyzed for each treatment and processing time. Percentages of chemical forms of myoglobin were compared for each treatment using the F-test (P<0.05) and Duncan's test. The parameters obtained from the mass spectra and the analyses of the MDSC thermograms were compared for each treatment through the analysis of variance (P<0.05) and Duncan's test.

Microbial growth and *E. coli* reduction were evaluated for 2 independent samples from 3 different batches of salami for each treatment and processing time. The growth parameters were compared for each treatment through the analysis of variance (P<0.05) and Tukey's test. The statistical software SAS 9.2 (32) was used (SAS Institute Inc., Cary, NC, USA).

3. Results and Discussion

3.1. Changes in the Total Fatty Acid (TFA) Profile.
The profiles of total fatty acids (TFA) during the processing of salami are expressed in mg/g of fat and are shown in Table 1. The concentration of the saturated fatty acids (SFA) myristic acid (C14:0) and palmitic acid (C16:0) during processing with the addition of the starter culture did not change significantly (P>0.05) as a function of the addition of $NaNO_2$. The SFA myristic and stearic were higher abundance in the starter culture treatments; this finding was in concordance with results reported by Aksu and Kaya [22] on Pastirma (Turkish Dry Meat Product) where the SFA concentration was lower in control group than in treatments with starter culture composed by *Staphylococcus xylosus + Lactobacillus sakei* and *Staphylococcus carnosus + Lactobacillus pentosus*.

Regarding the concentration of monounsaturated fatty acids (MUFA), palmitoleic acid (C16:1) exhibited higher concentration in the postfermentation and postripening phases for treatment C+. The MUFA oleic acid (C18:1) had the highest concentrations in the control treatments and no differences (P>0.05) were associated with the addition of $NaNO_2$. For the *Ls-Sc* and *Ls-Sc*+, there were no significant differences (P>0.05) in the concentration of the MUFA vaccenic acid (C18:1 trans-11) after the 7th day of ripening.

The polyunsaturated fatty acids (PUFA) linoleic acid and γ- linolenic acid were not identified in stages prior to prefermentation. During the ripening stage, *Ls-Sc*+ showed the highest concentration of this PUFA compared to treatment C, which could be correlated with the greater population density of the starter culture. According to Talon et al. [23] *S. carnosus* inhibited the oxidation of linoleic and linolenic acids in a pork fat but LAB such as *L. sakei, L. curvatus,* and *P. pentosaceus* did not present an antioxidant effect on linoleic acid *in vitro*. Zanardi et al. [24] did not find significant differences on

TABLE 1: Changes in total fatty acids (TFAs) during salami processing expressed in mg/g of fat. Treatments with (Ls-Sc) and without (C) starter addition; the superscript + indicates the addition of 50 ppm NaNO₂.

TFA	Treatments	Processing time (Days)					
		0	2	5	7	10	14
Myristic acid (C14:0)	Ls-Sc	$4.18 \pm 0.30^{a\,A}$	$1.62 \pm 0.19^{c\,D}$	$3.11 \pm 0.21^{a\,C}$	$3.71 \pm 0.30^{a\,B}$		$3.45 \pm 0.31^{a\,B,C}$
	Ls-Sc⁺	$3.68 \pm 0.14^{a\,A,B}$	$3.68 \pm 0.34^{b\,A,B}$	$3.20 \pm 0.00^{a\,B}$	$3.36 \pm 0.16^{a,b\,B}$		$4.24 \pm 0.81^{a\,A}$
	C	$3.63 \pm 0.14^{a\,A,B}$	$3.23 \pm 0.48^{b\,B,C}$		$2.41 \pm 0.72^{b,c\,C}$	$0.00 \pm 0.00^{b\,D}$	$4.09 \pm 0.37^{a\,A}$
	C⁺	$3.60 \pm 0.86^{a\,C}$	$5.54 \pm 0.00^{a\,A}$		$2.36 \pm 0.59^{c\,D}$	$3.38 \pm 0.69^{a\,C}$	$4.46 \pm 0.19^{a\,B}$
Palmitoleic acid (C16:1)	Ls-Sc	$3.72 \pm 0.24^{b,c\,A}$	$1.27 \pm 0.06^{c\,D}$	$3.09 \pm 0.06^{a\,B}$	$2.56 \pm 0.34^{a\,C}$		$2.75 \pm 0.24^{b\,B}$
	Ls-Sc⁺	$3.57 \pm 0.11^{c\,A}$	$2.73 \pm 0.60^{b\,B,C}$	$2.03 \pm 0.71^{b\,D}$	$3.31 \pm 0.11^{a\,A,B}$		$2.48 \pm 0.49^{b\,C,D}$
	C	$4.82 \pm 0.48^{a\,A}$	$2.31 \pm 0.39^{b\,B}$		$0.00 \pm 0.00^{d\,C}$	$0.00 \pm 0.00^{b\,C}$	$2.71 \pm 0.65^{b\,B}$
	C⁺	$4.35 \pm 0.40^{a,b\,B,C}$	$4.59 \pm 0.00^{a\,A,B}$		$1.26 \pm 0.00^{c\,D}$	$4.99 \pm 0.96^{a\,A}$	$3.94 \pm 0.35^{a\,C}$
Palmitic acid (C16:0)	Ls-Sc	$56.15 \pm 1.10^{b\,A}$	$56.33 \pm 2.85^{c\,A}$	$54.56 \pm 3.36^{b\,A}$	$57.25 \pm 4.95^{b\,A}$		$53.48 \pm 3.60^{b\,A}$
	Ls-Sc⁺	$58.60 \pm 1.26^{a,b\,B}$	$57.65 \pm 4.25^{c\,B}$	$63.77 \pm 4.37^{a\,A}$	$50.17 \pm 1.49^{c\,C}$		$56.05 \pm 2.54^{b\,B}$
	C	$57.19 \pm 2.94^{a,b\,C}$	$72.76 \pm 6.74^{b\,A,B}$		$68.94 \pm 3.97^{a\,B}$	$80.59 \pm 2.68^{a\,A}$	$71.60 \pm 5.82^{a\,B}$
	C⁺	$59.81 \pm 2.09^{a\,A}$	$87.72 \pm 1.48^{a\,B}$		$61.20 \pm 0.10^{d\,C}$	$64.54 \pm 3.00^{b\,D}$	$72.56 \pm 2.81^{a\,E}$
Oleic acid (C18:1)	Ls-Sc	$104.68 \pm 0.92^{a\,A}$	$91.62 \pm 8.03^{c\,C}$	$96.79 \pm 1.46^{a\,B}$	$92.63 \pm 6.54^{a\,B,C}$		$91.10 \pm 6.00^{b\,C}$
	Ls-Sc⁺	$107.12 \pm 3.28^{a\,A}$	$85.17 \pm 8.31^{c\,C}$	$86.94 \pm 8.17^{b\,B}$	$91.17 \pm 4.75^{a\,B}$		$90.24 \pm 3.66^{b\,B}$
	C	$104.49 \pm 4.21^{a\,C}$	$120.2 \pm 7.87^{b\,A}$		$123.25 \pm 3.70^{b\,D}$	$121.7 \pm 774^{a\,A}$	$110.28 \pm 9.05^{a\,B,C}$
	C⁺	$105.28 \pm 4.33^{a\,C}$	$125.8 \pm 1.40^{a\,A}$		$116.88 \pm 0.63^{b\,D}$	$114.9 \pm 3.37^{a\,B}$	$108.34 \pm 4.99^{a\,C}$
Vaccenic acid (C18:1 trans-11)	Ls-Sc	$4.41 \pm 0.68^{b\,A}$	$1.68 \pm 0.18^{b\,C}$	$1.42 \pm 0.00^{b\,C}$	$3.09 \pm 0.45^{b\,B}$		$3.14 \pm 0.30^{a\,B}$
	Ls-Sc⁺	$4.38 \pm 0.23^{b\,A}$	$2.76 \pm 0.00^{a\,A}$	$3.48 \pm 1.25^{a\,A,B}$	$3.50 \pm 0.03^{a\,A,B}$		$2.43 \pm 0.71^{a\,C}$
	C	$5.68 \pm 0.88^{a\,A}$	$2.71 \pm 0.13^{a\,B}$		$1.16 \pm 0.61^{a\,A}$	$4.45 \pm 0.10^{b\,B,C}$	$1.40 \pm 0.02^{b\,B,C}$
	C⁺	$4.42 \pm 0.08^{b\,C}$	$0.00 \pm 0.00^{c\,C}$		$1.55 \pm 0.83^{a\,A}$	$2.51 \pm 0.41^{a\,B}$	$1.37 \pm 0.00^{b\,C}$
Stearic acid (C18:0)	Ls-Sc	$22.40 \pm 1.38^{a\,A}$	$12.60 \pm 2.18^{a\,C}$	$14.59 \pm 1.41^{a\,C}$	$14.83 \pm 0.26^{a\,B}$		$16.18 \pm 0.33^{a\,B}$
	Ls-Sc⁺	$18.62 \pm 5.60^{a,b\,A}$	$11.12 \pm 1.80^{a\,C}$	$9.39 \pm 0.09^{b\,C}$	$13.79 \pm 0.81^{a\,B}$		$14.16 \pm 0.05^{c\,B}$
	C	$19.63 \pm 0.72^{b\,A}$	$12.48 \pm 0.55^{a\,B}$		$10.04 \pm 0.62^{b\,C}$	$7.21 \pm 1.87^{b\,D}$	$15.07 \pm 0.16^{b\,E}$
	C⁺	$19.74 \pm 0.28^{b\,A}$	$12.62 \pm 1.10^{a\,C}$		$11.59 \pm 0.51^{b\,D}$	$10.02 \pm 0.81^{a\,A}$	$15.75 \pm 0.43^{a\,B}$
Linoleic acid (C18:2)	Ls-Sc	0.00 ± 0.00	$2.07 \pm 0.32^{b\,B}$	$4.19 \pm 0.65^{b\,A}$	$1.05 \pm 0.06^{c\,C}$		$1.62 \pm 0.22^{c\,B}$
	Ls-Sc⁺	0.00 ± 0.00	$2.03 \pm 0.85^{b\,C}$	$7.75 \pm 2.94^{a\,A,B}$	$9.88 \pm 2.47^{a\,A}$		$6.65 \pm 0.17^{a\,B}$
	C	0.00 ± 0.00	$4.04 \pm 0.24^{a\,A}$		$1.74 \pm 0.60^{a\,A}$	0.00 ± 0.00	$0.00 \pm 0.00^{d\,C}$
	C⁺	0.00 ± 0.00	$4.25 \pm 0.98^{a\,A}$		$2.22 \pm 0.79^{a\,B}$	0.00 ± 0.00	$2.58 \pm 0.00^{b\,B}$
γ-linolenic acid (C18:3)	Ls-Sc	0.00 ± 0.00	0.00 ± 0.00	$1.49 \pm 0.30^{b\,C}$	$3.17 \pm 0.02^{b\,B}$		$26.62 \pm 1.29^{a\,A}$
	Ls-Sc⁺	0.00 ± 0.00	0.00 ± 0.00	$14.16 \pm 0.50^{a\,B}$	$14.30 \pm 0.37^{a\,C}$		$30.32 \pm 6.66^{a\,A}$
	C	0.00 ± 0.00	0.00 ± 0.00		$1.62 \pm 0.31^{c\,B}$	$3.91 \pm 0.21^{a\,A}$	$1.37 \pm 0.18^{b\,C}$
	C⁺	0.00 ± 0.00	0.00 ± 0.00		$0.00 \pm 0.00^{d\,B}$	$0.00 \pm 0.00^{b\,B}$	$4.94 \pm 0.00^{a\,A}$
Eicosadienoic acid (C20:2)	Ls-Sc	$1.20 \pm 0.00^{c\,C}$	$48.72 \pm 2.74^{a\,A}$	$20.91 \pm 4.18^{b\,B}$	$21.48 \pm 3.81^{b\,B}$		$0.00 \pm 0.00^{b\,C}$
	Ls-Sc⁺	$1.42 \pm 0.00^{b\,C}$	$39.21 \pm 1.67^{b\,C}$	$40.96 \pm 4.85^{a\,A}$	$27.80 \pm 3.69^{a\,B}$		$0.00 \pm 0.00^{b\,C}$
	C	$0.00 \pm 0.00^{d\,C}$	$0.00 \pm 0.00^{c\,C}$		$3.16 \pm 0.56^{c\,A}$	$0.00 \pm 0.00^{c\,C}$	$1.57 \pm 0.44^{a\,B}$
	C⁺	$2.56 \pm 0.00^{a\,B}$	$0.00 \pm 0.00^{c\,D}$		$3.78 \pm 0.19^{c\,A}$	$1.25 \pm 0.00^{a\,C}$	$0.00 \pm 0.00^{b\,D}$

Values are means ± SEM of three replicates.

a-d Different letters between rows (treatments) for each TFA show significant differences P<0.05 by Duncan.

A-D Different letters between columns (processing time) for each TFA show significant differences P<0.05 by Duncan.

TABLE 2: Onset temperature of the endothermic peak (T_o), maximum temperature (T_{max}), and enthalpy (ΔH) of the lipid fractions of salami treatment Ls-Sc^+ in the pre-fermentation and postripening stages.

Stage	Non-reversible heat flow			Total heat flow		
	T_o (°C)	T_{max} (°C)	ΔH (J/g)	T_o (°C)	T_{max} (°C)	ΔH (J/g)
Pre-fermentation	15.750	22.380	1.275	16.750	24.820	4.884
	32.700	35.780	0.371	37.980	43.290	0.881
	80.140	84.280	0.031			
Post - ripening	16.187	25.290	5.495	19.195	27.650	6.610
	40.105	45.080	1.230	39.805	45.330	2.211

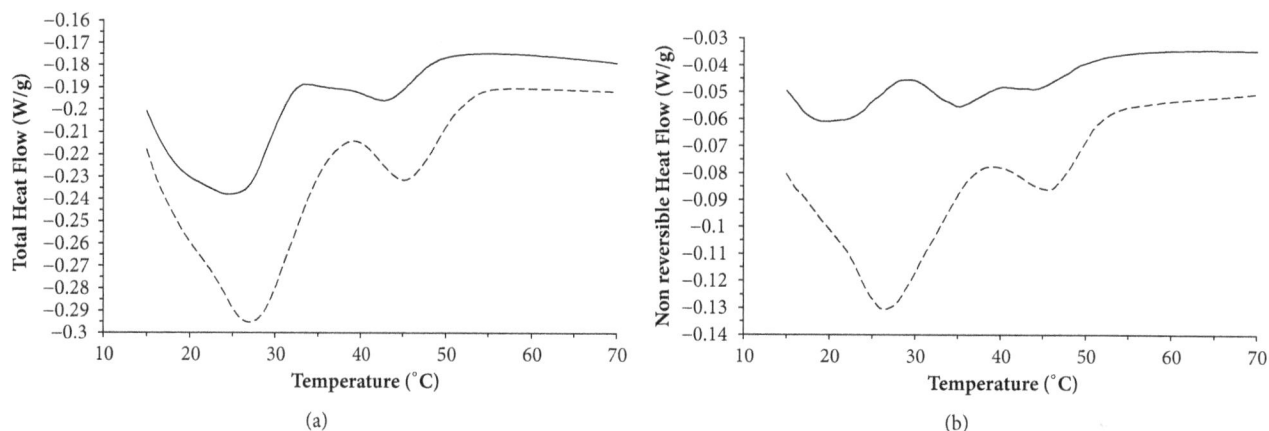

FIGURE 1: Modulated Differential Scanning Calorimetry (MDSC) curves indicating exothermic/endothermic total heat flow (a) and nonreversible heat flow (b) for salami lipid fraction in prefermentation stage (–) and postripening stage (- -).

total fatty acids profile for different fermented sausages with different concentrations for $NaNO_2$. However, these authors reported differences on polyunsaturated fatty acids (PUFAs) on the meat product associated with starter culture strain composition, where a starter culture composed by *Lactobacillus* spp. and *S. carnosus* presents lower PUFAs concentration than meat products with starter cultures composed by *L. curvatus* and *Micrococcus varians*; *L. sakei, P. pentosaceus, S. xylosus*, and *M. varians*; and *L. sakei, P. pentosaceus, S. xylosus, S. carnosus*, and *M. varians*.

In contrast, the PUFA eicosadienoic acid was not found during the postripening stage; this result is concordant with the results reported by Qiu et al. [25] where decrease the proportion of the long chain PUFAs such as C20:4, C20:2, and C20:1 during cantonese sausage processing. This behavior is due to the oxidation of polyunsaturated fatty acids releasing aldehydes and ketones that affect the flavor of the final product during ripening [26].

During the post-fermentation stage, treatments with and without starter culture (addition of 50 ppm $NaNO_2$) generated the highest concentration of SFA myristic acid and MUFA palmitoleic acid. These results could be related to the high μ_{max} values of CNS probably dominated by *S. carnosus* in treatments with the addition of $NaNO_2$. The SFA/ (MUFA + PUFA) ratio was higher during the entire process for treatments with the starter culture compared to the controls; however, this behavior was not associated with the addition of $NaNO_2$. The TFA ratio increased during the post-fermentation stage and decreased during ripening

for the Ls-Sc and Ls-Sc^+. The results previously reported by Marco et al. [27] are indicative that the total polyunsaturated fatty acid concentration change as a function of processing time on different lipid fraction (free fatty acids, phospholipids and triglycerides), where the free fatty acids within the total fatty acids present high increase on the concentration and proportion of PUFAs. Martín-Sánchez et al. [28] report PUFAs increase in dry-cured fermented Spanish sausage inoculated with *L. sakei* and *S. carnosus* as the starter culture during ripening, but during fermentation, there is a high oxidation rate of unsaturated triglycerides.

3.2. Modulated Differential Scanning Calorimetry (MDSC) of Salami Lipids between the Pre-Fermentation and Post-Ripening Stages for the Ls-Sc^+. The lipid thermograms of salami for the Ls-$Sc^+$are shown in Figure 1. Transitions in the total heat flow (Figure 1(a)) show combination on reversing and nonreversing curves (Figure 1(b)), which indicate that the transitions show a combination of melting and crystallization of fatty acids between the prefermentation and postripening stages; and their contribution can be determined by enthalpy [29]. For this reason, a higher enthalpy for transitions in the nonreversible heat flow in the prefermentation stage could be related to crystallization and fatty acids fusion events as the main energetic contribution in this processing stage.

For the nonreversible and total heat flow, an increase was observed ($P<0.05$) on T_o and ΔH in the post-ripening stage compared with prefermentation stage (Table 2). The

TABLE 3: Transitions of nonreversible and total heat flow, onset temperature of the endothermic peak (T_o), midpoint temperature (T_m), and specific heat (ΔCp) of lipid fractions of salami of treatment Ls-Sc$^+$ in the prefermentation and postripening stages.

Stage	Non-reversible heat flow			Total heat flow		
	T_o (°C)	T_m (°C)	ΔCp (J/g °C)	T_o (°C)	T_m (°C)	ΔCp (J/g °C)
Pre-fermentation	13.140	13.260	0.076	16.220	16.740	0.443
	16.430	16.860	0.098	27.510	28.670	0.538
	23.470	25.120	0.074	39.080	40.700	0.054
	28.637	29.033	0.020	44.110	46.450	0.227
	31.263	31.685	0.020	64.070	75.930	0.054
	43.695	47.340	0.166	81.410	93.180	0.096
	79.490	82.610	0.011			
	93.840	96.820	0.024			
Post-ripening	13.270	13.400	0.269	16.105	16.775	0.642
	17.380	16.650	0.134	29.230	31.150	0.759
	23.780	25.383	0.134	40.160	42.385	0.208
	27.445	29.130	0.305	47.105	48.905	0.360
	40.670	42.480	0.102	62.580	64.930	0.020
	47.075	49.535	0.262	82.845	92.950	0.046
	61.138	63.455	0.031			
	94.485	97.095	0.023			

increase on enthalpy (ΔH) can be related to the increase in the concentration of primary and secondary products of the lipid oxidation and free fatty acids [14] during the salami processing. Furthermore, in the nonreversible period three transitions were observed for prefermentation stage and two transitions for the postripening stage (Table 2). This behavior indicates a kinetic process on salami lipids, and it can be associated with changes on the enthalpy relaxation reflecting structural reorganization and modifications [29].

Transitions with similar T_o had an increment on Cp in the postripening stage respect to the pre-fermentation stage; this was more evident for nonreversible heat flow (Table 3). The phenomena observed is comparable to the results reported by Samyn et al. [29]; Hidalgo and Zamora [30]; Coupland and McClements [31], where the increment on Cp is in function of the increase of the degree of unsaturation and the lipid autoxidation reactions developed during salami ripening. This is also seen in the results for the treatment Ls-Sc$^+$ which are in concordance with the decreasing levels of SFA stearic acid, with significant differences (P<0.05) observed between the prefermentation and post-ripening periods. The concentration of PUFAs (linoleic and γ-linolenic acid) increased significantly (P<0.05) during the postripening period and the increment on lipid oxidation related to the relative abundance of MMb.

3.3. Changes in the Concentrations of Myoglobin Forms. The concentrations of DMB, OMB and MMB are shown in Table 4. According to Pérez-Alvarez et al. [32], the decrease in the concentration of OMB in the treatments with the addition of NaNO$_2$ was inverse to the formation of nitrosomyoglobin complexes. For treatments C$^+$, Ls-Sc, and C the decrease was more evident in the OMB related with the processing time. However, in the treatment with NaNO$_2$ the OMB decrease was perceived from day 7 of processing (ripening),

whereas for Ls-Sc and C this decrease was gradual from first day (fermentation). The oxidation of OMB increases proportionally at a lower pH, lower partial oxygen pressure, and lipid oxidation [8].

For all treatments, the concentration of MMB increased as a function of the processing time (Table 4). The highest concentration of MMB with significant differences (p<0.05) was for C$^+$ and C. Consistent with Tang et al. [33], these changes in myoglobin oxidation were directly related to lipid oxidation and its products (i.e., aldehydes) that initiate conformational changes in myoglobin and cause an increase in the oxidation of the haem group and, therefore, it is darkening [18]. Moreover, the lipid oxidation during salami processing could be related with an increase on NaCl concentration; according to Ying et al. [34], who reported in dry-cured goose sausages that this reaction during the ripening stage could be due to the pro-oxidative effect of salt. On the other hand, the increase in NaCl concentration increases the activity of endogenous enzyme lipoxygenase (LOX) involved in lipid oxidation.

The OMB oxidation to MMB generates reactive intermediates capable of enhancing further oxidation of unsaturated fatty acids; in this sense, a superoxide anion is achieved which one dismutase to hydrogen peroxide. The latter can react with MMB concurrently generated in this oxidation sequence to form an activated MMB complex capable of enhancing lipid oxidation that is attributed to ferryl myoglobin [8]. This phenomenon indicates that the starter culture addition could generate an antioxidant effect, associated with the S. carnosus capacity to consume the free oxygen and nitrate reductase, catalase, and SOD activities and for L. sakei the nitrite reductase and catalase activities. However, the 50 ppm of NaNO$_2$ addition did not accomplish a clear effect on the decrease of MMB formation. Furthermore, during

TABLE 4: Myoglobin species (%), deoxymyoglobin (DMB), oxymyoglobin, (OMB) and metmyoglobin (MMB). Treatments with (Ls- Sc) and without starter culture addition (C); the superscript $^+$ indicates the addition of 50 ppm $NaNO_2$.

Myoglobin Form (%)	Processing Time (Days)	Treatment			
		Ls-Sc$^+$	Ls-Sc	C$^+$	C
DMB	0	32.06 ± 0.42 a A	28.86 ± 0.84 c B	30.18 ± 0.65 a C	27.45 ± 0.65 c D
	1	29.18 ± 0.26 d A	29.04 ± 0.15 c A	26.04 ± 0.10 b C	27.54 ± 0.05 c B
	2	29.20 ± 0.52 d A	28.42 ± 0.20 c B	26.95 ± 0.99 c C	32.26 ± 0.50 a D
	3	29.84 ± 0.24 c A	30.80 ± 0.42 b B	24.14 ± 0.42 d C	27.89 ± 0.62 c D
	7	31.55 ± 0.12 b A	30.66 ± 0.75 b B	31.23 ± 0.70 e A,B	28.67 ± 0.68 b C
	14	28.75 ± 0.22 e A	33.20 ± 0.93 a B	28.61 ± 0.37 a A	27.09 ± 0.15 c C
OMB	0	18.56 ± 0.56 d A	25.43 ± 0.93 a B	20.58 ± 0.09 a C	22.47 ± 0.84 a D
	1	20.80 ± 0.87 b A	25.75 ± 0.43 a B	24.21 ± 0.40 b C	21.97 ± 0.48 a D
	2	21.75 ± 0.69 a A	22.60 ± 0.75 b B	25.78 ± 0.86 c C	17.80 ± 0.08 c D
	3	21.76 ± 0.55 a A	18.92 ± 0.24 d B	27.72 ± 0.61 d C	22.39 ± 0.67 a D
	7	17.94 ± 0.46 e B	19.94 ± 0.94 c A	16.60 ± 0.75 e C	19.84 ± 0.19 b A
	14	20.13 ± 0.51 c A	17.10 ± 0.79 e B	16.01 ± 0.92 e C	17.05 ± 0.67 d C
MMB	0	49.98 ± 0.90 b A	45.37 ± 0.59 e B	49.66 ± 0.19 b A	50.01 ± 0.68 c A
	1	50.34 ± 0.69 b A	45.17 ± 0.08 e C	49.57 ± 0.43 b B	50.70 ± 0.48 b A
	2	49.17 ± 0.21 c B	48.99 ± 0.23 d B	46.98 ± 0.50 d C	49.93 ± 0.72 c A
	3	48.60 ± 0.71 d A	50.67 ± 0.73 a B	47.65 ± 0.01 c C	49.58 ± 0.64 c D
	7	51.00 ± 0.86 a A	49.52 ± 0.55 c B	52.69 ± 0.39 a C	51.76 ± 0.69 a D
	14	51.09 ± 0.40 a A	50.14 ± 0.05 b B	52.32 ± 0.65 a C	51.70 ± 0.08 a D

Values are means ± SEM of nine samples with five replicates.

[a-e] Different letters between rows (processing time) for each Myoglobin species show significant differences P<0.05 by Duncan.

[A-D] Different letters between columns (treatment) for each processing time and Myoglobin species show significant differences P<0.05 by Duncan.

the salami processing the treatments with the addition of a starter culture acquired higher concentration of PUFAs than the control treatments; this behavior is also related to an antioxidant capacity from the starter culture, because primary and secondary products derived from unsaturated fatty acids oxidation can enhance myoglobin oxidation [8].

3.4. *Microbial Dynamics during Salami Processing.* The dynamics of microbial growth during the salami processes are shown in Figure 2. The population density of CNS was maintained for the treatment with the starter culture (Figures 2(a) and 2(b)), whereas for the control treatment a decrease in the microorganism population was observed during the first 6 hours of fermentation of the salami (Figures 2(c) and 2(d)). In accordance with previous reports by Hospital et al. [35], the addition of low levels of nitrite (<60ppm) can promote microbial growth under low levels of oxygen pressure (inner part of the meat product) also acting as a terminal acceptor of electrons, stimulating reductase activity. Microorganisms from the Staphylococcal genus can rapidly grow under aerobic conditions (surface), and their reductase activity decreases with high oxygen pressure; in this way, excess nitrite may cause the inhibition of growth associated with the interference of the electron transport chain. Furthermore, in the treatments Ls-Sc and Ls-Sc$^+$, oxygen consumption (reduction of redox potential) and nitrate reductase activity is expected, according to Tjener et al. [36]. This phenomenon generates a positive influence on color formation and a negative influence on lipid oxidation.

In all treatments of this study, there were no differences in the bacterial concentrations of BAL and CNS with the addition of 50 ppm $NaNO_2$ during the process of salami. The influence on changes in lipid fraction could be related to the fatty acid biosynthesis pathway by S. carnosus [37], where branched-chain fatty acids represent the majority of fatty acids produced by S. carnosus. However, for L. sakei were not found enzymes implicated with fatty acid biosynthesis or lipid transformation such as elongases or desaturases.

For the total LAB, there were no differences (P>0.05) in the μ_{max} between the Ls-Sc, Ls-Sc$^+$, and C$^+$ treatments; however, μ_{max} decreased to 0.003 h^{-1} following treatment C. In Ls-Sc$^+$, λ increased to 14.7 h (Table 5). The population density of LAB in all treatments increased until 72 h of salami processing to approximately of 7 log CFU.g^{-1} and the control treatments was maintained until the post-ripening stage, while for Ls-Sc and Ls-Sc$^+$ from 7th day (ripening) began to slight decrease, this behavior is concordant with results reported by Lorenzo et al. [38] where LAB population declines during ripening associated to the reduction of fermentable carbohydrates and a_w. The growth rate (μ_{max}) of aerobic mesophiles was determined as the total bacterial count. The C and C$^+$ treatments showed a higher μ_{max}; for these treatments, the dominant microbiota was native to the meat and adapted to the environment and therefore could present a higher growth rate. In contrast, no differences were associated with the addition of 50 ppm $NaNO_2$ for the evaluated treatments or the cell concentrations in the postripening stage (P>0.05).

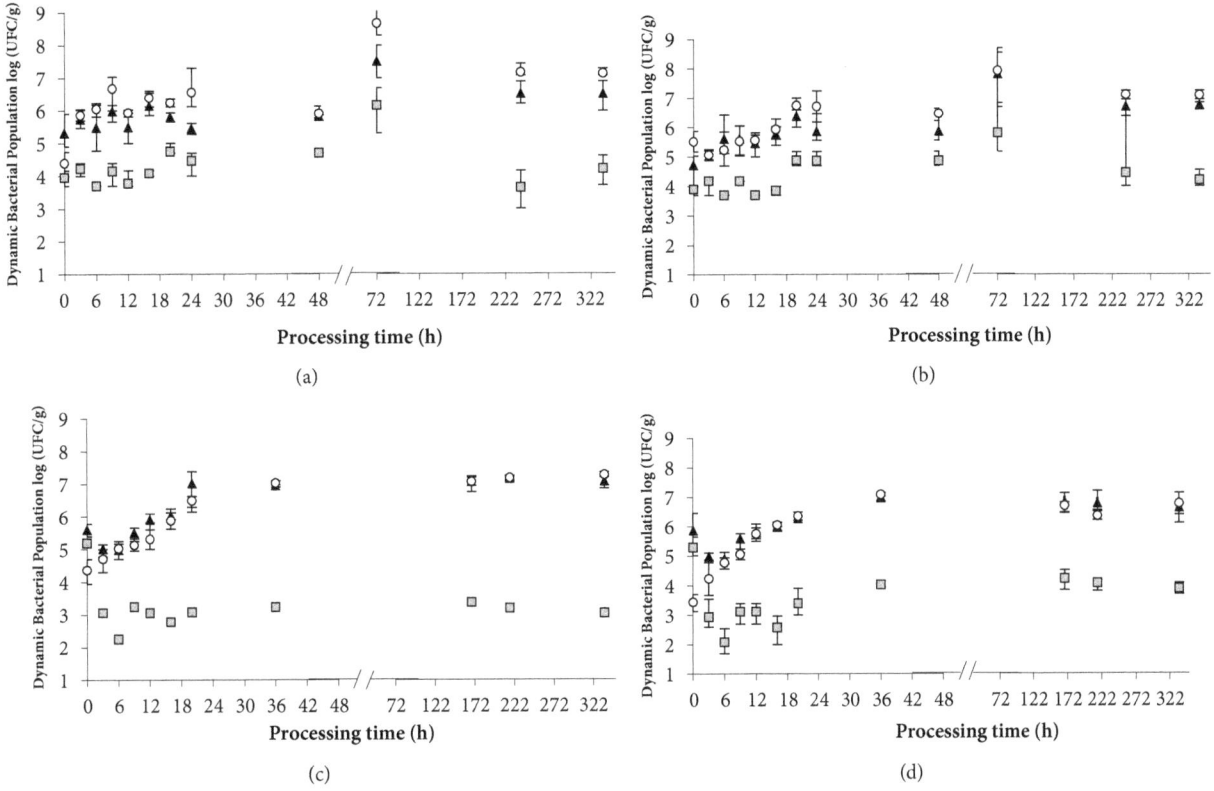

Figure 2: Dynamics of the microbial population during the processing of salami: fermentation at 21°C from 0 to 48 h and ripening at 17°C to 336 h (14 days). CNS (grey square), total bacterial count (black triangle), and LAB (white circle). Treatments *Ls-Sc* (a), *Ls-Sc⁺* (b), C (c), and C⁺ (d). The bars indicate the confidence intervals.

Table 5: Biokinetic parameter estimation using Baranyi and Roberts's model.

Microorganism	Treatment	μ_{max} (h⁻¹)		λ(h)
CNS	*Ls-Sc*	-0.0006 ± 0.0000	g	- *
	Ls-Sc⁺	0.037 ± 0.002	e,f,g	- *
	C	0.0007 ± 0.000	g	- *
	C⁺	0.164 ± 0.035	a	- *
LAB	*Ls-Sc*	0.029 ± 0.001	e,f,g	- *
	Ls-Sc⁺	0.067 ± 0.001	c,d,e	14.691 ± 0.200
	C	0.003 ± 0.0004	g	- *
	C⁺	0.052 ± 0.003	e,f	- *
Total count	*Ls-Sc*	0.019 ± 0.003	f,g	- *
	Ls-Sc⁺	0.034 ± 0.003	e,f,g	- *
	C	0.108 ± 0.007	b,c	- *
	C⁺	0.098 ± 0.012	c,d	- *

Values are means ± SEM with six replicates.
Different letters show significant differences P<0.05 by Tukey test.
-* Parameter not calculated (λ < 3h).

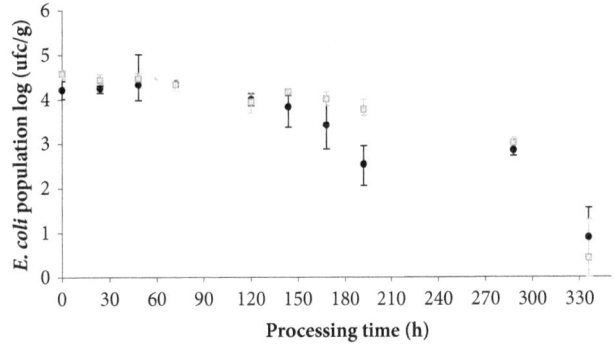

Figure 3: Dynamics of *E. coli* population during the processing of salami: fermentation at 21°C from 0 to 48 h and ripening at 17°C to 336 h (14 days). Treatments: *Ls-Sc* (grey square) and *Ls-Sc⁺* (black circle). The bars indicate the confidence intervals.

3.5. Validating the Reduction of E. coli during Salami Processing for Treatments with Starter Culture.

The *E. coli* reduction during salami processing is shown in Figure 3. *E. coli* was reduced in the treatment *Ls-Sc* in 4.15±0.75 log CFU/g (90.5%) and in *Ls- Sc⁺* in 4.22±0.81 log CFU/g (78.7%).

The inactivation parameters showed that (α) was 15.99 for *Ls-Sc* and 19.47 for *Ls- Sc⁺* implying greater resistance in the first phase of the salami processing for treatment with NaNO₂. The parameter (p) was 6 for *Ls-Sc* and 1.79 for *Ls-Sc⁺*, thus *E. coli* treatment resistance related with adaptation speed was reduced in the NaNO₂ addition. Inactivation rate (δ), for *Ls- Sc* (δ_1) was 269.55 h⁻¹ and (δ_2) was 544.31 h⁻¹, and for *Ls- Sc⁺* (δ_1) was 169.47 h⁻¹ and (δ_2) was 321.88 h⁻¹. As a consequence, the NaNO₂ addition reduces decimal inactivation time for both subpopulation of *E. coli*. Therefore, this behavior indicates the variability of the stress response of *E. coli* attributed to multiple factors combined as direct and indirect inhibition caused by the starter culture addition, $NaNO_2$, pH, and a_w. Holck et al. [39] mentioned that *E. coli* is generally inhibited in fermented sausages due to a combination of several hurdles: high salt, the presence of nitrite curing salt, low a_w, decrease of redox potential, the growth of competitive starter culture, and a decrease in pH. Wang et al. [40] found the growth inhibition of *E. coli* during fermentation and ripening of Chinese fermented sausages by *L. sakei* and 100 ppm of $NaNO_2$ addition. Results by Casey and Condon [41] reported that *E. coli* O157: H45 inhibition rate was faster with 50 ppm $NaNO_2$ addition than without $NaNO_2$ during processing of sausage fermented with *P. acidilactici*.

4. Conclusions

The addition of starter culture in salami processing increase the concentration of PUFAs (linoleic, γ- linolenic, and eicosadienoic), in contrast to the MMb concentration, which was lower. This latest concentration could be associated with an antioxidant effect produced by *S. carnosus* in the starter culture. On the other hand, the addition of 50 ppm $NaNO_2$ did not have influence in the concentrations of the TFAs and MMb formation. Moreover, the starter culture without and with 50 ppm $NaNO_2$ addition allowed the reduction of 4 log CFU/g of *E. coli*. However, the stress response of *E. coli* showed a broad variability because of bacteria inhibition (attributed to multiple factors combined). Therefore, these results allow analyzing changes during salami processing in a holistic form, where the application of starter culture with low nitrite concentrations could be a strategy for the meat industry to reduce this additive.

Conflicts of Interest

The authors declare that they have no conflicts of interest.

Acknowledgments

Authors acknowledge the financial support of Colciencias, Universidad de La Sabana and Vilaseca S.A., RC no. 221-2010, and the DGAPA-UNAM in PAPIIT Projects (key nos. IT202312 and IT203314) and the contribution in samples elaboration and comments of C. E. Juan Ramón Vilaseca and the MS- CG analysis of M.S. Luis Barbo Hernández Portilla.

References

[1] G. F. Cifuni, M. Contò, and S. Failla, "Potential use of visible reflectance spectra to predict lipid oxidation of rabbit meat," *Journal of Food Engineering*, vol. 169, pp. 85–90, 2016.

[2] J. D. Wood, R. I. Richardson, G. R. Nute et al., "Effects of fatty acids on meat quality: a review," *Meat Science*, vol. 66, no. 1, pp. 21–32, 2004.

[3] J. A. Ordóñez, E. M. Hierro, J. M. Bruna, and L. de La Hoz, "Changes in the components of dry-fermented sausages during ripening," *Critical Reviews in Food Science and Nutrition*, vol. 39, no. 4, pp. 329–367, 1999.

[4] A. Casaburi, R. Di Monaco, S. Cavella, F. Toldrá, D. Ercolini, and F. Villani, "Proteolytic and lipolytic starter cultures and their effect on traditional fermented sausages ripening and sensory traits," *Food Microbiology*, vol. 25, no. 2, pp. 335–347, 2008.

[5] V. Fuentes, M. Estévez, J. Ventanas, and S. Ventanas, "Impact of lipid content and composition on lipid oxidation and protein carbonylation in experimental fermented sausages," *Food Chemistry*, vol. 147, pp. 70–77, 2014.

[6] G. Jin, J. Zhang, X. Yu, Y. Zhang, Y. Lei, and J. Wang, "Lipolysis and lipid oxidation in bacon during curing and drying-ripening," *Food Chemistry*, vol. 123, no. 2, pp. 465–471, 2010.

[7] A. L. Alderton, C. Faustman, D. C. Liebler, and D. W. Hill, "Induction of redox instability of bovine myoglobin by adduction with 4-hydroxy-2-nonenal," *Biochemistry*, vol. 42, no. 15, pp. 4398–4405, 2003.

[8] C. Faustman, Q. Sun, R. Mancini, and S. P. Suman, "Myoglobin and lipid oxidation interactions: Mechanistic bases and control," *Meat Science*, vol. 86, no. 1, pp. 86–94, 2010.

[9] A. Roccato, M. Uyttendaele, F. Barrucci et al., "Artisanal Italian salami and soppresse: Identification of control strategies to manage microbiological hazards," *Food Microbiology*, vol. 61, pp. 5–13, 2017.

[10] E. Dalzini, E. Cosciani-Cunico, P. Monastero et al., "Reduction of Escherichia coli O157:H7 during manufacture and ripening of Italian semi-dry salami," *Italian Journal of Food Safety*, vol. 3, no. 2, pp. 137–139, 2014.

[11] M. Sánchez Mainar and F. Leroy, "Process-driven bacterial community dynamics are key to cured meat colour formation by coagulase-negative staphylococci via nitrate reductase or nitric oxide synthase activities," *International Journal of Food Microbiology*, vol. 212, pp. 60–66, 2015.

[12] S. Chaillou, M.-C. Champomier-Vergès, M. Cornet et al., "The complete genome sequence of the meat-borne lactic acid bacterium *Lactobacillus sakei* 23K," *Nature Biotechnology*, vol. 23, no. 12, pp. 1527–1533, 2005.

[13] W. P. Hammes, "Metabolism of nitrate in fermented meats: the characteristic feature of a specific group of fermented foods," *Food Microbiology*, vol. 29, no. 2, pp. 151–156, 2012.

[14] E. Chiavaro, M. T. Rodriguez-Estrada, C. Barnaba, E. Vittadini, L. Cerretani, and A. Bendini, "Differential scanning calorimetry: A potential tool for discrimination of olive oil commercial categories," *Analytica Chimica Acta*, vol. 625, no. 2, pp. 215–226, 2008.

[15] C. M. Blanco-Lizarazo, J. Arjona-Roman, A. Llorente-Bousquets, and I. Sotelo-Díaz, "Changes in myofibrillar and sarcoplasmic proteins in salami processing added with *Lactobacillus Sakei/Staphylococcus Carnosus* according to modulated differential scanning calorimetry and the color profile," *Journal of Food Process Engineering*, vol. 40, no. 1, Article ID e12330, 2017.

[16] P. W. Park and R. E. Goins, "In situ preparation of fatty acid methyl esters for analysis of fatty acid composition in foods," *Journal of Food Science*, vol. 59, no. 6, pp. 1262–1266, 1994.

[17] K. Sasaki, M. Mitsumoto, T. Nishioka, and M. Irie, "Differential scanning calorimetry of porcine adipose tissues," *Meat Science*, vol. 72, no. 4, pp. 789–792, 2006.

[18] American Meat Science Association, *Meat Color Measurement Guidelines*, merican Meat Science Association, Illinois, Ill, USA, 2012.

[19] J. Tang, C. Faustman, and T. A. Hoagland, "Krzywicki revisited: Equations for spectrophotometric determination of myoglobin redox forms in aqueous meat extracts," *Journal of Food Science*, vol. 69, no. 9, pp. C717–C720, 2004.

[20] J. Baranyi and T. A. Roberts, "A dynamic approach to predicting bacterial growth in food," *International Journal of Food Microbiology*, vol. 23, no. 3-4, pp. 277–294, 1994.

[21] L. Coroller, I. Leguerinel, E. Mettler, N. Savy, and P. Mafart, "General model, based on two mixed weibull distributions of bacterial resistance, for describing various shapes of inactivation curves," *Applied and Environmental Microbiology*, vol. 72, no. 10, pp. 6493–6502, 2006.

[22] M. I. Aksu and M. Kaya, "Effect of commercial starter cultures on the fatty acid composition of pastirma (Turkish dry meat product)," *Journal of Food Science*, vol. 67, no. 6, pp. 2342–2345, 2002.

[23] R. Talon, D. Walter, and M. C. Montel, "Growth and effect of staphylococci and lactic acid bacteria on unsaturated free fatty acids," *Meat Science*, vol. 54, no. 1, pp. 41–47, 2000.

[24] E. Zanardi, S. Ghidini, A. Battaglia, and R. Chizzolini, "Lipolysis and lipid oxidation in fermented sausages depending on different processing conditions and different antioxidants," *Meat Science*, vol. 66, no. 2, pp. 415–423, 2004.

[25] C. Qiu, M. Zhao, W. Sun, F. Zhou, and C. Cui, "Changes in lipid composition, fatty acid profile and lipid oxidative stability during Cantonese sausage processing," *Meat Science*, vol. 93, no. 3, pp. 525–532, 2013.

[26] A. Casaburi, M.-C. Aristoy, S. Cavella et al., "Biochemical and sensory characteristics of traditional fermented sausages of Vallo di Diano (Southern Italy) as affected by the use of starter cultures," *Meat Science*, vol. 76, no. 2, pp. 295–307, 2007.

[27] A. Marco, J. L. Navarro, and M. Flores, "The influence of nitrite and nitrate on microbial, chemical and sensory parameters of slow dry fermented sausage," *Meat Science*, vol. 73, no. 4, pp. 660–673, 2006.

[28] A. M. Martín-Sánchez, C. Chaves-López, E. Sendra, E. Sayas, J. Fenández-López, and J. Á. Pérez-Álvarez, "Lipolysis, proteolysis and sensory characteristics of a Spanish fermented dry-cured meat product (salchichón) with oregano essential oil used as surface mold inhibitor," *Meat Science*, vol. 89, no. 1, pp. 35–44, 2011.

[29] P. Samyn, G. Schoukens, L. Vonck, D. Stanssens, and H. Van Den Abbeele, "Quality of Brazilian vegetable oils evaluated by (modulated) differential scanning calorimetry," *Journal of Thermal Analysis and Calorimetry*, vol. 110, no. 3, pp. 1353–1365, 2012.

[30] F. Hidalgo and R. Zamora, "Fats: physical properties," in *Handbook of Food Science, Technology, and Engineering*, Y. H. Hui, Ed., CRC Press, Taylor and Francis, Boca Ratón, Fla, USA, 2005.

[31] J. N. Coupland and D. J. McClements, "Physical properties of liquid edible oils," *Journal of the American Oil Chemists' Society*, vol. 74, no. 12, pp. 1559–1564, 1997.

[32] J. A. Pérez-Alvarez, M. E. Sayas-Barberá, J. Fernández-López, and V. Aranda-Catalá, "Physicochemical characteristics of Spanish-type dry-cured sausage," *Food Research International*, vol. 32, no. 9, pp. 599–607, 1999.

[33] J. Tang, C. Faustman, T. A. Hoagland, R. A. Mancini, M. Seyfert, and M. C. Hunt, "Interactions between mitochondrial lipid oxidation and oxymyoglobin oxidation and the effects of vitamin E," *Journal of Agricultural and Food Chemistry*, vol. 53, no. 15, pp. 6073–6079, 2005.

[34] W. Ying, J. Ya-Ting, C. Jin-Xuan et al., "Study on lipolysis-oxidation and volatile flavour compounds of dry-cured goose with different curing salt content during production," *Food Chemistry*, vol. 190, pp. 33–40, 2016.

[35] X. F. Hospital, J. Carballo, M. Fernández, J. Arnau, M. Gratacós, and E. Hierro, "Technological implications of reducing nitrate and nitrite levels in dry-fermented sausages: Typical microbiota, residual nitrate and nitrite and volatile profile," *Food Control*, vol. 57, pp. 275–281, 2015.

[36] K. Tjener, L. H. Stahnke, L. Andersen, and J. Martinussen, "Growth and production of volatiles by Staphylococcus carnosus in dry sausages: Influence of inoculation level and ripening time," *Meat Science*, vol. 67, no. 3, pp. 447–452, 2004.

[37] R. Rosenstein, C. Nerz, L. Biswas et al., "Genome analysis of the meat starter culture bacterium *Staphylococcus carnosus* TM300," *Applied and Environmental Microbiology*, vol. 75, no. 3, pp. 811–822, 2009.

[38] J. M. Lorenzo, M. Gómez, and S. Fonseca, "Effect of commercial starter cultures on physicochemical characteristics, microbial counts and free fatty acid composition of dry-cured foal sausage," *Food Control*, vol. 46, pp. 382–389, 2014.

[39] A. L. Holck, L. Axelsson, T. M. Rode et al., "Reduction of verotoxigenic *Escherichia coli* in production of fermented sausages," *Meat Science*, vol. 89, no. 3, pp. 286–295, 2011.

[40] X. H. Wang, H. Y. Ren, D. Y. Liu, W. Y. Zhu, and W. Wang, "Effects of inoculating *Lactobacillus sakei* starter cultures on the microbiological quality and nitrite depletion of Chinese fermented sausages," *Food Control*, vol. 32, no. 2, pp. 591–596, 2013.

[41] P. Casey and S. Condon, "Synergistic lethal combination of nitrite and acid pH on a verotoxin-negative strain of Escherichia coli O157," *International Journal of Food Microbiology*, vol. 55, no. 1-3, pp. 255–258, 2000.

6

Evaluation of Catalytic Effects of Chymotrypsin and Cu^{2+} for Development of UV-Spectroscopic Method for Gelatin-Source Differentiation

Anis Hamizah,[1] **Ademola Monsur Hammed,**[1,2,3]
Tawakalit Tope Asiyanbi-H,[2] **Mohamed Elwathig Saeed Mirghani,**[1,3]
Irwandi Jaswir,[1,3] **and Nurrulhidayah binti Ahamad Fadzillah**[1]

[1]*International Institute for Halal Research and Training (INHART), International Islamic University Malaysia (IIUM),*
Gombak, Kuala Lumpur, Malaysia
[2]*Plant Sciences Department, North Dakota State University, Fargo, ND, USA*
[3]*Biotechnology Engineering Department, International Islamic University Malaysia (IIUM), Gombak, Kuala Lumpur, Malaysia*

Correspondence should be addressed to Ademola Monsur Hammed; demmarss@gmail.com

Academic Editor: Salam A. Ibrahim

The consumers interest in gelatin authentication is high due to allergic reactions and adoption of Halal and Kosher eating cultures. This research investigated browning development due to enzymatic hydrolysis and presence of Cu^{2+} during Maillard reaction of fish, porcine, and bovine gelatin. The rate of browning index samples showed two phases—rapid and slow—for all the gelatin samples and changes in browning index (ΔB_{index}) were increased (>100%) in presence of Cu^{2+}. ΔB_{index} of enzymatic hydrolysates were different among the gelatin species. Fish gelatin hydrolyzate displayed > 400% increase in browning in the first six hours compared to gelatin hydrolyzates from porcine (200%) and bovine (140%). The variation in ΔB_{index} of chymotrypsin digested gelatin in presence of Cu^{2+} could be valuable for the development of an efficient UV-spectroscopic method for gelatin differentiation.

1. Introduction

Maillard reaction is a nonenzymatic browning leading to formation of numerous compounds when proteins (such as gelatin) are heated in presence of sugar. Maillard reaction products (MRP) have fluorescence and browning capability suitable to measure reaction progress [1]. The degree of browning during Millard reaction depends on the type of sugar, temperature, pH, reaction time, concentration of reactants, presence of inorganic compounds, and most especially protein amino acids profiles [1–3]. The colour intensity of MRP from basic amino acids was reportedly greater than that of acidic amino acids, while nonpolar amino acids were of intermediate colour intensity. Also, browning was accelerated by the presence of metal ions (Fe^{2+} and Cu^{2+}) but not affected by Na$^+$ [3]. Study of Maillard reaction of essential amino acids with glucose revealed that amino acids were degraded differently and exhibited a varying browning degree [2].

Previous studies have shown that the amino acid profiles and film properties of different gelatin sources varied, most especially methionine and histidine [4]. The content of imino acids proline and hydroxyproline of fish collagen is generally lower than that of mammalian collagen [5]. Another report showed that properties of gelatin film depend on the source of the gelatin such that fish gelatin exhibited lowest water vapor permeability while pork gelatin exhibited least water solubility [6]. Therefore, there is a possibility that variation in the degree of browning in Millard reaction of different gelatin sources could be observed. The degree of browning can be detected by UV-spectroscopy and the readings can then be used to differentiate gelatin of varying sources.

We presume that distribution of aromatic amino acid on gelatin is species specific. Chymotrypsin cleaves peptide bond by aromatic residues, unlike trypsin that only cleaves peptide bond at lysine and arginine. Therefore, chymotrypsin digest of gelatin will possibly produce peptides and hydrolysates unique to different gelatin species. Previous studies have reported enzyme hydrolysis of gelatin yield peptides as a biomarker for gelatin-source differentiation [7]. Most of these studies made use of expensive and difficult procedures such as GC-MS, PCR, ELISA, and HPLC. This research is based on two principles: (i) the chymotrypsin hydrolysis of gelatin from a different source could produce peptides of varying properties that will yield a different degree of browning during Maillard reaction and (ii) the introduction of Cu^{2+} will expedite the rate of reaction and reduce the time for brown colour development.

Till date, there is only one study on the use of UV-spectroscopy reading of the degree of browning of MRP of gelatin for species-specific differentiation purpose. Bovine and porcine gelatin were successfully differentiated with UV-spectroscopy reading after ribose-induced Maillard reaction [8]. A literature search revealed a lack of studies on Maillard reaction of gelatin hydrolyzate and the effect of Cu^{2+} on the degree of browning of gelatin. In a preliminary study, Cu^{2+} has shown to improve browning index during Maillard reaction. Improvement of UV-spectroscopic measurement for differentiation of gelatin power could be achieved by considering effects of chymotrypsin and Cu^{2+} on the degree of browning after Maillard reaction process.

Hence, the aim of this study is to investigate the development of brown colour during Maillard reaction of fish, porcine, and bovine gelatin as affected by chymotrypsin hydrolysis and the presence of Cu^{2+}. It is hoped that detail understanding of reaction kinetics will be valuable towards the development of more efficient and reliable UV-spectroscopy based method for species-specific gelatin-source differentiation.

2. Methodology

2.1. Production of Gelatin Hydrolyzates.
Gelatin hydrolysates were produced using chymotrypsin to digest gelatin from fish, porcine, and bovine. The digestion was carried out for 4 h at 25°C with the enzyme-gelatin ratio of 1 : 250 (w : w). The reaction was stopped by heating the mixture at 100°C for 10 min. The solutions were centrifuged at 3000 rpm for 15 min and the supernatants were discounted off and referred to as gelatin hydrolyzates solution.

2.2. Nonenzymatic Browning of Gelatin and Gelatin Hydrolysate.
An equal volume of 0.25% xylose solutions and 1.0% of gelatin/hydrolyzates were thoroughly mixed to make a final solution containing 0.125 and 0.5% of xylose and gelatin/hydrolyzates, respectively. In another set of experiments, about 2 mM of $CuCl_2$ was added to the mixture. The nonenzymatic browning was carried out by heating the mixture at 95°C for 6, 12, and 24 h. The mixture was allowed to cool to room temperature before the determination of browning index.

2.3. Determination of Browning Index.
The browning index (B_{index}) of a cooled mixture of gelatin/hydrolyzate containing xylose with or without Cu^{2+} was measured at 420 nm using a microplate spectrophotometer. The change in the browning index (ΔB_{index}) was used to determine the effect of enzyme hydrolysis and presence of Cu2+. B_{index} and ΔB_{index} were determined using

$$B_{index} \text{ of sample} = A_{x\text{-}g} - A_g,$$

$$\Delta B_{index} \text{ of xylose-hydrolyzate} = \frac{\left(A_{x\text{-}h} - A_{x\text{-}g}\right) \times 100}{A_{x\text{-}g}},$$

ΔB_{index} of xylose-gelatin in presence of Cu^{2+}

$$= \frac{\left(A_{x\text{-}gCu} - A_{x\text{-}g}\right) \times 100}{A_{x\text{-}g}}, \tag{1}$$

ΔB_{index} of xylose-hydrolyzate in presence of Cu^{2+}

$$= \frac{\left(A_{x\text{-}hCu} - A_{x\text{-}g}\right) \times 100}{A_{x\text{-}g}},$$

where A_g is the absorbance of gelatin; $A_{x\text{-}g}$ is the absorbance of xylose-gelatin mixture; $A_{x\text{-}h}$ is the absorbance of xylose-hydrolyzate; $A_{x\text{-}gCu}$ is the absorbance of xylose-gelatin in presence of Cu^{2+}, and $A_{x\text{-}hCu}$ is the absorbance of xylose-hydrolyzate in presence of Cu^{2+}.

2.4. Data Analysis.
All data were collected in triplicate and their average and standard deviations were calculated. Diagrammatic representations were used to present the result for easy understanding.

3. Results and Discussion

3.1. Nonenzymatic Browning of Fish, Porcine, and Bovine.
Figure 1 shows that the browning index of gelatin from different samples increases with increase in heating time. There was an initial rise in the browning index for all the 3 gelatin samples in the first 6 hours followed by a steady increase. This is similar to previous reports that colour formation during Maillard reactions is usually rapid at the early reaction stage [9].

Several studies have shown that the amino acids compositions of gelatin vary according to the species [5]. It is possible that different regions/parts of the gelatin responded differently during the nonenzymatic browning. The variation in the reaction rate might be due to differences in the reactivity of amino acids present in the peptides. A study of Maillard reaction of essential amino acids with glucose revealed that amino acids were degraded differently and exhibited a varying degree of browning [2]. Compared to acidic amino acids, the basic amino acids contributed to browning intensity, while the nonpolar amino acids exhibited intermediate colour intensity [3]. Lysine participated in browning reaction induced by glucose compared to other amino acids and that threonine contributed very little towards browning [2].

This pattern can be explained that there are two regions in the gelatin polymer including the fast responding region and the slow responding region. The slow responding region

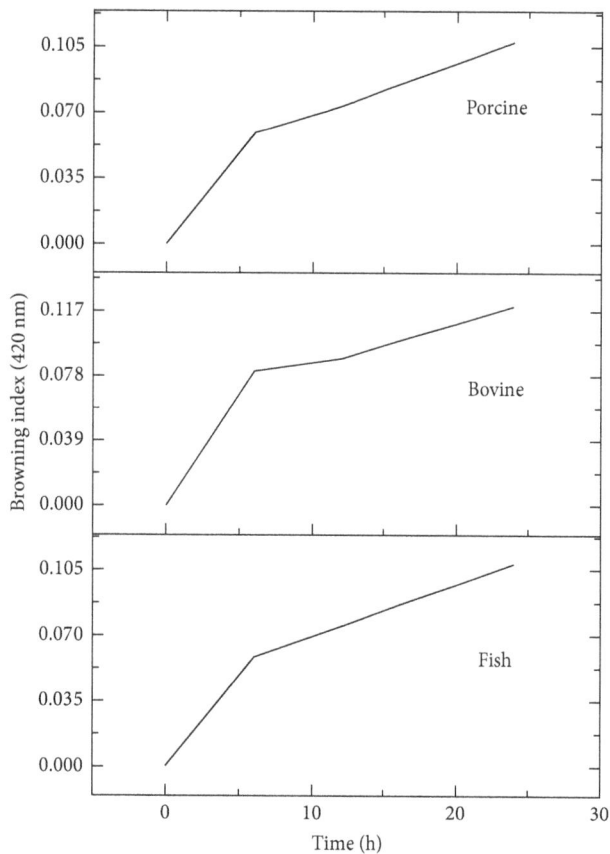

FIGURE 1: Browning index of xylose-induced nonbrowning of gelatin from different sources (fish, bovine, and porcine) at 95°C.

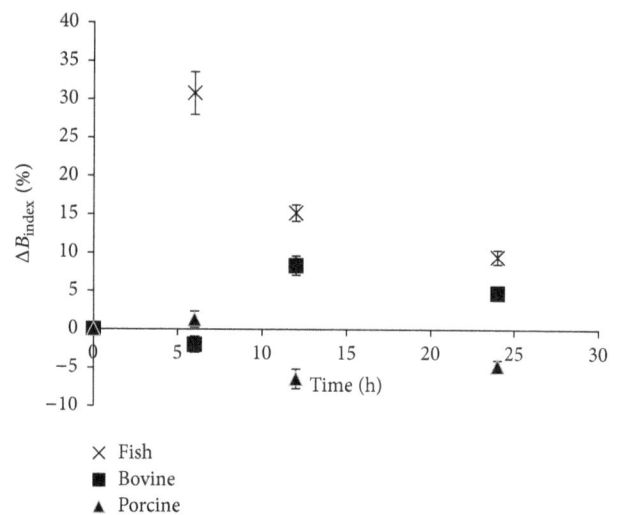

FIGURE 2: Change in the browning index due of chymotrypsin digestion on xylose-induced nonenzymatic browning of gelatin from different sources. ΔB_{index}: change in browning index.

might need to undergo initial reactions as a sequel to browning reaction. This surmised that gelatin of different sources has similar configurations which eventually reflect similarities in their Maillard reaction kinetic. However, the proportion of the two (slow and quick) regions might vary among the gelatin samples.

3.2. Effect of Chymotrypsin Hydrolysis of Gelatin on Degree of Xylose-Induced Nonenzymatic Browning. Structurally, collagen—the parent material of gelatin—is composed of nearly one-third of glycine and another 15 to 30% of proline and 4-hydroxyprolyl. 3-Hydroxyprolyl and 5-hydroxylysyl residues are also present in a smaller amount. Proline and hydroxyproline are responsible for the unique secondary structure of collagen as they limit rotation of the polypeptide backbone and thus contribute to the stability of triple helix [10]. Stabilization of collagen structure involves hydrogen bond (between glycine of the N terminal and proline of the adjacent chain), hydrophobic interaction, and Van der Walls interaction [11]. The collagen intermolar forces are likely inherited by gelatin, therefore, affecting the chemical properties of gelatin during Maillard reaction. The presence of intermolecular forces in gelatin is likely responsible for a reduction on the accessibility of amino acids to xylose during Maillard reaction. Hydrolysis of gelatin might result

into peptides that are more reactive and accessible during Maillard reaction.

In order to determine the effect of chymotrypsin digest of gelatin on contribution to browning index, the percentage difference in browning index of hydrolysates compared to their respective gelatin was estimated and referred to as a change in the browning index (ΔB_{index}). According to Figure 2, the effect of chymotrypsin digestion on ΔB_{index} varies among the gelatin samples. ΔB_{index} of fish gelatin hydrolysate was the highest among all the samples throughout the heating period. This suggests that digestion of fish gelatin produces peptides that contributed to ΔB_{index} compared to other samples.

Previous reports have shown that enzymatic digestion of gelatin produced different peptides fractions among species. The peptides might contribute differently towards the development of browning during Maillard reaction process and, hence, caused variation of ΔB_{index} among the samples.

The trends in browning development that follow were in 2 reaction phases, namely, increasing and decreasing phases. The increasing phase is early and high in fish hydrolysate compared to that of porcine and bovine. The increasing phase of bovine phase occurred only between 6 and 12 h. After 6 h reaction time, there was a decrease in browning in fish hydrolysate while the decreasing phase of porcine hydrolysate occurred after 12 h.

The reduction effect of chymotrypsin hydrolysis on ΔB_{index} observed in all the 3 samples might be explained by the lack of stability of their Maillard's reaction products. This reduction in browning occurred during the first six hours of heating in bovine gelatin hydrolysate, while that of fish hydrolysate occurred after the first six hours. Also, the rise in ΔB_{index} observed in all samples suggests that initial configuration of gelatin structure hindered the progress of Maillard reaction. Enzymatic hydrolysis did not only increase the surface area of gelatin and expose the amino acid but also

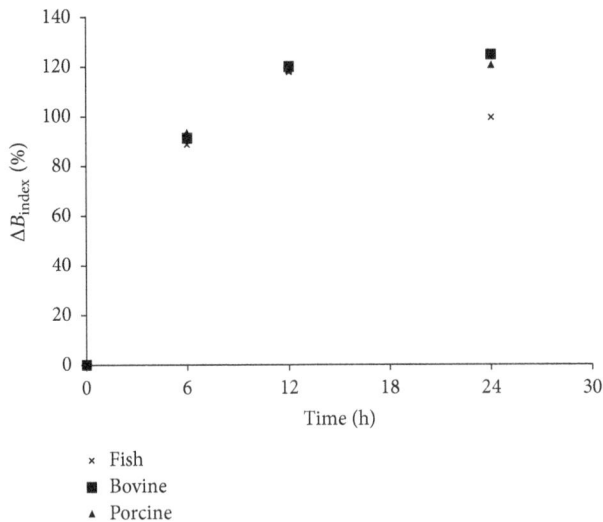

FIGURE 3: Change in the browning index due to the catalytic effect of Cu ion on the degree of xylose-induced nonenzymatic browning of gelatin from different sources. ΔB_{index}: change in browning index.

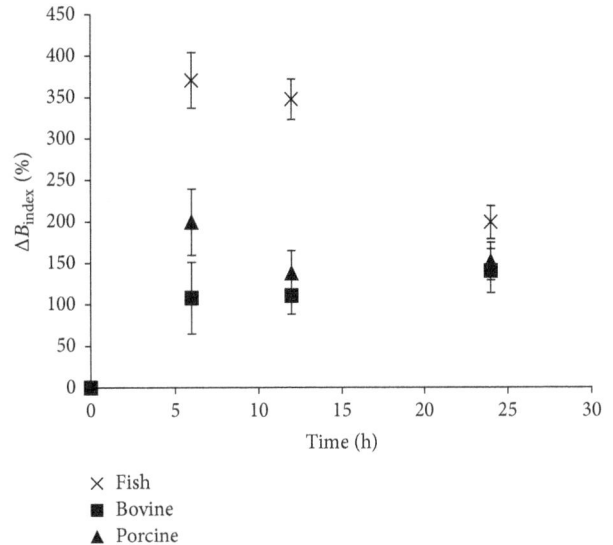

FIGURE 4: Change in the browning index due to combined catalytic effect of Cu^{2+} and chymotrypsin digestion on the degree of Maillard reaction of gelatin from different sources. ΔB_{index}: change in browning index.

reduce the hindering effect from gelatin configuration during Maillard reaction. In line with the findings of Su et al. [12], the smaller molecular peptides obtained from peanut protein exhibited higher reaction degree during Maillard reaction.

As stated earlier, the kinetic of nonenzymatic browning of gelatin of the three samples is similar and comprises fast and slow responding regions. However, enzymatic hydrolysis of gelatin resulted in hydrolysates that varied among the species. The rise in ΔB_{index} of fish hydrolysates might be because chymotrypsin hydrolysed the slow reactive regions of fish gelatin to produce more reactive peptides, compared to other gelatin samples. Another reason might be that chymotrypsin hydrolysed bovine gelatin might have produced peptides that participated in Maillard reaction after 6 h.

3.3. Catalytic Effect of Cu^{2+} on Degree of Xylose-Induced Nonenzymatic Browning of Gelatin from Different Sources.
The presence of Cu^{2+} during the nonenzymatic browning of gelatin samples causes an increase in the rate of more than 100% of B_{index}. The effect of Cu^{2+} on ΔB_{index} of the three gelatin samples was similar such that ΔB_{index} increased drastically until the 12 h of heating and then stabilized for bovine and porcine, while that of fish slightly decreased to about 100%. It is possible that Cu^{2+} enhanced formation of MRP and results in loss of slow reaction phase earlier observed in Figure 1. Also, Cu^{2+} might catalyse the formation of MRP from the majority of amino acids and reduced hindering effect of gelatin configuration during the reaction process. This observation agreed with the previous report that stated that Cu^{2+} contributed to the colour intensity of Millard reaction product [3]. The transition metals have been reported to have catalysed Maillard reaction by the oxidative pathway [3]. The presence of transition ions promotes the formation of chromophores and fluorophores during Maillard reaction of DNA with d-Fructose 6-Phosphate [13]. The

observed slight reduction in ΔB_{index} of fish might be due to loss of stability of MRP after extended heating.

3.4. Catalytic Effect of Cu^{2+} on Degree of Xylose-Induced Nonenzymatic Browning of Hydrolysates of Gelatin from Different Sources.
The combined effect of enzymatic hydrolysis and Cu^{2+} on the development of browning in Maillard reaction of the gelatin samples is shown in Figure 4. ΔB_{index} of fish hydrolysate was highest with a value of 400% followed by that of porcine (200%) and then bovine (140%). In the first 6 h, ΔB_{index} increased drastically in all samples and decreased steadily in porcine and fish while that of bovine kept increasing slightly.

Compared to that of the result in Figure 3, ΔB_{index} was increased in all samples. This suggests that Cu^{2+} catalysed the development of browning of peptides compared to gelatin. At this final stage, differentiating between gelatin from a different source can be achieved by comparing ΔB_{index} at different stages. In the early stage, fish gelatin exhibited highest ΔB_{index} followed by that of porcine and then bovine. The sharp reduction in ΔB_{index} observed in fish gelatin can be used to discriminate fish gelatin from mammalian gelatin.

4. Conclusion

Maillard reaction of gelatin from different sources exhibited similar reaction phases—the slow and the fast phases. Enzymatic degradation of gelatin prior to Maillard reaction caused a difference in the production of browning products among the species. Fish gelatin hydrolysate displayed multifold increase in browning in the first six hours compared to gelatin hydrolysates from porcine and bovine. Although the catalytic effect of Cu^{2+} during Maillard reaction was relatively similar in all the gelatin samples, Cu^{2+} affects gelatin hydrolysates differently. The variation in ΔB_{index} of chymotrypsin digested gelatin in presence of Cu^{2+} could be valuable for the development of an efficient UV-spectroscopy method for gelatin differentiation. Future works will investigate the effects of other reaction conditions on ΔB_{index} of enzymatic hydrolysates from gelatin.

Conflicts of Interest

The authors declare that there are no conflicts of interest regarding the publication of this paper.

References

[1] A. Etxabide, M. Urdanpilleta, P. Guerrero, and K. De La Caba, "Effects of cross-linking in nanostructure and physicochemical properties of fish gelatins for bio-applications," *Reactive and Functional Polymers*, vol. 94, article 3544, pp. 55–62, 2015.

[2] E. H. Ajandouz and A. Puigserver, "Nonenzymatic browning reaction of essential amino acids: effect of pH on caramelization and Maillard reaction kinetics," *Journal of Agricultural and Food Chemistry*, vol. 47, no. 5, pp. 1786–1793, 1999.

[3] E.-J. Kwak and S.-I. Lim, "The effect of sugar, amino acid, metal ion, and NaCl on model Maillard reaction under pH control," *Amino Acids*, vol. 27, no. 1, pp. 85–90, 2004.

[4] M. C. Gómez-Guillén, M. Pérez-Mateos, J. Gómez-Estaca, E. López-Caballero, B. Giménez, and P. Montero, "Fish gelatin: a renewable material for developing active biodegradable films," *Trends in Food Science and Technology*, vol. 20, no. 1, pp. 3–16, 2009.

[5] A. A. Karim and R. Bhat, "Gelatin alternatives for the food industry: recent developments, challenges and prospects," *Trends in Food Science and Technology*, vol. 19, no. 12, pp. 644–656, 2008.

[6] Z. A. Nur Hanani, Y. H. Roos, and J. P. Kerry, "Use of beef, pork and fish gelatin sources in the manufacture of films and assessment of their composition and mechanical properties," *Food Hydrocolloids*, vol. 29, no. 1, pp. 144–151, 2012.

[7] A. Rohman and Y. B. Che Man, "Analysis of pig derivatives for halal authentication studies," *Food Reviews International*, vol. 28, no. 1, pp. 97–112, 2012.

[8] T.-C. Tan, A. F. M. Alkarkhi, and A. M. Easa, "Assessment of the ribose-induced Maillard reaction as a means of gelatine powder identification and quality control," *Food Chemistry*, vol. 134, no. 4, pp. 2430–2436, 2012.

[9] M. Van Boekel, "Kinetic aspects of the Maillard reaction: a critical review," *Molecular Nutrition Food Research*, vol. 45, no. 3, pp. 150–159, 2001.

[10] R. Schrieber and H. Gareis, *GelaTine Handbook: Theory and Industrial Practice*, John Wiley & Sons, Hoboken, New Jersey, NJ, USA, 2007.

[11] T. Nogrady and D. F. Weaver, *Medicinal Chemistry: A Molecular and Biochemical Approach*, Oxford University Press, Oxford, UK, 2005.

[12] G. Su, L. Zheng, C. Cui, B. Yang, J. Ren, and M. Zhao, "Characterization of antioxidant activity and volatile compounds of Maillard reaction products derived from different peptide fractions of peanut hydrolysate," *Food Research International*, vol. 44, no. 10, pp. 3250–3258, 2011.

[13] J. Morita and N. Kashimura, "The Maillard reaction of DNA with D-Fructose 6-Phosphate," *Agricultural and Biological Chemistry*, vol. 55, no. 5, pp. 1359–1366, 1991.

7

Influence of Processing Methods on Proximate Composition and Dieting of Two *Amaranthus* Species from West Cameroon

Arnaud Landry Suffo Kamela,[1,2] **Raymond Simplice Mouokeu,**[3]
Rawson Ashish,[2] **Ghislain Maffo Tazoho,**[1] **Lamye Glory Moh,**[1]
Etienne Pamo Tedonkeng,[4] **and Jules-Roger Kuiate**[1]

[1]*Department of Biochemistry, Faculty of Science, University of Dschang, P.O. Box 67, Dschang, Cameroon*
[2]*Department of Food Engineering and Design, Indian Institute of Crop and Processing Technology, Thanjavur, India*
[3]*Institute of Fisheries and Aquatic Sciences, University of Douala, P.O. Box 7236, Douala, Cameroon*
[4]*Department of Animal Production, Faculty of Agronomy and Agricultural Science, University of Dschang, Dschang, Cameroon*

Correspondence should be addressed to Jules-Roger Kuiate; jrkuiate@yahoo.com

Academic Editor: Rosana G. Moreira

The effects of various processing methods on the proximate composition and dieting of *Amaranthus hybridus* and *Amaranthus cruentus* from West Cameroon were investigated in this study. Both amaranths leaves were subjected to same treatments (sun-dried and unsliced, sliced and cooked), milled, and analysed for their mineral and proximate composition. Thirty-Six *Wistar* albino rats of 21 to 24 days old were distributed in six groups and fed for 14 days with 10% protein based diets named D0 (protein-free diet), DI (egg white as reference protein), DII (sun-dried and unsliced *A. hybridus*), DIII (cooked and sliced *A. hybridus*), DIV (sun-dried and unsliced *A. cruentus*), and DV (cooked and sliced *A. cruentus*). The protein bioavailability and haematological and biochemical parameters were assessed in rats. The results showed that K, P, Mg, Zn, and Fe had the higher content in both samples regardless of processing method. The sun-dried and unsliced *A. cruentus* contained the highest value of crude protein 32.22 g/100 g DM (dry matter) while the highest crude lipid, 3.80 and 2.58%, was observed, respectively, in sun-dried and unsliced *A. hybridus* and cooked and sliced *A. cruentus*. Cooked and sliced *A. hybridus* and *A. cruentus* contained high crude fiber of 14 and 12.18%, respectively. Rats fed with diet DIII revealed the best protein bioavailability and haematological parameters whereas 100% mortality rate was recorded with group fed with diet DIV. From this study, it is evident that cooked and sliced *A. hybridus* and *A. cruentus* could play a role in weight reduction regimes.

1. Introduction

Vegetables and fruits offer the most rapid and cheapest sources of adequately supplied vitamins, minerals, and some essential amino acids [1]. They have the cheapest and most abundant sources of protein [2]. In Cameroon, as in many Africa countries, vegetables are very abundant immediately after the first rains, but they become scarce in the middle of the rainy season and more so in the dry season [3].

A great variety of local and introduced vegetable crops are grown in Cameroon and these crops together with a significant number of wild and semiwild plants like *Amaranthus hybridus* and *Amaranthus cruentus* form a valuable complementary food in the daily diet [4]. Most of them are consumed in the rural areas or in the communities where they are being planted [5]. They are underutilized when compared to the introduced varieties due to the flavor and unfamiliar taste impacted on the food [6, 7]. They are very perishable commodities with very high moisture contents; therefore, dehydration results in substantial reduction in weight and bulk with consequent savings in storage costs [8]. As green leafy vegetables are edible parts of the plants and are usually cooked before consumption, cooking causes significant changes in the nutritional properties of food as well as gelatinization of starches and coagulation of proteins to improve their digestibility and sensory properties [9].

TABLE 1: Ingredient and composition of diets.

Ingredients (g/100 g)	Diets					
	D0	DI	DII	DIII	IV	DV
Corn starch	15	15	15	15	15	15
Corn oil	5	5	5	5	5	5
Mineral complex	4	4	4	4	4	4
Vitamin complex	1	1	1	1	1	1
Cellulose	7.9	7.9	7.9	7.9	7.9	7.9
Protein source	—	13.37	32.3	31.4	30.1	38.9
Sucrose	67.1	53.73	34.8	35.7	37	28.2
Total	100	100	100	100	100	100

D0: protein-free diet, DI: egg white (protein reference) diet, DII: sun-dried and unsliced *A. hybridus* diet, DIII: cooked and sliced *A. hybridus*, DIV: sun-dried and unsliced *A. cruentus*, and DV: cooked and sliced *A. cruentus*.

Udosen and Ukpanah [10] observed that processing causes losses in some of the antinutritional factors and some good nutrients as well.

In many homes in Cameroon, the outstanding preservative method practiced on green leafy vegetables is slicing or not, followed by sun drying which is often combined with cooking. These methods are found to be effective in improving digestibility and increasing nutrient bioavailability and also minimize foodborne diseases [11]. Through boiling, some antinutrients contents of the leaves can be reduced [12]. However, information appears scanty on the nutritional composition of the vegetable harvested in Cameroon. The aim of this work was to determine the mineral and proximate composition of two leafy vegetables consumed in West Cameroon which undergo slicing, sun drying, and cooking processing techniques as well as protein bioavailability using rats' model. The awareness will encourage the consumption and exploitation of these cheap sources of not well-known food item at any time and throughout the year.

2. Materials and Methods

2.1. Plant Leaves Collection and Processing. The leaves of *Amaranthus hybridus* and *Amaranthus cruentus* were obtained from cultivated farmlands located at Foto, Dschang City of the West Region of Cameroon, from April 2013 to September 2013. The botanical identification was done at the National Herbarium in Yaounde (Cameroon) by referring to the voucher specimen number 15630 HNC.

The collected samples were thoroughly mixed, their stalks and dust discarded, and they were divided into two groups. The first group was sun dried without slicing with frequent turning till leaves were crumbly, while the second group (100 g) was cooked after slicing for 15 min following bleaching in boiling water (1000 mL) for 10 minutes. Cooked and sliced samples were cooled at ambient temperature (20°C ± 2), water was removed, and the sample sun dried for 3 days.

2.2. Assessment of Minerals and Heavy Metals Content. Minerals and heavy metals including Na, K, Ca, P, Mg, Mn, Fe, and Zn and As, Hg, Pb, and Cd, respectively, were analysed using ICP-OES (Perkin-Elmer; Model Optima™ 2000 DV,

Schwerzenbach, Switzerland) [13]. Analyses were performed in triplicate.

2.3. Proximate Composition Analysis. The methods adopted for the proximate composition analysis were those recommended by the Association of Official Analytical Chemists [14]. Analyzed parameters included dry mater, ash content, crude fat, crude fibre, and proteins.

2.4. Evaluation of Protein Bioavailability

2.4.1. Experimental Animals. Thirty-six *Wistar* albinos rats of both sexes, 21 to 24 days old, weighing 34–43 g were used in the experiment. These animals were bred in the Animal House of the Department of Biochemistry, University of Dschang, Cameroon, and were fed with a standard rat diet. Food and water were given to all animals used for the experiments ad libitum. Animals were maintained at room temperature (22 ± 2°C) and were handled according to standard protocols for the use of laboratory animals. The studies were conducted according to the ethical guidelines of Committee for Control and Supervision of Experiments on Animals (Registration number 173/CPCSEA, dated 28 January, 2000), Government of India, on the use of animals for scientific research.

2.4.2. Experimental Design. The bioassay experiments were carried out according to AOAC [14] protocol. The animals were weighed, divided into six groups (D0, DI, DII, DIII, DIV, and DV) of six animals each, and housed individually in stainless steel screening bottom cages (permitting free dropping of faeces). Highly absorbent paper was placed under the cages to catch spilled food and to minimize contamination of faeces with urine.

The experimental diets (Table 1) containing 10% protein were prepared following ICN protocol [15]. The first group (D0) was given the N-free basal diet and the second group (DI) was fed with reference diet (egg white), while the test groups (DII, DIII, DIV, and DV) were randomly allocated to the diets containing test ingredients (sun-dried and unsliced *A. hybridus*, cooked and sliced *A. hybridus,* sun-dried and unsliced *A. cruentus*, and cooked and sliced *A. cruentus.*

TABLE 2: Minerals and heavy metals composition (mg/Kg) of two differently processed *amaranths* species.

Processing	Na	K	Ca	P	Mg	Mn	Fe	Zn	As	Hg	Pb	Cd
						A. hybridus						
SDU	6.185	697.3	307.7	111.5	111.8	0.117	0.624	3.027	0.024	0.109	0.007	0.002
CS	3.549	298.7	535.5	110.4	102.9	0.914	2.912	10.04	0.026	0.137	0.018	0.001
						A. cruentus						
SDU	3.368	796	422.5	173.6	106.5	0.462	3.402	48	0.025	ND	0.024	ND
CS	3.913	323.2	576.8	79.56	88.47	0.597	5.685	38.22	0.037	ND	0.022	ND

SDU: sun-dried and unsliced, CS: cooked and sliced, and ND: not detected.

TABLE 3: Proximate composition (g/100 g dry matter) of protein of reference (egg white) and differentially processed *A. hybridus* and *A. cruentus*.

Diet	Dry matter	Crude protein	Crude lipid	Crude fibre	Ash
Egg white	89.95 ± 0.99^{ab}	80.05 ± 0.85^{a}	1.07 ± 0.74^{c}	1.73 ± 0.20^{e}	6.89 ± 0.05^{c}
			A. hybridus		
SDU	89.58 ± 0.025^{ab}	30.95 ± 0.07^{dc}	3.80 ± 0.39^{a}	6.02 ± 0.65^{d}	15.43 ± 0.07^{a}
CS	90.38 ± 0.08^{a}	31.79 ± 0.04^{c}	1.21 ± 0.03^{c}	14.00 ± 0.08^{a}	10.45 ± 0.12^{bd}
			A. cruentus		
SDU	88.487 ± 0.1^{b}	32.22 ± 0.02^{b}	1.64 ± 0.09^{bc}	8.76 ± 0.15^{c}	9.67 ± 0.04^{d}
CS	90.02 ± 0.8^{ab}	25.65 ± 0.01^{e}	2.58 ± 0.07^{b}	12.18 ± 0.66^{b}	9.27 ± 0.60^{d}

Results are expressed as mean ± standard deviation. Means in the same column followed by the same letters are not significantly different at 5%. DI: egg white (standard protein), SDU: sun-dried and unsliced, and CS: cooked and sliced.

2.4.3. Nutritional Evaluations. The absorbed nitrogen, true digestibility (TD), feed efficiency (FE), efficiency of food utilization (EFU), protein efficiency ratio (PER), and net protein ratio (NPR) were all computed as described by Pirman et al. [16].

2.4.4. Biochemical Assay. At the end of the experimental period (14 days), blood samples were collected following overnight fasting by cardiac puncture from chloroform anaesthetized rats into heparinised and nonheparinised tubes. The nonheparinised tubes were allowed to clot and were centrifuged at 3000 rpm for 15 min to obtain the serum. These sera were assayed for alanine aminotransferase (ALAT), aspartate aminotransferase (ASAT), creatinine, albumin, and urea using SPINREACT kit. The heparinised blood was used for determination of some haematological parameters including red blood cells, white blood cells, platelets, and their different indices [17, 18].

2.5. Statistical Analysis. Statistical analyses were performed using graph pad prism version 5.00 software. Data were analysed by one-way analysis of variance (ANOVA), followed by Bonferroni post hoc test. Results are expressed as mean ± standard deviation of replicated samples. Differences were considered significant at $P < 0.05$.

3. Results and Discussion

3.1. Results

3.1.1. Minerals and Heavy Metal Content. The effects of processing treatments on mineral and heavy metals contents of two amaranths species are given in Table 2. Regardless of the processing methods, the level of K, P, and Mg remained high in both samples of sun-dried and unsliced amaranthus leaves. In both amaranths, increase of Ca, Mn, and Fe content was observed after cooking and slicing. Regardless of the amaranth, the levels of Zn and Na were relatively affected by processing methods. The levels of Hg and Cd were relatively high in *A. hybridus* regardless of the processing method.

3.1.2. Proximate Composition of Leafy Vegetables. The proximate composition of protein reference (egg white) and differentially processed *A. hybridus* and *A. cruentus* are presented in Table 3. The protein content of cooked and sliced *A. cruentus* had the least value (25.65%) while the highest value (32.22%) was recorded in sun-dried and unsliced *A. cruentus* leaves sample. The various food processing techniques caused significant differences ($P < 0.05$) between sun-dried and unsliced and cooked and sliced *A. hybridus*. Cooking improved the proximate composition of the vegetables relative to the sun drying and unslicing method. However, cooking and slicing significantly decreased protein level in *A. cruentus* as well as the total lipid of *A. hybridus*. The ash value for *A. hybridus* was influenced by the processing method, with a significant increase ($P < 0.05$) in sun-dried and unsliced sample (15.43%), relative to the cooked and sliced sample (10.45 ± 0.12); meanwhile, there was no processing-related significant variation ($P > 0.05$) in the ash content of *A. cruentus*.

3.1.3. Growth Performance and Protein Digestibility of Rats Fed with Processed Amaranths Leaves. The results of growth performance and protein quality on rats fed with different

TABLE 4: Growth performance and protein quality of rats fed with diets containing processed *A. hybridus* and *A. cruentus* leaves.

Parameters	Standards		A. hybridus-based diets		A. cruentus-based diets	
	D0	DI	DII	DIII	DIV	DV
Food intake (g)	21.92 ± 2.92^c	68.56 ± 9.04^a	54.70 ± 5.35^b	73.46 ± 5.57^a	RNA	31.63 ± 5.77^c
Protein consumed (g)	RNA	6.74 ± 1.06^{ab}	5.47 ± 0.53^b	7.34 ± 0.55^a	RNA	3.16 ± 0.57^c
weight gain (g)	-2.80 ± 0.83^d	24.4 ± 3.71^a	6.00 ± 1.58^c	13.60 ± 1.67^b	RNA	-2.25 ± 2.06^d
Feed efficiency (g/g)	RNA	2.78 ± 0.41^b	9.57 ± 2.47^a	5.50 ± 1.13^b	RNA	RNA
Faeces weight (g)	2.20 ± 0.44^d	7.40 ± 1.67^c	14.00 ± 1.00^{ab}	16.90 ± 1.94^a	RNA	11.75 ± 1.70^b
EFU (%)	RNA	36.5 ± 5.30^a	10.93 ± 2.39^c	18.68 ± 3.18^b	RNA	RNA
PER	RNA	3.65 ± 0.53^a	1.09 ± 0.23^c	1.86 ± 0.31^b	RNA	RNA
NPR	RNA	3.27 ± 0.14^a	0.63 ± 0.23^c	1.52 ± 0.31^b	RNA	RNA
TD (%)	RNA	89.84 ± 7.47^a	84.42 ± 8.94^a	92.00 ± 8.43^a	RNA	57.23 ± 7.06^b

Results are expressed as mean ± standard deviation. Means in the same line followed by the same letters are not significantly different at 5%. D0: protein-free diet, DI: egg white (standard protein), DII: sun-dried and unsliced *A. hybridus*, DIII: cooked and sliced *A. hybridus*, DIV: sun-dried and unsliced *A. cruentus*, and DV: cooked and sliced *A. cruentus*. TD: true digestibility, FE: feed efficiency, EFU: efficiency of food utilization, PER: protein efficiency ratio, NPR: net protein ratio, and RNA: results not available. All the group IV animals died before the end of experimental period.

processed diets of *A. hybridus* and *A. cruentus* leaves are presented in Table 4. The rats fed with standard protein diet (DI) and cooked *A. hybridus* diet (DIII), respectively, had the highest food (68.56 and 73.46 g) and protein (6.74 and 7.34 g) intake and, consequently, showed the highest level of weight gain (24.4 g and 13.60 g) while those fed with cooked *A. cruentus* diet (DV) and protein-free diet (D0) significantly ($P < 0.05$) lost weight. Animals fed with sun-dried *A. hybridus* diet (DII) revealed the best feed efficiency while the highest value (16.90 g) of faeces weight was observed in animals fed with DIII. Weight of animals fed with diet DIV kept decreasing during experimental period and at the end, all the animals died. Animals fed with diet DIII significantly ($P < 0.05$) had the highest efficiency food utilization (EFU), protein efficiency ratio (PER), and net protein ratio (NPR) compared to animals fed with DI. With the death of all animals fed with DIV diet, no data about the protein quality was recorded while neither feed efficiency, EFU, and PER nor NPR were available for these animals.

3.1.4. Biochemical Parameters and Haematological Indices. The results of the change in some serum enzyme activities and metabolites of rats fed with diets containing green leafy vegetables are shown in Table 5. Apart from DIII and DV, there was a significant decrease in serum activity of aspartate aminotransferase (AST) and alanine aminotransferase (ALT) of animals fed with formulated diets. The formulated diets did not influence the level of blood creatinine as compared to the control. However, these diets significantly increased ($P < 0.05$) the serum level of urea with diet DV having the highest value. On the contrary, these diets significantly decreased serum concentration of albumin with the effect of the diets being comparable. The results of the haematological parameters indicate that white blood cells (WBC), red blood cells (RBC), haemoglobin (HB), mean cell haemoglobin (MCH), mean cell haemoglobin concentration (MCHC), mean platelet volume (MPV), and plateletcrit (PCT) were not significantly different ($P > 0.05$) between diet DI and

diets DII, DIII, and DV. Animals fed with diets DII, DIII, and DV showed significant decrease ($P < 0.05$) in levels of lymphocytes (LYM) and platelets (PLT) when compared to rats fed with DI. In general, there was a significant ($P < 0.05$) increase of mid-cell (MID) and granulocytes (GRAN) in groups fed with the formulated diets compared to the standard diet (DI).

4. Discussion

4.1. Mineral and Heavy Metals. The mineral analysis of treated leaves powder revealed the presence of eight (8) minerals in both samples, revealing them as excellent sources of macro- and micronutrients. The level of mineral content (K, P, and Mg) was lower in cooked and sliced leaves when compared to sun-dried and unsliced samples. This is in line with the observations of Bakr and Gawish [19], Shahnaz et al. [20], and Oboh [21] that various conventional food processing techniques (cutting, bleaching, cooking, etc.) cause a decrease in the mineral content of vegetables. Losses of the mineral elements during boiling or cooking are generally attributed to the leaching of the cell content including minerals [19]. The result of mineral analysis of the vegetables suggests consumption of enough quantities to meet Recommended Daily Allowance (RDA). The value of Na in both sun-dried and unsliced and cooked and sliced samples was low compared to the value (300.06–600.83 mg/kg) obtained by Makobo et al. [22] for sun-dried and bleached *Amaranthus cruentus*. Na is required for maintenance of fluid balance and osmotic pressure in the body for cellular activities [23, 24]. The K content found in both vegetables regardless of the processing method was higher than 241.88 mg/kg reported by Ogbadoyi et al. [25] for *Amaranthus cruentus*. However, the result of this study reveals that K content dropped below those reported in the available literature and this suggests the consumption of large quantities of these vegetables in order to meet the RDA for minerals. For instance, adult minimum K requirement for

TABLE 5: Biochemical parameters and haematological indices of rats fed with different processed *A. hybridus* and *A. cruentus* meal.

Parameters	Standard	A. hybridus-based diets		A. cruentus-based diets	
	DI	DII	DIII	DIV	DV
AST (U/L)	105.30 ± 6.10^a	52.50 ± 1.61^b	35.53 ± 2.70^c	RNA	35.04 ± 0.83^c
ALT (U/L)	27.25 ± 0.59^a	17.85 ± 0.10^c	21.00 ± 1.01^b	RNA	21.29 ± 0.80^b
Creatinine (mg/dL)	1.41 ± 0.64^a	0.99 ± 0.26^a	1.24 ± 0.31^a	RNA	0.58 ± 0.38^a
Urea (mg/dL)	34.01 ± 0.58^d	38.81 ± 1.09^c	43.37 ± 3.72^b	RNA	114.10 ± 0.70^a
Albumin (g/dL)	4.07 ± 0.11^a	3.08 ± 0.19^b	3.08 ± 0.19^b	RNA	3.44 ± 0.79^{ab}
WBC ($10^3/\mu L^{-1}$)	3.50 ± 2.32^a	4.24 ± 1.41^a	6.14 ± 2.48^a	RNA	2.90 ± 1.37^a
LYM (%)	84.24 ± 1.80^a	70.72 ± 4.95^b	76.80 ± 4.92^{ab}	RNA	69.27 ± 5.15^b
MID (%)	5.46 ± 0.34^c	11.38 ± 2.95^b	9.30 ± 0.95^b	RNA	16.30 ± 2.55^a
GRAN (%)	3.95 ± 1.15^b	17.9 ± 2.48^a	16.73 ± 6.65^a	RNA	13.73 ± 2.95^a
RBC ($10^6/\mu L$)	5.31 ± 0.94^a	4.66 ± 2.68^a	5.30 ± 1.13^a	RNA	6.69 ± 1.05^a
HB ($g \cdot dL^{-1}$)	11.46 ± 1.91^a	12.94 ± 1.06^a	13.76 ± 1.83^a	RNA	15.2 ± 2.25^a
HCT (%)	32.08 ± 4.06^a	33.7 ± 2.37^a	34.46 ± 3.79^a	RNA	44.27 ± 1.70^b
MCV (fL)	60.96 ± 3.46^{ab}	58.38 ± 4.64^b	68.28 ± 3.92^a	RNA	66.3 ± 5.11^{ab}
MCH (pg)	21.56 ± 0.92^a	21.80 ± 1.36^a	24.20 ± 4.80^a	RNA	22.67 ± 0.23^a
MCHC ($g \cdot dL^{-1}$)	35.64 ± 2.69^a	37.50 ± 0.95^a	36.33 ± 2.07^a	RNA	34.93 ± 5.67^a
PLT ($10^3/\mu L$)	581.20 ± 4.49^a	436.40 ± 5.22^d	486.80 ± 4.14^c	RNA	13.00 ± 2.64^b
MPV (fL)	7.78 ± 1.51^a	8.12 ± 0.53^a	7.92 ± 0.50^a	RNA	7.80 ± 0.61^a
PCT (%)	0.41 ± 0.22^a	0.35 ± 0.28^a	0.37 ± 0.10^a	RNA	0.39 ± 0.11^a

Results are expressed as mean ± standard deviation. Means in the same line followed by the same letters are not significantly different at 5%. DI: egg white (standard protein), DII: sun-dried and unsliced *A. hybridus*, DIII: cooked and sliced *A. hybridus*, DIV: sun-dried and unsliced *A. cruentus*, and DV: cooked and sliced *A. cruentus*. WBC: white blood cell, LYM: lymphocytes, MID: mid-cells, GRAN: granulocyte, RBC: red blood cell, HB: haemoglobin, HCT: haematocrit, MCV: mean cell volume, MCH: mean cell haemoglobin, MCHC: mean cell haemoglobin concentration, PLT: platelets, MPV: mean platelet volume, PCT: plateletcrit, and RNA: results not available. All the group IV animals died before the end of experimental period.

health set by the 1989 RDA is 2000 mg daily [9]. Then to meet the RDA for this important mineral involved in cellular metabolism, water used in the boiling must be included in the meal preparations [25]. The level of Ca, Mn, and Fe observed in the cooked two species of vegetables clearly indicates that these vegetables are good sources of those minerals when compared to the values obtained for cereals [26].

Calcium and phosphorus are associated with each other for the growth and maintenance of bones, teeth, and muscles [27] while Mg is an important cofactor of enzymes involved in cell respiration, glycolysis, and transmembrane transporter [23]. Iron is an essential trace element for haemoglobin formation, normal function of central nervous system, and energy metabolism [28]. The highest value of zinc (48 mg/kg) was found in sun-dried and unsliced *A. cruentus*. Therefore, the consumption of cooked and sliced *A. hybridus* in developing countries could correct Zn deficiency which is related to decreased growth in infants and children [29].

The values of As, Cd, Pb, and Hg obtained in *A. hybridus* were lower than those obtained in the same species of amaranths by Oti Wilberforce and Nwabue [30] in the state of Nigeria. The vegetables analysed regardless of the treatment methods showed lower levels of the minerals and it can be suggested that the consumption of average amounts of these vegetables could not pose a health risk for the consumers as the values obtained are far below the permissible limits of 0.2 mg/kg (cadmium), 0.1 mg/kg (arsenic), and 0.01 mg/kg (mercury) [31].

4.2. Proximate Composition. The results of the proximate analysis showed that sun-dried and unsliced *A. cruentus* was the richest source of crude protein, while cooked and sliced *A. hybridus* had the highest value of crude fibre. Studies have shown that these leaves are usually cooked before consumption [32]. The cooked values were therefore of great importance. The cooked and sliced sample had the lower value than sun-dried and unsliced sample. The reduction could have been due to the different levels of heat treatment and the severity of thermal process during cooking where some nutrients were leached off by water during the process [33]. Krauss et al. [34] indicated that plant foods providing more than 12% protein calorific values are good sources of protein. Therefore, the leafy vegetables studied are good sources of protein and can be used as diet supplements for people suffering from undernutrition diseases if and only if those proteins could be bioavailable. The crude fibre content in both samples ranged from 6.02 to 14% with the highest level in cooked and sliced *A. hybridus*. These values are higher than those obtained by Mensah et al. [35] in *A. cruentus* (1.8%) and Asaolu et al. [36] in *A. hybridus* (8.05%). The plants can serve as good roughage in the intestine for better functioning of the alimentary system [37] since it had been reported that food fibre aids absorption of trace elements in the gut and reduces absorption of cholesterol. Besides, vegetables rich in fibre are natural broom for the body which help to prevent constipation, bowel problems, and piles [38].

4.3. Growth Performance and Protein Digestibility of Rats Fed with Processed Amaranths Leaves. Weight loss was observed with the animals fed with basal diets (D0) and cooked and sliced *A. cruentus* (DV) while 100 percent of mortality rates were recorded in the group fed with sun-dried and unsliced *A. cruentus*-based diet (DIV). For the basal diet, it is not surprising since it lacks protein as Cameron and Eshelman [39] showed that deficiencies in dietary protein slow growth and delay maturation. The loss in weight observed with rats fed with sun-dried and unsliced *A. hybridus* (DII), cooked and sliced *A. hybridus* (DIII), and cooked and sliced *A. cruentus* (DV) is then a consequence of lack of full utilization and poor protein quality indices such as low values of PER, NPR, and TD. Furthermore, it could be either due to the fact that plant protein can be encased in cellulose walls, which are hard to penetrate making proteins less accessible to digestive enzymes [40], or attributed to the rich sources of dietary fibre contained in those vegetables which are known to decrease protein utilization [41]. This finding is consistent with earlier report. Agbede et al. [42] reported that vegetables proteins have lower quality protein than the animal protein, and such a diet needs to be supplemented with another protein source relatively rich in the essential amino acids. The reduction in weight gain also implies that cooked and sliced *A. cruentus* can be used in weight reduction regimens. Studies have shown that weight reduction is one of the ways of reducing coronary risk incidence, as well as managing diabetes mellitus, dyslipidemia, hypertension, and obesity [34], and is one of the strategies for improving low high density lipoprotein cholesterol (HDL-C) levels. Those results are in line with the faeces. Indeed, the faeces bulk of animals fed with these vegetables based diets was higher than those of animals fed with the standard and nitrogen-free diets. This could be explained by the abundance of fibre content in the vegetables. Cummings [43] and EFSA [44] stated that dietary fibres contribute to an increase in faecal bulk with some beneficial physiological effect like the laxative effect, capacity to decrease gastrointestinal transit time, loose stools, bloating and distension, borborygmi, abdominal discomfort, and flatus [45]. Furthermore, faeces weight is inversely related to colon cancer [43].

The death of animal fed with sun-dried and unsliced *A. cruentus*-based diet could be probably due to the presence of high amount of antinutritional factors such as trypsin inhibitors, phytate, and polyphenols that could have interfered with metabolic processes by reducing the bioavailability of nutrients [46] and limiting the digestibility of plant protein [47] and can also provoke deleterious effects on many organs [48].

4.4. Biochemical Parameters and Haematological Indices. There were significant increases in serum alanine aminotransferase (ALT), albumin, and aspartate aminotransferase (AST) in rats fed with diet containing cooked and sliced *A. cruentus* and sun-dried and unsliced *A. hybridus*. This indicates possible damage of some organ such as liver and heart by those vegetables. ALT is regarded to be more specific indicator of liver inflammation, while AST may be elevated in diseases of other organs such as heart and muscle diseases

[49]. There was also a significant increase in the serum albumin in rats fed with diet containing cooked and sliced *A. cruentus*; this clearly indicates that *A. cruentus* may not cause liver damage since albumin is produced mainly in the liver. Except for the group fed with cooked and sliced *A. cruentus* $(0.58 \pm 0.038 \, \text{mg·dL}^{-1})$, the level of creatinine in other groups was higher than the normal values $(0.2–0.8 \, \text{mg·dL}^{-1})$ [50]. This is an indication that the samples contained considerable amount of phytochemical compounds that may cause kidney related malfunctions.

The best RBC and HB values were observed with animals fed with diet containing cooked and sliced *Amaranthus* species. This shows that these processed leaves help in blood formation due to availability of crude protein and iron, meaning that there is no risk relative to the anaemia and related diseases with the consumption of the studied amaranth species. White blood cell count and MID cell are related to immune system and bone marrow and are indicators of the ability of an organism to eliminate infection [51]. The white blood cell of animal fed with sun-dried and unsliced and cooked and sliced *A. hybridus*-based diet was not significantly different to the reference diet, showing that the feed does not affect the immune systems of animal.

5. Conclusion

The present study clearly indicates that cooking method remains the best way for good utilization of these green leafy vegetables. It was also found that cooked and sliced *A. hybridus* compared to *A. cruentus* can better support growth performance. However, cooked and sliced *A. cruentus* and *A. hybridus* could be used in weight reduction regimes.

Competing Interests

The authors declare that they have no competing interests.

Acknowledgments

The authors acknowledged the financial assistance of the Department of Science & Technology and the Ministry of External Affairs, Government of India, and FICCI for CV Raman International Fellowship for African Researchers grant offer.

References

[1] T. C. Mosha and H. E. Gaga, "Nutritive value and effect of blanching on the trypsin and chymotrypsin inhibitor activities of selected leafy vegetables," *Plant Foods for Human Nutrition*, vol. 54, no. 3, pp. 271–283, 1999.

[2] A. O. Fasuyi, "Nutritional potentials of some tropical vegetable leaf meals: chemical characterization and functional properties," *African Journal of Biotechnology*, vol. 5, no. 1, pp. 49–53, 2006.

[3] A. I. Ihekoronye and P. O. Ngoddy, *Integrated Food Science and Technology for the Tropics*, Macmillan Publishers, London, UK, 2nd edition, 1985.

[4] J. M. C. Stevels, *Légumes traditionnels du Cameroun, une étude agrobotanique*, Wageningen Agricultural University Papers, Wageningen, The Netherlands, 1990.

[5] A. R. Olaposi and A. O. Adunni, "Chemical composition of three traditional vegetables in Nigeria," *Pakistan Journal of Nutrition*, vol. 9, no. 9, pp. 858–860, 2010.

[6] F. O. Orech, T. Akenga, J. Ochora, H. Friis, and J. Aagaard-Hansen, "Potential toxicity of some traditional leafy vegetables consumed in Nyang'oma Division, Western Kenya," *African Journal of Food, Agriculture, Nutrition and Development*, vol. 5, no. 1, 2005.

[7] F. I. Smith and P. Eyzaguirre, "African leafy vegetables: their role in the World Health Organization's global fruit and vegetables initiatives," *African Journal of Food, Agriculture, Nutrition and Development*, vol. 7, pp. 1–9, 2007.

[8] S. S. Sobowale, O. P. Olatidoye, O. O. Olorode, and O. K. Sokeye, "Effect of preservation methods and storage on nutritional quality and sensory properties leafy vegetables consumed in Nigeria," *Journal of Medical and Applied Biosciences*, vol. 2, pp. 46–56, 2010.

[9] H. D. Mepba, L. Eboh, and D. B. Banigo, "Effect of processing treatments on the nutritional composition and consumer acceptance of some Nigeria edible leafy vegetables," *African Journal of Biotechnology*, vol. 4, pp. 157–159, 2007.

[10] E. O. Udosen and U. M. Ukpanah, "The toxicants and phosphorus content of some Nigerian vegetables," *Plant Foods for Human Nutrition*, vol. 44, no. 3, pp. 285–289, 1993.

[11] P. Fellows, *Food Processing Technology: Principles and Practice*, Ellis Harwood Limited, West Sussex, UK, 1990.

[12] S. C. Noonan and G. P. Savage, "Oxalate content of foods and its effect on humans," *Asia Pacific Journal of Clinical Nutrition*, vol. 8, no. 1, pp. 64–74, 1999.

[13] B. Du, F.-M. Zhu, and F.-Y. Li, "Measurement and analysis of mineral components in grape wine by inductively coupled plasma-optical emission spectrometer," *Advance Journal of Food Science and Technology*, vol. 4, no. 5, pp. 277–280, 2012.

[14] AOAC, *Association of Official Analytical Chemists. Official Method of Analysis of AOAC International*, AOAC, Washington, DC, USA, 16th edition, 1997.

[15] ICN, International conference of nutrition, biochemicals catalog no. 103312 and no. 960219, Rome, Cleavaland, 1992.

[16] T. Pirman, M. Mari, and A. Orenik, "Changes in digestibility and biological value of pumpkin seed Cake protein after limiting amino acids supplementation," *Izvorni znanstveni članak*, vol. 49, no. 2, pp. 95–102, 2007.

[17] J. P. Benson and B. Cales, "Animal anatomy and physiology," in *Laboratory Text Book*, pp. 325–341, Wm.C. Brown Communication, Dubuque, Iowa, USA, 1992.

[18] H. Theml, *Atlas de Poche d'Hématologie*, Flammarion Médecine-Science, Paris, France, 2000.

[19] A. A. Bakr and R. A. Gawish, "Trials to reduce nitrate and oxalate content in some leafy vegetables. 2. Interactive effects of the manipulating of the soil nutrient supply, different blanching media and preservation methods followed by cooking process," *Journal of the Science of Food and Agriculture*, vol. 73, no. 2, pp. 169–178, 1997.

[20] A. Shahnaz, K. M. Khan, A. Munirm, and S. Muhammed, "Effect of peeling and cooking on nutrient in vegetables," *Pakistan Journal of Nutrition*, vol. 2, no. 3, pp. 189–191, 2003.

[21] G. Oboh, "Effect of blanching on the antioxidant properties of some tropical green leafy vegetables," *LWT—Food Science and Technology*, vol. 38, no. 5, pp. 513–517, 2005.

[22] N. D. Makobo, M. D. Shoko, and T. A. Mtaita, "Nutrient content of amaranth (*Amaranthus cruentus* L.) under different processing and preservation methods," *World Journal of Agricultural Sciences*, vol. 6, pp. 639–643, 2010.

[23] N. W. Tietz, A. B. Carl, and R. A. Edward, *Test Book of Clinical Chemistry*, W.B. Saunders Company, London, UK, 2nd edition, 1994.

[24] H. M. Aliyu and A. I. Morufu, "Proximate analysis of some leafy vegetables (Roselle, jute and bitter leaf)," *International Journal of Food and Agricultural Economics*, vol. 3, pp. 194–198, 2006.

[25] E. O. Ogbadoyi, A. Musa, J. A. Oladiran, M. I. S. Ezenwa, and F. H. Akanya, "Effect of processing methods on some nutrients, antinutrients and toxic substances in *Amaranthus cruentus*," *International Journal of Applied Biology and Pharmaceutical Technology*, vol. 2, pp. 487–502, 2011.

[26] H. Oumarou, R. A. Ejoh, R. Ndjouenkeu, and A. Tanya, "Nutrient content of complementary foods based on processed and fermented sorghum, groundnut, spinach, and mango," *Food and Nutrition Bulletin*, vol. 26, no. 4, pp. 385–392, 2005.

[27] M. Turan, S. Kordali, H. Zengin, A. Dursun, and Y. Sezen, "Macro and micro mineral content of some wild edible leaves consumed in eastern Anatolia," *Acta Agriculturae Scandinavica Section B: Soil and Plant Science*, vol. 53, no. 3, pp. 129–137, 2003.

[28] H. Ishida, H. Suzuno, N. Sugiyama, S. Innami, T. Tadokoro, and A. Maekawa, "Nutritive evaluation on chemical components of leaves, stalks and stems of sweet potatoes (*Ipomoea batatas poir*)," *Food Chemistry*, vol. 68, no. 3, pp. 359–367, 2000.

[29] K. H. Brown, J. M. Peerson, J. Rivera, and L. H. Allen, "Effect of supplemental zinc on the growth and serum zinc concentrations of prepubertal children: a meta-analysis of randomized controlled trials," *American Journal of Clinical Nutrition*, vol. 75, no. 6, pp. 1062–1071, 2002.

[30] J. O. Oti Wilberforce and F. I. Nwabue, "Heavy metals effect due to contamination of vegetables from Enyigba Lead Mine in Ebonyi State, Nigeria," *Environmental and Pollution*, vol. 2, no. 1, 2013.

[31] WHO/FAO, *Food Standard Programme Codex Alimentarius Commission 13th Session. Report of the thirty eight session of the codex committee on food hygiene: Houston, United States of America*, ALINORM, 2007.

[32] W. S. Jansen Van Rensburg, S. L. Venter, T. R. Netshiluvhi, E. Van Den Heever, H. J. Vorster, and J. A. De Ronde, "Role of indigenous leafy vegetables in combating hunger and malnutrition," *South African Journal of Botany*, vol. 70, no. 1, pp. 52–59, 2004.

[33] J. O. Olusanya, *Essentials of Food and Nutrition*, Apex Books, Lagos, Nigeria, 2008.

[34] R. M. Krauss, P. J. Blanche, R. S. Rawlings, H. S. Fernstrom, and P. T. Williams, "Separate effects of reduced carbohydrate intake and weight loss on atherogenic dyslipidemia," *The American Journal of Clinical Nutrition*, vol. 83, no. 5, pp. 1025–1031, 2006.

[35] J. K. Mensah, R. I. Okoli, J. O. Ohaju-Obodo, and K. Eifediyi, "Phytochemical, nutritional and medical properties of some leafy vegetables consumed by Edo people of Nigeria," *African Journal of Biotechnology*, vol. 7, no. 14, pp. 2304–2309, 2008.

[36] S. S. Asaolu, O. S. Adefemi, I. G. Oyakilome, K. E. Ajibulu, and M. F. Asaolu, "Proximate and mineral composition of nigerian leafy vegetables," *Journal of Food Research*, vol. 1, no. 3, pp. 214–218, 2012.

[37] G. M. Wardlaw and A. M. Smith, *Contemporary Nutrition*, McGraw-Hill International, New York, NY, USA, 7th edition, 2009.

[38] J. Pallavi and M. Dipika, "Effect of dehydration on the nutritive value of drumstick leaves," *Molecular Systems Biology*, vol. 1, pp. 5–9, 2010.

[39] G. N. Cameron and B. D. Eshelman, "Growth and reproduction of hispid cotton rats (*Sigmodon hispidus*) in response to naturally occurring levels of dietary protein," *Journal of Mammalogy*, vol. 77, no. 1, pp. 220–231, 1996.

[40] V. R. Young and P. L. Pellette, "Plant proteins in relation to human protein and amino acid requirements," *The American Journal of Clinical Nutrition*, vol. 59, pp. 1203–1212, 1994.

[41] R. Modgil and M. Modgil, "Effect of feeding Chayote (*Sechum edule*) and Bottle Gourd (*Lageneria siceraria*) as a source of fiber on biological utilization of diet in rats," *Journal of Human Ecology*, vol. 15, pp. 109–111, 2004.

[42] J. O. Agbede, M. Adegbenro, O. Aletor, and A. Mohammed, "Evaluation of the nutrition value of *Vernonia amygdalina* leaf protein concentrates for infant weaning foods," *Acta Alimentaria*, vol. 36, no. 3, pp. 387–393, 2007.

[43] J. H. Cummings, "The effect of dietary fiber on fecal weight and composition," in *CRC Handbook of Dietary Fiber in Human Nutrition*, G. A. Spiller, Ed., pp. 263–349, CRC Press, Boca Raton, Fla, USA, 2nd edition, 1993.

[44] EFSA Panel on Dietetic Products Nutrition and Allergies, "Scientific Opinion on the substantiation of health claims related to wheat bran fibre and increase in faecal bulk (ID 3066), reduction in intestinal transit time (ID 828, 839, 3067, 4699) and contribution to the maintenance or achievement of a normal body weight (ID 829) pursuant to Article 13(1) of Regulation," *EFSA Journal*, vol. 8, pp. 1817–1835, 2010.

[45] M. T. Flood, M. H. Auerbach, and S. A. S. Craig, "A review of the clinical toleration studies of polydextrose in food," *Food and Chemical Toxicology*, vol. 42, no. 9, pp. 1531–1542, 2004.

[46] P. O. Agbaire and O. O. Emoyan, "Nutritional and antinutritional levels of some local vegetables from Delta State, Nigeria," *African Journal of Food Science*, vol. 6, no. 1, pp. 8–11, 2012.

[47] V. A. Aletor, "Cyanide in Garri 2. An assessment of some aspects of the nutrition fed garri containing various residual cyanide levels," *International Journal of Food Sciences and Nutrition*, vol. 4, pp. 289–295, 1993.

[48] C. N. Esenwah and M. J. Ikenebomeh, "Processing effects on the nutritional and anti-nutritional contents of African locust bean (*Parkia biglobosa* Benth.) seed," *Pakistan Journal of Nutrition*, vol. 7, no. 2, pp. 214–217, 2008.

[49] D. E. Johnston, "Special considerations in interpreting liver function tests," *American Family Physician*, vol. 59, pp. 2223–2230, 1999.

[50] C. Johnson-Delaney, *Exotic Animal Companion Medicine Handbook for Veterinarians*, Zoological Education Network, Lake Worth, Fla, USA, 2005.

[51] J. A. Saliu, O. O. Elekofehinti, K. Komolafe, and G. Oboh, "Effects of some green leafy vegetables on the hematological parameters of diabetic rats," *Journal of Natural Product and Plant Resources*, vol. 2, pp. 482–485, 2012.

8

Evaluation of the Starch Quantification Methods of *Musa paradisiaca, Manihot esculenta,* and *Dioscorea trífida* using Factorial Experiments

J. J. Lafont-Mendoza ⓘ,[1] **C. A. Severiche-Sierra ⓘ,**[2,3] **and J. Jaimes-Morales**[2]

[1] *Universidad de Cordoba, Monteria, Colombia*
[2] *Universidad de Cartagena, Cartagena de Indias, Colombia*
[3] *Corporacion Universitaria Minuto de Dios (UNIMINUTO), Barranquilla, Colombia*

Correspondence should be addressed to J. J. Lafont-Mendoza; jenniferlafontmendoza@gmail.com

Academic Editor: Salam A. Ibrahim

Background. Starch and its products are used in a variety of ways for both the food and nonfood industries. A factorial experiment is carried out with two factors to explain the behavior of the percentage of starch, where the factors correspond to the extraction method and to the raw material. *Method.* Three methods were used in triplicate: the first followed the official technique of the Association of Official Analytical Chemists (AOAC), to perform acid hydrolysis and quantification of starch by Titulation; the second method involved the colorful reaction with iodine using the UV equipment to measure the absorbance and calculate the percentage of starch; as a third method the FTIR was used, through which the concentration of the starch was calculated by the area under the curve obtained from the spectrum. *Results.* there is an effect of both the method and the raw material on the percentage of starch, while there was no effect of the interaction; the Tukey test indicates that the highest average percentage of extraction occurs with the extraction method by Titulation and with the starch of *Manihot esculenta. Conclusion.* It is used as raw material. The method of quantification of starch by UV-VIS spectroscopy was the best for the study samples because it presented less deviation in relation to the FTIR and Titulation methods.

1. Introduction

The starch is a food reserve polysaccharide predominant in plants, it is the most important and abundant from the commercial point of view, and it is the most common way to include carbohydrates in our diet; the foods rich in it are a good source of energy [1]; starch has been a fundamental part of the diet of man for many years; in addition to this, it has been given a large number of industrial uses so it is considered, after cellulose, the most important polysaccharide from the commercial point of view. This carbohydrate is found in various sources such as cereals, tubers, and some fruits, and although its composition does not change the properties, if it does so, this depends on the source from which it is extracted [2–4].

The starch comes from different sources with different crystalline structures; cereal grains such as corn, wheat, or rice are sources of starch, such as roots and tubers; for example, tapioca, cassava root, and potatoes are frequently used in the preparation of gluten-free foods; the reversible transformations between the starch and glucose that intervene in the maturation and after the harvest have a remarkable influence on the quality, and the concentration of the starch varies according to the state of maturity [5, 6]. Starch is also derived from legumes such as soybeans and chickpeas; starch granules form different grains that differ in size, ranging from 2 to 150 microns, and in shape, which can be round or polygonal [7].

Starch and its products are used in a variety of ways in both the food and nonfood industries. In food, it is used as an ingredient in different preparations and in the nonfood industry as a raw material for the elaboration of a wide range of products. The consumption of starch is destined approximately 75 percent to the industrial sector and 25 percent to the food sector [8–10].

In this work three raw materials are studied for the quantification of their starch. Among these is *Musa paradisiaca*, which belongs to the Musaceae family, grows abundantly in many developing countries, and as food is considered one of the most important sources of energy for people who live in the humble regions of many countries [11]. Next, the *Manihot esculenta* is detailed, it is a perennial woody plant; its stem is cylindrical and formed by knots (point where the leaf joins the stem) and internodes (portion of the stem between two knots). It can be multiplied better in a vegetative way; therefore the stems are important because when they are mature they are cut into stakes of 7 to 30 centimeters in length with which the plant propagates [12]. Finally we have *Dioscorea trifida*; in this the leaves are heart shaped, alternate or opposite, long stalked; their stems are winged or oval cross section with small flowers in clusters or in panicles of three sepals and three stamens; although they are food species, they are characterized by poor flowering [13].

Based on these three agricultural raw materials outlined above, it is necessary to quantify their starch by the three analytical methods most used in these cases: initially the Titulation (acid hydrolysis) typical of traditional analytical chemistry; UV spectroscopic methods; IR studied by instrumental analysis. To compare the methods of quantification, the support of statistical tools is necessary; however, a descriptive study is not sufficient for this purpose, due to the inherent variations in said process; therefore, the support of an inferential tool is necessary to identify the importance of the effect that these factors have on % starch. The experimental design is a very useful tool in the comparison of processes [14], while the factorial experiments allow you to observe the effect of several factors on a variable response, as well as the interaction of them [15].

Taking into account the fact that it is very important to know the properties of starch and determine how these vary depending on the source from which it is obtained, by means of this work we quantify *Paradisiaca Musa*, *Manihot esculenta*, and *Dioscorea trifida*, by means of Titulation, UV-VIS spectroscopy, and FTIR in order to demonstrate which of these raw materials provides the highest percentage of starch and which of the three techniques is most appropriate for each one.

2. Materials and Methods

2.1. Starch Removal. The method of wet starch extraction consisted basically of grinding the pulp and removing in liquid medium those components that are relatively larger, such as fibers and proteins, using sieves; subsequently, the elimination of the water by decantation is facilitated and the sedimented material was washed to eliminate the last different fractions of the starch and finally it was subjected to drying at room temperature. For the extraction of the starch, 2000 grams of each raw material was used, which was acquired from the market of the city of Monteria (Cordoba, Colombia). The method developed consisted of the following stages [16].

Washing: It was carried out using potable water.

Peeling: Separation of the pulp shell manually with knife. In the case of *Musa paradisiaca*, after peeling, it was disinfected with a solution prepared with 1% sodium hypochlorite for 10 minutes.

Maceration: The samples were divided into slices, and water was added to then decrease size to the maximum in a blender for two minutes for the case of *Musa paradisiaca* and *Dioscorea trifida*; for *Manihot esculenta* this had a grating process.

Sieving: The product that was obtained is a mixture of starch, water, proteins, minerals, and impurities. To separate them, this slurry was passed through sieves.

Decantation: The suspension obtained was deposited in a plastic container and left for 4 hours, then the supernatant was removed to ensure that all impurities had been removed, and the wet starch was washed and sieved three more times.

Drying: The wet starch was dried in the environment for a day and then steamed to refine it.

2.2. Quantification of Starch. The quantification of the starch content of the three raw materials studied was carried out using the Titulation methods (acid hydrolysis), spectroscopic analysis (UV-VIS), and IR with Fourier transform (FTIR), which are described below.

Titulation (acid hydrolysis): The technique 920.44 of the (AOAC, 1995) [17] was used for the acid hydrolysis of the starch; for this determination 5 g of macerated sample was used; the method of acid hydrolysis was tested, by means of which 20 mL of concentrated hydrochloric acid and 200 mL of distilled water were added to the sample of starch and heated for 2.5 h in a ball fitted with a condenser to avoid evaporation. It was then cooled and neutralized with sodium hydroxide in beads, transferred to a 250 mL flask, made up to volume with distilled water, and filtered. To 25 mL of the liquid obtained were added equal volumes of alkaline tartrate, copper sulfate prepared according to method 923.09 (AOAC, 2000) [18], and distilled water, giving a final volume of 100 mL. This solution was carefully heated for 4 minutes, followed by 2 minutes of boiling; when hot, it was filtered under vacuum; after filtering the entire sample, it was washed with water at 60°C and with ethyl alcohol, to accelerate the subsequent drying process. It was dried in an oven at 100°C until constant weight; then the filter paper was weighed obtaining the content of copper oxide (I) and, based on it, the grams of glucose and subsequently the percentage of starch; the procedure was repeated by triplicate.

Spectrophotometric analysis (UV-VIS): UV-Vis spectroscopy was one of the first physical methods applied to quantitative analysis and the determination of molecular structures. The technique of UV- Vis spectroscopy is widely used in quantitative analysis, although in qualitative analysis, in the determination of structures, it is surpassed by other techniques, such as infrared spectroscopy and nuclear magnetic resonance. The absorption of UV-Vis radiation by molecules and compounds is due to the electronic transitions of certain groups that make them called chromophobes. These are unsaturated groups, with a large number of electrons responsible for the absorption of UV and Vis radiation. This absorption of energy causes electronic transitions and the usual formation of excited states. Based on the type of electronic transitions that occur, the absorbing species can be classified and, in addition, it is possible to correlate the spectral behavior of a certain species with its chemical characterics. The electronic transitions that originate from the absorption of UV-Vis radiation can be divided into electronic transitions between orbitals, charge transfer transitions, and ligand field transitions. This procedure of colorful reaction with iodine can only be used when the starch is completely dissolved (gelatinized). For this procedure, 3 mL of the starch solution was placed in a test tube and 1 mL of an I / KI solution was added; then the intensity of the blue color produced in a 600 nm spectrophotometer was measured against a reagent blank.; the amount of starch present in the sample was calculated from a standard curve prepared in the range of 10-50 mg of soluble starch / mL, treated in the same manner as the test sample. For each sample of *Manihot esculenta* starch, *Musa paradisiaca*, and *Dioscorea trífida*, the procedure was repeated in triplicate [19].

IR spectroscopic analysis with Fourier transformed (FTIR): The basis of near infrared spectroscopy consists essentially in the emission of a monochromatic beam of light on the sample, which, depending on its composition and the nature of the bonds present in its molecules, will perform a selective absorption of energy and it will reflect another determined quantity, which is quantified by some detectors present in the NIR instrument and will be used to indirectly quantify the amount of infrared energy absorbed. In this way, the spectra collected in the infrared region are represented graphically as the energy absorbed as a function of the wavelength. In this regard it should be noted that only molecules or part of molecules that vibrate with a frequency similar to that of the incident energy will absorb infrared radiation, so that their vibrational and rotational states are modified, with vibrations of light atoms taking place with strong molecular links. These characteristics correspond to the functional groups C-H, O-H, and N-H of the organic compounds that are part of both plant and animal tissues. For this reason, NIR spectroscopy is practically oriented to the determination and quantification of organic compounds that present the functional groups described above; consequently, from the spectra obtained in the infrared region, information about the chemical composition of the sample analyzed can be obtained. Therefore, the relationship of the energy absorbed with the analytical composition or known characteristics of the calibration samples allows obtaining prediction models for the automatic and instantaneous analysis of thousands of samples. In this sense it can be said that in the last ten years some analytical services based on NIRS technology have been consolidated within the agri-food industry, and many others have been created, both public and private, since NIRS technology or spectroscopy in near infrared is an analytical technique widely implemented in agri-food industries as a support tool for the quality control of raw materials and products, mainly due to its rapid analysis, low cost per sample, simplicity of handling, similar precision to the reference method, no or little preparation of the sample

for analysis, etc. For the quantification of starch we proceeded to make a calibration curve and analyzed the spectra of *Musa paradisiaca* starch, *Manihot esculenta*, and *Dioscorea trífida* samples, using an infrared spectrophotometer with transformed Fourier equipped with a KBr tablet system and total reflectance system attenuated at a temperature of 25 ± 2°C; the solid sample of starch will be introduced in KBr tablets with different concentrations of starch, and finally it will be measured in the equipment in the region of 400 to 4,000 cm-1. We calculate the problem concentration from a calibration curve prepared in the range of 20-100 mg of starch/mg of KBr. For each sample of *Manihot esculenta* starch, *Musa paradisiaca*, and *Dioscorea trífida*, the procedure was repeated in triplicate [20].

2.3. Statistical Analysis. Initially, the descriptive statistics of the percentage of starch are made, with respect to the raw material used and the extraction method; subsequently, an analysis of variance is carried out using a factorial experiment 3^2 with three repetitions, to determine the effect of the extraction method, the raw material, and the interaction between raw material and extraction. Finally, the Tukey test and the media graphs are performed to observe the behavior of the factors, a strategy of data analysis similar to that carried out by Baldiris et al. (2017) [21].

3. Results and Discussion

Next, we present the analysis of the results obtained by the procedures used to quantify the starch content of the three raw materials studied (*Musa paradisiaca, Manihot esculenta,* and *Dioscorea trífida*) which were carried out using the Titulation methods (acid hydrolysis), analysis spectroscopic (UV- VIS), and IR with Fourier transform (FTIR).

3.1. Descriptive Statistics. According to the results obtained in Table 1, we proceed to calculate descriptive statistics as average, standard deviation, and coefficient of variation (CV).

The results of Table 2 show that the Titulation method is the one that provides the highest averages of percentage of starch in the different raw materials, while the percentages of average starch extracted from the *Manihot esculenta* are those that show the highest values. With respect to the variations, low dispersions are observed in the different methods and raw materials (CV <12%); however, the greatest variations are observed in the method by Titulation and with the raw material of the *Dioscorea trífida*.

3.2. Analysis of Variance. Before performing experimental design for starch percentages, the diagnostic graphs of the assumptions of the model are made as illustrated in Figure 1.

Figures 1(a) and 1(b) correspond to the dispersion graph by raw material and by method, respectively; in them it is observed that the behavior of the variations of the groups is very similar, which corroborates the observed deviations obtained in the descriptive measures. With respect to the independence of the data, the chronological sequence graph, Figure 1(c) shows subgroups between the data, which can be

TABLE 1: Starch percentages for the different raw materials using the different methods.

Raw material	Method		
	Titulation (%)	UV (%)	IR (%)
Musa paradisiaca	17,33	15,63	16,85
	20,15	16,97	17,47
	16,69	16,27	16,7
Manihot esculenta	46,95	43,82	43,44
	50,43	43,45	43,74
	47,95	43,28	43,62
Dioscorea trífida	33,16	29,52	29,76
	28,35	29,77	30,99
	35,94	29,48	29,15

TABLE 2: Descriptive statistics for the percentage of starch with respect to the method and the raw material.

Raw material	Statistics	Method		
		Titulation (%)	UV (%)	IR (%)
Musa paradisiaca	Average	18,06	16,29	17,01
	Deviation	1,84	0,67	0,41
	CV	10,20%	4,11%	2,40%
Manihot esculenta	Average	48,44	43,52	43,6
	Deviation	1,79	0,28	0,15
	CV	3,70%	0,63%	0,35%
Dioscorea trífida	Average	32,48	29,59	29,97
	Deviation	3,84	0,16	0,94
	CV	11,82%	0,53%	3,13%

TABLE 3: Analysis of variance for the percentage of starch depending on the method and the raw material.

Source of variation	Sum of squares	DF	F-Value	P-Value
Method	54.7	2	107.142	0.0008613∗ ∗ ∗
Raw material	3546.7	2	6.950.566	< 2.2e-16∗ ∗ ∗
Method: Raw material	12.6	4	12.394	0.3295978
Residuals	45.9	18		

∗ ∗ ∗: Highly significant difference (P-Value <0.01).

TABLE 4: Tukey test among raw materials.

	Difference	lwr	upr	P-Value
Musa paradisiaca -Dioscorea trífida	-1.356.222	-1.548.394	-1.164.050	0,0000∗ ∗ ∗
Manihot esculenta- Dioscorea trífida	1.450.667	1.258.494	1.642.839	0,0000∗ ∗ ∗
Manihot esculenta- Musa paradisiaca	2.806.889	2.614.717	2.999.061	0,0000∗ ∗ ∗

∗ ∗ ∗: Highly significant difference (P-Value <0.01).

explained by a possible effect of the factors considered in the study, and finally Figure 1(d) shows the normal probability graph, where it is observed that the data conform to this behavior. The above indicates that the analysis of variance can be performed since the data meet the specifications for it, as shown in Table 3.

Table 3 shows the analysis of variance for the percentage of starch, where a factorial experiment is carried out under a completely randomized design. The P-Values of the test indicate that there is an effect both of the extraction method and of the raw material used (P-Values less than 0.01); with

respect to the interaction, it is observed that it was not significant, since the P-Value is 0.32. The Tukey test is then carried out, both for the method and for the raw material, to observe the behavior of the said differences.

The means test for the different raw materials shown in Table 4 indicates that there is a highly significant difference with respect to the percentage of starch extracted; the average chart, Figure 2(a), shows that for *Manihot esculenta* there is a higher average extraction of starch percentage.

For the methods, there is a highly significant difference between the Titulation method and the UV and IR methods

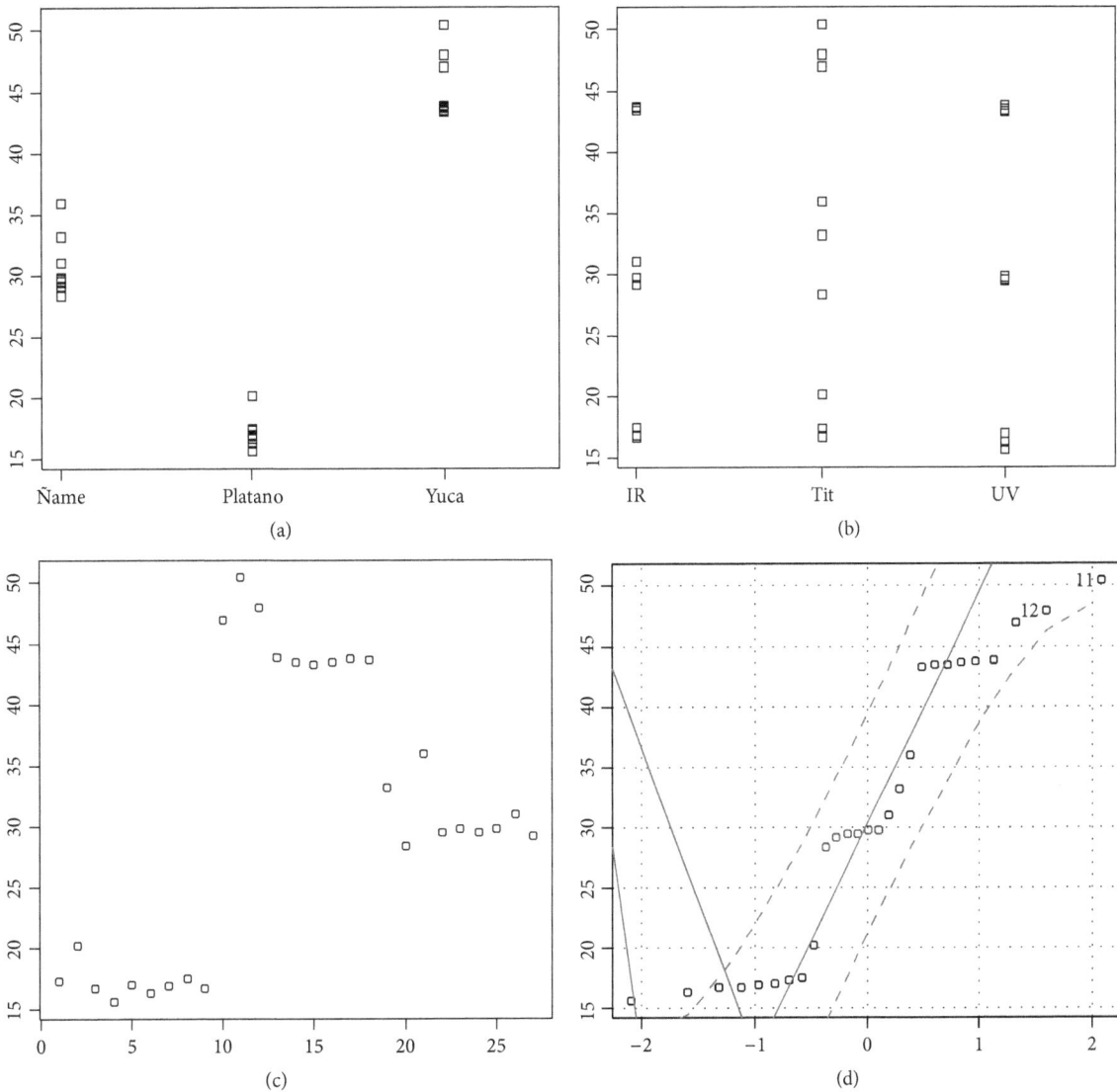

FIGURE 1: Diagnostic graphs of the assumptions of the variance analysis model.

TABLE 5: Tukey test among raw materials.

	Difference	lwr	upr	P-Value
Tit-IR	28.033.333	0.8816116	4.725.055	0.0042229∗ ∗ ∗
UV-IR	-0.3922222	-23.139.440	1.529.500	0.8621970
UV-Tit	-31.955.556	-51.172.773	-1.273.834	0.0013494∗ ∗ ∗

∗ ∗ ∗: Highly significant difference (P-Value <0.01).

(P-Values <0.01) as noted in Table 5; the said method has the highest average value with respect to the percentage of starch; see Figure 2(b).

4. Conclusions

According to the review of the literature, the results shown, and their discussion, the following conclusions can be obtained: *Manihot esculenta* presented a higher yield of starch extraction, turning it into a potential source for this product

as an alternative in the food industry with respect to the other raw materials studied which presented a lower extraction yield. The assumptions of normality, homogeneity of variance, and independence are fully met. As for the experimental design, it indicates that there is a highly significant effect of the raw material and the extraction method on the percentage of starch, but not on the interaction. Finally, the means tests show that the highest average starch percentages are obtained for the extraction method by Titulation and for the starch based on *Manihot esculenta*.

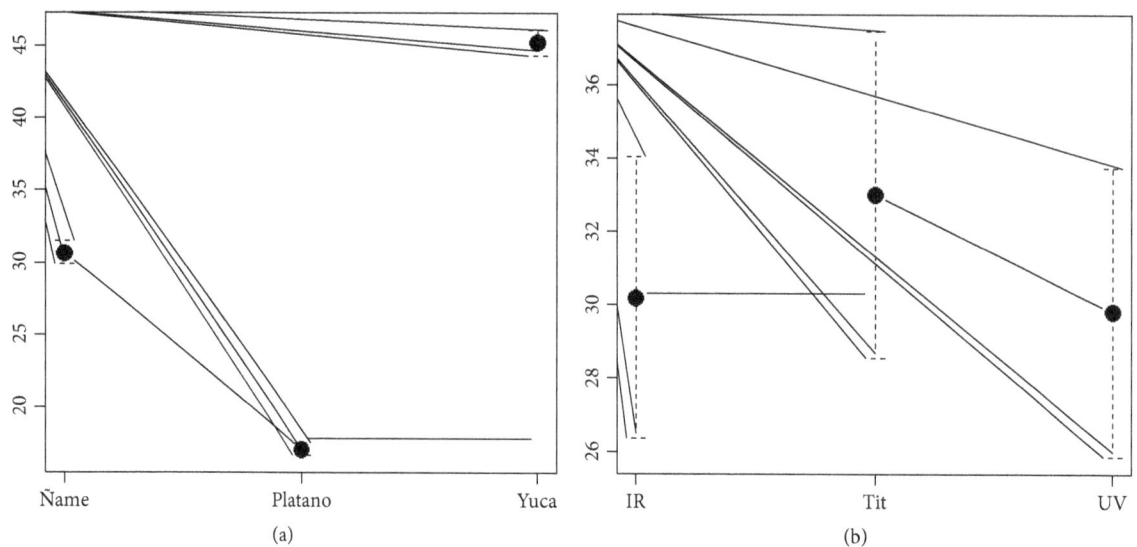

FIGURE 2: Average chart for the percentage of starch by raw material and by extraction method.

Conflicts of Interest

The authors declare that they have no conflicts of interest.

Authors' Contributions

J. J. Lafont-Mendoza and C. A. Severiche-Sierra conceived and designed the research; J. Jaimes-Morales participated in the sampling, the experimental work, the data collection, and the bibliographic search. J. J. Lafont-Mendoza, C. A. Severiche-Sierra, and J. Jaimes-Morales analyzed and interpreted the data and wrote and reviewed the final document. All authors read and approved the manuscript.

Acknowledgments

The authors are grateful for the financial support provided by the Universidad de Cordoba and the cooperation in scientific-technical and laboratory topics by the Universidad de Cartagena.

References

[1] J. J. Lafont, n. S. Páez, and A. A. Portacio, "Extracción y Caracterización Fisicoquímica del Aceite de la Semilla (Almendra) del Marañón (Anacardium occidentale L)," Información tecnológica, vol. 22, no. 1, pp. 51–58, 2011.

[2] J. Jaimes, I. Rios, and C. Severiche, "Nanotecnología y sus aplicaciones en la industria de alimentos," in Alimentos Hoy, vol. 25, pp. 51–76, 2017.

[3] M. Häkkinen, "Musa chunii Häkkinen, a new species (Musaceae) from Yunnan, China and taxonomic identity of Musa rubra," Journal of Systematics and Evolution, vol. 47, no. 1, pp. 87–91, 2009.

[4] G. Aurore, B. Parfait, and L. Fahrasmane, "Bananas, raw materials for making processed food products," Trends in Food Science & Technology, vol. 20, no. 2, pp. 78–91, 2009.

[5] S. Washaya, J. F. Mupangwa, and E. Muranda, "Nutritional Value and Utilization of Yams (Dioscorea steriscus) by Residents of Bindura Town High Density Suburbs, Zimbabwe," Advances in Agriculture, vol. 2016, Article ID 5934738, 7 pages, 2016.

[6] M. Hernández-Medina, J. G. Torruco-Uco, L. Chel-Guerrero, and D. Betancur-Ancona, "Caracterización fisicoquímica de almidones de tubérculos cultivados en Yucatán, México," Ciência e Tecnologia de Alimentos, vol. 28, no. 3, pp. 718–726, 2008.

[7] A. Kumar, P. Ramakumar, A. A. Patel, V. K. Gupta, and A. K. Singh, "Influence of drying temperature on physico-chemical and techno-functional attributes of elephant foot yam (Amorphophallus paeoniifolius) var. Gajendra," Food Bioscience, vol. 16, pp. 11–16, 2016.

[8] L. Shamla and P. Nisha, "Acrylamide formation in plantain (Musa paradisiaca) chips influenced by different ripening stages: A correlation study with respect to reducing sugars, amino acids and phenolic content," Food Chemistry, vol. 222, pp. 53–60, 2017.

[9] F. Zhu, "Structure, properties, and applications of aroid starch," Food Hydrocolloids, vol. 52, pp. 378–392, 2016.

[10] J. D. Hoyos-Leyva, L. Alonso-Gomez, J. Rueda-Enciso, H. Yee-Madeira, L. A. Bello-Perez, and J. Alvarez-Ramirez, "Morphological, physicochemical and functional characteristics of starch from Marantha ruiziana Koern," LWT- Food Science and Technology, vol. 83, pp. 150–156, 2017.

[11] M. Rivas, P. Zamudio, and L. Bello, "Efecto del grado de acetilación en las características morfológicas y fisicoquímicas del almidón de plátano," Revista mexicana de ingeniería química, vol. 8, pp. 291–297, 2009.

[12] L. Suarez and V. Mederos, "Apuntes sobre el cultivo de la yuca (Manihot esculenta Crantz). Tendencias actuales," in Cultivos Tropicales, vol. 32, pp. 27–35, 2011.

[13] E. Pacheco and N. Techeira, "Propiedades químicas y funcionales del almidón nativo y modificado de ñame (Dioscorea alata)," in *Revista Interciencia*, vol. 34, pp. 280–285, 2009.

[14] A. Kitsche and F. Schaarschmidt, "Analysis of Statistical Interactions in Factorial Experiments," *Journal of Agronomy and Crop Science*, vol. 201, no. 1, pp. 69–79, 2015.

[15] F. C. Zimmer, A. H. P. Souza, A. F. C. Silveira et al., "Application of factorial design for optimization of the synthesis of lactulose obtained from whey permeate," *Journal of the Brazilian Chemical Society*, vol. 28, no. 12, pp. 2326–2333, 2017.

[16] J. J. Lafont, A. A. Espitia, and J. R. Sodré, "Potential vegetable sources for biodiesel production: Cashew, coconut and cotton," *Materials for Renewable and Sustainable Energy*, vol. 4, no. 1, 2015.

[17] Association of Official Analytical Chemists (AOAC), *Official Methods of Analysis of the Association of Official Analytical Chemists*, AOAC, Arlington, Va, USA, 15th edition, 1995.

[18] Association of Official Analytical Chemists AOAC, "Official Method 925.09 Sucrose in fruits and fruit products read with AOAC official method 923.09," in *Lane and Eynon general volumetric method*, John Wiley & Sons Inc., Arlington, TX, USA, 2005.

[19] V. G. Uarrota, R. Moresco, B. Coelho et al., "Metabolomics combined with chemometric tools (PCA, HCA, PLS-DA and SVM) for screening cassava (*Manihot esculenta* Crantz) roots during postharvest physiological deterioration," *Food Chemistry*, vol. 161, pp. 67–78, 2014.

[20] A. Rohman, D. L. Setyaningrum, and S. Riyanto, "FTIR Spectroscopy Combined with Partial Least Square for Analysis of Red Fruit Oil in Ternary Mixture System," *International Journal of Spectroscopy*, vol. 2014, Article ID 785914, 5 pages, 2014.

[21] I. Baldiris, Y. Marrugo, C. Severiche et al., "Delayed Organoleptic Maturation of Tomato Variety Milano (Lycopersicum esculentum Mill) Using Giberelina," *International Journal of ChemTech Research*, vol. 10, pp. 1032–1037, 2017.

9

Fractal Dimension Analysis of Texture Formation of Whey Protein-Based Foods

Robi Andoyo [iD]**, Vania Dianti Lestari, Efri Mardawati** [iD]**, and Bambang Nurhadi**

Department of Food Industrial Technology, Faculty of Agro-Industrial Technology, Universitas Padjadjaran, Jl. Raya Bandung Sumedang Km. 21, Jatinangor, Sumedang 40600, Indonesia

Correspondence should be addressed to Robi Andoyo; r.andoyo@unpad.ac.id

Academic Editor: Salam A. Ibrahim

Whey protein in the form of isolate or concentrate is widely used in food industries due to its functionality to form gel under certain condition and its nutritive value. Controlling or manipulating the formation of gel aggregates is used often to evaluate food texture. Many researchers made use of fractal analysis that provides the quantitative data (i.e., fractal dimension) for fundamentally and rationally analyzing and designing whey protein-based food texture. This quantitative analysis is also done to better understand how the texture of whey protein-based food is formed. Two methods for fractal analysis were discussed in this review: image analysis (microscopy) and rheology. These methods, however, have several limitations which greatly affect the accuracy of both fractal dimension values and types of aggregation obtained. This review therefore also discussed problem encountered and ways to reduce the potential errors during fractal analysis of each method.

1. Introduction

The network system of proteins is often applied to food, one of which is whey protein. This is because whey may affect the structure or texture as well as viscosity of foods depending on the processing applied [1]. Whey protein is also a high-quality protein that provides essential amino acids and has a high bioavailability (protein efficiency ratio or PER) compared to the other sources of protein [2]. This protein functionality is influenced by the ability of whey protein to form an aggregated network that can act as gelling agent in protein-fortified food. The textural properties of gel-type foods are principally determined by the combination of structural properties of the gel matrix and filler particles. The filler particles can be classified as "active" or "inactive" based on the resulting effects of these particles in rheological properties of gel. The "inactive" filler has a low chemical affinity on the polymer matrix so it cannot strengthen the resulting gel. The "active" filler has a strong interaction with the polymer matrix and therefore can strengthen the resulting gel structure [3]. It is important to predict or manipulate the ability of whey protein in forming gel in order to produce foods with desirable texture. Gelling properties of whey protein are necessary in determining consumer acceptance

in many kinds of food products, such as processed meat, milk, bread, and cakes, and improving product appearance by preventing surface moisture in yogurt [1, 4].

The application of whey protein as both structure builder and structure breaker in food should therefore be conducted by manipulating the aggregation process of whey-based particles network to produce food products with controlled texture. This accounts for a fundamental and quantitative understanding of the aggregation process. Many studies had observed the kinetics and aggregation process of particle network dispersions both theoretically and experimentally. Whey protein can theoretically form or alter the food texture because it can undergo conformational changes and form network or gel aggregate under certain and continuous treatment. Whey protein aggregation process consists of three stages that often occur subsequently, including conformational changes, chemical reactions, and physical interactions [5], and can be done through heat or cold gelation methods. The rheological properties of protein gels produced from the process vary greatly, depending on pH, ionic strength, temperature and heating rate, and gelation methods used [6, 7].

Whey protein gels under certain length scales were proven to have self-similar structure; this structure can be described and quantified experimentally by fractal concept.

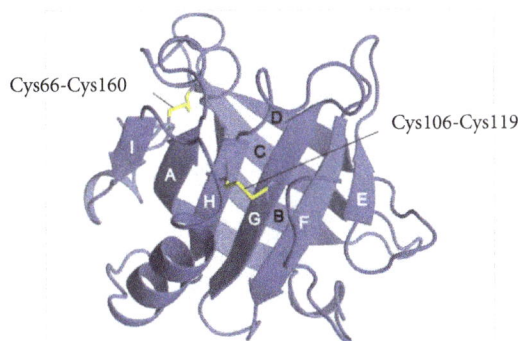

FIGURE 1: *Structure of β-Lg.* Yellow lines represent disulfide bonds (adapted from Ikeguchi [89]).

The fractal concept can be used to describe and quantify the structure of aggregated particles by using fractal dimension D or D_f—this shows the relationship between the number of particles on aggregate and their size, $N \sim R^D$. D_f values of 1.7–1.8 represent diffusion-limited cluster-cluster aggregation (DLCA) while D_f value of 2.0–2.2 represents reaction-limited cluster-cluster aggregation (RLCA)—the higher D_f value also indicates a denser aggregate structure while lower D_f value indicates a more tenuous aggregate. Observation of the fractal aggregation process can further be used to control whey protein-based aggregation so that foods with uniform and appropriate texture can be produced [8].

There are several techniques that can be used to analyze fractal structure in aggregates. These experimental methods were widely used to analyze and evaluate texture characteristics of whey protein-based foods which are carried out by forming a model of whey protein gel system that can be used as medium. The measurement of this physical quantity is related to mass distribution in space and can be done through various techniques, such as scattering, settling, microscopy (image analysis), and rheology [9, 10]. This paper will only limit the discussion on two methods: microscopy and rheology. These methods, however, have several limitations which might affect the fractal dimension values and type of aggregations obtained. The review is therefore divided into three sections; the first section discussed the structure of whey protein and its functionality; the second section is about fractal theories in relation to food structure, methods for fractal analysis, and their limitations; the third section is a compilation on fractal dimension values and the types of aggregation of whey protein aggregates gathered from several researches. This was done to finally understand how the texture of whey protein-based food is formed despite the limitations available on each method.

2. Structure and Functionality of Whey Protein

2.1. Composition and Structure. Proteins provide various functions in food and also maintain the stability of food structure. Whey proteins can interact with each other and with other types of protein to make networks associated with gels. Whey protein comprises a globular structure as shown

in Figure 1. Globular proteins are usually curled up so that the hydrophobic R regions are centered in the molecule to avoid the polar environment around them, while the hydrophilic R group is located on the surface of the molecule. This makes globular proteins (i.e., whey protein) soluble in water. β-Lg as a major component in whey protein determines the properties of the overall whey protein. Each β-Lg molecule has 5 cysteine (amino acid) residues: Cys-66, Cys-106, Cys-119, Cys-121, and Cys-160. Cys-66/Cys-160 and Cys-106/Cys-119 are connected by disulfide bonds (-SS) to form oxidized form of cysteine, that is, cystine, while Cys-121 is available as a free thiol (-SH) group. One molecule of β-Lg (whey protein), therefore, consists of one free thiol group and two disulfide bonds [11, 12]. The free thiol group is entrapped in the three-dimensional structure of *native* whey protein and exposed during denaturation. These structures are important in terms of protein aggregation [13]. Heat treatment, high pressure, mechanical stress, and oil-in-water exposure to air during food processing can cause changes in the tertiary (globular) structure of the whey protein so that the thiol and cystine groups become exposed to the solvent and become chemically reactive [12].

2.2. Whey Protein as Gelling Agent. Whey protein can enhance texture formation in food due to one of its functionalities to form gel. This is an important functional attribute to many food applications, such as meat and milk processing, and bakery, and also to improving appearance of food products such as yogurt [1, 4, 14].

Rheological properties of protein gel vary depending on the type of protein, pH, ionic strength, and rate of heating. Gel of globular protein is classified into two distinct types of morphology: particulate and fine-stranded gels. Extensive reviews on these morphologies were done by Bryant and Julian McClements [15] and Nicolai and Durand [6]. In general, particulate gels are turbid gels that can be formed by modifying the pH close to the isoelectric point or at high ionic strength (low electrostatic repulsion force), while fine-stranded gels are transparent filamentous gels that can be formed at pH far from the isoelectric point in the absence of salt or low ionic strength (high electrostatic repulsion force). Figure 2 showed the microstructure on each type of gel.

Verheul and Roefs [5] stated that there are three phenomena involved (often subsequently) in aggregation process of

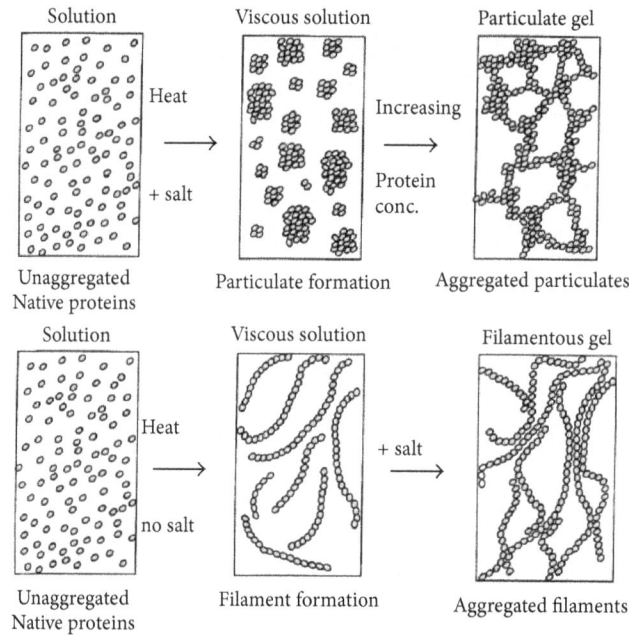

FIGURE 2: *Illustration of gel formation* (adapted from Bryant and Julian McClements [15]).

globular protein, that is, conformational changes, chemical reaction, and physical interaction. Conformational changes are related to denaturation process and consist of two steps. The first step is *unfolding* of globular protein which is then followed by the second step: aggregation [1]. Proteins in the native form initially undergo a process of structural conformation changes (denaturation) due to modification of the external environment [16]. The *unfolding* of globular whey protein causes the exposure of hydrophobic, free thiol, and disulfide groups which are initially present in the interior of globular protein, thus allowing inter- and intramolecular interactions among whey proteins [15]. This exposure leads to chemical reactions and aggregation through covalent and noncovalent bonds [5, 7, 8, 11, 15, 17–19]. The aggregation process itself is divided into two stages: primary aggregation that leads to the formation of spherical particles or flexible strands (filament) and secondary aggregation that occurs as protein and salt concentration increase. This secondary aggregation finally leads to formation of gel, precipitate, or fractal clusters [6]. A further stage of aggregate formation is physical interaction or commonly noticed as cluster-cluster aggregation. Particles in this stage diffuse and stick together randomly in certain media with certain probability and form larger clusters (gel network) [20]. The process of gel formation by whey protein itself can be done in two ways, namely, heat and cold gelation.

Heat Gelation. Heat gelation is done by heating whey protein solution sample at gelation temperature until denaturation and aggregation occur. The gel is prepared by heating the native whey protein solution (C_g = 2–200 g/L, temperature > 60°C) for 10 minutes to 24 hours at various pH (2.5–9.0) with certain salt concentration (~20 mM–1.0 M NaCl) [15, 18, 21–28] followed by rapid cooling at room temperature [1].

McClements and Keogh [29] observed that the process of gel formation in heat gelation is influenced by temperature and heating time. The gelation temperature of aggregated whey protein was 48°C while for native whey protein it was 77°C. This is because the aggregated whey protein has more exposed hydrophobic groups to the whey protein solution. Hydrophobic interactions that occur can therefore overcome the electrostatic repulsion forces at lower temperature so the gelation temperature is also lowered. On the other hand, the gel will not form on native whey protein solution until the temperature reaches a point where unfolding occurs. This effect of heating is also influenced by pH, ionic strength, and the concentration of salt, protein, sugar, and fat [1, 14]. Aggregation and gelation in heat gelation also occur simultaneously [13].

The formation of covalent bonds in the form of additional disulfide bonds from the oxidation of thiol groups or thiol-disulfide exchange reaction occurs in both gelation types. The formation of additional disulfide bonds stabilizes some weak clusters and decreases the potential for restructuring during gel formation [30]. Hoffmann and van Mil [31] examined the role of free thiol groups and disulfide bonds in the heat gelation of β-Lg; the result showed that β-Lg is dispersed at neutral pH and heating at 65°C enhanced the formation of aggregate mainly due to presence of intermolecular disulfide crosslink.

Cold Gelation. Cold gelation consists of two stages: the preparation of heat-denatured whey protein solution and gelation at low (or ambient) temperature. In contrast to Verheul and Roefs [5], research by Alting et al. [32] indicated that the initial stage in cold gelation process was physical interaction. This stage was then followed by an increase in hardness and stiffness of the gel due to a covalent reaction between structural elements of the gel. Aggregation and gelation process occur separately throughout this process. This causes the properties of the gel in cold gelation to be readily adjusted in the early stages of aggregation before the gelation [13, 15].

The aim of preparing a heat-denatured whey protein solution is to produce a solution containing filamentous type of protein aggregate that does not gel [15]. This is done by heating the native whey protein solution at low concentration (C_g = 6–10% (w/v)) [29, 32–38], at temperatures between 70 and 90°C for 5–60 minutes [29, 30, 35], at pH distant from the isoelectric point (~2 units above the pI of protein), and at low ionic strength (<50 mM NaCl or <10 mM $CaCl_2$ at pH 7 [29, 35, 36, 39–43]) to form soluble aggregates. The process of gel formation is influenced by a combination of protein concentration, pH, ionic strength, temperature, and heating time. The gel is formed on a critical gel concentration (C_g), where at a concentration higher than C_g the gel network will not flow if tilted but will flow at a lower concentration. C_g will decrease with increasing salt concentration and decrease in pH, while aggregate size will increase [30]. The next step is to induce gelation at room temperature by adding salt (salt-induced) NaCl [8, 38] or $CaCl_2$ [8, 35, 38, 42]. Higher NaCl concentrations are needed to induce gel formation from cold gelation process than $CaCl_2$ concentrations [44, 45]; this is because calcium has a specific ion bridge effect that contributes to gel formation. The protein concentration which is typically used for salt-induced cold gelation is 100 g/L, where the rate of aggregation after salt addition increases sharply with increasing salt and protein concentration. The rate of aggregation also increases with increasing temperature due to exposure of hydrophobic groups that lead to hydrophobic interactions [29, 30, 44]. Kuhn et al. [46] suggested that the use of different types of salt encourages the formation of different gel structures. Gelation with $CaCl_2$ produced irregular microstructure (particulate) resulting in stronger, more elastic, and turbid gels, whereas gelation with NaCl produced gel with a more-ordered microstructure but more brittle.

Gelation also can be performed by adjusting pH of the heat-denatured solution to the isoelectric point of protein. This is commonly called acid-induced cold gelation, one type of reagents which were used frequently is glucono-δ-lactone (GDL) [32–34, 47–50]. The addition of acid leads to a decrease in electrostatic forces on proteins so that aggregates are formed. Stronger gel is generally produced by acid gelation rather than by calcium gelation [48, 49, 51], where maximum strength is achieved at pH ~ 5 which is the isoelectric point of β-Lg [13, 47, 49]. The morphological properties of the initial gel network of acid-induced cold gelation are formed as a result of noncovalent bonding [52]. The formation of additional disulfide bonds stabilizes some of the weak clusters and decreases the potential restructuring during gel formation. The formation of disulfide bonds in acid-induced cold gelation had also been studied by Alting et al. [32, 33, 52]; the formation of disulfide bonds in this type of cold gelation was surprising since oxidation of thiol groups or thiol-disulfide exchange reaction generally occurs under alkaline conditions [15]. The results showed

that the formation of additional disulfide bonds may occur, depending on the pH used during the process, in which the disulfide bonds cannot form at pH 2.5–3.5. The rate of acid gelation also controls the level of structural rearrangements during gel formation and finally affects the properties of the gel. The acid gel texture after gel formation tends to be unstable, and hardening gradually occurs due to thiol-disulfide exchange reaction during storage at a pH higher than 3.9 [53].

There were different insights regarding the role of thiol-disulfide exchange reaction in acid-induced cold gelation. Famelart et al. [54] also stated that thiol-disulfide exchange reaction during acid-induced cold gelation also prevented the formation of large covalent structures during gelation of acid under pH 5; the result also suggested that the role of thiol-disulfide exchange reaction during acid gelation at various pH was almost insignificant. This was demonstrated by the insignificant difference of elastic modulus of milk gel formed by heating reconstituted milk with and without addition of thiol-blocking agent (N-ethyl-maleimide or NEM). This was in contrast with Vasbinder et al. [55], Lucey et al. [56], and Lucey et al. [57] that showed significant differences in elastic modulus and gel hardness made with or without the addition of thiol-blocking agent, in which the gel with the addition of thiol-blocking agent has a slightly modulated (~20%) elastic modulus and gel hardness (~30%) after gelation. This clearly showed the presence of additional disulfide bonds during and after gelation. Relatively different results were shown by Cavallieri et al. [47], in which the disulfide bond is associated with the internal stabilization of whey protein aggregates formed only during the initial heating in acid-induced cold gelation.

3. Fractal Concept and Quantification Methods

Many foods, like many other natural materials, are inherently irregular in conformation. Food has a complex geometry in which a large category of structural irregularities exists, including pores (bread, snack, and cereals), protuberances (cauliflower), and replicating structures (broccoli). Such attributes may exist over wide levels of magnification. Fractal concept was first introduced by Mandelbrot [58] to describe dimensions between regular or conventional dimensions 1, 2, and 3. Fractal dimension indicates the degree to which an image or object deviates from regularity (Figure 3). A feature of mathematically constructed fractal objects is self-similarity: the attribute of having the same appearance at all magnifications or length scales. Natural objects like food therefore can be characterized quantitatively in terms of their fractal dimension, which may serve as an index of irregularities [59]. The fractal model may describe particle aggregation in a mixture of both particles, from the point when acidification renders the particles reactive [21] to the point when clusters have grown enough to come in close contact, interpenetrate, and percolate. The fractal dimension (D_f) is used to describe the occupancy of the structure in the volume of the gel, while the scaling behavior can give insight

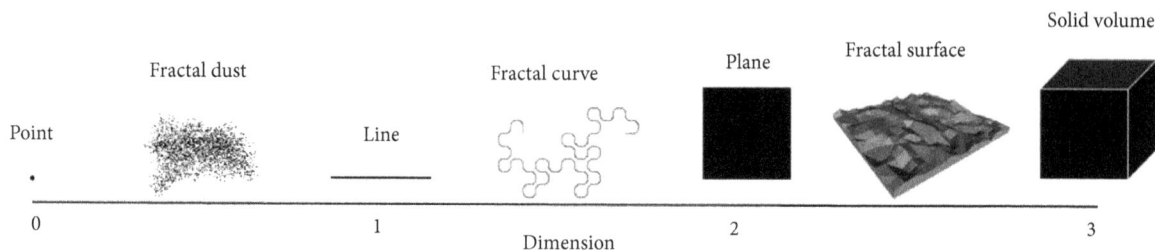

FIGURE 3: Illustration for conventional (Euclidean) and fractal dimensions.

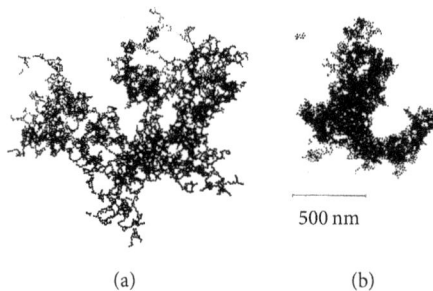

500 nm

(a)   (b)

FIGURE 4: *DLCA (a); dan RLCA (b)* (adapted from Weitz et al. [61]).

into the mechanism of assembly of the particles. Fractal analysis explains the quantitative analytical method that characterized disorder shape of particles network of colloidal system [58]. The fractal concept has led to considerable advances of our understanding of colloid aggregation process including the structure and size distribution of the aggregates as well as the kinetics [60].

3.1. Fractal Dimension in relation to Types of Aggregation.
Gel is an elastic network formed by collection of crosslinked macromolecular chains that interact with each other. Physical interaction processes (cluster-cluster aggregation) which occur in gel formation are greatly affected by one of the main characteristics of particle, namely, Brownian motion. This motion occurs since the small size of particles causes unequal collisions. As a result, the particles change direction and produce a zigzag random motion. The Brownian motion allows particles to overcome the effect of gravity so that the particles do not settle out or separate from the dispersing medium.

The growth model or ideal aggregate formation of particles network based on Brownian motion is divided into two limiting regimes: diffusion-limited (DLCA) and reaction-limited cluster-cluster aggregation (RLCA) (Figure 4) [8]. The fractal concept can be used to explain the fractal aggregation regimes of DLCA and RLCA by quantifying the structure of aggregated particles by using a parameter, namely, fractal dimension (D_f). Aggregation on DLCA regime is very rapid and the collisions among particles are limited only by Brownian motion. The aggregate formed on this regime is indicated by D_f value of 1.7–1.8. Meanwhile, aggregation in RLCA regime occurs more slowly due to the electrostatic repulsions between approaching particles.

Aggregates formed under these conditions are marked with slightly higher D_f (2.0–2.2). Weitz et al. [61] also stated that aggregate formation with D_f value above 2.0 was irreversible so that it could be best described as in RLCA regime. The D_f values represented in both regimes were also proven to have important effects on the mechanical properties of aggregated network [8, 10, 61–63].

3.2. Scaling Theory.
The macroscopic nature of gel is somehow affected by hierarchical level of various factors, which can be attributed to the properties of each individual protein molecule. The importance of this structural hierarchy cannot be ignored (Figure 5). They physicochemical properties of individual protein molecules as well as environmental conditions (pH, temperature, ionic strength, etc.) will affect the type of primary particles formed, interactions between particles, and ultimately the final three-dimensional network structure. The observed macroscopic properties of the system are influenced by various levels of structure in the hierarchy; the most influencinglevel is definitely the one closest to the macroscopic level, that is, the microstructure level. Prediction and manipulation of the macroscopic nature of the gel therefore require an understanding of the effect of properties present at the microstructure level on the macroscopic properties of gel. Not involving the microstructure level will undoubtedly lead to failure in predicting and manipulating macroscopic properties of gels [38].

There are several developing methods for determining fractal dimension values of complex material such as whey protein gels, namely, scattering, settling, image analysis, and rheology. This review will only limit its discussion to two methods: image analysis and rheology. This is also based on

Molecules (Primary, Secondary, Tertiary, Quaternary structure)

Heat, Mass, Momentum Transfer and Molecular
Thermodynamics/kinetic

Primary Colloidal Particles (sub-micron)

Heat, Mass, Momentum Transfer

Microstructural Elements (micron)

Heat, Mass, Momentum Transfer

Microstructures (sub-millimeter)

Heat, Mass, Momentum Transfer

Macroscopic (gel)

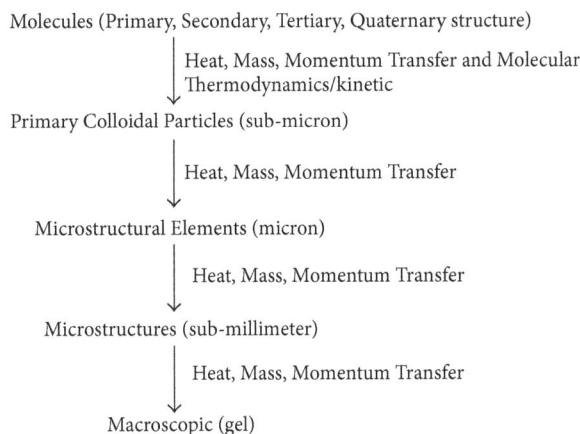

FIGURE 5: *Mechanism of protein gel formation* (adapted from Marangoni et al. [38]).

the fact that the gel aggregate is a quantifiable fractal structure from rheological and optical measurements [38].

3.3. Fractal Structure Measurement

3.3.1. Microscopy (Image Analysis). Interpretation of rheological and mechanical properties depends on the structural information derived primarily from microscopic methods, image analysis, and the use of computer simulations to test the model of food structure, in which rheological techniques cannot provide direct information about the underlying food structure at molecular level. This is because most of the important elements affecting physical and rheological properties as well as texture and sensory properties are available in sizes below 100 μm. The development of the first (optical) microscope opens a new way to visualize and describe material structure at the molecular level. Microscopic techniques have the same significance as rheology techniques and are available for analyzing food structure at different hierarchical levels. The use of microscopic methods to study food can reveal additional structural information and new insights and applications in food science and technology [9, 64, 65].

Image analysis works well for large particles with high contrast and low dimensionality. High contrast and large particles are the attributes needed to produce a good and clear image, so that the important structural information can be extracted from the image. The whey protein aggregate of 2D images is obtained from several types of microscope: transition electron microscope (TEM) [34, 38], scanning electron microscope (SEM) [46], and confocal scanning laser microscope (CSLM) [21, 32, 34, 60, 66–69]. Image taken from CSLM should be in 2D format, meaning that the Z-stack should not be applied. The images will further be processed by using image processing software (*ImageJ*, *Fiji*, etc.). Through the software, images with certain resolution are converted to grayscale format and then to binary color (black and white). Determination of threshold value (T_L) is therefore important as a first step before the image is converted to binary color. Thresholding is performed to determine whether a pixel of a given intensity (0–255; 0 = black, 255 = white) in the image is considered as an object or background. For the image taken from CSLM, the concentration of protein label, that is, Rhodamine B, should be added by considering the total protein concentration in the sample; otherwise the unbound Rhodamine B will appear as an artefact.

Several methods to determine T_L of aggregate images were proposed by Kuhn et al. (2010), Pugnaloni et al. (2005), and Thill et al. (1998) [46, 70, 71]. After conversion to binary image, the D_f value from the image is then obtained by using box counting method (BCM), in which various sizes of grids are placed on the image and the number of boxes containing the pixels of object (N) is calculated on each grid size (L). The log-log graph between N as ordinate and L as abscissa is then plotted and the slope of the graph is a fractal dimension value in two-dimensional space (D_{BCM}). The D_f value in three-dimensional space which represents the gel aggregate structure is determined as follows: $D_f = D_{\text{BCM}} + 1$ [9, 72].

Limitations on Image Analysis Technique. Andoyo et al. [34] observed the microstructure of mixture of whey protein aggregates and native micellar casein by using both CSLM and TEM. The results showed that the D_f value from CSLM image is slightly lower (2.64) than the TEM image (2.69), although both results were still in the same aggregation regime (RLCA). Hagiwara et al. [60] observed images of β-Lg and BSA gels made by heat gelation method in the absence of salt at pH 7.0 by using CSLM. The resulting aggregate image is not clear; the researchers proposed that it occurred because the size of the aggregate was too small to be observed by microscope. This also happened in the research done by Andoyo et al. [34] which used TEM to observe acid-induced cold-set gel in whey protein aggregate system—the D_f value cannot be extracted from the image since the floc was so small that the image had a low contrast and T_L cannot be determined. The advantages and disadvantages of image analysis from different types of microscope for dairy gels had been explicitly described by Ercili-Cura [73]. In general, CSLM, which was widely used to analyze the whey protein aggregate images, is capable of projecting three-dimensional structure of gel unto two-dimensional plane (image). The initial sample handling for CSLM is also easier compared to the other microscope, so it does not trigger the structural changes of the gel. CSLM can also provide a more homogeneous luminous flux on each observed image so that the intensity of the background and object colors can be clearly distinguished. On the other hand, SEM and TEM produced a higher resolution image compared to CSLM, but initial sample preparation is so rough and complicated that it might trigger structural and morphological changes of gel aggregates.

In addition, determination of T_L also affects the D_f value obtained from an image. Ako et al. [66] applied different methods of image analysis. Various gray level intensities as T_L in β-Lg gel image, higher T_L, resulted in lower D_f value and vice versa. In higher T_L, more pixels of the image are considered as part of the background. The effect of changing thresholding led to a change in the contrast and thus change of the D_f value. The D_f value was also determined by the rhodamine concentration; it should be sufficient to give a proper signal. However, too much rhodamine added could change the structure. Furthermore, it was not satisfied with the self-similar concept. The exact T_L value therefore greatly determines the accuracy of D_f value and aggregation regime obtained. In addition to taking a long time and inconsistency, different results are likely to be obtained on different time or by different operators. Manual thresholding errors cause more problems during analysis compared to the other causes [74]. This was also supported by Andoyo et al. [34] who used three different manual thresholding methods [46, 70, 71] on mixture of whey protein aggregates and native micellar casein image, in which the result showed that different proposed manual thresholding methods resulted in relatively distinct values of T_L and D_f. Russ [74] suggested that the use of automated thresholding methods is therefore more advisable. These automated thresholding methods are readily available in the image processing software based on several algorithms proposed by many image processing researchers [75, 76]. These algorithms are based on the information or knowledge about the subject and images and how the images are acquired. However, the different types of automated thresholding algorithms that were previously proposed by many researchers also provide different binary images. Further objective and quantitative evaluation is needed to determine the exact algorithm for each type of image. Sezgin and Sankur [77] performed a quantitative evaluation based on the average value of 5 criteria against 40 algorithms for automated thresholding. These algorithms were applied to document images and also to the nondestructive testing images (NDT), in which there were 6 images produced from light microscope with bimodal histogram distribution [78]. The result showed that the minimum error method algorithm proposed by Kittler and Illingworth [75] produced the most uniform and accurate binary image among other methods. This result was also supported by previous evaluation done by Glasbey [79]. To the authors' knowledge, this automated thresholding algorithm has been applied to images of whey protein gel but it should be optimized; furthermore the use of this algorithm for whey protein aggregate image with bimodal and multimodal [80] histogram distributions is quite promising for certain type of gels. For images with unimodal histogram distribution, statistical analysis of geometric parameters can

be done prior to image processing as stated in Chu et al. [81] or Silva et al. [82]. Thresholding of the images can be optimized by using automated methods for certain types of gel. For a certain type of gel, the calculation of D_f is sensitive to the threshold value. This is the reason why an optimized method was used. Variations were also observed when manual thresholding of images with a wide range of threshold values was applied. Nevertheless, changing the threshold level or using different thresholding method does not really change the message.

3.3.2. Rheology. Determination of D_f value using rheological properties of gel requires a model that covers the relationship between the structure of the gel with its rheological properties; the most appropriate model for gel aggregate structures with self-similar pattern is based on scaling theory. The early development of scaling theory to explain the elasticity of gel was proposed by Buscall et al. [83] who proposed that aggregates network is fractal on a scale greater than their primary particle size and formulated the power-law relationship of elastic modulus (E) to solid volume fractions. The value of E is equal to particle concentration of φ^A and the value of strain at limit of linearity (γ_0) is equal to particle concentration of φ^B which are then linearized to

$$\log E = A \log \varphi$$
$$\log \gamma_0 = B \log \varphi, \tag{1}$$

where A and B are the slopes of log-log plot between rheological parameters and particle concentration. Exponents A and B will vary with different gel systems [32]. The calculation of fractal dimensions by means of rheological method is generally applicable to various rheological parameters of a material, including strain at limit of linearity (γ_0) [20, 32, 34, 38, 60, 68]; storage modulus (G') [8, 21, 32, 34, 84]; elasticity (E) [38, 60, 68]; and shear stress (σ) [8]. These various rheological properties of gel aggregate can be determined by using rheometer or texture analyzer. The power-law theory was further verified by many researchers, so three models of scaling theory were presented: Bremer [85], Shih et al. [63], Wu and Morbidelli [10]—these three scaling models are basically used to find the relationship between rheological properties of a gel and its network structure. Bremer [85] classified the particles of gel network into two types: straight strands and curved strands (Table 1). Shih et al. [63] classified the gel network into strong-link and weak-link regime (strong-link: the extent of viscoelastic linear region decreased as increasing protein concentration and vice versa (Table 1)), based on the strength of inter- and intrafloc interactions, while Wu and Morbidelli [10] added another regime, that is, transition regime to describe the intermediate situation where both inter- and intrafloc interactions give similar effect to the gel elasticity. Furthermore, Narine and Marangoni [86] proposed the mechanical models for colloidal aggregates that relate the fractal dimension of a colloidal aggregate to mechanical properties, as follows:

$$G' \sim \frac{mA}{6c\pi\sigma\xi d_0^3} \Phi^{1/(d-D)}, \tag{2}$$

TABLE 1: Scaling models for determining fractal dimension.

A	B	References	Gel classification
$2/(3 - D_f)$		Bremer [85]	(Straight strands)
$3/(3 - D_f)$	$1/(D_f - 3)$	Bremer [85]	(Curved strands)
$(3 + x)/(3 - D_f)$	$-1(1 + x)/(3 - D_f)$	Shih et al. [63]	(Strong-link regime)
$1/(3 - D_f)$	$1/(D_f - 3)$	Shih et al. [63]	(Weak-link regime)
$\beta/(3 - D_f)$	$(3 - \beta - 1)/(3 - D_f)$	Wu & Morbidelli [10]	$\beta = 1 + (2 + x)(1 - \alpha)$ $\alpha = 0 \to$ strong-link regime $\alpha = 1 \to$ weak-link regime $0 < \alpha < 1 \to$ transition regime

(Adapted from Alting et al. [32] and Wu and Morbidelli [10]).

where G' is elastic modulus, m is spring constant, c is proportionality constant, A is Hamaker's constant, ξ is diameter of the microstructure, σ is the diameter of a microstructural element assumed to be spherical, Φ is the volume fraction, d is the Euclidean dimension of the network, usually 3, and D is the fractal dimension of the network. They found that the model successfully identifies key parameters that are important in determining the fractal dimension value. The model relates the values of Hamaker's constants and size of microstructural elements with the composition of the particles network. Furthermore, this model follows the weak-link theory in which the rheological parameter of the network is dependent on the nature of the link between microstructure as opposed to the strength of the microstructures themselves and only valid for the relatively high percentages of solid content (60–100%).

Mellema et al. [87] proposed five types of gel structures, namely, random, curved, hinged, straight, and rigid, which could be derived using the framework of Kantor and Webman. This model is generalized by introducing a scaling parameter, ξ as percolation length, δ as a measure of the number of deformable links in a strand, and a parameter ε as a measure of the bendability of the link. The model is shown as follows:

$$G' \propto \xi^{-(1+2\varepsilon+\delta)}$$

$$\gamma_0 \propto \xi^{2\varepsilon+\delta-1} \qquad (3)$$

$$\sigma \propto \xi^{-2}.$$

This model can at least cover a wide range of gel types by using three rheological parameters; the storage modulus, the yield stress, and the maximum linear strain as a function of the volume fraction. Table 1 summarizes these three scaling models.

Limitations on Rheological Technique. Ikeda et al. [20] used the scaling model of Shih et al. [63] that showed that the value of D_f for heat-induced WPI gel produced at 25 mM, 100 mM, and 500 mM NaCl decreased, except at 100 mM NaCl. This suggested that WPI gel at given NaCl concentration (100 mM NaCl) cannot be predicted appropriately by the fractal model. The study also showed that limit of linearity cannot be used to determine the value of D_f since it produced unreasonable value (D_f between 0.2 and 0.7).

The limit of linearity parameter in this study was only used to determine the regime classification of the resulting gel in this study. Research done by Alting et al. [32] on acid-induced cold-set WPI gel using scaling model of Shih et al. [63] and Bremer [85] also yielded ambiguous D_f values. This ambiguity also occurred in heat-induced β-Lg gel at various NaCl concentrations in the research done by de Kruif et al. [21]. The use of different scaling models can also result in different D_f values. This significantly affects the interpretation of the resulting aggregation regime. Although rheological technique is relatively easy to apply and can be performed at higher range of particle concentration, this method has limitation in terms of appropriate scaling model and rheological parameters. The use of scaling models and rheological parameters under different gel conditions determines the accuracy and consistency of the resulting D_f values.

Another limitation includes the following: rheological data from various instruments such as rheometer or texture analyzer cannot directly generate strain values at limit of linearity; this leads to the need for further processing of data manually to determine the value. Hagiwara et al. [60, 67, 68] defined limit of linearity as the value of strain in which there was a deviation of 5% (see (2)) between the ordinate value (σ) of stress-strain curve with γ (strain) $\times E$ (elasticity). The elasticity value itself was the slope value of linear part of stress-strain curve at $\gamma < 0.01$. Andoyo et al. [34] also defined limit of linearity the same way but formulated the calculation for deviation (see (5)) in a relatively different way, in which there was a deviation of 5% between slope at the end of linear part of stress-strain curve ($R^2 \sim 0.99$) ($slope_o$) with slope at certain point ($slope_n$) after the end of linear point of stress-strain curve.

$$\text{Deviation}_H = \frac{\sigma - (\gamma \times E)}{(\gamma \times E)} \times 100\% \qquad (4)$$

$$\text{Deviation}_A = \frac{slope_o - slope_n}{slope_o} \times 100\% \qquad (5)$$

Limit of linearity

$$\qquad (6)$$

= strain (γ) value at 5% deviation.

Each method used to determine the value of deviation and limit of linearity is relatively subjective; therefore each produces different value of limit of linearity. The use of 5% deviation value is also determined subjectively by the researchers so that different percentages of deviation will also result in different limit of linearity values. This was supported by Andoyo et al. [34], where the use of 5% and 10% deviations in determining the strain at limit of linearity yielded relatively different values. Hagiwara et al. [67] also indicated that the slope value of $\log \gamma_0$ and $\log \varphi$ could not be used to find D_f values of the observed gel system; this might due to the nongenerality of method used to determine the value of limit of linearity or the 5% deviation value used was less accurate to determine the value. In some cases, the fractal assembly of whey protein gels only occurred in the suspensions at the early stages of the acid gelation/aggregation, until the flocs started to come in contact, interpenetrate, and eventually percolate into a gel [34, 88]. Therefore, the range of length scales where acid gelation self-similarity investigated was ranging from ~0.1 to ~10 μm. Fractality may also exist in whey protein-containing samples below the 0.1 μm length scale and it is possible that the flocculation mechanism somewhat differed among different samples tested [88]. In another case, acid cold gelation probably starts off as a fractal process but is rapidly taken over by another mechanism at larger length scales (>100 nm) [32].

4. D_F Value of Whey Protein Aggregates

4.1. D_f from Microscopy. This section will explain all the D_f value compilations from various researchers by using microscopy method. de Kruif et al. [21] examined the gel from the same solution but with addition of 0.1 M and 0.5 M NaCl heated at 68.5°C. Hagiwara et al. [60, 67, 68] analyzed gels made from BSA solution with the addition of 0.1 M NaCl (pH 5.1), 30 mM CaCl$_2$ (pH 7.0), and 5 mM CaCl$_2$ (pH 7.0) which were heated at 50°C for 60 minutes followed by 95°C for 10 minutes. In addition, observations were also made on the gel made from the β-Lg solution with the addition of 1.0 M NaCl at pH 7.0 and 30 mM CaCl$_2$ at pH 7.0 which were heated at 40°C for 60 minutes and then 95°C for 10 minutes. All gels formed from those various researches were then observed by microscopy method. The D_f values generated from this method for heat gelation of whey protein and its components are in the range of 2.2–2.81. The addition of salt and adjustment of pH of the protein solution prior to heating process to regulate the ionic strength of the solution also affected the value of D_f, where D_f decreased with increasing ionic strength. Higher D_f values may occur due to restructuring and micro-phase separation process occurred in the protein gel during aging (storage) [6, 32, 68]. The image from CSLM also showed that the gel structure tended to be more homogeneous at NaCl concentrations below 0.2 M and more heterogeneous at higher ionic strengths. Gel formed from whey protein solution with salt concentrations higher than 0.2 M had a more heat-sensitive structure, whereas at lower ionic strength the heating temperature only affected the kinetics of gelation [21, 66]. The D_f value derived from the microscopy method for whey protein gel made from the heat gelation method lay between 2.2 and 2.7 in various types of conditions so that it could be ideal for reaction controlled gel

formation. However, such high range of D_f values could lead to a different structure formation mechanism as it is not fully complying with well-defined RLCA model.

Marangoni et al. [38] studied salt-induced cold-set WPI gels with 10% (w/v) protein concentration induced by CaCl$_2$ (10, 30, and 120 mM) and 9, 10, 11, and 12% (w/v) protein concentration induced by 300 mM NaCl. The heating was carried out at temperature of 80°C for 30 minutes before gelation at room temperature. Kuhn et al. [46] also studied gel with the same gelation process in 10% (w/w) WPI concentration which was heated at 90°C for 30 minutes and then diluted to 5, 6, 7, 8, and 9% (w/w) WPI concentrations. These solutions were then induced by 150 mM NaCl or 150 mM CaCl$_2$ to form gel. The D_f values for these salt-induced cold-set whey protein gels were in the range of 2.45 and 2.81 for NaCl-induced and 2.63 and 2.82 for CaCl$_2$-induced gels. The difference in values resulting from both studies can be due to differences in rate of aggregation, time required to achieve equilibrium conditions, or heating temperatures [46]. Marangoni et al. [38] also stated that microscopy method yielded D_f values similar to rheological technique at NaCl and CaCl$_2$ concentrations above 30 mM. The D_f values obtained for whey protein gel prepared by the salt-induced cold gelation method from various studies under various conditions were between 2.25 and 2.82.

The images from SEM showed that the cold-set gel induced by CaCl$_2$ had a thinner, more compact microstructure, with a less porous network. As WPI concentration increased, the number of pores decreased and the gel network formed had a denser structure; this could be associated with an increase in water-holding capacity (WHC). Cold-set WPI gels induced by NaCl represented a more porous network but had a higher WHC than gels induced by CaCl$_2$; this might be because the WHC depended not only on the porosity of the gel, but also was influenced by the polymer characteristics (its availability in water binding) which was highly dependent on the types of salt added. NaCl-induced gels were transparent (fine-stranded) while CaCl$_2$-induced gels were turbid (particulate). Fine-stranded gels tended to have higher WHC since they contained smaller and more homogeneous pore sizes that can bind water more strongly [15, 46, 47]. The increased salt concentration also caused the gel to become more turbid; this suggested an increase in the size of the particle diameter as the salt concentration increased at constant D_f value. Increased protein concentrations led to more transparent gels; this indicated a decrease in particle size as the concentration of protein increased at constant D_f value [38, 50].

Alting et al. [32] compared gels formed from 9% (w/w) WPI heated at 68.5°C and induced by GDL with and without addition of thiol-blocking agent NEM. Andoyo et al. [34] observed D_f values in WPA (whey protein aggregates) made from 90 g protein/kg WPI solution heated at 68.5°C, pH 7.5, for 2 hours and then standardized to 70 g protein/kg. D_f observations were also performed on mixture of WPA and NMC (native micellar casein) with a ratio of 20 : 80. Both types of solution were then set to achieve certain concentrations (14–62 g protein/kg for WPA and 15–90 g protein/kg for WPA and NMC mixtures). The gel formation

TABLE 2: Fractal dimension (D_f) values of different types of gels, measured by using microscopic method.

Reference	Gel type	D_f
de Kruif et al. (1995)	*Salt-induced cold gelation* β-Lg [NaCl] = 0.1 or 0.5 M	2.2
Hagiwara et al. (1997a)	*Heat-induced* β-Lg, pH = 7.0, [CaCl$_2$] = 30 mM	2.7
Hagiwara et al. (1997b)	*Heat induced* β-Lg, pH = 7.0, [NaCl] = 1.0 M	2.68
Hagiwara et al. (1998)	*Heat-induced* BSA, I = 50 mM buffer pH 5.1, 0.1 M NaCl, II = 50 mM buffer pH 7.0, 30 mM CaCl$_2$, III = 50 mM buffer pH 7.0, 5 mM CaCl$_2$	I = 2.81 II = 2.81 III = 2.68
Marangoni et al. (2000)	*Heat-induced* WPI (5% w/v); *salt-induced cold gelation*	~2.45 for 300 mM NaCl at different protein concentration
Alting et al. (2003)	*Acid-induced cold gelation* WPI 9%	2.3
Kuhn et al. (2010)	*Heat-induced* WPI + 150 mM NaCl or CaCl$_2$	CaCl$_2$ = 2.82 NaCl = 2.81
Torres et al. (2012)	Yogurt with substituted microparticulate whey protein (MWP)	1.4–2.6
Andoyo et al. (2015)	*Acid-induced cold gelation* whey protein aggregates (WPA) & WPA + native micellar casein (NMC), pH 4.5	CSLM: WPA = 2.15 WPA + NMC = 2.64 TEM: WPA + NMC = 2.69
Eissa, Khan (2005)	Whey protein solution: 3% and 7.5% (heat with/without transglutaminase, TG), *low pH cold-set whey protein gel*, *final pH 4.0*	CSLM: 1.96–1.98 Rheology: 2.05–2.09

was then induced by using GDL. The D_f values generated from microscopy method for acid-induced cold-set gels were in the range of 2.15–2.69. The results were consistent with other studies that suggested that D_f values for gels induced by using GDL or microorganisms were in the range of 2.3–2.4 [85]. Eissa and Khan [90] compared D_f values obtained from gels made from transglutaminase-modified WPI and WPI without modification. WPI solution at various concentrations and pH 7.0 was initially heated at 80°C for 1 hour and transglutaminase enzyme was added at 50°C while cooling, stirred for 20 minutes, and incubated at the same temperature for 10 hours. The gel formation was then induced by using GDL at room temperature until pH increased to 4.0. The resulting gel based on the CSLM image showed that both types of gel had fine-stranded morphology with D_f values of 1.96–1.98.

The microscopy method has even been used not only for whey protein-based gel models, but also for comparing the structure of protein-based food products. Torres et al. [69] substituted the use of fats in yogurt by using several types of microparticulate whey protein (MWP) with different nutrient compositions (including different amount of native and denatured whey protein). The results showed that the D_f values were in the range of 1.4–2.6, in which lower amount of native whey protein would form less interconnected network with low self-similarity. MWP with higher amount of native whey protein produced yogurt with characteristics similar to that of high-fat yogurt. This suggested that the denatured whey protein might act as a structure breaker because of its inability to form a cohesive network. The appropriate MWP types therefore could substitute the use of fats in yogurt production. Detailed data presented above was collected in Table 2.

4.2. D_f from Rheology. This section will discuss all the D_f value compilations from various researchers by using rheology method. Stading et al. [84] conducted a study of β-Lg gels made by heat gelation method. The β-Lg solution at pH 5.3 or 7.5 was heated at a rate of 0.1°C/minute or 5°C/minute from 30 to 90°C and was held constant for an hour. Microstructures formed on rapid-heated gels were more open and inhomogeneous and had a higher fracture stress; meanwhile the slow-heated gels had more homogeneous and compact microstructure. This was also supported by D_f value generated from the same study by using scaling model of Bremer [85] with rheological parameter G', where gels formed at pH 5.3 had a particulate morphology with D_f value of 2.46 for 0.1°C/minute heating rate and 2.46 for 5°C/minute. Gels at pH 7.5 were fine-stranded with D_f value of 2.94 for 0.1°C/minute and 2.91 for 5°C/minute. Hagiwara et al. [60, 67, 68] also examined the β-Lg and BSA gels made by heat gelation by using scaling model of Shih et al. [63] and elasticity as rheological parameter to determine D_f. BSA and β-Lg solutions with various protein concentrations were initially added with 50 mM buffer, pH 7.0, or 50 mM acetate buffer, pH 5.1, and then varied with or without salt addition (0.1 M NaCl or 5 and 30 mM CaCl$_2$). The solutions were then heated at 40–50°C for 60 minutes, followed by a temperature of 95°C for 10 minutes, and cooled to 25°C and stored for 24 hours. The addition of NaCl and CaCl$_2$ salts to the protein solution prior to heating caused the gel to be classified in the weak-link regime with D_f values between 2.61 and 2.82, whereas the gel formed without the addition of salt at pH 7.0 fell into the classification of strong-link regime with D_f values between 2.00 and 2.20. Hagiwara et al. [67] also more specifically suggested that the

D_f values resulting from BSA gel aggregates were greater than D_f values of aggregates in the aqueous solution (results gathered from scattering method) [91]. This might be due to the interpenetration effect between aggregates at high protein concentration to form a compact gel (higher D_f value). de Kruif et al. [21] examined the gel from WPI at various concentrations added with 0.1 and 0.5 M NaCl prior to heating (heat gelation). Heating was done at a temperature of 68.5°C for 20 hours. The log-log graph between the concentrations of WPI and G' showed that the value of slope decreased with increasing NaCl concentration; therefore, the D_f value also decreased as NaCl concentration increased. This was in contrast to the observations done using permeability method, where the D_f value increased along with the increase of NaCl concentration. These statements differed from those of Ikeda et al. [20] and Vreeker et al. [8] who stated that the increase of NaCl concentration added in heat gelation method caused the decrease in D_f value. Despite the ambiguity, authors tried to extract the D_f value from the study. We observed that the gel belonged to the weak-link regime based on Shih et al. [63] with D_f 2.81 for the gel with addition of 0.1 M NaCl and 2.78 for 0.5 M NaCl. The result was in line with microscopy method (D_f = 2.20). Ikeda et al. [20] also examined WPI gel made by heat gelation method. The WPI solution was initially stirred for 1 hour in 25–1000 mM NaCl, adjusted to pH 7.0, diluted to the desired protein concentration, then heated from 25 to 90°C at a rate of 2.5°C/minute, and held for 1 hour. Gel produced based on Shih et al. [63] and strain as rheological parameter belonged to the strong-link regime, while, based on stress and storage modulus, the gel under this study was included in the weak-link regime. The D_f values determined in this study were based on the strong-link regime with D_f 2.2 (25 mM NaCl), 1.5 (100 mM NaCl), and 1.8 (500 mM NaCl). The difference of regime classification on some rheological parameters was then clarified by Wu and Morbidelli [10] who proposed that the gels were on the transition regime with D_f values of 2.56 (25 mM NaCl), 2.22 (100 mM NaCl), and 2.20 (500 mM NaCl). The results made more sense because the increase of NaCl concentration should accelerate the aggregation process, causing a decrease in D_f value [8]. This was also supported by research done by Foegeding et al. [92] which showed that microstructure image of gel at 100 mM NaCl comprised a mixture of gel with a fine-stranded and particulate morphologies, therefore indicating that, at 100 mM NaCl, a transition of microstructural changes of gel occurred. Verheul and Roefs [93] also showed that WPI gel formed with NaCl concentration of 0.4–3.0 M resulted in D_f value of 2.20 by using permeability method; nevertheless the fractal concept cannot simply be applied to WPI gels. The type of aggregation on heat gelation by using rheology method in general followed the reaction controlled aggregation with some limitations as already explained above. The resulting aggregation type was similar to microscopy method and even other methods.

Marangoni et al. [38] and Kuhn et al. [46] also conducted a comparison of D_f values resulting from the microscopy and rheology method in salt-induced cold gelation system. The resulting D_f values were 2.45 and 2.62 for the NaCl-induced gels and 2.63 and 2.66 for the CaCl$_2$-induced gels. It was also observable that the gel produced by salt-induced cold gelation of whey protein followed the weak-link regime, although both researchers used different scaling models and rheological parameters to determine the regime and D_f values of gels. The D_f values generated from the salt-induced cold gelation method lied between 2.62 and 2.66; the results were similar to those observed using the microscopy method.

Vreeker et al. [8] examined acid-induced cold-set gel prepared from WPI solution with 1 and 10% protein (w/w) at pH 6.7 which was then heated at 70 and 90°C for 60 minutes. The solution was then cooled to 20°C and gel formation was induced by using 0.1 M HCl to a pH of 5.4. The results showed that the gel belonged to the weak-link regime with D_f value 2.0–2.5. Alting et al. [32] also examined the WPI gel by same gelation method, but using a WPI concentration of 9% (w/w) which was heated at 68.5°C and diluted to 0.5–9% (w/w). The solution was then stored at 40°C and GDL was added to a pH of 5. D_f values from various methods and scaling models were not obtained; this might be because gel made by acid-induced cold gelation process at a certain level (>100 nm) did not generate fractal structure, although the aggregate image generated from CSLM could give a D_f value of 2.2 on the same gel system. Andoyo et al. [34] observed the D_f value by using rheology method on the same acid-induced cold gelation system with observation by microscopy method. The scaling model used was that of Shih et al. [63] with storage modulus and limit of linearity as rheological parameters. Both rheological parameters produced similar D_f values, although for WPA system, D_f values from rheology method (1.15–1.7) were slightly different compared to the microscopy method (2.15). WPA gel belonged to strong-link regime and gel made from mixture of WPA and NMC belonged to weak-link regime with D_f value 2.29–2.6. The values were not of much difference with microscopy method (D_f 2.64–2.69). Eissa and Khan [90] also examined acid-induced cold-set gels made from WPI with or without enzyme modification. The results showed that the D_f value obtained from rheology (2.05–2.09 for strong-link regime) did not vary much with the microscopy method (1.96–1.98). Based on log-log graph from limit of linearity, the resulting gel belonged to strong-link regime, but based on log-log graph of elastic modulus, the resulting gel belonged to weak-link regime. Authors attempted to obtain D_f values of gels based on weak-link regime; results obtained were relatively larger (D_f 2.77–2.78). The observations also showed that microstructure and D_f values for both gels were similar, but the fracture stress and strain values were relatively different; this might be because several factors that affect the fracture properties of gel were not observable at microstructure level. The rate of aggregation resulting from acid-induced cold gelation method by using rheology method in general was varied among different researchers, ranging from 1.15 to 2.85. However, we could conclude that the gelation process was not ideal RLCA. The pretreatment of whey proteins can be done to modify the aggregation process so as to alter the D_f value, thus changing the type of aggregation [34, 94]. Detailed data presented above was collected in Table 3.

The values of D_f obtained from rheology method from most of the studies were not of much difference with the

TABLE 3: Fractal dimension (D_f) values of different types of gels, measured by using rheological method.

Reference	Gel type	D_f
Hagiwara et al. (1997a)	*Heat induced:* BSA I = pH 7.0 II = 50 mM buffer pH 7.0, 30 mM CaCl₂ III = 50 mM buffer pH 7.0, 5 mM CaCl₂ β-Lg IV = pH 7.0 V = pH 7.0 + 30 mM CaCl₂	I = strong-link, 2.00–2.07 II = weak-link, 2.82 III = weak-link, 2.61 IV = strong-link, 2.14–2.20 V = weak-link, 2.69
Hagiwara et al. (1997b)	*Heat induced β-Lg* pH = 7.0 [NaCl] = 1.0 M	Weak-link, 2.62
Hagiwara et al. (1998)	*Heat-induced BSA* I = 50 mM buffer pH 5.1, 0.1 M NaCl II = 50 mM buffer pH 7.0, 30 mM CaCl₂ III = 50 mM buffer pH 7.0, 5 mM CaCl₂	I = weak-link, 2.82 II = weak-link, 2.82 III = weak-link, 2.61
Vreeker et al. (1992)	*Acid-induced cold gelation WPI* (1 or 10% w/w) + 0.1 M NaCl or CaCl₂ at pH 5.4	G' Shih et al. (1990) = 2.0 Bremer (1992) = 2.3 σ: Bremer (1992) = 2.4–2.5
Ikeda et al. (1999)	*Heat-induced WPI* pH 7.0 + 25–1000 mM NaCl	25 mM NaCl = 2.2 100 mM NaCl = 1.5 (*incorrectly predicted*) 500 mM NaCl = 1.8 50, 80, 500, and 1000 mM NaCl = DLCA
Alting et al. (2003)	*Acid-induced cold gelation WPI 9%*	D_f 2.3, acid-induced cold gelation probably starts off as a fractal process but is rapidly taken over by another mechanism at larger length scales (>100 nm)
de Kruif et al. (1995)	*Salt-induced cold gelation β-Lg* [NaCl] = 0.1 or 0.5 M	After passing the gelation threshold, gel with more or less fractal-like structure was formed and it coarsens with increasing salt concentration. Nevertheless, gels properties could not completely be described using the scaling laws as explained by many authors
Marangoni et al. (2000)	*Heat-induced WPI (5% w/v)* *Salt-induced cold gelation*	CaCl₂ = weak-link, 2.63 NaCl = weak-link, 2.45
Stading et al. (1993)	*Heat-induced β-Lg at pH 5.3 with different heating rate*	Microstructure of gels showed straight-strand type pH 5.3 = 2.46–2.47 pH 5.7 = 2.91–2.94

TABLE 3: Continued.

Reference	Gel type	D_f
Kuhn et al. (2010)	*Heat-induced* WPI + 150 mM NaCl or CaCl$_2$	CaCl$_2$ = weak-link, 2.66 NaCl = weak-link, 2.62
Andoyo et al. (2015)	*Acid-induced cold gelation* WPA & WPA + NMC, pH 4.5	WPA gels: strong-link $G' = 1.6$–1.7 $\gamma_0 = 1.15$ WPA + NMC gels: weak-link $G' = 2.6$ $\gamma_0 = 2.29$
Hagiwara et al. (1996)	BSA dissolved in HEPES buffer of pH 7.0 and acetate buffer of pH 5.1 to 0.1% and 0.001% solutions, heated at 95°C, varying the heating time	BSA at pH 7.0 were about 2.1 and 1.5; D_f of heat-induced aggregates at pH 5.1 was about 1.8
Foegeding et al. (1995)	A fine-stranded matrix formed in protein suspensions contained monovalent cation (Li$^+$, K$^+$, Rb$^+$, and Cs$^+$) chlorides, sodium sulfate, or sodium phosphate at ionic strengths ≤ 0.1 mol/dm^3. This matrix varies in stress and strain at fracture at different salt concentrations	Protein-specific factors can affect the dispersibility of proteins and thereby determine the microstructure and fracture properties of globular protein gels
Verheul & Roefs (1998)	Gels were made at near-neutral pH. Protein concentration (35–89 g/l) and NaCl concentration (0.1–3 mol/dm^3) were systematically varied	Gel structure did not change much after gel formation, while gel rigidity continued to increase, and at the gel point only part of the protein in the dispersion contributes to the gel network. The fractal concept cannot simply be applied to WPI gels
Eissa, Khan (2005)	Whey protein solution: 3% and 7.5% (heat with/without transglutaminase, TG), *low pH cold-set whey protein gel, final pH 4.0*	CSLM: 1.96–1.98 Rheology: 2.05–2.09

values of D_f obtained from microscopy, or even other, methods. This indicated that the various rheological properties of gel aggregates were reflection of the fractal structure of gel aggregate [34, 38, 46, 60, 68]. The salt-induced cold gelation process generally resulted in weak-link gels [38, 46, 51], whereas acid-induced cold gelation generally produced strong-link gel when the initial pH of protein solution was above 7.0 and weak-link gel when pH is below 7.0 [8, 32, 34, 51, 90, 95]. The strong-link gels were generally transparent (fine-stranded; gel formed at low ionic strength and pH far from pI), while weak-link gels were turbid (particulate; gel formed at high ionic strength and pH close to pI). Hagiwara et al. [60] therefore suggested that the transparency and turbidity of gel aggregates were universal characteristics for gels with both strong- and weak-link regime, respectively. Aggregation for all three types of gelation was reaction limited, but addition of other components or pretreatment of the protein prior to processing could be done to modify the aggregation process so as to produce structure, D_f values, regime classification, and different aggregation types.

From the above description regarding D_f value by microscopy and rheology, in general there were a variation of D_f values among researchers and differences of D_f values within the same gel system. These may occur due to several reasons; for example, different length scale used among researcher means that different resolution of the measurements and dynamic changes inside the gel make the scaling laws not able to be applied completely as explained by several authors. The fractal gel model assumes that the gel consists of crosslinked flocs that fill up the space. The flocs are supposed to be fractal structures formed by aggregated particles. The model allows for the possibility that different types of bonds are formed between particles within the flocs and between particles of different flocs. One way to obtain such a structure is by random aggregation of particles, which leads to fractal aggregates (flocs) that grow until they fill up the space and interconnect into a space filling network. Furthermore, the volume fraction of the particles used by authors was small so that the fractal character of the flocs can be expressed before they fill up the space. Different thresholding methods used for D_f values by microscopy lead to varied values of D_f. This is one of the critical steps in using digital imaging for fractal analysis and different thresholding value can lead to different fractal dimensions. Thresholding is not straightforward and changing the threshold value can slightly change the fractal dimension. This is probably the reason why fractal dimensions calculated using different rheological and microscopy parameter are varied. However, the results are still on the same correspondence aggregation regime.

5. Conclusion

Results confirmed the applicability of fractal analysis by macroscopic or microscopic methods in describing whey protein-based gels, in which viscoelastic measurements correlated well with the microstructure of gels. This can be shown by the values of D_f obtained from rheological measurement

which agreed to some extent with those from image analysis, an indication that the rheological behavior of the aggregate gels is a reflection of fractal structure of the aggregates in gels. Both methods tended to produce D_f values with same aggregation types for similar gel systems and even yielded D_f values similar to other methods. Results from numerous studies confirmed that fractal analysis from macroscopic and microscopic methods were applicable to quantify whey protein-based gels despite the presence of several limitations in both methods.

Conflicts of Interest

The authors declare that there are no conflicts of interest.

References

[1] S. Jovanović, M. Barać, and O. Maćej, "Whey Proteins-Properties And Possibility of Application," *Journal for Dairy Production and Processing Improvement*, vol. 55, pp. 215–233, 2005.

[2] C. I. Onwulata and P. J. Huth, *Whey Processing, Functionality and Health Benefits*, Wiley-Blackwell, Oxford, UK, 2008.

[3] J. Chen and E. Dickinson, "Viscoelastic properties of heat-set whey protein emulsion gels," *Journal of Texture Studies*, vol. 29, no. 3, pp. 285–304, 1998.

[4] L. M. Huffman and L. De Barros Ferreira, "Whey-Based Ingredients," *Dairy Ingredients for Food Processing*, pp. 179–198, 2011.

[5] M. Verheul and S. P. F. M. Roefs, "Structure of particulate whey protein gels: effect of NaCl concentration, pH, heating temperature, and protein composition," *Journal of Agricultural and Food Chemistry*, vol. 46, no. 12, pp. 4909–4916, 1998.

[6] T. Nicolai and D. Durand, "Controlled food protein aggregation for new functionality," *Current Opinion in Colloid & Interface Science*, vol. 18, no. 4, pp. 249–256, 2013.

[7] P. Walstra, "Dispersed Systems: Basic Considerations," in *Food Chemistry*, pp. 95–155, Marcel Dekker, Inc, New York, NY, USA, 3rd edition, 1996.

[8] R. Vreeker, L. L. Hoekstra, D. C. den Boer, and W. G. M. Agterof, "Fractal aggregation of whey proteins," *Topics in Catalysis*, vol. 6, no. 5, pp. 423–435, 1992.

[9] G. C. Bushell, Y. D. Yan, D. Woodfield, J. Raper, and R. Amal, "On techniques for the measurement of the mass fractal dimension of aggregates," *Advances in Colloid and Interface Science*, vol. 95, no. 1, pp. 1–50, 2002.

[10] H. Wu and M. Morbidelli, "Model relating structure of colloidal gels to their elastic properties," *Langmuir*, vol. 17, no. 4, pp. 1030–1036, 2001.

[11] T. Croguennec, S. Bouhallab, D. Mollé, B. T. O'Kennedy, and R. Mehra, "Stable monomeric intermediate with exposed Cys-119 is formed during heat denaturation of β-lactoglobulin," *Biochemical and Biophysical Research Communications*, vol. 301, no. 2, pp. 465–471, 2003.

[12] R. W. Visschers and H. H. J. De Jongh, "Bisulphide bond formation in food protein aggregation and gelation," *Biotechnology Advances*, vol. 23, no. 1, pp. 75–80, 2005.

[13] A. C. Alting, M. Weijers, E. H. de Hoog et al., "Acid-induced cold gelation of globular proteins: effects of protein aggregate characteristics and disulfide bonding on rheological properties,"

Journal of Agricultural and Food Chemistry, vol. 52, no. 3, pp. 623–631, 2004.

[14] J. E. Kinsella and D. M. Whitehead, "Proteins in Whey: Chemical, Physical, and Functional Properties," *Advances in Food and Nutrition Research*, vol. 33, no. C, pp. 343–438, 1989.

[15] C. M. Bryant and D. Julian McClements, "Molecular basis of protein functionality with special consideration of cold-set gels derived from hat-denatured whey," *Trends in Food Science & Technology*, vol. 9, no. 4, pp. 143–151, 1998.

[16] H. H. J. de Jongh, "Proteins in food microstructure formation," in *Understanding and Controlling Microstructure of Complex Foods*, pp. 40–66, Woodhead Publishing Limited &CRC Press LLC, FLorida, Fla, USA, 2007.

[17] A. Brodkorb, T. Croguennec, S. Bouhallab, and J. J. Kehoe, "Heat-Induced Denaturation, Aggregation and Gelation of Whey Proteins," in *Advanced Dairy Chemistry*, P. L. H. McSweeney and J. A. O'Mahony, Eds., pp. 155–178, Springer, New York, NY, USA, 2016.

[18] K. Katsuta, D. Rector, and J. e. Kinsella, "Viscoelastic properties of whey protein gels: mechanical model and effects of protein concentration on creep," *Journal of Food Science*, vol. 55, pp. 516–521, 1990.

[19] R. N. W. Zeiler and P. G. Bolhuis, "Numerical study of the effect of thiol-disulfide exchange in the cluster phase of β-lactoglobulin aggregation," *Faraday Discussions*, vol. 158, pp. 461–477, 2012.

[20] S. Ikeda, E. A. Foegeding, and T. Hagiwara, "Rheological study on the fractal nature of the protein gel structure," *Langmuir*, vol. 15, no. 25, pp. 8584–8589, 1999.

[21] K. G. de Kruif, M. A. M. Hoffmann, M. E. Van Marle et al., "Gelation of proteins from milk," *Faraday Discussions*, vol. 101, pp. 185–200, 1995.

[22] L. Donato, C. Schmitt, L. Bovetto, and M. Rouvet, "Mechanism of formation of stable heat-induced β-lactoglobulin microgels," *International Dairy Journal*, vol. 19, no. 5, pp. 295–306, 2009.

[23] C. Le Bon, T. Nicolai, and D. Durand, "Kinetics of aggregation and gelation of globular proteins after heat-induced denaturation," *Macromolecules*, vol. 32, no. 19, pp. 6120–6127, 1999.

[24] M. McSwiney, H. Singh, O. Campanella, and L. K. Creamer, "Thermal gelation and denaturation of bovine β-lactoglobulins A and B," *Journal of Dairy Research*, vol. 61, no. 2, pp. 221–232, 1994.

[25] M. Paulsson, P. Dejmek, and T. Van Vliet, "Rheological Properties of Heat-Induced β-Lactoglobulin Gels," *Journal of Dairy Science*, vol. 73, no. 1, pp. 45–53, 1990.

[26] M. Paulsson, P-O. Hegg, and H. B. Castberg, "Heat-induced gelation of individual whey proteins a dynamic rheological study," *Journal of Food Science*, pp. 87–90, 1986.

[27] M. Stading and A.-M. Hermansson, "Viscoelastic behaviour of β-lactoglobulin gel structures," *Topics in Catalysis*, vol. 4, no. 2, pp. 121–135, 1990.

[28] M. Verheul, S. P. F. M. Roefs, and K. G. De Kruif, "Kinetics of Heat-Induced Aggregation of βLactoglobulin," *Journal of Agricultural and Food Chemistry*, vol. 46, no. 3, pp. 896–903, 1998.

[29] D. J. McClements and M. K. Keogh, "Physical properties of cold-setting gels formed from heat-denatured whey protein isolate," *Journal of the Science of Food and Agriculture*, vol. 69, no. 1, pp. 7–14, 1995.

[30] T. Nicolai, M. Britten, and C. Schmitt, "β-Lactoglobulin and WPI aggregates: Formation, structure and applications," *Food Hydrocolloids*, vol. 25, no. 8, pp. 1945–1962, 2011.

[31] M. A. Hoffmann and P. J. van Mil, "Heat-Induced Aggregation of β-Lactoglobulin: Role of the Free Thiol Group and Disulfide Bonds," *Journal of Agricultural and Food Chemistry*, vol. 45, no. 8, pp. 2942–2948, 1997.

[32] A. C. Alting, R. J. Hamer, C. G. De Kruif, and R. W. Visschers, "Cold-set globular protein gels: Interactions, structure and rheology as a function of protein concentration," *Journal of Agricultural and Food Chemistry*, vol. 51, no. 10, pp. 3150–3156, 2003.

[33] A. C. Alting, H. H. J. De Jongh, R. W. Visschers, and J.-W. F. A. Simons, "Physical and chemical interactions in cold gelation of food proteins," *Journal of Agricultural and Food Chemistry*, vol. 50, no. 16, pp. 4682–4689, 2002.

[34] R. Andoyo, F. Guyomarc'h, A. Burel, and M.-H. Famelart, "Spatial arrangement of casein micelles and whey protein aggregate inacid gels: Insight on mechanisms," *Food Hydrocolloids*, vol. 51, pp. 118–128, 2015.

[35] S. Barbut and E. A. Foegeding, "Ca2+-Induced Gelation of Preheated Whey Protein Isolate," *Journal of Food Science*, vol. 58, pp. 867–871, 1993.

[36] C. Elofsson, P. Dejmek, M. Paulsson, and H. Burling, "Characterization of a cold-gelling whey protein concentrate," *International Dairy Journal*, vol. 7, no. 8-9, pp. 601–608, 1997.

[37] P. Hongsprabhas and S. Barbut, "Ca2+-induced gelation of whey protein isolate: Effects of pre-heating," *Food Research International*, vol. 29, no. 2, pp. 135–139, 1996.

[38] A. G. Marangoni, S. Barbut, S. E. McGauley, M. Marcone, and S. S. Narine, "On the structure of particulate gels - The case of salt-induced cold gelation of heat-denatured whey protein isolate," *Food Hydrocolloids*, vol. 14, no. 1, pp. 61–74, 2000.

[39] S. Barbut, "Effects of calcium level on the structure of pre-heated whey protein isolate gels," *LWT- Food Science and Technology*, vol. 28, no. 6, pp. 598–603, 1995.

[40] E. Doi, "Gels and gelling of globular proteins," *Trends in Food Science & Technology*, vol. 4, no. 1, pp. 1–5, 1993.

[41] P. Hongsprabhas and S. Barbut, "Effect of gelation temperature on Ca2+-induced gelation of whey protein isolate," *LWT- Food Science and Technology*, vol. 30, no. 1, pp. 45–49, 1997.

[42] P. Hongsprabhas and S. Barbut, "Protein and salt effects on Ca2+-induced cold gelation of whey protein isolate," *Journal of Food Science*, vol. 62, no. 2, pp. 382–385, 1997.

[43] S. P. Roefs and K. G. De Kruif, "A Model for the Denaturation and Aggregation of beta-Lactoglobulin," *European Journal of Biochemistry*, vol. 226, no. 3, pp. 883–889, 1994.

[44] K. Ako, T. Nicolai, and D. Durand, "Salt-induced gelation of globular protein aggregates: Structure and kinetics," *Biomacromolecules*, vol. 11, no. 4, pp. 864–871, 2010.

[45] C. M. Bryant and D. J. McClements, "Influence of NaCl and CaCl2 on cold-set gelation of heat-denatured whey protein," *Journal of Food Science*, vol. 65, no. 5, pp. 801–804, 2000.

[46] K. R. Kuhn, Â. L. F. Cavallieri, and R. L. da Cunha, "Cold-set whey protein gels induced by calcium or sodium salt addition," *International Journal of Food Science & Technology*, vol. 45, no. 2, pp. 348–357, 2010.

[47] A. L. F. Cavallieri, A. P. Costa-Netto, M. Menossi, and R. L. Da Cunha, "Whey protein interactions in acidic cold-set gels at different pH values," *Dairy Science & Technology*, vol. 87, no. 6, pp. 535–554, 2007.

[48] Z. Y. Ju and A. Kilara, "Gelation of pH-Aggregated Whey Protein Isolate Solution Induced by Heat, Protease, Calcium Salt, and Acidulant," *Journal of Agricultural and Food Chemistry*, vol. 46, no. 5, pp. 1830–1835, 1998.

[49] Z. Y. Ju and A. Kilara, "Textural properties of cold-set gels induced from heat-denatured whey protein isolates," *Journal of Food Science*, vol. 63, no. 2, pp. 288–292, 1998.

[50] Z. Y. Ju and A. Kilara, "Properties of Gels Induced by Heat, Protease, Calcium Salt, and Acidulant from Calcium Ion-Aggregated Whey Protein Isolate," *Journal of Dairy Science*, vol. 81, no. 5, pp. 1236–1243, 1998.

[51] Z. Y. Ju and A. Kilara, "Effects of Preheating on Properties of Aggregates and of Cold-Set Gels of Whey Protein Isolate," *Journal of Agricultural and Food Chemistry*, vol. 46, no. 9, pp. 3604–3608, 1998.

[52] A. C. Alting, R. J. Hamer, C. G. De Kruif, and R. W. Visschers, "Formation of disulfide bonds in acid-induced gels of preheated whey protein isolate," *Journal of Agricultural and Food Chemistry*, vol. 48, no. 10, pp. 5001–5007, 2000.

[53] A. C. Alting, E. T. Van Der Meulena, J. Hugenholtz, and R. W. Visschers, "Control of texture of cold-set gels through programmed bacterial acidification," *International Dairy Journal*, vol. 14, no. 4, pp. 323–329, 2004.

[54] M.-H. Famelart, N. H. T. Le, T. Croguennec, and F. Rousseau, "Are disulphide bonds formed during acid gelation of preheated milk?" *International Journal of Food Science & Technology*, vol. 48, no. 9, pp. 1940–1948, 2013.

[55] A. J. Vasbinder, A. C. Alting, R. W. Visschers, and C. G. De Kruif, "Texture of acid milk gels: Formation of disulfide cross-links during acidification," *International Dairy Journal*, vol. 13, no. 1, pp. 29–38, 2003.

[56] J. A. Lucey, M. Tamehana, H. Singh, and P. A. Munro, "Effect of interactions between denatured whey proteins and casein micelles on the formation and rheological properties of acid skim milk gels," *Journal of Dairy Research*, vol. 65, no. 4, pp. 555–567, 1998.

[57] J. A. Lucey, C. T. Teo, P. A. Munro, and H. Singh, "Rheological properties at small (dynamic) and large (yield) deformations of acid gels made from heated milk," *Journal of Dairy Research*, vol. 64, no. 4, pp. 591–600, 1997.

[58] B. Mandelbrot, *The Fractal Geometry of Nature*, Times Books, New York, NY, USA, 1982.

[59] A. H. Barrett and M. Peleg, "Applications of fractal analysis to food structure," *LWT- Food Science and Technology*, vol. 28, no. 6, pp. 553–563, 1995.

[60] T. Hagiwara, H. Kumagai, and T. Matsunaga, "Fractal Analysis of the Elasticity of BSA and β-Lactoglobulin Gels," *Journal of Agricultural and Food Chemistry*, vol. 45, no. 10, pp. 3807–3812, 1997.

[61] D. A. Weitz, J. S. Huang, M. Y. Lin, and J. Sung, "Limits of the fractal dimension for irreversible kinetic aggregation of gold colloids," *Physical Review Letters*, vol. 54, no. 13, pp. 1416–1419, 1985.

[62] L. G. B. Bremer, B. H. Bijsterbosch, P. Walstra, and T. van Vliet, "Formation, properties and fractal structure of particle gels," *Advances in Colloid and Interface Science*, vol. 46, no. C, pp. 117–128, 1993.

[63] W.-H. Shih, W. Y. Shih, S.-I. Kim, J. Liu, and I. A. Aksay, "Scaling behavior of the elastic properties of colloidal gels," *Physical Review A: Atomic, Molecular and Optical Physics*, vol. 42, no. 8, pp. 4772–4779, 1990.

[64] J. M. Aguilera, "Why food micro structure?" *Journal of Food Engineering*, vol. 67, no. 1-2, pp. 3–11, 2005.

[65] V. J. Morris and K. Groves, "Food microstructures: Microscopy, measurement and modelling," *Food Microstructures: Microscopy, Measurement and Modelling*, pp. 1–438, 2013.

[66] K. Ako, D. Durand, T. Nicolai, and L. Becu, "Quantitative analysis of confocal laser scanning microscopy images of heat-set globular protein gels," *Food Hydrocolloids*, vol. 23, no. 4, pp. 1111–1119, 2009.

[67] T. Hagiwara, H. Kumagai, and K. Nakamura, "Fractal analysis of aggregates in heat-induced BSA gels," *Food Hydrocolloids*, vol. 12, no. 1, pp. 29–36, 1998.

[68] T. Hagiwara, H. Kumagai, T. Matsunaga, and K. Nakamura, "Analysis of aggregate structure in food protein gels with the concept of fractal," *Bioscience, Biotechnology, and Biochemistry*, vol. 61, no. 10, pp. 1663–1667, 1997.

[69] I. C. Torres, J. M. Amigo Rubio, and R. Ipsen, "Using fractal image analysis to characterize microstructure of low-fat stirred yoghurt manufactured with microparticulated whey protein," *Journal of Food Engineering*, vol. 109, no. 4, pp. 721–729, 2012.

[70] L. A. Pugnaloni, L. Matia-Merino, and E. Dickinson, "Microstructure of acid-induced caseinate gels containing sucrose: Quantification from confocal microscopy and image analysis," *Colloids and Surfaces B: Biointerfaces*, vol. 42, no. 3-4, pp. 211–217, 2005.

[71] A. Thill, S. Veerapaneni, B. Simon, M. Wiesner, J. Y. Bottero, and D. Snidaro, "Determination of structure of aggregates by confocal scanning laser microscopy," *Journal of Colloid and Interface Science*, vol. 204, no. 2, pp. 357–362, 1998.

[72] R. Quevedo, L.-G. Carlos, J. M. Aguilera, and L. Cadoche, "Description of food surfaces and microstructural changes using fractal image texture analysis," *Journal of Food Engineering*, vol. 53, no. 4, pp. 361–371, 2002.

[73] D. Ercili-Cura, "Imaging of Fermented Dairy Products," in *Imaging Technologies and Data Processing for Food Engineers*, N. Sozer, Ed., Food Engineering Series, pp. 99–128, Springer International Publishing, 2016.

[74] J. C. Russ, *The Image Processing Handbook*, CRC Press, 7th edition, 2016.

[75] J. Kittler and J. Illingworth, "Minimum error thresholding," *Pattern Recognition*, vol. 19, no. 1, pp. 41–47, 1986.

[76] N. Otsu, "A threshold selection method from gray-level histograms," *IEEE Transactions on Systems, Man, and Cybernetics*, vol. 9, no. 1, pp. 62–66, 1979.

[77] M. Sezgin and B. Sankur, "Survey over image thresholding techniques and quantitative performance evaluation," *Journal of Electronic Imaging*, vol. 13, no. 1, pp. 146–168, 2004.

[78] Z.-K. Huang and K.-W. Chau, "A new image thresholding method based on Gaussian mixture model," *Applied Mathematics and Computation*, vol. 205, no. 2, pp. 899–907, 2008.

[79] C. A. Glasbey, "An analysis of histogram-based thresholding algorithms," *CVGIP: Graphical Models and Image Processing*, vol. 55, no. 6, pp. 532–537, 1993.

[80] J. H. Xue and Y.-J. Zhang, "Ridler and Calvard's, Kittler and Illingworth's and Otsu's methods for image thresholding," *Pattern Recognition Letters*, vol. 33, no. 6, pp. 793–797, 2012.

[81] C. P. Chu, D. J. Lee, and J. H. Tay, "Bilevel thresholding of floc images," *Journal of Colloid and Interface Science*, vol. 273, no. 2, pp. 483–489, 2004.

[82] J. V. C. Silva, D. Legland, C. Cauty, I. Kolotuev, and J. Floury, "Characterization of the microstructure of dairy systems using automated image analysis," *Food Hydrocolloids*, vol. 44, pp. 360–371, 2015.

[83] R. Buscall, P. D. A. Mills, J. W. Goodwin, and D. W. Lawson, "Scaling behaviour of the rheology of aggregate networks formed from colloidal particles," *Journal of the Chemical Society,*

Faraday Transactions 1: Physical Chemistry in Condensed Phases, vol. 84, no. 12, pp. 4249–4260, 1988.

[84] M. Stading, M. Langton, and A.-M. Hermansson, "Microstructure and rheological behaviour of particulate β-lactoglobulin gels," *Topics in Catalysis*, vol. 7, no. 3, pp. 195–212, 1993.

[85] L. G. B. Bremer, *Fractal Aggregation in Relation to Formation and Properties of Particle Gels*, Wageningen University, 1992.

[86] S. S. Narine and A. G. Marangoni, "Mechanical and structural model of fractal networks of fat crystals at low deformations," *Physical Review E: Statistical, Nonlinear, and Soft Matter Physics*, vol. 60, no. 6, pp. 6991–7000, 1999.

[87] M. Mellema, J. H. J. Van Opheusden, and T. Van Vliet, "Categorization of rheological scaling models for particle gels applied to casein gels," *Journal of Rheology*, vol. 46, no. 1, pp. 11–29, 2002.

[88] N. Mahmoudi, S. Mehalebi, T. Nicolai, D. Durand, and A. Riaublanc, "Light-scattering study of the structure of aggregates and gels formed by heat-denatured whey protein isolate and β-lactoglobulin at neutral pH," *Journal of Agricultural and Food Chemistry*, vol. 55, no. 8, pp. 3104–3111, 2007.

[89] M. Ikeguchi, "Transient non-native helix formation during the folding of β-lactoglobulin," *Biomolecules*, vol. 4, no. 1, pp. 202–216, 2014.

[90] A. S. Eissa and S. A. Khan, "Acid-induced gelation of enzymatically modified, preheated whey proteins," *Journal of Agricultural and Food Chemistry*, vol. 53, no. 12, pp. 5010–5017, 2005.

[91] T. Hagiwara, H. Kumagai, and K. Nakamura, "Fractal analysis of aggregates formed by heating dilute bsa solutions using light scattering methods," *Bioscience, Biotechnology, and Biochemistry*, vol. 60, no. 11, pp. 1757–1763, 1996.

[92] E. A. Foegeding, E. L. Bowland, and C. C. Hardin, "Factors that determine the fracture properties and microstructure of globular protein gels," *Topics in Catalysis*, vol. 9, no. 4, pp. 237–249, 1995.

[93] M. Verheul and S. P. F. M. Roefs, "Structure of whey protein gels, studied by permeability, scanning electron microscopy and rheology," *Food Hydrocolloids*, vol. 12, no. 1, pp. 17–24, 1998.

[94] E. A. Foegeding, J. P. Davis, D. Doucet, and M. K. McGuffey, "Advances in modifying and understanding whey protein functionality," *Trends in Food Science & Technology*, vol. 13, no. 5, pp. 151–159, 2002.

[95] F. Li, X. Kong, C. Zhang, and Y. Hua, "Gelation behaviour and rheological properties of acid-induced soy protein-stabilized emulsion gels," *Food Hydrocolloids*, vol. 29, no. 2, pp. 347–355, 2012.

10

Extending Shelf Life of Indonesian Soft Milk Cheese (Dangke) by Lactoperoxidase System and Lysozyme

Ahmad Ni'matullah Al-Baarri [ID],[1,2] **Anang Mohamad Legowo,**[1]
Septinika Kurnia Arum,[3] **and Shigeru Hayakawa**[4]

[1]*Department of Food Technology, Faculty of Animal and Agricultural Sciences, Diponegoro University, Semarang 50275, Indonesia*
[2]*Laboratory of Food Technology, Integrated Laboratory, Diponegoro University, Semarang 50275, Indonesia*
[3]*Department of Animal Science, Faculty of Animal and Agricultural Sciences, Diponegoro University, Semarang 50275, Indonesia*
[4]*Department of Applied Biological Sciences, Faculty of Agriculture, Kagawa University, Miki-cho 761-0795, Japan*

Correspondence should be addressed to Ahmad Ni'matullah Al-Baarri; albari@live.undip.ac.id

Academic Editor: Salam A. Ibrahim

Dangke, a type of fresh soft cheese made of bovine and buffalo milk, is a traditional dairy product used in South Sulawesi, Indonesia. It is prepared from fresh milk using the conventional method, which easily destroys the quality. This study was conducted to assess whether using lactoperoxidase system and lysozyme as preservative agents could suppress the growth of bacteria in dangke. The pH value, total microbial count, and hardness of dangke were determined to measure the quality. Lactoperoxidase and lysozyme were purified from fresh bovine milk, and their purity was confirmed using SDS-PAGE. The combination of lactoperoxidase system and lysozyme was able to remarkably suppress the total microbial count in dangke from 7.78 ± 0.67 to 5.30 ± 0.42 log CFU/ml during 8 h of storage at room temperature. Preserving dangke in this enzyme combination affected its hardness, but there was no remarkable change in the pH value. Results of this study may provide knowledge to utilize a new method to preserve the quality of dangke.

1. Introduction

Dangke, a type of fresh soft cheese, is a traditional dairy product available in Enrekang Regency, South Sulawesi province, Indonesia. The nutrient content of dangke in %w/w comprises 55% water, 23.8% protein, 14.8% fat, and 2.1% ash [1]. It is produced by heating fresh milk and then adding papaya latex to precipitate casein. Commonly, local people used papaya latex from unripe papaya fruit, thus keeping the slightly bitter taste [2]. Traditionally, curd and whey are separated using a coconut shell, which is a process involved in the shaping stage in the preparation of dangke. After the shaping process, dangke is packed in a banana leaf and is ready to be consumed. The conventional method of producing dangke does not involve high food hygiene standards, resulting in an increased possibility for contamination with bacteria. Dangke is usually preserved using salt, though there is the problem of a relatively short shelf life (±2 days) by storing at room temperature [1].

Today, the production of dangke has increased along with the increase in consumer demand [2]. The distribution of dangke has been reported to reach out of the province, including to other countries such as Brunei Darussalam and Malaysia. Nationally, dangke is being already distributed to Java and Sumatra islands, consistent with the increase in the number of tourism activities. Therefore, preservation is an important factor to maintain the quality of dangke.

Lactoperoxidase (LPO) is a heme-containing glycoprotein of 608 amino acids with a molecular mass of 78 kDa and has already been known as a natural enzyme found in plants, animals, and humans. LPO is abundantly found in milk, saliva, and tear glands [3–5] and can serve as a natural antimicrobial in combination with thiocyanate (SCN^-) and hydrogen peroxide (H_2O_2), which is known as the lactoperoxidase system (LPOS) [5–7]. LPO catalyzes the oxidation of thiocyanate (SCN^-) by hydrogen peroxide (H_2O_2), resulting in the production of hypothiocyanite ($OSCN^-$). Hypothiocyanite is a compound that is responsible for killing bacteria, fungi, and viruses by destroying the sulfhydryl groups (SH groups) of the cell membrane, resulting in damage to the vital cell membrane, which leads to cell death [8–12]. LPOS has been used as a natural preservative in some foods, such as milk [13, 14], fruits, chicken, and vegetables [15]. LPOS is effective in suppressing the growth of *Pseudomonas*, *Escherichia coli*, and *Salmonella typhimurium* on cottage cheese [16].

Lysozyme (1,4-beta-N-acetylmuramidase, 14.4 kDa) is a hydrophilic protein that has been widely used as a natural preservative. It is naturally found in egg white and milk [17, 18]. Lysozyme hydrolyzes 1,4-β-linkages between N-acetylmuramic acid and N-acetylglucosamine present in the peptidoglycan. Gram-positive bacteria are highly susceptible to lysozyme because of the presence of peptidoglycan in their cell walls, but lysozyme is not effective in killing Gram-negative bacteria [19–21], which indicates the need for a combination with other compounds. Lysozyme has been used as an antimicrobial and an antiviral in food and pharmaceutical industries [22], where it causes inhibition of the growth of pathogenic bacteria and could thus extend the shelf life of food. It is also used in the preservation of fruits, vegetables, beans, tofu, curd, meat, sausages, salads, and semi-hard-type cheese such as Edam, Gouda, and some Italian cheese. It has also been reported to have protective effects against pathogenic bacteria such as *Bacillus cereus* in cheese [23]. On the other hand, lysozyme has also been added to infant formulas to achieve the similarity to human milk [24, 25].

Previous research has shown that weak inhibition by LPOS in dangke could result in the extension of shelf life for only 6 h at room temperature [2]. Therefore, a synergistic effect of LPOS to inhibit bacteria may be useful to solve this problem. Thus, in this study, lysozyme was added to LPOS to extend the shelf life of dangke. This experiment might provide knowledge to utilize a new method for extending the shelf life of dangke using natural LPOS and lysozyme.

2. Materials and Methods

2.1. Materials. Fresh bovine milk samples were provided by a campus farm. Fresh duck eggs were purchased from a local farm. Latex from young papaya was used to obtain papain enzyme to precipitate the protein. SP-Sepharose Fast Flow (SP-FF) (Lot No. 10072021) was used for lysozyme purification. LPO from bovine whey was obtained from the Chemical and Food Nutrition Laboratory, Food Technology Department, Faculty of Animal and Agricultural Sciences, Diponegoro University. Hydrogen peroxide (H_2O_2) and potassium thiocyanate (KSCN) were used as LPO substrate. A 0.2 μm syringe filter was used to sterilize the enzyme.

2.2. Lysozyme Purification. Lysozyme purification was carried out following the method described by Naknukool et al. [26]. Duck egg white was mixed with 3-fold volume sodium acetate buffer (0.05 M, pH 5.0). The mixture was centrifuged at 6000 rpm for 15 min to separate the supernatant, and then the supernatant was applied in an SP-FF column for lysozyme purification. Then, 500 ml of sodium acetate

buffer (0.05 M, pH 5.0) was subsequently eluted through the column. Lysozyme was obtained using serial dilution with 300 ml of 0.1, 0.3, and 0.5 M NaCl in sodium phosphate buffer (0.05 M, pH 9.0). The eluate was then collected in 10 ml tubes. The purity of the eluate was determined using sodium dodecyl sulfate polyacrylamide gel electrophoresis (SDS-PAGE).

2.3. Preparation of Dangke. The preparation of dangke was adopted from the traditional method that has been followed by the local people in South Sulawesi. Fresh bovine milk was heated at 75°C for 20 min, and then latex from young papaya (0.3% w/v) was subsequently added to the milk. The curd that formed was separated using a clean filter cloth and then pressed to produce dangke. Using this method, 680 g of dangke could be produced from 1 l of milk.

2.4. Preparation of LPOS Solution. LPOS solution was prepared using a mixture of 300 μL of LPO, 300 μL of 0.9 mM H_2O_2, and 300 μL of 0.9 mM KSCN. Prior to application, this mixture was filtered using a 0.2 μm syringe filter, placed in a microtube, and left to stand for 1 h at 30°C.

2.5. Preservation of Dangke. A total of 1 g of dangke was used for the evaluation of total microbial count and the pH value, while for the evaluation of hardness, dangke was cut into a rectangular shape measuring 2.5 × 1 × 1 cm. The dangke was then stored at 30°C for 18 h for the calculation of total microbial count, while the dangke stored at 30°C for 8 h was used to analyze the pH value and hardness. Prior to evaluation, dangke was immersed in various preservation solutions (LPOS, lysozyme, and LPOS + lysozyme) at 30°C for 4 h. Dangke immersed in sterile pure water was used as a control.

2.6. Microbial Count. 3 M Petrifilm Aerobic Count Plates (3 M Microbiology, St. Paul, Minn., USA) were used for assessing the microbial count of dangke following a previous method described by Rasbawati [2], with a minor modification. Briefly, dangke was subjected to serial dilutions of sterile 0.88% NaCl solution to enumerate the bacteria. The diluted mixture (1000 μl) was spread onto the plates and incubated at 37°C for 48 h. The CFUs of the microbes in the sample solution were counted on the plates.

2.7. Hardness Measurement. Dangke samples measuring 2.5 × 1 × 1 cm were analyzed for hardness. Texture analyses were conducted using Brookfield Texture Analyzer (CT3) under the following conditions: a ø12.7 mm ball probe was penetrated to a depth of 4 mm into the sample at a speed of 1 mm/s, and the textural hardness was measured in triplicate and expressed in Newton.

2.8. Statistical Analysis. The total microbial count was analyzed descriptively with two replications. The pH value and hardness were analyzed using ANOVA with three replications. Statistical analyses were performed using R software for Macintosh. Duncan's multiple range test ($P < 0.05$) was used to calculate the significance among values.

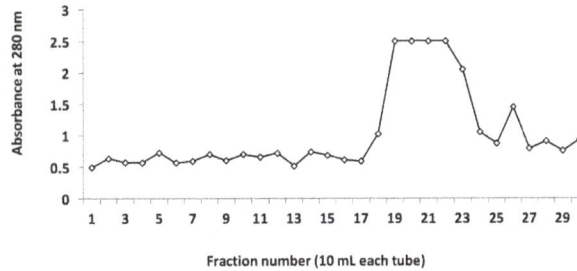

FIGURE 1: Absorbance at 280 nm of the eluate from SP Sepharose Fast Flow column (10 ml each tube) containing a high concentration of protein. Fraction numbers 1–10, 11–20, and 21–30 were obtained from elution with 0.1, 0.3, and 0.5 M NaCl in sodium phosphate buffer (0.05 M, pH 9.0), respectively.

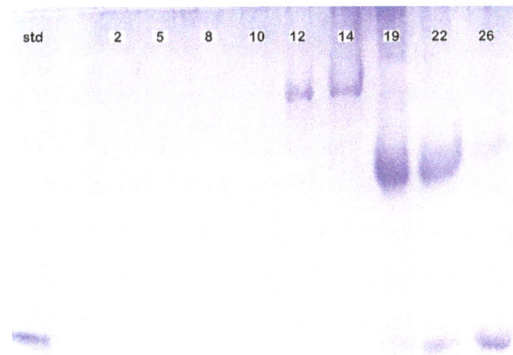

FIGURE 2: Sodium dodecyl sulfate polyacrylamide gel electrophoresis (SDS-PAGE) profile of eluate through SP Sepharose Fast Flow. Lane from left to right: standard (std) using α-lactalbumin (a 14 kDa protein), fraction numbers 2, 5, 8, 10, 12, 14, 19, 22, and 26. Fraction number 26 showed a single band, indicating that pure lysozyme was detected. Thus, this fraction was used for the entire research.

3. Results and Discussion

3.1. Purification of Lysozyme.
Three-step dilutions with various concentrations of NaCl were carried out to obtain the lysozyme. Figure 1 shows the absorbance at 280 nm of the elution from each step of dilution. Fractions numbers 1–10, 11–20, and 21–30 were obtained from the elution against phosphate buffer (pH 9.0) containing 0.1, 0.3, and 0.5 M NaCl, respectively. A high peak of protein concentration activity was detected from fractions numbers 19–22. However, the elution from these fractions showed more than one band (Figure 2), whereas fraction number 26 showed a single band representing pure lysozyme with a molecular weight of 14 kDa. Therefore, fraction number 26 was used for the entire study. The elution was then mixed, and the protein concentration was determined using the Lowry method, resulting in a value of 0.10%. This value was comparable to that reported in another study that showed that the protein concentration from purified protein determined using a similar method was almost 0.1% [25].

3.2. Total Microbial Count.
Figure 3 shows the total microbial count in the dangke samples that were immersed in sterile pure water, LPOS, lysozyme, and a combination of LPOS + lysozyme for 18 h at room temperature. It can be seen that the total microbial count in dangke has increased by storage time. Immersing in sterile pure water at 0 h showed the highest bacterial count (4.15±0.21 CFU/ml) compared to those with other treatments, whereas immersing dangke in lysozyme resulted in the lowest total number of bacteria (2.07±0.32 log CFU/ml). Immersing dangke in LPOS and the combination of LPOS + lysozyme resulted in a total bacterial number of 2.95±0.91 and 2.39±0.54, respectively. Immersing dangke for 8 h increased the total bacterial count in all treatments, ranging from 5.30 ± 0.42 to 7.78 ± 0.67 log CFU/ml, and a longer immersion time of up to 18 h resulted in further increase in the bacterial count, ranging from 8.11 ± 0.37 to 8.71 ± 0.57 log CFU/ml (Figure 3).

Immersing dangke in LPOS and lysozyme or its combination reduced the total microbial count, as shown in Figure 3. LPOS, lysozyme, and the combination of LPOS + lysozyme were able to decrease the population of bacteria at 0 h of storage to almost 1.20, 2.08, and 1.76 log CFU/ml, respectively, when compared to the total bacterial count in dangke immersed in sterile pure water as control. Among all the treatments, lysozyme exhibited the strongest antibacterial activity, whereas LPOS exhibited the weakest antimicrobial activity.

The antibacterial activity of LPO is due to hypothio-cyanite production from the enzymatic reaction between

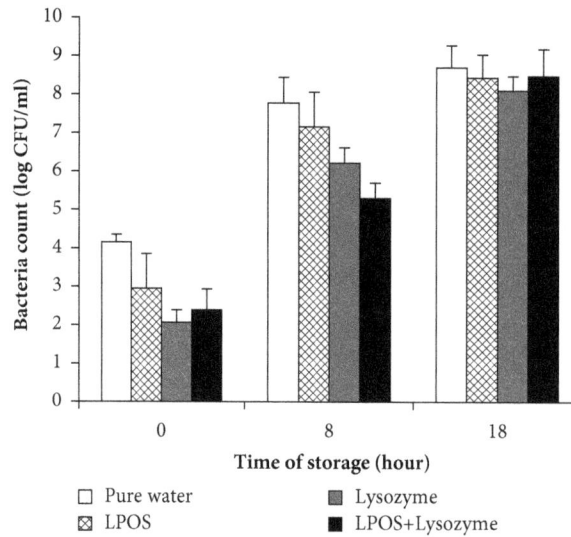

FIGURE 3: Total microbial count in dangke after immersing for 10 min in solutions containing LPOS, lysozyme, and the combination of LPOS + lysozyme. Dangke immersed in pure water was used as control. Values are the mean from three replicates of the experiment, and error bars represent standard error.

hydrogen peroxide and thiocyanate. Hypothiocyanite is a short-lived product that is responsible for killing bacteria, fungi, and viruses by destructing the sulfhydryl (SH) groups of the cell membrane [3, 8, 26–28]. Lysozyme is known to exert its antimicrobial activity against bacteria, fungi, protozoa, and viruses by destroying the structural components on the cell walls of bacteria and fungi [29–31]. Lysozyme catalyzes the β1–4 bonds between N-acetylmuramic acid and N-acetylglucosamine in the peptidoglycan, resulting in bacterial death. Gram-positive bacteria are highly susceptible to lysozyme as they contain 90% peptidoglycan in their cell walls, whereas the peptidoglycan content in Gram-negative bacteria is only 5%–10% [19, 32]. It has been well documented that several bacteria found in raw milk might also be found in cheese due to the handling process prior to cheese-making [33]. Among these bacteria, the Gram-positive bacteria such as *Enterococcus, Pediococcus, Aerococcus, Staphylococcus* [33], and *Bacillus* spp. [34] are commonly found in milk. The dominance of Gram-positive bacteria may provide an answer for the high antimicrobial activity of lysozyme in cheese.

In the present study, all the preservatives were unable to inhibit the growth of bacteria in dangke stored for 18 h because of the high total microbial count (from 8.11 ± 0.37 to $8.71 \pm 0.57 \log$ CFU/ml). This result is consistent with [35] that showed that the hypothiocyanite generated from limited amount of substrates (0.3 mM H_2O_2 and 0.3 mM SCN^-) was able to kill the total bacteria in milk if the initial population of bacteria did not exceed $8.00 \log$ CFU/ml. Furthermore, [2] reported that the LPOS was unable to reduce the total microbial count in dangke stored for 12 h with a total microbial count of 10^{10} CFU/ml.

The combination LPOS + lysozyme was unable to suppress the growth of bacteria in dangke at the maximum storage time; however, the synergistic effect of this combination could be observed at 8 h of storage of dangke, resulting in the least total bacterial count of 5.30 ± 0.42 CFU/ml compared to that with other treatments. Since the Indonesian National Standard (2008) has stated that the maximum allowed limit of total bacteria in cheese is $6 \log$ CFU/ml, the combination of LPOS + lysozyme may be applied to meet the requirement of the maximum allowed amount of total bacteria in cheese.

3.3. pH Value. The development of appropriate pH and texture is required to produce the preferred cheese by storage during a period of time [36]. Based on the data shown in Table 1, the pH value of dangke stored at room temperature for 8 h varied from 6.22 ± 0.30 to 6.77 ± 0.02. Dangke immersed in sterile pure water showed a significant increase in pH value, ranging from 6.22 ± 0.30 to 6.54 ± 0.05, whereas immersing dangke in LPOS, lysozyme, and the combination of LPOS + lysozyme did not show a significant change in the pH value.

It has been reported that the increase in pH value was due to the process of deamination of amino acids resulting in the production of NH_3 and the metabolism of lactic acid bacteria to produce CO_2 [37]. This reason is in agreement with the result of total bacteria shown in Figure 3, where the total bacterial count was found to be decreased along with treatments in the preservative solutions. The decreased number of live bacteria contributed to the decreased production of CO_2, resulting in less change in the pH value.

The initial pH value of dangke was detected to be 6.22 ± 0.30, while [2] stated that the initial pH value of dangke was 7.17. Another study reported an initial pH value of 6.40 [38]. It has been recognized that the initial pH value of dangke was relatively similar to the pH of fresh milk. The variation in the initial pH value of dangke may be explained by the wide variation in the pH value of papaya latex. It has been documented that the pH of papaya latex ranged from 6.00 to 8.75 [38, 39], thus probably resulting in the alteration of initial pH value of dangke from the initial pH value of fresh milk.

TABLE 1: pH value of dangke immersed in pure water, LPOS, lysozyme, and the combination of LPOS + lysozyme.

Storage period (h)	Dangke pH value			
	Pure water	LPOS[ns]	LZ[ns]	LPOS + LZ[ns]
0	6.22 ± 0.30^b	6.59 ± 0.01	6.72 ± 0.01	6.71 ± 0.01
1	6.46 ± 0.13^a	6.48 ± 0.09	6.54 ± 0.04	6.62 ± 0.01
2	6.54 ± 0.06^a	6.64 ± 0.07	6.71 ± 0.01	6.62 ± 0.01
3	6.52 ± 008^a	6.68 ± 0.10	6.72 ± 0.05	6.64 ± 0.05
4	6.43 ± 0.03^{ab}	6.56 ± 0.11	6.71 ± 0.03	6.64 ± 0.08
5	6.54 ± 0.05^a	6.75 ± 0.07	6.69 ± 0.03	6.71 ± 0.08
6	6.53 ± 0.04^a	6.64 ± 0.05	6.73 ± 0.09	6.68 ± 0.03
7	6.46 ± 0.02^a	6.58 ± 0.11	6.69 ± 0.01	6.64 ± 0.02
8	6.54 ± 0.05^a	6.66 ± 0.01	6.77 ± 0.02	6.70 ± 0.03

The superscript letters indicate significant difference among the storage periods; "ns" means not significant. Data are the average values from triplicate of the experiment ± standard error.

TABLE 2: Hardness (N) of dangke after immersing in pure water, LPOS, lysozyme, and the combination of LPOS + lysozyme.

Dangke	Pure water[ns]	LPOS	LZ	LPOS + LZ[ns]
Initial	1.984 ± 0.75	2.110 ± 0.56^b	2.734 ± 0.47^a	2.035 ± 0.69
Final	1.535 ± 1.03	3.620 ± 0.90^a	1.750 ± 0.32^b	2.798 ± 0.73

The superscript letters indicate significant difference among the storage periods; "ns" means not significant. Data are the hardness values at 8 h storage at 30°C.

3.4. Hardness. Table 2 shows the results of the measurement of hardness of dangke immersed in sterile pure water, LPOS, lysozyme, and the combination of LPOS + lysozyme at 0 h of storage time (initial) and 8 h of storage time (final). Based on the statistical analysis, sterile pure water and the combination of LPOS + lysozyme had no significant effects on the hardness of dangke; however, LPOS increased the hardness of dangke to a value of 71.6% from the initial point, resulting in final textural hardness of 3.62 ± 0.90 N. The hardness of dangke immersed in lysozyme was found to be significantly decreased. Based on the results shown in Table 2, the decrease in hardness of dangke immersed in lysozyme was 36%, resulting in a final hardness value of 1.750 ± 0.32 N. The increase in hardness of dangke immersed in LPOS may be explained by the generation of hypothiocyanite and hypothiocyanous acid by the enzymatic reaction between KSCN and H_2O_2 using LPO as a catalyzer. Reference [40] stated that hypothiocyanite is an anion and the conjugate base of hypothiocyanous acid which is an organic compound and a part of thiocyanate containing the functional group SCN^-. Hypothiocyanous acid is a fairly weak acid with an acid dissociation constant of 5.3 [41]. It has been recognized that some factors, including pH, can affect the rheological properties of dangke. For instance, a decrease in pH of Gouda cheese resulted in an increase in hardness [42] and vice versa, which is similar to the result of the present study.

The measurement of hardness is necessary to determine the quality of rheological properties. Since dangke is commonly consumed after deep frying or is served with other food products, a hard-texture-dangke is commonly preferred.

Therefore, based on this reason, the LPOS treatment might be an appropriate method to preserve dangke and strengthen its hardness.

4. Conclusions

LPOS, lysozyme, and the combination of LPOS + lysozyme were able to inhibit the growth of microbes in dangke stored for 8 h. The highest antimicrobial activity was found in dangke preserved in the combination of LPOS + lysozyme immersion. The change in pH value was also maintained by immersing dangke in all treatments. The hard texture of dangke was found in dangke immersed in LPOS; therefore the treatment with the combination of LPOS and lysozyme was suggested to retain the softness of dangke.

Conflicts of Interest

The authors declare that there are no conflicts of interest regarding the publication of this article.

Acknowledgments

The corresponding author is highly indebted to the Ministry of Research, Technology, and Higher Education of the

Republic of Indonesia for entirely supporting the financial requirement for this study.

References

[1] W. Hatta, M. Sudarwanto, I. Sudirman, and R. Malak, "Prevalence and sources of contamination of Escherichia coli and Salmonella spp. in cow milk dangke, Indonesian fresh soft cheese," *Global Veterinaria*, vol. 11, no. 3, pp. 352–356, 2013.

[2] A. N. Mukhlisah, I. I. Arief, and E. Taufik, "Physical, microbial, and chemical qualities of dangke produced by different temperatures and papain concentrations," *Media Peternakan*, vol. 40, no. 1, pp. 63–70, 2017.

[3] Rasbawati, B. Dwiloka, A. N. Al-Baarri, A. M. Legowo, and V. P. Bintoro, "Total bacteria and pH of Dangke preserved using natural antimicrobial lactoferrin and lactoperoxidase from bovine whey," *International Journal of Dairy Science*, vol. 9, no. 4, pp. 116–123, 2014.

[4] L. M. Wolfson and S. S. Sumner, "Antibacterial activity of the lactoperoxidase system: a review," *Journal of Food Protection*, vol. 56, no. 10, pp. 887–892, 1993.

[5] K. D. Kussendrager and A. C. M. van Hooijdonk, "Lactoperoxidase: physico-chemical properties, occurrence, mechanism of action and applications," *British Journal of Nutrition*, vol. 84, no. 1, pp. S19–S25, 2000.

[6] J.-W. Boots and R. Floris, "Lactoperoxidase: From catalytic mechanism to practical applications," *International Dairy Journal*, vol. 16, no. 11, pp. 1272–1276, 2006.

[7] E. Seifu, E. M. Buys, and E. F. Donkin, "Significance of the lactoperoxidase system in the dairy industry and its potential applications: A review," *Trends in Food Science & Technology*, vol. 16, no. 4, pp. 137–154, 2005.

[8] A. N. Al-Baarri, M. Hayashi, M. Ogawa, and S. Hayakawa, "Effects of mono-and disaccharides on the antimicrobial activity of bovine lactoperoxidase system," *Journal of Food Protection*, vol. 74, no. 1, pp. 134–139, 2011.

[9] E. Borch, C. Wallentin, M. Rosén, and L. Björck, "Antibacterial effect of the lactoperoxidase/thiocyanate/hydrogen peroxide system against strains of campylobacter isolated from poultry," *Journal of Food Protection*, vol. 52, no. 9, pp. 638–641, 1989.

[10] M. Hernandez, B. Van Markwijk, and H. Vreeman, "Isolation and properties of lactoperoxidase from bovine milk," *Netherlands Milk and Dairy Journal*, vol. 44, pp. 213–231, 1990.

[11] S. Modi, S. S. Deodhar, D. V. Behere, and S. Mitra, "Lactoperoxidase-catalyzed oxidation of thiocyanate by hydrogen peroxide: nitrogen-15 nuclear magnetic resonance and optical spectral studies," *Biochemistry*, vol. 30, no. 1, pp. 118–124, 1991.

[12] J. Lu, N. Argov-Argaman, J. Anggrek et al., "The protein and lipid composition of the membrane of milk fat globules depends on their size," *Journal of Dairy Science*, vol. 99, no. 6, pp. 4726–4738, 2016.

[13] A. Naidu, *Natural Food Antimicrobial Systems*, A. S. Naidu, Ed., CRC Press, Boca Raton, Fla, US, 2000.

[14] E. Seifu, E. M. Buys, E. F. Donkin, and I.-M. Petzer, "Antibacterial activity of the lactoperoxidase system against food-borne pathogens in Saanen and South African Indigenous goat milk," *Food Control*, vol. 15, no. 6, pp. 447–452, 2004.

[15] N. O. Asaah, F. Fonteh, P. Kamga, S. Mendi, and S. Imele, "Activation of the lactoperoxidase system as a method of preserving raw milk in areas without cooling facilities," *African Journal of Food, Agriculture, Nutrition and Development*, vol. 7, pp. 1–15, 2007.

[16] V. Touch, S. Hayakawa, S. Yamada, and S. Kaneko, "Effects of a lactoperoxidase-thiocyanate-hydrogen peroxide system on *Salmonella enteritidis* in animal or vegetable foods," *International Journal of Food Microbiology*, vol. 93, no. 2, pp. 173–183, 2004.

[17] R. G. Earnshaw, J. G. Banks, D. Defrise, and C. Francotte, "The preservation of cottage cheese by an activated lactoperoxidase system," *Food Microbiology*, vol. 6, no. 4, pp. 285–288, 1989.

[18] V. A. Proctor and F. E. Cunningham, "The chemistry of lysozyme and its use as a food preservative and a pharmaceutical," *Critical Reviews in Food Science and Nutrition*, vol. 26, no. 4, pp. 359–395, 1988.

[19] D. E. Conner, "Naturally occurring compounds," in *Antimicrobials in Foods*, P. M. Davidson and A. L. Branen, Eds., pp. 441–468, Marcel Dekker Inc., New York, NY, USA, 1993.

[20] J. N. Losso, S. Nakai, and E. A. Charter, *Lysozyme, Natural Food Antimicrobial System*, CRC Press, Boca Raton, Fla, USA, 2000.

[21] L. Vannini, R. Lanciotti, D. Baldi, and M. E. Guerzoni, "Interactions between high pressure homogenization and antimicrobial activity of lysozyme and lactoperoxidase," *International Journal of Food Microbiology*, vol. 94, no. 2, pp. 123–135, 2004.

[22] A. Bera, S. Herbert, A. Jakob, W. Vollmer, and F. Götz, "Why are pathogenic staphylococci so lysozyme resistant? The peptidoglycan O-acetyltransferase OatA is the major determinant for lysozyme resistance of *Staphylococcus aureus*," *Molecular Microbiology*, vol. 55, no. 3, pp. 778–787, 2005.

[23] P. S. Davidson and S. Zivanovic, *The Use of Natural Antimicrobial, in Food Preservation Techniques*, Woodhead Publishing Ltd, Cambridge, UK, 2003.

[24] E. C. Scharfen, D. A. Mills, and E. A. Maga, "Use of human lysozyme transgenic goat milk in cheese making: Effects on lactic acid bacteria performance," *Journal of Dairy Science*, vol. 90, no. 9, pp. 4084–4091, 2007.

[25] F. E. Cunningham, V. A. Proctor, and S. T. Goetsch, "Egg-White Lysozyme As A Food Preservative: An Overview," *World's Poultry Science Journal*, vol. 47, no. 2, pp. 141–163, 1991.

[26] S. Naknukool, S. Hayakawa, T. Uno, and M. Ogawa, "Antimicrobial Activity of Duck Egg Lysozyme against Salmonella enteritidis," in *Proceedings of the 13th World Congress of Food Science & Technology*, vol. 3, pp. 1783–1794, Nantes, France, September 2006.

[27] E. Seifu, E. F. Donkin, and E. M. Buys, "Potential of lactoperoxidase to diagnose subclinical mastitis in goats," *Small Ruminant Research*, vol. 69, no. 1-3, pp. 154–158, 2007.

[28] F. Bafort, O. Parisi, J.-P. Perraudin, and M. H. Jijakli, "Mode of Action of Lactoperoxidase as Related to Its Antimicrobial Activity: A Review," *Enzyme Research*, vol. 2014, Article ID 517164, 13 pages, 2014.

[29] A. N. Al-Baarri, M. Ogawa, and S. Hayakawa, "Application of lactoperoxidase system using bovine whey and the effect of storage condition on lactoperoxidase activity," *International Journal of Dairy Science*, vol. 6, no. 1, pp. 72–78, 2011.

[30] C. C. Fuglsang, C. Johansen, S. Christgau, and J. Adler-Nissen, "Antimicrobial enzymes: Applications and future potential in the food industry," *Trends in Food Science & Technology*, vol. 6, no. 12, pp. 390–396, 1995.

[31] K. V. R. Reddy, R. D. Yedery, and C. Aranha, "Antimicrobial peptides: premises and promises," *International Journal of Antimicrobial Agents*, vol. 24, no. 6, pp. 536–547, 2004.

[32] S. Wang, T. B. Ng, T. Chen et al., "First report of a novel plant lysozyme with both antifungal and antibacterial activities,"

Biochemical and Biophysical Research Communications, vol. 327, no. 3, pp. 820–827, 2005.

[33] H. R. Ibrahim, T. Aoki, and A. Pellegrini, "Strategies for new antimicrobial proteins and peptides: Lysozyme and aprotinin as model molecules," *Current Pharmaceutical Design*, vol. 8, no. 9, pp. 671–693, 2002.

[34] I. Verdier-Metz, G. Gagne, S. Bornes et al., "Cow teat skin, a potential source of diverse microbial populations for cheese production," *Applied and Environmental Microbiology*, vol. 78, no. 2, pp. 326–333, 2012.

[35] D. Samaržija, Š. Zamberlin, and T. Pogačić, "Psychrotrophic bacteria and their negative effects on milk and dairy products quality," *Mljekarstvo*, vol. 62, no. 2, pp. 77–95, 2012.

[36] V. Y. Villa, A. M. Legowo, V. P. Bintoro, and A. N. Al-Baarri, "Quality of fresh bovine milk after addition of Hypothiocyanite-rich-solution from Lactoperoxidase system," *International Journal of Dairy Science*, vol. 9, no. 1, pp. 24–31, 2014.

[37] Indonesian National Standard, *SNI 01-6366-2000 on Microbial Contamination Limit and Limit Maximum Residues in Foodstuffs of Animal Origin*, National Standardization Agency (BSN), Jakarta, Indonesia, 2008.

[38] M. El-Hofi, A. Ismail, F. A. Rabo, S. El-Dieb, and O. Ibrahim, "Studies on acceleration of ras cheese ripening by aminopeptidase enzyme from buffaloes' pancreas II- utilization of buffaloes' pancreas aminopeptidase in acceleration of ras cheese ripening," *New York Science Journal*, vol. 3, no. 9, pp. 91–96, 2010.

[39] M. M. El-Sheikh, M. H. El-Senaity, Y. B. Youssef, N. M. Shahein, and S. N. Abd Rabou, "Effect ripening conditions on properties of blue cheese produced from cow's and goat's milk," *American Journal of Science*, vol. 7, no. 1, pp. 485–490, 2011.

[40] P. Chaiwut, S. Nitsawang, L. Shank, and P. Kanasawud, "A comparative study on properties and proteolytic components of papaya peel and latex proteases," *Chiang Mai Journal of Science*, vol. 34, no. 1, pp. 109–118, 2007.

[41] P. G. Furtmüller, M. Zederbauer, W. Jantschko et al., "Active site structure and catalytic mechanisms of human peroxidases," *Archives of Biochemistry and Biophysics*, vol. 445, no. 2, pp. 199–213, 2006.

[42] E. L. Thomas, K. A. Pera, K. W. Smith, and A. K. Chwang, "Inhibition of Streptococcus mutans by the lactoperoxidase antimicrobial system," *Infection and Immunity*, vol. 39, no. 2, pp. 767–780, 1983.

11

Underwater Shockwave Pretreatment Process to Improve the Scent of Extracted *Citrus junos* Tanaka (Yuzu) Juice

Eisuke Kuraya, Akiko Touyama, Shina Nakada, Osamu Higa, and Shigeru Itoh

National Institute of Technology, Okinawa College, 905 Henoko, Nago City, Okinawa 905-2192, Japan

Correspondence should be addressed to Eisuke Kuraya; kuraya@okinawa-ct.ac.jp

Academic Editor: Mitsuru Yoshida

Citrus junos Tanaka (yuzu) has a strong characteristic aroma and thus its juice is used in various Japanese foods. Herein, we evaluate the volatile compounds in yuzu juice to investigate whether underwater shockwave pretreatment affects its scent. A shockwave pretreatment at increased discharge and energy of 3.5 kV and 4.9 kJ, respectively, increased the content of aroma-active compounds. Moreover, the underwater shockwave pretreatment afforded an approximate tenfold increase in the scent intensity of yuzu juice cultivated in Rikuzentakata. The proposed treatment method exhibited reliable and good performance for the extraction of volatile and aroma-active compounds from the yuzu fruit. The broad applicability and high reliability of this technique for improving the scent of yuzu fruit juice were demonstrated, confirming its potential for application to a wide range of food extraction processes.

1. Introduction

Citrus fruits are widely cultivated in regions between tropical and temperate zones and include some of the most important commercial crops. *Citrus junos* Tanaka (yuzu), a sour fruit, is cultivated mainly in Japan and Korea. In Japan, its annual production amounted to approximately 22,900 tons in 2013. Yuzu produced in Rikuzentakata (Iwate prefecture) is known as the "Northern Limit Yuzu" (NLY) and is renowned for its pleasant aroma [1]. NLY is produced in small quantities and as a consequence, it tends to be expensive. In addition, Rikuzentakata sustained significant damage during the Great Tohoku Earthquake and Tsunami in 2011. Thus, NLY is currently being promoted as part of the reconstruction plan.

Almost all parts of the yuzu fruit, including its peel, juice, and seeds, are utilized. Compared to other citrus fruits, yuzu has a strong characteristic aroma and is well known for the pleasant fragrance of its outer rind. Therefore, yuzu is used industrially in the production of sweets, beverages, cosmetics, perfumes, and aromatherapy products [2, 3]. Important bioactive components present in yuzu fruits include vitamin C, β-carotene, flavonoids, limonoids, and dietary fiber. The predominant citrus fruit-derived flavonoids are glycosides that also function as antioxidants [4, 5]. Citrus limonoids

such as limonin and nomilin are responsible for the bitter taste of citrus fruits and are characterized by their substituted furan moiety [5]. Owing to these characteristics, yuzu juice is commonly used in Japanese cooking. In the standard juicing process, high pressure is applied to the flavedo and/or albedo of the yuzu fruit, resulting in the extraction of ascorbic acid, flavanone glycosides, and odorants present in the juice. Owing to the high pressure employed, overextraction of bitter components (e.g., naringin, neohesperidin, and limonoids) [6] often occurs and for this reason, the pressure must be carefully controlled. As a consequence, it is difficult to obtain fruit juice with a strong aroma using the standard juicing processes.

Shockwaves propagate in plant media at rates exceeding the speed of sound, dividing into penetration and reflected waves upon a change in density. Underwater shockwaves cause instantaneous high pressure, splitting open cell structures and instantly generating multiple cracks on cell wall [7, 8]. The shockwave pretreatment liquefies vegetable material immediately (to a consistency similar to that of carrot juice) and simultaneously destroys cell structures. In previous studies, the effect of instantaneous high pressure produced by a conventional mixer blending extraction method contributed to increased yields in the tomato saponin, esculeoside A, from tomatoes [9] and an increased extraction efficiency of lipophilic gingerols and shogaols from ginger [10]. In a recent study, we have shown that multiple cracks generated by underwater shockwaves act as permeation pathways, increasing the extraction ratio of essential oils in steam distillation processes [7]. This observation indicates that the implementation of underwater shockwave treatment as a preprocessing step can be useful in the extraction of functional components from food materials.

Ultrasound-assisted extraction is also known to have a significant effect on the rates of various processes in chemical and food industries. Using ultrasound, full extractions can be completed in minutes with high reproducibility; moreover, solvent consumption is reduced and a final product with higher purity is yielded [11]. The effects of ultrasound propagation in solid/liquid media have been described in the literature previously [12, 13]. Pulsed electric field (PEF) processing is a nonthermal food processing technology based on the application of short pulses of high voltage to the food product. This process has been investigated for its suitability to enhance the extractability of fruit and vegetable juices as well as intracellular compounds [14]. In one such application, a higher concentration of bioactive compounds, for example, polyphenols in tomatoes, was reported after the PEF treatment [15]. Boussetta et al. investigated the effects of applying PEF and high voltage electrical discharges on the efficiency of the extraction of total soluble matter and polyphenols from grape skins (*Vitis vinifera* L.), and observed large differences in the total amount of polyphenols before and after the treatment [16, 17]. However, before the extraction of intracellular metabolites in vegetables and fruits can take place, these processing methods require liquefaction. This allows the rapid fragmentation of the raw material and generation of cavitation bubbles that increase the electrical conductivity and permeability of the whole plant tissue sample in solid/liquid media.

Following our shockwave pretreatment method, the cell walls disintegrated and the fruit softened, thus allowing access to the juice. This pressurization method also facilitated efficient extraction of functional compounds from the fruit tissue. We have recently reported that the multiple cracks generated by underwater shockwaves act as permeation pathways, thereby increasing the extraction ratio of essential oils in steam distillation processes [7]. Moreover, we observed that the flavanone glycoside content and oxygen radical absorbance capacity value of yuzu juice increased ca. 1.7-fold following the underwater shockwave pretreatment of the whole fruit with pilot-scale processing equipment [8]. These results indicate that efficient extraction of ascorbic acid, flavanone glycosides, and limonoids from yuzu fruit can be achieved using underwater shockwave treatment as a preprocessing step. Moreover, the dynamic control of instantaneous high pressure by underwater shockwaves proves its feasibility as a preprocessing step in the extraction of functional components from food materials.

In this study, we introduce an innovative application for this pretreatment process, aimed at improving the strength of scent extracted from *Citrus junos* Tanaka (yuzu) fruit juice. We also evaluate the content of volatile compounds in yuzu fruit juice to investigate whether such a pretreatment method affects its characteristics.

2. Material and Methods

2.1. Plant Material. Fruits of NLY produced in Rikuzentakata (Iwate prefecture) and yuzu produced in Kochi (KY; grown in Kami city, Kochi prefecture; Kumon variety) were provided by the Iwate Agricultural Research Center. All fruits were fully ripe and firm when harvested in November 2013. For both types of fruit, the juice was extracted by hand pressing using a hand-operated citrus juicer (Nanyo LLC., Tokushima, Japan). The samples of yuzu juice examined were comprised of mixtures of 30 and 12 fruit samples from KY and NLY, respectively. The juice samples were filtered through a nonwoven fabric net and the extract was immediately prepared for analysis.

2.2. Chemicals. n-Hexane, sodium sulfate, and 1-pentanol (internal standard, special grade reagent) were purchased from Nacalai Tesque, Inc. (Kyoto, Japan). Distilled and deionized water was used in all aqueous solutions.

2.3. Underwater Shockwave Pretreatment. Pilot-scale processing equipment was developed for the pretreatment of the whole yuzu fruit with continuous underwater shockwaves [8]. The high voltage supply generated a voltage of 2.5–3.5 kV and the underwater shockwaves were generated temporarily in the vessel by electrical discharge. A whole piece of yuzu fruit was placed in a silicone tube (ID = 114 mm, OD = 124 mm), separated from the shockwave generation source, and subjected to the instantaneous high-pressure load produced by the underwater shockwave. On the basis of a previous study [18], we assumed that the pressure produced by the shockwave generator at 3.5 kV and 4.9 kJ was ~40 MPa. The volatile compound content was evaluated to determine whether the pretreatment affects the characteristics of yuzu juice.

2.4. Isolation of Volatile Compounds Present in Yuzu Juice. The juice sample (1 mL) was extracted with distilled n-hexane (1 mL). After adding the extraction solvent (n-hexane), the juice sample was sonicated for 5 min and allowed to stand at 4°C for 24 h. The hexane layer was separated and dried over anhydrous sodium sulfate. Aromatic characterization of the volatile compounds was carried out by gas chromatography-mass spectrometry (GC/MS).

2.5. Analysis of Volatile Components in Extracted Juice. The juice aroma profiles were evaluated to determine whether the pretreatment affected the characteristics of yuzu juice. Two sampling methods were employed, namely, the n-hexane extraction method and the headspace method. The juice samples extracted with n-hexane were analyzed using a gas chromatograph coupled to a mass spectrometer (QP-2010 Plus; Shimadzu Co., Kyoto, Japan) and equipped with an autoinjector (AOC-20i, Shimadzu). Quantitative determinations of volatile components were based on the analyses of

TABLE 1: Production districts, cultivars, and average weights of tested yuzu fruits.

Sample	Producing district	Cultivar	Weight of fruit (g)
KY[a]	Kami City, Kochi pref.	Kumon	129.6 ± 8.7
NLY[b]	Rikuzentakata, Iwate pref.	No name	81.3 ± 14.0

[a]$n = 30$; [b]$n = 12$.

peak areas. Furthermore, the aroma characteristics of the yuzu juice sample (close to sensory evaluation) were analyzed directly by headspace (HS)-GC/MS. In qualitative analysis, 0.2% (v/v) 1-pentanol was used as the internal standard. The internal standard solution (100 μL) was added to the yuzu juice sample (diluted 100-fold; 1.9 mL) in a 20-mL gas-tight vial sealed with a septum. Using a headspace sampler (TurboMatrix HS-40, PerkinElmer, Inc., MA, USA), the sample vial was pressurized above the capillary column head pressure with a carrier gas (helium) and heated at 60°C for 35 min to equilibrate with the vapor-phase extraction. The pressurized vapor phase, including the volatile compounds, was subsequently transferred to the GC/MS system with an injection time of 0.05 min.

GC/MS analyses were performed using a ZB-WAX Plus column (length = 60 m, ID = 0.32 mm, and thickness = 0.5 μm; Phenomenex Inc., Torrance, CA, USA). The GC oven temperature program was set as follows: 40°C held for 1 min, increased at 10°C/min to 160°C, increased at 5°C/min to 200°C, and held for 4 min. The injector and detector temperatures were set at 200°C. The mass range was scanned from 30 to 600 amu. Control of the GC/MS system and data peak processing were performed using Shimadzu GC/MS solution software (Version 2.7). The volatile components were identified by comparing their retention indices and mass fragmentation patterns with MS libraries (NIST05 and FFNSC Library ver. 1.2; Shimadzu Co., Kyoto, Japan). The volatile components were identified based on their linear retention indices (RIs) and by comparison of their mass spectra with the MS data of reference compounds. The linear RIs were determined for all constituents using a homologous series of n-alkanes (C8–C24), injected under the same chromatographic conditions as the samples. The odor-activity of all yuzu oil compounds was evaluated based on their flavor dilution (FD) factor values obtained using GC-olfactometry and aroma extract dilution analysis (AEDA) by Lan-Phi et al. [3].

3. Results and Discussion

We employed an underwater shockwave treatment method as a preprocessing step for the extraction of volatile components from yuzu samples. By comparing the results obtained from the analyses of two different yuzu varieties, we elucidated successfully the effect of shockwave treatment on NLY and established how it compares to its effect on KY. The weight of the fruit, production district, and cultivar are listed in Table 1. The weights of KY and NLY fruits were calculated using the average of the weights recorded for 30 and 12 fruits, respectively. For both varieties, the juice samples consisted of

a mixture of 30 (for KY) and 12 (for NLY) fruit samples. In general, Kochi is the largest yuzu-producing district owing to its mild climate. Although it is possible to grow yuzu in Rikuzentakata, the fruit is generally smaller as a result of the colder climate of the Tohoku district. The weights of the KY and NLY fruits were calculated to be 129.6 ± 8.7 and 81.3 ± 14.0 g/fruit, respectively. As predicted, the average weight of the NLY fruit was lower than that associated with the KY fruit.

In this study, pilot-scale processing equipment was developed for the pretreatment of yuzu fruits with underwater shockwaves. It was possible to continuously feed yuzu fruits into the underwater shockwave treatment vessel [8]. Following exposure of the whole fruit to the shockwave treatment, the oil glands were crushed and volatile oils and juice were released from the fruit. This was attributed to the spalling destruction caused by the shockwave treatment, whereby the resulting cracks could act as permeation pathways for the aroma compounds during the extraction process. These observations prove that the underwater shockwaves reached the entire cell and readily destroyed the fruit oil glands and cell walls.

The volatile compounds were analyzed by GC/MS and identified by their peak areas and comparison to mass spectrometry libraries. We identified 19 compounds in the n-hexane juice extract, including limonene, γ-terpinene, myrcene, α-pinene, β-pinene, β-phellandrene, p-cymene, and α-terpinolene (Table 2). Notably, the volatile contents in the extracted KY and NLY juice samples were distinctly different. The shockwave (SW) pretreatment afforded negligible changes in the relative concentrations of the two major volatile components in KY limonene (control = 86.8%; with pretreatment = 83.5–85.3%) and γ-terpinene (control = 7.19%; with pretreatment = 7.90–8.49). Negligible changes were also observed in the main volatile compounds of NLY extracts (limonene; control = 72.5%; with pretreatment = 71.4–71.7%). Interestingly, the relative concentrations of sabinene and caryophyllene were not detected in the untreated and SW 2.5 kV-treated NLY juice extracts but increased after a shockwave treatment of 3.5 kV was employed. These results indicate that the oil glands distributed in the yuzu peel are effectively destroyed by the SW treatment of 3.5 kV (but not 2.5 kV), and the multiple cracks created by the shockwave processing act as permeation pathways, increasing the extraction ratio of volatile compounds obtained during extraction.

The volatile content in treated and untreated yuzu fruits also differed, suggesting that, in addition to increasing the content of volatile compounds in the juice samples, the shockwave treatment also changed the scent intensity of the juice. The aroma characteristics of the citrus juice samples were analyzed by HS-GC/MS; this is close to sensory evaluation. Figure 1 illustrates the contents of volatile compounds calculated from the ratio of the integrated area of the aroma component relative to the area arising from the internal standard. The contents of volatile components of untreated KY and NLY fruits were determined as 0.287 and 0.329, respectively, indicating that NLY fruit juice exhibits stronger scent intensity than KY juice. The volatile content of KY juice increased with an increase in applied shockwave energy (~ ×1.3 at 3.5 kV). However, for NLY, a more marked increase

TABLE 2: Volatile composition of yuzu juices.

| Compound | RI | Relative concentration (%) | | | | | |
| | ZB-WAX Plus | KY | | | NLY | | |
		Control	SW 2.5 kV	SW 3.5 kV	Control	SW 2.5 kV	SW 3.5 kV
α-Pinene	990	1.51	1.52	1.72	1.81	1.89	2.15
α-Thujene	993	—	—	—	0.08	0.16	0.11
β-Pinene	1077	0.43	0.48	0.50	1.11	1.13	1.15
Sabinene	1090	—	—	—	—	—	0.12
Myrcene	1127	2.44	2.23	2.20	1.86	1.91	1.94
α-Phellandrene	1135	0.10	0.33	0.17	0.75	0.79	0.78
α-Terpinene	1149	0.17	0.16	0.15	0.32	0.29	0.32
Limonene	1169	86.77	83.50	85.26	72.46	71.42	71.70
β-Phellandrene	1180	0.40	1.59	0.91	4.06	4.55	4.43
γ-Terpinene	1215	7.19	8.49	7.90	14.65	14.85	15.14
p-Cymene	1242	0.32	0.42	0.42	0.80	0.97	0.78
α-Terpinolene	1253	0.29	0.27	0.26	0.44	0.43	0.42
Copaene	1462	0.01	0.04	0.02	0.01	0.03	0.02
Linalool	1495	0.06	0.46	0.28	1.06	0.87	0.35
Caryophyllene	1573	—	—	—	—	—	0.07
(E)-β-Farnesene	1613	0.04	0.22	0.08	0.30	0.25	0.30
γ-Cadinene	1682	0.16	0.11	0.08	0.07	0.09	0.05
Bicyclogermacrene	1706	—	—	—	0.07	0.06	0.07
δ-Cadinene	1721	0.02	0.06	0.04	0.03	0.06	0.03
Total		99.89	99.88	99.97	99.89	99.77	99.92

RI: Retention Index values were determined relative to n-alkanes on the ZB-WAX Plus column. The correct isomer was not identified. SW represents shockwave treatment.

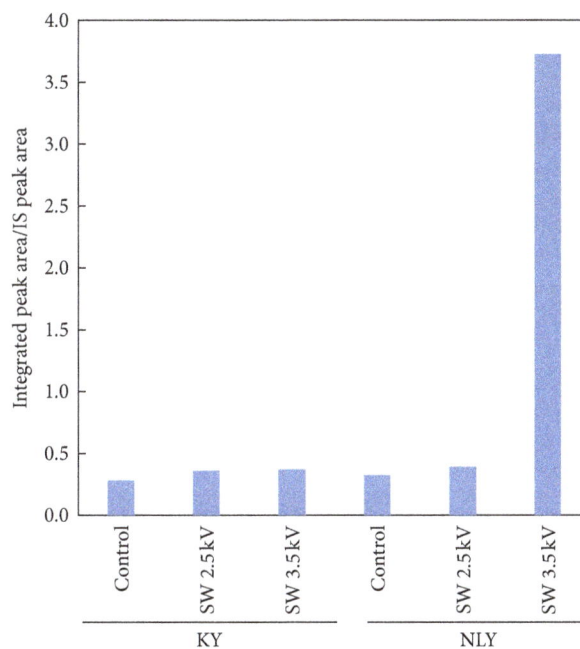

FIGURE 1: Changes in the content of volatile components in samples of yuzu juice as a function of treatment (SW represents shockwave treatment). The contents of volatile compounds were calculated from the ratio of the integrated areas of aroma components to the internal standard area.

TABLE 3: Overview of major volatile compounds in yuzu juice.

| Compound | FD Value[a] | | Intensity ratio of volatile component to internal standard | | | | | |
| | | | KY | | | NLY | | |
	$\log_3$ (FD-factor)[b]	Control	SW[c] 2.5 kV	SW[c] 3.5 kV	Control	SW[c] 2.5 kV	SW[c] 3.5 kV
Limonene	9	0.25	0.30	0.32	0.24	0.28	2.67
β-Phellandrene	8	0.001	0.006	0.003	0.013	0.018	0.165
γ-Terpinene	6	0.020	0.031	0.030	0.048	0.059	0.565

[a]Reference: Lan-Phi et al., 2009 [3]. [b]The base-3 logarithm of flavor dilution factor value on DB-WAX column. [c]SW represents shockwave treatment.

TABLE 4: Overview of trace volatile compounds in yuzu juice.

| Compound | FD Value[a] | | Intensity ratio of volatile component to internal standard ($\times 10^{-3}$) | | | | | |
| | | | KY | | | NLY | | |
	$\log_3$ (FD-factor)[b]	Control	SW[c] 2.5 kV	SW[c] 3.5 kV	Control	SW[c] 2.5 kV	SW[c] 3.5 kV
α-Pinene	8	4.32	5.56	6.43	5.94	7.51	80.1
Myrcene	8	7.00	8.15	8.27	6.13	7.59	72.0
p-Cymene	8	0.89	1.52	1.56	2.63	3.84	29.3
α-Terpinolene	9	0.81	1.00	0.96	1.44	1.72	15.7
Linalool	7	0.18	1.66	1.04	3.48	3.44	12.8
(E)-β-Farnesene	9	0.11	0.81	0.31	1.00	1.00	11.1
Bicyclogermacrene	8	—	—	—	0.22	0.26	2.81

[a]Reference: Lan-Phi et al., 2009 [3]. [b]The base-3 logarithm of flavor dilution factor value on DB-WAX column. [c]SW represents shockwave treatment.

(maximum concentration ratio = 3.72; ~ ×11.3 increase) was observed when SW 3.5 kV was applied.

The odor-active components in yuzu oils, determined on the basis of the flavor dilution (FD) factor value, are important in the evaluation of the aroma characteristics of juice samples. The dilution of odor-active compound was performed until no odor was detected in the most diluted sample. The highest dilution at which an individual component could be detected was defined as the FD factor for that odorant [3, 19]. Thus, volatile compounds with high FD values contribute to the yuzu flavor. These compounds include limonene, β-phellandrene, and γ-terpinene (major volatile compounds) together with α-pinene, myrcene, p-cymene, α-terpinolene, linalool, (E)-β-farnesene, and bicyclogermacrene (trace volatile compounds). In this study, the contents of these components were evaluated to determine whether the shockwave pretreatment affects the characteristics of yuzu juice. The contents of major volatile components of the examined juice samples are listed in Table 3. Limonene, β-phellandrene, and γ-terpinene represent the odor-active components associated with yuzu juices and are described as having citrus-like aroma characteristic to the Kumon cultivar. Employing shockwave pretreatment at SW 3.5 kV increased the content of all of these components and thus, the odor-active content ratio of NLY fruit increased substantially (≥ ×10).

The contents of trace volatile components of juice samples are listed in Table 4. Although the contents of these compounds were low, their FD values were high, thus contributing strongly to the aroma of yuzu juice. Employing

shockwave pretreatment at 3.5 kV increased the contents of all of these components. Notably, the odor-active content ratio of NLY increased substantially and the content of each compound was ca. ×10 greater than the increase observed for the major components. These results indicate that shockwave processing leads to the destruction of fruit oil glands and formation of multiple cracks, thus resulting in improved passage of volatile components from the yuzu fruit and more efficient extraction of volatile compounds, as well as a substantial increase in the odor of yuzu juice.

4. Conclusion

This study demonstrates the applicability and reliability of the underwater shockwave preprocessing treatment to improving the characteristics of odorants extracted from yuzu juice. This technique is reliable and exhibits excellent results in the extraction of volatile and aroma components from yuzu fruit prior to the extraction process. This underwater shockwave pretreatment technique can be also applied to other plant materials through the same mechanism, even if the odorants may be stored in different parts of the biological material. Moreover, this innovative process has the potential to find application in a wide range of extraction processes such as juice extraction or extraction of components from other naturally occurring foods and medicinal plants. Thus, the dynamic control of instantaneous high pressure by underwater shockwaves should prove valuable in many industrial applications.

Conflicts of Interest

The authors declare that there are no conflicts of interest regarding the publication of this paper.

Acknowledgments

The authors are very grateful to Dr. Katsuya Higa (Deceased 17 February 2015), a professor at the National Institute of Technology, Okinawa College, for his invaluable advice on this research. They also gratefully acknowledge the Iwate Agricultural Research Center and Ms. Megumi Kubota for sample preparation and GC/MS analyses. This research was conducted based on "A Scheme to Revitalize Agriculture and Fisheries in Disaster Area through Deploying Highly Advanced Technology" by the Ministry of Agriculture, Forestry and Fisheries.

References

[1] E. Kuraya, S. Nakada, M. Kubota, T. Hasegawa, and S. Itoh, "Chemical and aroma profiles of Northern Limit Yuzu (Citrus junos) peel oils from different producing districts," Natural Volatiles & Essential Oils, vol. 2, no. 3, 84 pages, 2015.

[2] M. Sawamura, Ed., Citrus Essential, Oils: Flavor and Fragrance, Wiley, Hoboken, NJ, USA, 2010.

[3] N. T. Lan-Phi, T. Shimamura, H. Ukeda, and M. Sawamura, "Chemical and aroma profiles of yuzu (Citrus junos) peel oils of different cultivars," Food Chemistry, vol. 115, no. 3, pp. 1042–1047, 2009.

[4] Z. Zou, W. Xi, Y. Hu, C. Nie, and Z. Zhou, "Antioxidant activity of Citrus fruits," Food Chemistry, vol. 196, pp. 885–896, 2016.

[5] A. Sawabe, H. Kumamoto, and Y. Matsubara, "Bioactive glycosides in citrus fruit peels," Bulletin of the Institute for Comprehensive Agricultural Sciences Kinki University, vol. 67, no. 6, pp. 57–67, 1998.

[6] A. Shimada, "Antioxidant activity and lipase and alpha-glucosidase inhibitory activities of yuzu juice (Citrus junos Tanaka)," Journal of Yasuda Women's University, vol. 43, pp. 351–357, 2015.

[7] E. Kuraya, Y. Miyafuji, A. Takemoto, and S. Itoh, "The effect of underwater shock waves on steam distillation of Alpinia zerumbet leaves," Transactions of the Materials Research Society of Japan, vol. 39, no. 4, pp. 447–449, 2014.

[8] E. Kuraya, S. Nakada, A. Touyama, and S. Itoh, "Improving the antioxidant functionality of Citrus junos Tanaka (yuzu) fruit juice by underwater shockwave pretreatment," Food Chemistry, vol. 216, pp. 123–129, 2017.

[9] H. Manabe, A. Takemoto, H. Maehara et al., "Efficient improved extraction of tomato saponin using shock waves," Chemical and Pharmaceutical Bulletin, vol. 59, no. 11, pp. 1406–1408, 2011.

[10] H. Maehara, T. Watanabe, A. Takemoto, and S. Itoh, "A new processing of ginger using the underwater shock wave," Materials Science Forum, vol. 673, pp. 215–218, 2011.

[11] F. Chemat, N. Rombaut, A. Sicaire, A. Meullemiestre, A. Fabiano-Tixier, and M. Abert-Vian, "Ultrasound assisted extraction of food and natural products. Mechanisms, techniques, combinations, protocols and applications. A review," Ultrasonics Sonochemistry, vol. 34, pp. 540–560, 2017.

[12] T. J. Mason, L. Paniwnyk, and J. P. Lorimer, "The uses of ultrasound in food technology," Ultrasonics Sonochemistry, vol. 3, no. 3, pp. S253–S260, 1996.

[13] K. S. Suslick and G. J. Price, "Applications of ultrasound to materials chemistry," Annual Review of Materials Science, vol. 29, pp. 295–326, 1999.

[14] S. Asavasanti, W. Ristenpart, P. Stroeve, and D. M. Barrett, "Permeabilization of plant tissues by monopolar pulsed electric fields: Effect of frequency," Journal of Food Science, vol. 76, no. 1, pp. E98–E111, 2011.

[15] A. Vallverdú-Queralt, I. Odriozola-Serrano, G. Oms-Oliu, R. M. Lamuela-Raventós, P. Elez-Martínez, and O. Martín-Belloso, "Changes in the polyphenol profile of tomato juices processed by pulsed electric fields," Journal of Agricultural and Food Chemistry, vol. 60, no. 38, pp. 9667–9672, 2012.

[16] N. Boussetta, E. Vorobiev, T. Reess et al., "Scale-up of high voltage electrical discharges for polyphenols extraction from grape pomace: Effect of the dynamic shock waves," Innovative Food Science and Emerging Technologies, vol. 16, pp. 129–136, 2012.

[17] N. Boussetta, N. Lebovka, E. Vorobiev, H. Adenier, C. Bedel-Cloutour, and J.-L. Lanoisellé, "Electrically assisted extraction of soluble matter from chardonnay grape skins for polyphenol recovery," Journal of Agricultural and Food Chemistry, vol. 57, no. 4, pp. 1491–1497, 2009.

[18] O. Higa et al., "Mechanism of the shock wave generation and energy efficiency by underwater discharge," The International Journal of Multiphysics, vol. 6, no. 2, pp. 89–98, 2012.

[19] N. Miyazawa, N. Tomita, Y. Kurobayashi et al., "Novel character impact compounds in yuzu (Citrus junos sieb. ex Tanaka) peel oil," Journal of Agricultural and Food Chemistry, vol. 57, no. 5, pp. 1990–1996, 2009.

12

Effects of Roasting Temperature and Time on the Chemical Composition of Argan Oil

Rahma Belcadi-Haloui,[1] **Abderrahmane Zekhnini** (ID),[2]
Yassine El-Alem,[3] **and Abdelhakim Hatimi**[1]

[1]*Laboratory of Plants Biotechnologies, Faculty of Sciences, BP 8016, Agadir 80 000, Morocco*
[2]*Laboratory of Aquatic Systems, Faculty of Sciences, BP 8016, Agadir 80 000, Morocco*
[3]*Autonomous Establishment of Control and Coordination of Exports, 23 E, Industrial Zone of Tassila, Agadir 80 000, Morocco*

Correspondence should be addressed to Abderrahmane Zekhnini; a.zekhnini@uiz.ac.ma

Academic Editor: Rosana G. Moreira

This work aimed at assessing the effects of roasting temperature and duration on chemical composition of argan oil. Thus, argan oils extracted from almonds roasted at different temperatures (75-175°C) and times (10-30 min) were analyzed and compared to a control. The physicochemical parameters (acidity, peroxide value, and absorbance at 232, 270 nm) increased slightly and the fatty acid composition did not show significant variation, regardless of roasting temperature and duration. The browning index increased significantly for temperatures greater than or equal to 100°C. The tocopherols content significantly decreased with roasting temperature and time (from 977.9 to 305.2 mg/kg after roasting at 175°C for 10 min). However, fluctuations are noted as a function of temperature. The phospholipids content increased with roasting temperature and time (from 0.198 % to 1.370 % after roasting at 175°C for 30 min). The decrease in the tocopherols content would be due to their thermolability. The increase in phospholipids and tocopherols content could be explained by better extractability. The results obtained make it possible to conclude that a roasting at 125-150°C / 10 min would allow the development of the organoleptic properties of the oil, notably its hazelnut flavour, without compromising its oxidative stability.

1. Introduction

Argan oil extracted from the fruits of *Argania spinosa* L. (Sapotaceae) has many nutritional, cosmetic, and therapeutic properties thanks to its richness in unsaturated fatty acids (UFA) and bioactive substances such as tocopherols and phytosterols [1]. The cosmetic oil is extracted by mechanical press from unroasted almonds. As for the edible oil, it is extracted by press or in the traditional way by hand kneading roasted almonds. Roasting is usually carried out on the wood fire by constantly stirring the almonds until browning. It constitutes an important step in the development of the organoleptic characteristics of the oil, in particular the typical hazelnut aroma appreciated by the consumer.

In industry, heating of oilseeds is carried out with the aim of increasing oil extraction yield, reducing seed moisture, and deactivating lipases and lipoxygenases which could induce fatty acids (FA) oxidation [2]. The oxidation could lead to loss of nutritional value and decreased shelf life [3]. The roasting of oilseeds could also improve the stability of vegetable oils by increasing the extraction yield of antioxidant substances [4, 5]. As regards the effect of roasting on the content of tocopherols which represent the major antioxidants of vegetable oils, the data from the literature are inconsistent. Depending on the nature and variety of oilseeds, roasting may lead to a decrease [5–7] or an increase in tocopherols level in oil [8–10]. Since tocopherols are thermosensitive, their final concentration in the oil will depend essentially on the temperature and the duration of the roasting.

In the case of argan oil, the literature does not report studies of the impact of roasting on the chemical composition of the oil, with the exception of the study by Harhar et al. [11] in which only the duration of the roasting was taken into consideration for a constant temperature of 110°C. The objective of this work was to evaluate the concomitant effects of the temperature and the duration of the roasting of the *A. spinosa* almonds on the chemical composition of the oil. To do this, the roasting was carried out by combining temperatures of 75, 100, 125, 150, and 175°C with times of 10, 20, and 30 min. The chemical composition of the oil related to the content of fatty acids, tocopherols, and phospholipids. The physicochemical parameters considered were acidity, peroxide value (PV), absorbance at 232 nm (K232) and 270 nm (K270), and browning index (assessed by absorbance at 420 nm).

2. Materials and Methods

2.1. Argan Fruits and Reagents. The fruits of argan tree were collected in the forest of Admine located at 15 km in the east of Agadir city (south of Morocco). Tocopherols and fatty acids homologues were purchased from Sigma-Aldrich (St Louis, USA). Hexane, tetrahydrofuran, chloroform, and cyclohexane were obtained from Merck (Darmstadt, Germany). All other chemicals and solvents were of analytical grade.

2.2. Preparation of Argan Oil. Argan fruits were depulpated and obtained seeds were crushed manually. Then, almonds were distributed in batches of 300 g, uniformly spread on a stainless steel plate, and placed in an electric convection oven (Binder ED 115, USA) under continuous aeration. Roasting temperatures of 75, 100, 125, 150, and 175°C were applied for duration of 10, 20, or 30 min. After cooling, the almonds were subjected to a mechanical press for cold extraction of the oil (Komet S87G press, Germany). A control sample of oil was obtained from unroasted almonds.

2.3. Determination of Physicochemical Parameters. The acid value (AV) and the peroxide value (PV) were determined according to IUPAC 2.201 [12] and IUPAC 2.501 [13], respectively.
The absorbance at 233 and 270 nm was determined using a Varian DMS 80 spectrophotometer after diluting oil in cyclohexane (1/100, v/v).
The color of oils was evaluated as described by Yoshida et al. [8]. Solutions of 5% of oils in chloroform (w/v) were prepared to determine absorbance at 420 nm in a Varian DMS 80 spectrophotometer.

2.4. Fatty Acids Analysis. Fatty acid methyl esters (FAME) were prepared according to AOCS method Ce 1K-07 [14]. FAME were analyzed by gas chromatography (Agilent 6890, USA). The apparatus was equipped with a flame ionisation detector and a BPX70 column (60 m length, 0.32 mm internal diameter, and 0.25 μm film; SGE Europe). Helium was used as the carrier gas at a flow rate of 1 ml/min. The temperature of the column oven was 170°C. The temperature of the injector and the detector was 220°C.

2.5. Tocopherols Analysis. The analysis of tocopherols was carried out using high performance liquid chromatography (Flexar, Perkinelmer, USA). The apparatus was equipped with a C18 column (250 mm length, 4 mm internal diameter, and 5 μm particle size; Varian Inc.) and a fluorescence detector. The excitation and detection wavelengths were 290 and 330 nm, respectively. The mobile phase consisted of 2:98 tetrahydrofuran/hexane (v/v) mixture and the flow rate was 1 ml/min. Oil samples were diluted in hexane (0.1g/ml) before analysis.

2.6. Determination of Phospholipids Content. The phospholipids content was estimated by measuring phosphorous according to AOCS recommended method Ca 12-5515 [15]. The method is based on ashing the oil sample in the presence of zinc oxide. The phosphorus is then determined by a spectrophotometric measurement as a blue phosphomolybdic acid complex (Varian DMS 80 spectrophotometer). The phospholipids content is obtained by multiplying the phosphorus content by 25.

2.7. Statistical Analysis. The Statistica 6 software was used for processing data from three replicates (n = 3). Significant differences from the control were determined using the analysis of variance (ANOVA) followed by Newman-Keuls student method.

3. Results and Discussion

3.1. Physicochemical Parameters. The effect of roasting on the initial physicochemical characteristics of the argan oil is presented in Table 1. The acidity of 0.18 for the control oil increased significantly under the effect of the roasting whatever the temperature and the duration. The maximum values (0.28-0.3) were recorded for the temperature of 175°C. This result is consistent with previous works on pine nuts, colza, and sesame oils [16–18]. The increase in acidity could be explained by the release of FA following the hydrolysis of the triglycerides which constitute the major components of vegetable oils. It could trigger reactions of oxidative degradation of the oil due to the prooxidative action of free FA [19].

The PV was 0.1 meq O_2/Kg for the control oil. It significantly increased in oils extracted from almonds roasted at 125°C for 30 min, 150°C for 10 min, and at 175°C for 10, 20, and 30 min durations. PV is a parameter commonly used to evaluate the primary oxidation state of vegetable oils during technological treatments and storage. It reflects the content of hydroperoxides formed by oxidation of the UFA. In this regard, previous studies of rapeseed, sesame, and walnut seeds oils reported an increase in the PV of the oil with an increase in the roasting temperature of the seeds [16, 18, 20]. Our results showed that roasting at elevated temperatures resulted in oil oxidation which could affect its stability during storage.

K232 reflects the content of conjugated dienes (CD) produced from the hydroperoxides of the UFA. K270 permits to evaluate the secondary oxidation products resulting from the decomposition of hydroperoxides [3]. In our study, the K232 value of 0.844 for the control oil increased to 0.878, 0.896, and 0.906, respectively, under the 175°C/10 min, 75°C/30 min, and 100°C/10 min conditions and decreased to 0.814, 0.806, and 0.793, respectively, for the couples temperature (°C)/time (min) of 175/30, 100/20, and 75/10. The absorption at 270 nm also showed variations according to the roasting conditions. Overall, it increased from 0.041 to 0.062, respectively, at

TABLE 1: Effects of temperature and duration of roasting on the physicochemical characteristics of argan oil.

Roasting conditions	Peroxide value (meq O_2/Kg)	Acidity (%)	K232	K270
Control	0.10 ± 0.01	0.180 ± 0.020	0.844 ± 0.021	0.054 ± 0.001
75°C				
10 min	0.11± 0.01	0.230 ± 0.010 *	0.793 ± 0.042 *	0.057 ± 0.003*
20 min	0.11± 0.01	0.240 ± 0.010 *	0.815 ± 0.014 *	0.054 ± 0.001*
30 min	0.11± 0.01	0.240 ± 0.010 *	0.896 ± 0.075 *	0.062 ± 0.002*
100°C				
10 min	0.12 ± 0.02	0.207 ± 0.015 *	0.906 ± 0.062 *	0.052 ± 0.001*
20 min	0.12 ± 0.01	0.210 ± 0.010 *	0.806 ± 0.097 *	0.048 ± 0.001*
30 min	0.12 ± 0.01	0.217 ± 0.015 *	0.846 ± 0.061 *	0.051 ± 0.001*
125°C				
10 min	0.11 ± 0.02	0.250 ± 0.010 *	0.858 ± 0.092 *	0.053 ± 0.001*
20 min	0.13 ± 0.02	0.240 ± 0.010 *	0.824 ± 0.063 *	0.050 ± 0.001*
30 min	0.14 ± 0.01 *	0.220 ± 0.010 *	0.836 ± 0.095 *	0.052 ± 0.003*
150°C				
10 min	0.14 ± 0.02 *	0.223 ± 0.012 *	0.860 ± 0.033 *	0.050 ± 0.002*
20 min	0.13 ± 0.02	0.260 ± 0.017 *	0.834 ± 0.034 *	0.044 ± 0.004 *
30 min	0.12 ± 0.02	0.270 ± 0.010 *	0.824 ± 0.025*	0.041 ± 0.003
175°C				
10 min	0.15 ± 0.02 *	0.280 ± 0.010 *	0.878 ± 0.021 *	0.055 ± 0.002 *
20 min	0.15 ± 0.02 *	0.283 ± 0.015 *	0.827 ± 0.031 *	0.053 ± 0.003 *
30 min	0.16 ± 0.02 *	0.300 ± 0.010 *	0.814 ± 0.097*	0.052 ± 0.004 *

Values present means ± standard deviations from triplicate measurements.
* Significantly different from control ($p < 0.05$).

150°C/30 min and 75°C/30 min conditions, compared to the control (0.054). Previous studies reported that increasing temperature and time of seed roasting led to an increase in K232 and K270 [18, 21]. The fluctuations of the CD contents recorded in this work would be due to the instability of these compounds which transform into secondary oxidation products. The increase in the K270 value indicates the formation of undesirable oxidation products such as carbonyl compounds. However, the values of acidity, PV, and K270 encountered under the various conditions of our work remained below the limits set by the Moroccan Standard NM 08.5.090 [22] characterizing the "extra virgin" quality of argan oil (acidity < 0.8; PV < 15 meq O_2/kg, K270 < 0.35).

The development of the color of argan oil prepared from almonds roasted at different temperatures and times is shown in Figure 1. The absorbance at 420 nm increased significantly from 100°C with both time and temperature. These results are in agreement with data previously reported for other edible oils [9, 10, 17]. They reflect the formation of browning substances from nonenzymatic Maillard-type reactions between reducing sugars and free amino acids or amides [23]. The products of the Maillard reaction are also responsible for the development of the flavour of the edible argan oil. For this reason, roasting constitutes an important step in the edible oil production. Some derivatives of the Maillard reaction also afford numerous food safety benefits [24]. However, others exhibited toxicity [25, 26].

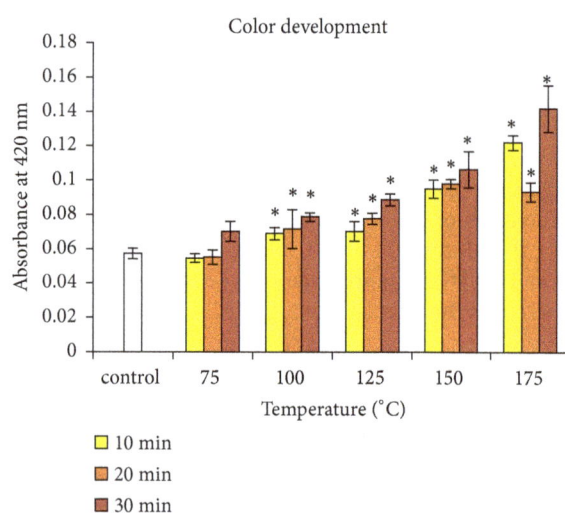

FIGURE 1: Changes in absorbance (at 420 nm) of argan oil according to roasting temperature and time. * Significantly different from control ($p < 0.05$).

3.2. Fatty Acid Composition.

The FA composition of oils obtained from unroasted and roasted almonds is shown in Table 2. Argan oil from unroasted almonds consisted of 47.01% oleic and 32.66% linoleic acids as UFA and 13.59%

TABLE 2: Effect of temperature and duration of roasting on the content of the argan oil in principal fatty acids.

Roasting conditions	Fatty acid (%)				
	C16:0	C18:0	C18:1	C18:2	C18:3
Control	13.59 ± 0.10	5.49 ± 0.04	47.00 ± 0.05	32.66 ± 0.02	0.085 ± 0.001
75°C					
10 min	12.89 ± 0.09	5.61 ± 0.07	46.79 ± 0.06	33.44 ± 0.04	0.084 ± 0.003
20 min	13.17 ± 0.03	5.44 ± 0.06	47.75 ± 0.05	32.38 ± 0.11	0.082 ± 0.001
30 min	13.17 ± 0.03	5.34 ± 0.05	47.73 ± 0.02	32.50 ± 0.07	0.083 ± 0.002
100°C					
10 min	13.41 ± 0.03	5.46 ± 0.07	46.71 ± 0.08	33.23 ± 0.06	0.090 ± 0.001
20 min	13.50 ± 0.04	5.52 ± 0.05	46.39 ± 0.07	33.30 ± 0.11	0.091 ± 0.001
30 min	13.54 ± 0.02	5.51 ± 0.07	46.21 ± 0.02	33.41 ± 0.06	0.092 ± 0.001
125°C					
10 min	13.48 ± 0.08	5.50 ± 0.05	46.11 ± 0.04	33.61 ± 0.12	0.092 ± 0.001
20 min	13.41 ± 0.03	5.48 ± 0.10	45.88 ± 0.03	33.91 ± 0.06	0.095 ± 0.001
30 min	13.31 ± 0.10	5.52 ± 0.09	46.16 ± 0.17	33.66 ± 0.09	0.100 ± 0.013
150°C					
10 min	13.58 ± 0.01	5.52 ± 0.12	46.08 ± 0.07	33.46 ± 0.03	0.094 ± 0.002
20 min	13.31 ± 0.01	5.36 ± 0.21	46.24 ± 0.14	33.72 ± 0.04	0.094 ± 0.004
30 min	13.36 ± 0.01	5.51 ± 0.10	45.82 ± 0.04	33.96 ± 0.01	0.096 ± 0.003
175°C					
10 min	13.33 ± 0.02	5.45 ± 0.10	46.06 ± 0.03	33.80 ± 0.02	0.095 ± 0.002
20 min	13.16 ± 0.02	5.28 ± 0.16	47.16 ± 0.13	33.03 ± 0.03	0.091 ± 0.003
30 min	13.28 ± 0.01	5.47 ± 0.15	47.48 ± 0.41	32.12 ± 0.17	0.086 ± 0.004

Values present mean ± standard deviations from triplicate measurements.
No significant variation was recorded compared to the control.

palmitic and 5.49% stearic acids as saturated fatty acids (SFA). These results are consistent with previous works [27, 28]. The roasting did not lead to significant variations of FA composition. Similar results were reported for safflower and rice germ oils [9, 10]. This result would be due to the protective effect of almonds tissues by limiting the contact between UFA and oxygen which is responsible for initiating the oxidative processes [29]. In addition, natural antioxidants such as polyphenols and tocopherols present in almonds tissues could exert a protective effect against UFA oxidation. In general, vegetable oils are estimated by their FA composition. Their nutritional value is related to the richness of essential FA ($\omega 3$ and $\omega 6$ families). Argan oil is produced for food and cosmetic purposes. In the process of extracting the edible oil, the roasting of the almonds is systematically carried out for the development of the organoleptic characteristics such as flavour and color. As for the cosmetic oil, it is cold extracted from unroasted almonds. Our results showed that both types of extraction processes result in the same FA profile.

3.3. Tocopherols Composition. The effect of roasting on tocopherols content is shown in Figure 2. The control oil exhibited a total tocopherols level of 977.9 mg/kg with 46.1, 875.5, and 56.3 mg/kg of α-, γ-, and δ-tocopherol. Thus, γ-tocopherol represented 89.5% of the total tocopherols of argan oil,

α-tocopherol 4.7%, and δ-tocopherol 5.8%. These results are consistent with data from the literature and confirm the richness of argan oil in γ-tocopherol [1]. Roasting caused a significant decrease in tocopherols content. The lowest levels were recorded in the condition 175°C/10 min with values of 32.1, 604.1, and 44.9 mg/kg for α-, γ-, and δ-tocopherol, respectively. The loss was about 69% for the three forms. This showed that the thermal sensitivity is comparable for the three isomers of tocopherols during argan almonds roasting. Our results also exhibited fluctuations of the tocopherols content with increasing temperature and time of roasting. Considering the effect of the roasting time, it appears that the content of the three forms of tocopherols was comparable at the roasting temperature of 100°C. On the other hand, the values were higher for the duration of 10 min when the roasting temperature was 125 and 150°C. Some studies carried out on pumpkin, rapeseed, and sesame seed oils recorded successive decreases and increases in the tocopherols content according to the increase of the roasting temperature and duration [29, 30]. Other works reported either an increase or a decrease in the level of tocopherols as a function of the roasting temperature. As an example, Lee et al. [10] noted an increase in tocopherols content in safflower oil with increasing roasting temperature, and Eitenmiller [31] reported a decrease in content of tocopherols with increasing

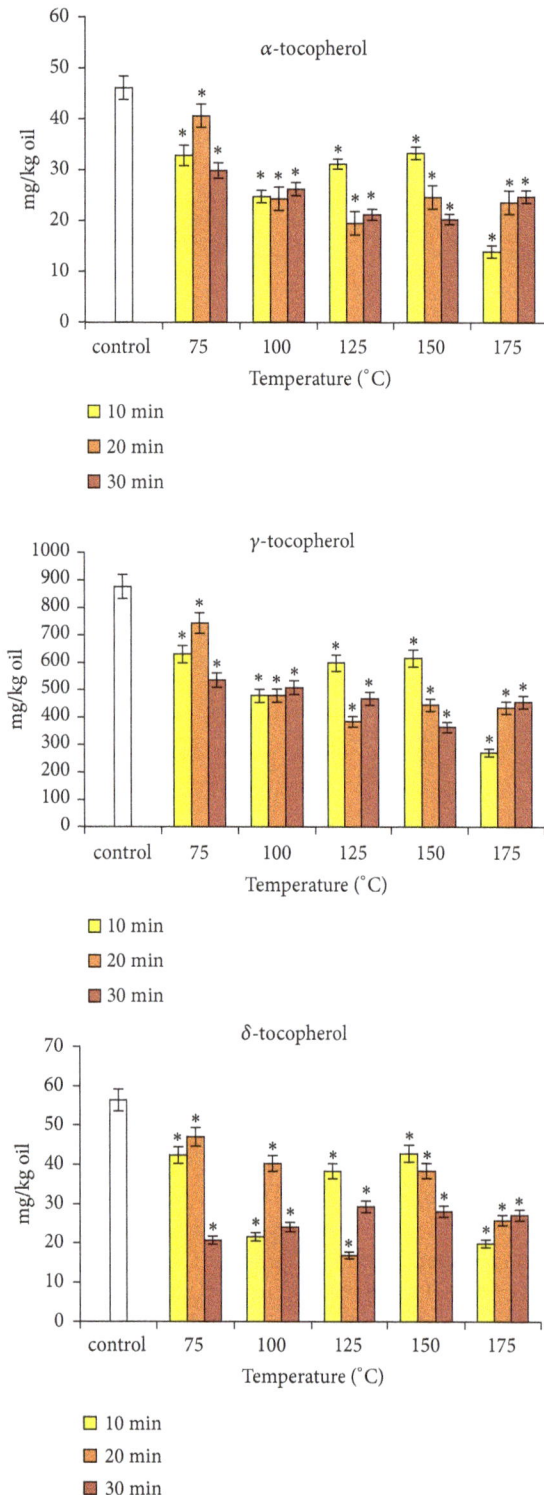

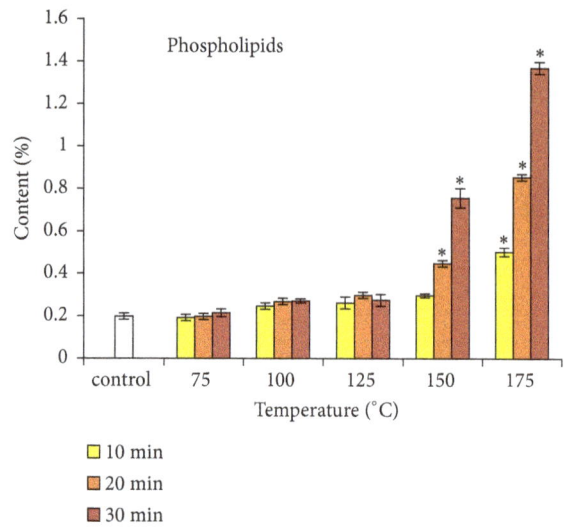

FIGURE 3: Changes in phospholipids content of argan oil according to roasting temperature and time. * Significantly different from control (p < 0.05).

extraction due to the destruction of the tocopherol-retaining cellular structures and rupture of their binding to membrane proteins and/or phospholipids [4, 31, 32]. The decrease in the tocopherols level following the almonds roasting constitutes a significant loss of the nutritional value of the edible argan oil, in particular for the high temperatures which are widely used during the artisanal extraction process. In order to preserve the nutritional properties and stability of the argan oil, it would be advisable to avoid high roasting temperatures. This will make it possible to minimize the destruction of tocopherols which are endowed with a high antioxidant power.

3.4. Phospholipids Content. The variation of the phospholipids content under the effect of roasting is represented by Figure 3. The control oil showed a content of 0.198%. The values obtained at the roasting temperatures of 75, 100, and 125°C did not differ significantly from that of the control oil. They increased significantly after roasting at temperatures of 150 (20 and 30 min) and 175°C (10, 20, and 30 min). The values also increased with the roasting time (0.447 and 0.756%, respectively, after 20 and 30 min at 150°C, and at 0.503, 0.855, and 1.370%, respectively, after 10, 20, and 30 min at 175°C). These results are in agreement with previous works carried out on safflower, rice germ, argan, pumpkin, and mustard seeds reporting an increase in phospholipids content as a function of temperature and/or duration of roasting [9–11, 33, 34]. As for tocopherols, the increase in the phospholipids level could be explained by a better extractability following the destruction of almonds tissues by roasting.

Several studies have focused on the effects of roasting conditions on the physicochemical characteristics of edible oils. Significant changes in some characteristics such as color, aroma, fatty acid profile, and bioactive compounds have been reported [9, 16]. The heating of oil seeds also leads to the initiation of the oxidation of UFA and the loss of natural antioxidants. The natural antioxidants content strongly influences

FIGURE 2: Changes in tocopherols content of argan oil according to roasting temperature and time. * Significantly different from control (p < 0.05).

temperature and roasting time for peanut oil. To explain these variations, two major hypotheses were adopted: (i) the decrease in tocopherols level would be due to their thermolability; (ii) the increase would result from a better the oxidation stability of the oil during storage. In this regard, the tocopherols present the most effective antioxidants for reducing lipid oxidation in oils [35]. The α-tocopherol also has vitamin activity. Therefore, the roasting process at uncontrolled temperatures could induce a reduction in oxidative stability and a loss of nutritional value of the oil.

The studies on the effects of roasting on the chemical composition and the oxidative stability of argan oil are not numerous. In addition, the temperature used by the authors remains relatively low (110°C) [11, 36, 37] compared to works done on other oilseeds [6, 7, 29, 38], but also with regard to the temperatures applied for the traditional extraction of argan oil intended for commerce or for a familiar use. Indeed, the extraction of argan oil is mainly done using the artisanal process, especially in rural areas of southern Morocco. Almonds from the argan tree are roasted over a wood fire using a stove. They are then ground using a rotary arm grinding stone and the resulting paste is crushed by hand. The addition of small amounts of warm water facilitates the extraction of the oil [39]. The argan oil obtained by this process is reputed to be poorly preserved [40]. Indeed, high roasting temperatures, contact with the metal, and the addition of water are all factors for initiating the oxidation of the oil. In this regard, Mathaüs et al. reported that the traditional oil was characterized by negative sensory attributes after 12 weeks of storage [36]. This result has been attributed in particular to the conditions of manual extraction practiced in the artisanal process. More recently, cooperatives producing argan oil use electric ovens for roasting and the extraction is carried out by press. This allowed an improvement in the quality of the oil compared to the artisanal process [39]. Mathaüs et al. also noted an improvement in the oxidative stability and sensory quality of oil extracted from roasted almonds at 110°C for 30 min [36]. These results were explained by a better extraction of antioxidants and the formation of Maillard reaction products during roasting. In fact, for the same temperature (110°C), other authors reported a significant increase in phospholipids content [11, 37] and a significant decrease in moisture [11]. As for the content of FA and tocopherols, Harhar et al. did not report significant changes [11]. However, the roasting time of 45 min (at 110°C) induced a reduction in the oxidative stability of the oil [11]. This allowed these authors to conclude that 110°C/30 min (temperature/time) represents an optimal roasting condition for preserving the nutritional and the organoleptic properties of argan oil.

In order to better reproduce the roasting conditions used in the two argan oil extraction processes (artisanal and mechanical), we used temperatures between 75 and 175°C for periods of 10 to 30 min. Our results showed that PV and acidity increased with temperature and roasting time with maximum values reaching 175°C/30 min. With regard to tocopherols, the lowest levels were obtained after roasting at 175°C/10 min. The temperatures of 125° and 150°C induced a lower increase in PV and a lower reduction in the tocopherols content especially when the duration of roasting was limited to 10 min. These same conditions caused a significant increase in the browning index. From these results, it can be concluded that heating argan almonds at high temperatures is not recommended, even for short periods.

Since the roasting of argan almonds is an important step for the development of the organoleptic characteristics of edible argan oil, particularly its aroma and color, it would be desirable to use temperatures between 125 and 150°C for duration of 10 min. Nevertheless, our study requires additional investigations to evaluate the oxidative stability of oils extracted from roasted almonds at different temperatures and times.

4. Conclusion

This study showed that the temperature and the duration of roasting (75 to 175°C, and 10 to 30 min, respectively) had little effect on the physicochemical characteristics of the argan oil. Similarly, the composition of FA was not modified by roasting whatever the temperature and the time used. However, the browning index increased significantly from 100°C, important decrease in the tocopherols content was recorded at 175°C, and significant increase in the phospholipids level was obtained from 150°C/20 min condition. These results make it possible to conclude that roasting at temperatures of 125-150°C for 10 min would allow the development of organoleptic characteristics without compromising the nutritional value and the oxidative stability of the oil.

Conflicts of Interest

The authors declare that there are no conflicts of interest regarding the publication of this article.

Acknowledgments

The authors would like to thank Ms. Jamila Idbourrous, Director of UCFA Tissaliwine (argan products cooperative), Agadir, Morocco, for the extraction of argan oil.

References

[1] A. El Abbassi, N. Khalid, H. Zbakh, and A. Ahmad, "Physicochemical Characteristics, Nutritional Properties, and Health Benefits of Argan Oil: A Review," *Critical Reviews in Food Science and Nutrition*, vol. 54, no. 11, pp. 1401–1414, 2014.

[2] G. Durmaz and V. Gökmen, "Impacts of roasting oily seeds and nuts on their extracted oils," *Lipid Technology*, vol. 22, no. 8, pp. 179–182, 2010.

[3] E. Choe and D. B. Min, "Mechanisms and factors for edible oil oxidation," *Comprehensive Reviews in Food Science and Food Safety*, vol. 5, no. 4, pp. 169–186, 2006.

[4] B. Vaidya and E. Choe, "Effects of seed roasting on tocopherols, carotenoids, and oxidation in mustard seed oil during heating," *Journal of the American Oil Chemists' Society*, vol. 88, no. 1, pp. 83–90, 2011.

[5] B. Jannat, M. R. Oveisi, N. Sadeghi et al., "Effect of roasting process on total phenolic compounds and γ-tocopherol contents of Iranian sesame seeds (*Sesamum indicum*)," *Iranian Journal of Pharmaceutical Research*, vol. 12, no. 4, pp. 751–758, 2013.

[6] H. Yoshida and S. Takagi, "Effects of seed roasting temperature and time on the quality characteristics of sesame (Sesamum indicum) oil," *Journal of the Science of Food and Agriculture*, vol. 75, no. 1, pp. 19–26, 1997.

[7] E. Şimşek, M. M. özcan, D. Arslan, A. ünver, and G. Kanbur, "Changes in chemical composition of some oils extracted from seeds roasted at different temperatures," *Quality Assurance and Safety of Crops & Foods*, vol. 7, no. 5, pp. 801–808, 2015.

[8] H. Yoshida, S. Takagi, and S. Mitsuhashi, "Tocopherol distribution and oxidative stability of oils prepared from the hypocotyl of soybeans roasted in a microwave oven," *Journal of the American Oil Chemists' Society*, vol. 76, no. 8, pp. 915–920, 1999.

[9] I.-H. Kim, C.-J. Kim, J.-M. You et al., "Effect of roasting temperature and time on the chemical composition of rice germ oil," *Journal of the American Oil Chemists' Society*, vol. 79, no. 5, pp. 413–418, 2002.

[10] Y.-C. Lee, S.-W. Oh, J. Chang, and I.-H. Kim, "Chemical composition and oxidative stability of safflower oil prepared from safflower seed roasted with different temperatures," *Food Chemistry*, vol. 84, no. 1, pp. 1–6, 2004.

[11] H. Harhar, S. Gharby, B. Kartah, H. El Monfalouti, D. Guillaume, and Z. Charrouf, "Influence of argan kernel roasting-time on virgin argan oil composition and oxidative stability," *Plant Foods for Human Nutrition*, vol. 66, no. 2, pp. 163–168, 2011.

[12] C. Paquot, "2.201 - Determination of the Acid Value (A. V.) and the Acidity," in *Standard Methods for the Analysis of Oils, Fats and Derivatives*, pp. 52–55, Pergamon, 6th edition, 1979.

[13] C. Paquot, "2.501 - Determination of the Peroxide Value (P.V.)," in *Standard Methods for the Analysis of Oils, Fats and Derivatives*, pp. 138-139, Pergamon, 6th edition, 1979.

[14] AOCS, "Official Method Ce 1k-07, Direct Methylation of Lipids for the Determination of Total Fat, Saturated, cis-Monounsaturated cispolyunsaturated, and trans fatty acids by chromatography," in *Official method and recommended practices of the American Oil Chemists Society*, American Oil Chemists Society, USA, 5th edition, 2007.

[15] AOCS, *Official Method Ca 12-55: Phosphorus in Oils, Colorimetric*, American Oil Chemists' Society, 2009.

[16] H. Yoshida, "Composition and quality characteristics of sesame seed (Sesamum indicum) oil roasted at different temperatures in an electric oven," *Journal of the Science of Food and Agriculture*, vol. 65, no. 3, pp. 331–336, 1994.

[17] L. Cai, A. Cao, G. Aisikaer, and T. Ying, "Influence of kernel roasting on bioactive components and oxidative stability of pine nut oil," *European Journal of Lipid Science and Technology*, vol. 115, no. 5, pp. 556–563, 2013.

[18] A. Rękas, M. Wroniak, and K. Krygier, "Effects of different roasting conditions on the nutritional value and oxidative stability of high-oleic and yellow-seeded Brassica napus oils," *Grasas y Aceites*, vol. 66, no. 3, 2015.

[19] C. Scrimgeour, "Chemistry of Fatty Acids," in *Bailey's Industrial Oil and Fat Products*, John Wiley & Sons, Inc., 2005.

[20] B. Vaidya and J.-B. Eun, "Effect of roasting on oxidative and tocopherol stability of walnut oil during storage in the dark," *European Journal of Lipid Science and Technology*, vol. 115, no. 3, pp. 348–355, 2013.

[21] K. Kraljić, D. Škevin, M. Pospišil, M. Obranović, S. N. D. Signeral, and T. Bosolt, "Quality of rapeseed oil produced by conditioning seeds at modest temperatures," *Journal of the American Oil Chemists' Society*, vol. 90, no. 4, pp. 589–599, 2013.

[22] SNIMA, "Norme marocaine NM 08.5.090," in *Huiles d'argane. Spécifications*, Service de normalisation industrielle marocaine, Rabat, Maroc, 2003.

[23] P. E. Koehler and G. V. Odell, "Factors Affecting the Formation of Pyrazine Compounds in Sugar-Amine Reactions," *Journal of Agricultural and Food Chemistry*, vol. 18, no. 5, pp. 895–898, 1970.

[24] D. J. Liska, C. M. Cook, D. D. Wang, and J. Szpylka, "Maillard reaction products and potatoes: have the benefits been clearly assessed?" *Food Science & Nutrition*, vol. 4, no. 2, pp. 234–249, 2016.

[25] J. O'Brien, P. A. Morrissey, and J. M. Ames, "Nutritional and toxicological aspects of the maillard browning reaction in foods," *Critical Reviews in Food Science and Nutrition*, vol. 28, no. 3, pp. 211–248, 1989.

[26] N. Tamanna and N. Mahmood, "Food processing and maillard reaction products: Effect on human health and nutrition," *International Journal of Food Science*, vol. 2015, 2015.

[27] R. Belcadi-Haloui, A. Zekhnini, and A. Hatimi, "Comparative study on fatty acid and tocopherol composition in argan oils extracted from fruits of different forms," *Acta Botanica Gallica*, vol. 155, no. 2, pp. 301–305, 2008.

[28] R. Belcadi-Haloui, A. Zekhnini, and A. Hatimi, "Comparative Study on Output and Chemical Composition of Argan Oils Extracted from Fruits of Different Regions in the Southwest of Morocco," *Food Science and Quality Management*, vol. 30, pp. 1–6, 2014.

[29] R. Chirinos, D. Zorrilla, A. Aguilar-Galvez, R. Pedreschi, and D. Campos, "Impact of Roasting on Fatty Acids, Tocopherols, Phytosterols, and Phenolic Compounds Present in Plukenetia huayllabambana Seed," *Journal of Chemistry*, vol. 2016, 2016.

[30] M. Murkovic, V. Piironen, A. M. Lampi, T. Kraushofer, and G. Sontag, "Changes in chemical composition of pumpkin seeds during the roasting process for production of pumpkin seed oil (Part 1: non-volatile compounds)," *Food Chemistry*, vol. 84, no. 3, pp. 359–365, 2004.

[31] R. R. Eitenmiller, "Effects of Dry Roasting on the Vitamin E Content and Microstructure of Peanut (Arachis hypogae)," *Journal of Agriculture & Life Sciences*, vol. 45, pp. 121–133, 2011.

[32] R. A. Moreau, K. B. Hicks, and M. J. Powell, "Effect of heat pretreatment on the yield and composition of oil extracted from corn fiber," *Journal of Agricultural and Food Chemistry*, vol. 47, no. 7, pp. 2869–2871, 1999.

[33] V. Vujasinovic, S. Djilas, E. Dimic, Z. Basic, and O. Radocaj, "The effect of roasting on the chemical composition and oxidative stability of pumpkin oil," *European Journal of Lipid Science and Technology*, vol. 114, no. 5, pp. 568–574, 2012.

[34] K. Shrestha, F. G. Gemechu, and B. De Meulenaer, "A novel insight on the high oxidative stability of roasted mustard seed oil in relation to phospholipid, Maillard type reaction products, tocopherol and canolol contents," *Food Research International*, vol. 54, no. 1, pp. 587–594, 2013.

[35] S.-W. Huang, E. N. Frankel, and J. B. German, "Antioxidant Activity of α-and γ-Tocopherols in Bulk Oils and in Oil-in-Water Emulsions," *Journal of Agricultural and Food Chemistry*, vol. 42, no. 10, pp. 2108–2114, 1994.

[36] B. Matthäus, D. Guillaume, S. Gharby, A. Haddad, H. Harhar, and Z. Charrouf, "Effect of processing on the quality of edible argan oil," *Food Chemistry*, vol. 120, no. 2, pp. 426–432, 2010.

[37] I. Zaanoun, S. Gharby, I. Bakass, E. Ait Addi, and I. Ait Ichou, "Kinetic parameter determination of roasted and unroasted argan oil oxidation under Rancimat test conditions," *Grasas y Aceites*, vol. 65, no. 3, 2014.

[38] A. A. Mariod, S. Y. Ahmed, S. I. Abdelwahab et al., "Effects of roasting and boiling on the chemical composition, amino acids

and oil stability of safflower seeds," *International Journal of Food Science & Technology*, vol. 47, no. 8, pp. 1737–1743, 2012.

[39] Z. Charrouf, D. Guillaume, and A. Driouich, "The Argan tree, an asset for Morocco," *Biofutur*, no. 220, pp. 54–56, 2002.

[40] M. Hilali, Z. Charrouf, A. E. A. Soulhi, L. Hachimi, and D. Guillaume, "Influence of origin and extraction method on argan oil physico-chemical characteristics and composition," *Journal of Agricultural and Food Chemistry*, vol. 53, no. 6, pp. 2081–2087, 2005.

13

Effects of Blanching and Natural Convection Solar Drying on Quality Characteristics of Red Pepper (*Capsicum annuum* L.)

James Owusu-Kwarteng,[1] **Francis K. K. Kori,**[2] **and Fortune Akabanda**[1]

[1]*Department of Applied Biology, Faculty of Applied Sciences, University for Development Studies, P.O. Box 24, Navrongo, Ghana*
[2]*Department of Applied Physics, Faculty of Applied Sciences, University for Development Studies, P.O. Box 24, Navrongo, Ghana*

Correspondence should be addressed to James Owusu-Kwarteng; jowusukwarteng@uds.edu.gh

Academic Editor: Pankaj B. Pathare

The objective of this work was to determine the effects of blanching and two drying methods, open-sun drying and natural convection solar drying, on the quality characteristics of red pepper. A 2 × 3 factorial design with experimental factors as 2 drying methods (open-sun drying and use of solar dryer) and 3 levels of pepper blanching (unblanched, blanched in plain water, and blanched in 2% NaCl) was conducted. Dried pepper samples were analysed for chemical composition, microbial load, and consumer sensory acceptability. Blanching of pepper in 2% NaCl solution followed by drying in a natural convection solar dryer reduced drying time by 15 hours. Similarly, a combination of blanching and drying in the solar dryer improved microbial quality of dried pepper. However, blanching and drying processes resulted in reduction in nutrients such as vitamin C and minerals content of pepper. Blanching followed by drying in natural convection solar dryer had the highest consumer acceptability scores for colour and overall acceptability, while texture and aroma were not significantly ($p > 0.05$) affected by the different treatments. Therefore, natural convection solar dryer can be used to dry pepper with acceptable microbial and sensory qualities, as an alternative to open-sun drying.

1. Introduction

Red pepper (*Capsicum annuum* L.) is used as a spice or major ingredient in dishes around worldwide and perhaps considered the first spice to have been used by man with archaeological evidence of pepper and other fossil foods dating back to 6000 years old [1]. Red pepper is generally known to be cholesterol free, have low sodium and caloric contents, and serve as good source of vitamins A and C [2]. Additionally, red pepper is known to possess antimicrobial activity [3] and reduces the risk of diseases such as arthritis, cancer, and diabetes [4–8]. In food processing, red pepper is also used as colouring and flavouring agent in sauces, soups, pickles, and pizzas [9].

Ghana contributes about 1% of world pepper production and was ranked the eleventh largest producer of pepper in the world and second in Africa with an estimated total production of 88,000 metric tons [10]. Like other fresh fruits and vegetables, fresh pepper is a perishable produce and deteriorates within a few days after harvest without proper storage

or preservation measures. The perishable nature of pepper can lead to economic losses which is further aggravated by storage and marketing problems and lack of appropriate processing technologies [11]. To prevent losses due to postharvest deterioration, dehydration is one important method often adopted for preservation and value addition to perishable agricultural products through moisture control. The major goal in drying fruits and vegetables such as pepper is to reduce the moisture content to desirable levels, usually 5–10%, which allows for safe storage over an extended period of time [12].

Unlike many industrialized countries where mechanized drying of fruits and vegetables is practiced, traditional open-sun drying of fruits and vegetables is a common practice in developing countries. However, traditional open-sun drying of fruits and vegetables can be time-consuming and less hygienic. Although modern mechanized drying of food commodities is faster than open-sun drying and uses much less land space, the equipment cost as well as the continued recurring cost of fuel or energy to operate these systems is very high. Therefore, small-scale farmers and other players in

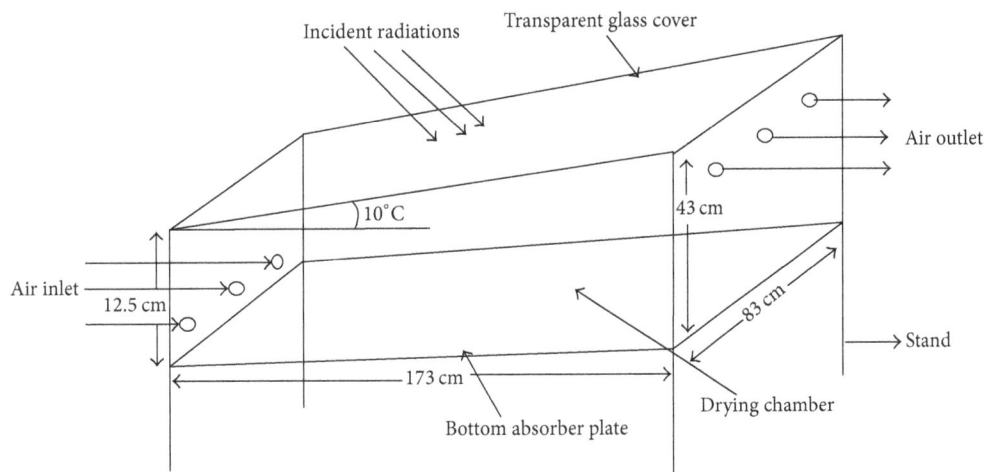

FIGURE 1: Schematic diagram of solar dryer.

the pepper value chain in rural communities in Ghana resort to the traditional open-sun drying method of preservation which results in low-quality products. As an alternative to mechanized and open-sun drying methods, solar dryers are being investigated and used for drying various fruits and vegetables [12], especially at geographic locations where there is enough sunshine during the harvest season [13]. Further advantages of solar drying over conventional open sun drying are the improvement in hygienic quality of the dried products, safe moisture content, colour and taste, and the protection of produce from rain, dust, and insects [14].

To obtain dried pepper with the best organoleptic and nutritional qualities, pretreatment such as blanching may be applied prior to drying [15–17]. Blanching is considered a pretreatment or unit operation prior to freezing, canning, or drying in which fruits or vegetables are heated for the purpose of inactivating enzymes; modifying texture; preserving colour, flavour, and nutritional value. Therefore, the objective of this work was to determine the effects of aquathermal blanching and natural convection solar drying methods on the quality characteristics of pepper.

2. Materials and Methods

2.1. Design and Fabrication of Natural Convection Solar Dryer. A direct natural convection solar dryer used in this study was designed and fabricated as shown in Figure 1. It was constructed mainly with wood on the sides and a transparent glass (166 cm × 76 cm) covering the top inclined at an angle of 10° to face the equator. A metal plate, painted black to improve absorption, was fitted at the bottom of the box with cardboard (insulator) beneath it. The inner side walls of the box were lined with aluminum foil for the reflection of radiation back to the interior space of box. Three inlet and outlet vents were created to allow air convectional current take place.

Further details and measurements of the solar collector are shown in Table 1. The constructed direct solar drying system was installed in an open place at the Navrongo Campus of the University for Development Studies. The drier

TABLE 1: Details of solar collector assembly material.

Parameter	Value/description
Collector area	1.44 m2 (0.83 m × 1.73 m)
Absorber plate	Galvanized steel 0.67 mm thick
Absorber plate surface treatment	Black paint coating
Glazing	Transparent glass of surface area 1.26 m^2 and 5 mm thick
Bottom insulation	Cardboard 20 mm thickness
Inside reflector	Aluminium foil of 0.5 mm thickness
Casing	Hard wood of 2.5 cm thick
Collector slope	10°

was positioned to face the equator to maximize the collection of solar radiation throughout the day. Navrongo is located at 10.8940°N, 1.0921°W.

2.2. Experimental Design. A 2 × 3 factorial design with experimental factors as two pepper drying methods (open-sun drying and use of solar dryer) and 3 levels of pepper blanching (unblanched, blanched in plain water, and blanched in 2% NaCl) were conducted. The samples were analysed for their chemical composition including moisture, crude protein, ash, and fibre contents [18]. Additionally, mineral content, microbial counts, and consumer sensory analysis of dried pepper samples were carried out following standard methods.

2.2.1. Sample Treatments. Fresh *Capsicum annuum* L. (red pepper) were purchased from the local open market in Navrongo. The pepper was first washed thoroughly under running tap water and then blanched at 93°C in plain or in 2% NaCl solution for 4 min. Prior to drying experiments, the initial moisture content of fresh untreated pepper was determined by drying in an oven set at 120°C until constant mass was observed. The initial and final masses of the red pepper samples were recorded with an electronic balance.

Initial moisture content of red pepper was found to be about 75% on wet basis.

2.2.2. Drying of Pepper.

2.2.2. Drying of Pepper. Differently treated pepper samples were loaded onto wire-mesh trays at $5 \, kg/m^2$ and dried by suspending them on weighing balance in the drying chamber of the direct solar dryer and in the open sun simultaneously. The samples were spread evenly in a single layer on the stainless-steel wire-mesh trays and kept inside the drying chamber of the natural convection solar dryer or in the open sun. The drying experiments were carried out in the month of May, 2016, at University for Development Studies, Navrongo Campus, Ghana. Each experiment started at 09:00 a.m. and continued until 5:00 p.m. daily. Relative humidity during the experimental period ranged between 55% and 65%. During the experimental period, ambient temperature, temperatures within the dryer, and the absorber plate temperatures were measured at 1 h intervals. Additionally, the mass of pepper was measured at 1 h intervals throughout the experiments until a final moisture content of 5% (ideal for long-term storage of pepper) was achieved [12].

2.2.3. Chemical Analysis of Dried Pepper. Crude protein, fibre, and ash were determined following the procedures by AOAC methods 970.22, 985.29, and 972.15, respectively [18]. Mineral analyses were carried out using AOAC procedure of Atomic Absorption Spectrophotometer (AAS) [18]. Briefly, one (1) ml aliquots of the digest from pepper sample were used to determine Ca and Fe contents of pepper using Spectra AA 220FS Spectrophotometer (Varian Co., Mulgrave, Australia) with an acetylene flame.

2.2.4. Microbiological Analysis. Serial 10-fold dilutions were made by weighing 10 g of dried pepper samples into 90 ml of Butterfield's phosphate buffer (Hardy Diagnostics, CA). Appropriate dilutions were surface-plated onto Tryptic Soy Agar (TSA; Difco) supplemented with cycloheximide (50 mg/l) to inhibit the growth of mold and onto CHRO-Magar ECC plates and incubated at 37°C for 24 h. Following incubation, all visible colonies on TSA were enumerated as Aerobic Mesophilic count (AMC), and all pink (coliforms) and blue (presumptive *E. coli*) colonies on CHROMagar ECC plates were enumerated. Microbial counts were reported as CFU/g of pepper sample.

2.2.5. Consumer Sensory Evaluation of Dried Pepper. Consumer sensory quality of the dried pepper was carried out to assess attributes including colour, texture, aroma, and overall acceptability by 65 volunteered untrained panelists drawn from the Faculty of Applied Sciences of the University for Development Studies. The panelist were adults between the ages of 18 and 40 years who are familiar with whole dried pepper. The panelist independently, in separate sensory evaluation booths, assessed the various products for sensory qualities using a nine-point hedonic scale with 1, 5, and 9 representing "dislike extremely," "neither like nor dislike," and "like extremely," respectively [19]. All products were presented to the panelists randomly and placed side by side, with each panelist receiving two rounds of each product.

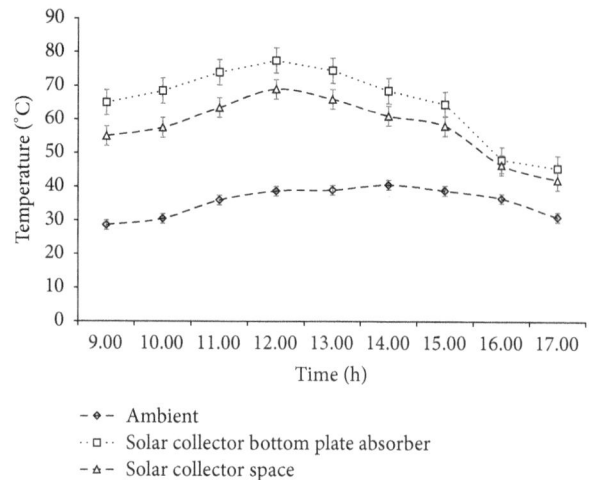

FIGURE 2: Ambient and collector temperatures on sunny hours of the day during drying experiments.

2.3. Statistical Analysis. All experiments were carried out in triplicate and values are presented as means with standard deviations (SD) where applicable. Data obtained were subjected to analysis of variance (ANOVA) and Least Significant Difference (LSD) was used to separate means at $p < 0.05$ using the MINITAB statistical software package (MINITAB Inc. Release 14 for Windows, 2004).

3. Results and Discussion

3.1. Ambient and Solar Dryer Temperatures. Variations in ambient temperature and temperatures in the drying chamber of solar dryer and absorber plate during the experimental period are shown in Figure 2. The average ambient temperature increased from 28.6°C at 9:00 GMT to 40.5°C at 14:00 GMT and decreased thereafter to 31°C at 17:00 GMT. In a similar trend, temperature of the drying chamber within the solar dryer increased from an average 55.0°C at 9:00 to 69°C at 12:00 GMT and decreased thereafter to average of 42°C at 17:00 GMT. The highest average temperatures of drying chamber and bottom absorber plate of the solar dryer, 69.0°C and 77.5°C, respectively, were attained at 12:00 GMT.

3.2. Drying Characteristics of Pepper. The reduction in moisture contents of pepper samples during drying in the open sun and in solar dryer is shown in Figure 3. Fresh pepper samples had an average moisture content of 75% prior to pretreatment and drying. Pepper samples that were blanched in 2% NaCl solution and subsequently dried in the solar dryer had moisture content reduced to 5% after 13 h of drying. Pepper samples that were blanched in plain water and unblanched pepper attained moisture contents of 5% after drying in a period of 16 and 17 h, respectively, in solar dryer. On the other hand, all pepper samples that were dried in the open sun, irrespective of pretreatment, attained moisture contents of 5% after 28 h of drying. Thus, blanching in 2% NaCl followed by use of the natural convection solar dryer reduced the drying time for pepper by 15 hours. In general,

TABLE 2: Nutrient composition of dried pepper[*].

Sample	Protein (g/100 g)	Ash (g/100 g)	Fibre (g/100 g)	Vitamin C (mg/100 g)	Ca (mg/100 g)	Fe (mg/100 g)
Fresh pepper	2.8 ± 0.04^a	2.6 ± 0.07^a	4.8 ± 1.00^a	175.6 ± 1.5^a	18.5 ± 0.9^a	2.1 ± 0.10^a
UBOS	2.7 ± 0.10^a	1.5 ± 0.05^b	4.7 ± 0.81^a	103.0 ± 0.8^b	13.0 ± 1.1^b	0.9 ± 0.05^b
UBSC	2.8 ± 0.06^a	1.4 ± 0.01^{bc}	4.8 ± 0.50^a	99.5 ± 1.2^c	12.3 ± 0.8^{cb}	1.1 ± 0.08^b
BWOS	2.7 ± 0.05^a	1.2 ± 0.08^{cd}	4.8 ± 1.02^a	101.0 ± 0.9^{bc}	14.5 ± 0.9^{bd}	0.8 ± 0.06^{cb}
BSOS	2.8 ± 0.12^a	1.5 ± 0.06^b	4.6 ± 0.65^a	100.3 ± 1.5^{bc}	14.3 ± 1.0^{bd}	0.7 ± 0.07^{cd}
BWSC	2.7 ± 0.08^a	1.1 ± 0.10^d	4.7 ± 0.50^a	98.8 ± 1.1^c	15.0 ± 0.5^{bd}	0.9 ± 0.12^b
BSSC	2.7 ± 0.05^a	1.3 ± 0.05^{bc}	4.8 ± 0.50^a	109.9 ± 0.8^{bc}	15.2 ± 1.2^{bd}	0.7 ± 0.05^{cd}

[*]Means ($\pm$SD) values with different letters as superscripts in a column are significantly different ($p < 0.05$). UBOS: unblanched open-sun drying; UBSC: unblanched solar collector drying; BWOS: blanched in plain water open-sun drying; BSOS: blanched in 2% NaCl solution open-sun drying; BWSC: blanched in plain water solar collector drying; BSSC: blanched in 2% NaCl solution solar collector drying.

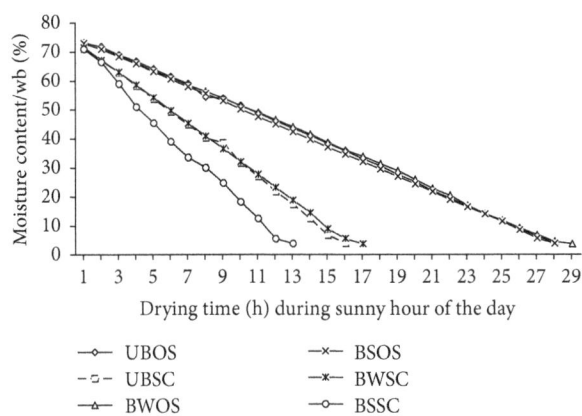

FIGURE 3: Moisture content of pepper during drying in open sun and in solar dryer. Drying was stopped when moisture content reduced to 5%. UBOS: unblanched open-sun drying; UBSC: unblanched solar collector drying; BWOS: blanched in plain water open-sun drying; BSOS: blanched in 2% NaCl solution open-sun drying; BWSC: blanched in plain water solar collector drying; BSSC: blanched in 2% NaCl solution solar collector drying.

average drying time for pepper samples that were dried in the natural convection solar dryer was significantly lower ($p < 0.05$) than pepper samples dried in the open sun. Blanching either in 2% NaCl solution or in water significantly ($p < 0.05$) improved drying rate of pepper (Figure 3). In general, drying pepper in the solar dryer reduced drying time by about 49–54% depending on pretreatment method applied to the pepper.

In a review of various types of solar drying systems for agricultural commodities, Fudholi et al. [20] reported that the moisture content of fresh chili decreased from 80% to 5% under solar drying in 48 h. Improvement in the drying rates of blanched pepper has previously been reported [5, 16, 21–23], and the observation has been attributed to possible rupturing of cell membrane making pepper tender and thus facilitating faster removal of moisture in the drying process [22].

3.3. Effect of Blanching and Drying Method on Nutrient Composition of Pepper.
The nutrient composition of pepper following blanching treatments and drying is shown in

Table 2. Total proteins, ash, and fibre contents of fresh pepper were 2.8 ± 0.04, 2.6 ± 0.10, and 4.8 ± 1.00 (g/100 g), respectively, while vitamin C, calcium (Ca), and iron (Fe) contents were 175.6 ± 1.5, 18.5 ± 0.9, and 2.1 ± 0.10 (mg/100 g), respectively. Generally, crude proteins and fibre contents were not significantly affected by the blanching treatments and drying methods. However, crude ash content significantly reduced following blanching and drying. Similarly, vitamin C, Ca, and Fe contents reduced significantly ($p < 0.05$) after blanching treatments and drying (Table 2).

Generally, vegetables and fruits serve as good sources of energy, minerals, and vitamins. However, during dehydration processes, changes in nutritional quality of heat sensitive vitamins and other nutrients occur [12, 24]. It has previously been reported that significant losses in the content of vitamin C, minerals, and polyphenols in vegetables occur after aquathermal blanching [25, 26], an observation which may be due to their sensitivity to heat and/or leaching of these compounds in water [24, 27]. However, steam blanching was reported to retain higher amounts of vitamin C in spinach compared with hot water blanching [28]. The present results show that neither blanching in NaCl solution nor plain water could significantly retain the vitamin C and mineral contents of pepper as there were significant losses in these nutrients after blanching and drying.

3.4. Microbial Load in Dried Pepper.
Mean aerobic mesophilic counts (AMC) and E. coli/coliform (ECC) counts in dried pepper are shown in Figure 4. Aerobic mesophilic counts ranged between 4.6 ± 1.2 and 6.7 ± 0.8 cfu/g, while ECC counts ranged between 1.8 ± 0.5 and 3.5 ± 1.0 cfu/g of dried pepper samples.

Generally, pepper samples that were blanched and dried in solar collector had significantly ($p < 0.05$) lower microbial load when compared to samples that were dried in the open sun with or without blanching (Figure 4). Thus, a combination of blanching and the solar collector drying processes can significantly reduce the microbial load in dried pepper. Through blanching and drying to control moisture levels (water activity, a_w) in pepper, microbial load can be controlled although spores may not be killed. These may bring microbial load of dried pepper to allowable safe limits while supporting long shelf life of these products.

TABLE 3: Consumer sensory evaluation of dried pepper.

Sensory attribute	BWOS	BSOS	UBOS	BWSC	BSSC	UBSC
Colour	5.8 ± 1.9^a	5.9 ± 2.0^a	3.8 ± 2.1^b	6.8 ± 2.0^c	7.0 ± 1.7^c	4.1 ± 1.9^b
Aroma	6.4 ± 2.2^a	6.1 ± 1.9^a	6.3 ± 1.9^a	6.0 ± 1.8^a	6.3 ± 1.6^a	6.2 ± 1.8^a
Texture	5.9 ± 1.8^a	5.7 ± 1.5^a	6.1 ± 2.0^a	5.7 ± 1.7^a	5.8 ± 1.7^a	6.1 ± 1.7^a
Overall acceptability	6.2 ± 1.4^a	6.4 ± 1.2^a	4.6 ± 1.7^b	6.9 ± 1.7^c	7.1 ± 1.9^c	4.8 ± 2.0^b

UBOS: unblanched dried in open sun; UBSC: unblanched dried in solar collector; BWOS: blanched in plain water and dried in open sun; BSOS: blanched in 2% salt solution and dried in open sun; BWSC: blanched in plain water and dried in solar collector; BSSC: blanched in 2% salt solution and dried in solar collector. Values with same superscript in a row are not significantly different ($p > 0.05$).

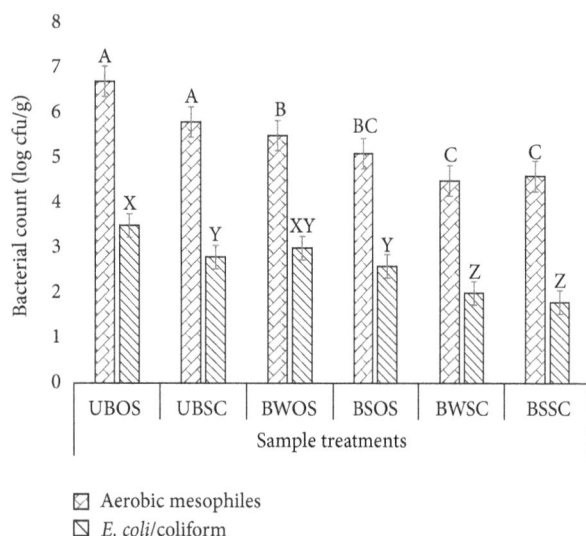

FIGURE 4: Mean aerobic mesophilic and *E. coli*/coliform counts in dried pepper. For AMC or ECC, means (±SD) with different letters are significantly different ($p < 0.05$). UBOS: unblanched open-sun drying; UBSC: unblanched solar collector drying; BWOS: blanched in plain water open-sun drying; BSOS: blanched in 2% NaCl solution open-sun drying; BWSC: blanched in plain water solar collector drying; BSSC: blanched in 2% NaCl solution solar collector drying.

3.5. Consumer Sensory Evaluation of Dried Pepper. Consumer sensory evaluation of the differently treated and dried pepper samples assessed using a 9-point hedonic scale is shown in Table 3. Consumer evaluation of dried pepper covered sensory characteristics such as colour, aroma, texture, and overall acceptability. Results of consumer evaluation generally showed no significant difference ($p > 0.05$) in aroma and texture between the different pepper samples. However, significant differences ($p < 0.05$) were observed in colour and overall acceptability between samples. For colour, pepper samples that were blanched either in plain water or in 2% NaCl solution and subsequently dried using the solar dryer scored significantly higher (7.0), whereas pepper samples that were not blanched and subsequently dried in the open sun scored significantly ($p < 0.05$) lower (3.8). For overall acceptability, the highest score of 7.1 was attained for pepper samples that were blanched in 2% NaCl solution and subsequently dried in the solar dryer, indicating the panelist likeness for the product.

It has been suggested that inactivating the enzymes responsible for browning (polyphenoloxidase, lipoxygenase, and peroxidase), during blanching improves both colour and flavour of vegetables [29]. However, if not carefully managed, blanching may also be accompanied by a reduction in sensory and nutrient qualities in many foods, principally due to Maillard reaction [30].

4. Conclusions

Blanching of pepper in 2% NaCl solution followed by drying in a natural convection solar dryer reduced drying time by 15 hours when compared to drying in the open sun which takes at least 28 hours. Similarly, a combination of blanching and drying in the solar dryer improved microbial quality of the dried pepper. However, the blanching and drying processes resulted in reduction of nutrients such as vitamin C and minerals contents. Consumer sensory analysis showed that blanching, followed by drying in natural convection solar dryer, had the highest scores for colour and overall acceptability, while texture and aroma were not significantly ($p > 0.05$) affected by the various treatments. Therefore, natural convection solar dryer has the potential for commercial small-medium scale industrial application for drying pepper with acceptable microbial and sensory qualities, especially in developing rural communities where equipment cost as well as recurring cost of fuel or energy to operate mechanical dryers can prove to be high. This in turn would ensure reduction in postharvest losses of pepper and thus better economic returns for farmers and processors.

Conflicts of Interest

The authors declare that they have no conflicts of interest.

Acknowledgments

The authors acknowledge the contributions of Lorlorli Blemayi-Honya, Ramatu Alhassan, and Hafiz Mutalib during our data collection.

References

[1] T. A. Hill, H. Ashrafi, S. Reyes-Chin-Wo et al., "Characterization of *Capsicum annuum* genetic diversity and population structure based on parallel polymorphism discovery with a 30K unigene pepper GeneChip," *PLoS ONE*, vol. 8, no. 2, article e56200, 2013.

[2] P. A. Luning, D. Yuksel, R. Vries, and J. P. Roozen, "Aroma changes in fresh bell peppers (Capsicum annuum) after hot-air

drying," *Journal of Food Science*, vol. 60, no. 6, pp. 1269–1276, 1995.

[3] N. M. Wahba, A. S. Ahmed, and Z. Z. Ebraheim, "Antimicrobial effects of pepper, parsley, and dill and their roles in the microbiological quality enhancement of traditional Egyptian Kareish cheese," *Foodborne Pathogens and Disease*, vol. 7, no. 4, pp. 411–418, 2010.

[4] Z. DongLin and H. Yasunori, "Phenolic compounds: ascorbic acid cartoneoids and antoxidant, carotenoids and antioxidant properties of green, red and yellow bell pepper," *Journal of Food Agriculture and Environment*, vol. 2, pp. 22–27, 2003.

[5] A. Eleyinmi, S. Akoja, N. Ilelaboye, and F. Aiyeleye, "Effect of different pre-drying operations on some nutritionally valuable minerals, ascorbic acid and rehydration index of capsicum species," *Tropical Agricultural Research and Extension*, vol. 5, no. 1, pp. 57–61, 2002.

[6] Y. Lee, L. R. Howard, and B. Villalón, "Flavonoids and antioxidant activity of fresh pepper (Capsicum annuum) cultivars," *Journal of Food Science*, vol. 60, no. 3, pp. 473–476, 1995.

[7] H. Nishino, M. Murakoshi, H. Tokuda, and Y. Satomi, "Cancer prevention by carotenoids," *Archives of Biochemistry and Biophysics*, vol. 483, no. 2, pp. 165–168, 2009.

[8] M. Ozgur, T. Ozcan, A. Akpinar-Bayizit, and L. Yilmaz-Ersan, "Functional compounds and antioxidant properties of dried green and red peppers," *African Journal of Agricultural Research*, vol. 6, no. 25, pp. 5638–5644, 2011.

[9] A. M. Chuah, Y.-C. Lee, T. Yamaguchi, H. Takamura, L.-J. Yin, and T. Matoba, "Effect of cooking on the antioxidant properties of coloured peppers," *Food Chemistry*, vol. 111, no. 1, pp. 20–28, 2008.

[10] FAO, "Chillis, peppers and greens." Food and Agriculture Organization, 2011. Available online at: http://faostat.fao.org/site/291/default.aspx.

[11] M. Kaushal, V. Joshi, and R. Sharma, "Preparation and evaluation of value added products from bell pepper," *Indian Food Packer*, vol. 65, no. 6, pp. 159–165, 2011.

[12] A. R. Eswara and M. Ramakrishnarao, "Solar energy in food processing—a critical appraisal," *Journal of Food Science and Technology*, vol. 50, no. 2, pp. 209–227, 2013.

[13] D. R. Pangavhane and R. L. Sawhney, "Review of research and development work on solar dryers for grape drying," *Energy Conversion and Management*, vol. 43, no. 1, pp. 45–61, 2002.

[14] V. Gupta, L. Sunil, A. Sharma, and N. Sharma, "Construction and performance analysis of an indirect solar dryer integrated with solar air heater," in *Proceedings of the International Conference on Modelling Optimization and Computing*, pp. 3260–3269, April 2012.

[15] I. Doymaz and M. Pala, "Hot-air drying characteristics of red pepper," *Journal of Food Engineering*, vol. 55, no. 4, pp. 331–335, 2002.

[16] T. Y. Tunde-Akintunde, "Effect of pretreatment on drying time and quality of chilli pepper," *Journal of Food Processing and Preservation*, vol. 34, no. 4, pp. 595–608, 2010.

[17] P. Wiriya, T. Paiboon, and S. Somchart, "Effect of drying air temperature and chemical pretreatments on quality of dried chilli," *International Food Research Journal*, vol. 16, no. 3, pp. 441–454, 2009.

[18] AOAC, *Official Methods of Analysis*, AOAC International Press, Washington, DC, USA, 2005.

[19] H. T. Lawless and H. Heymann, *Sensory Evaluation of Food: Principles and Practices*, Springer Science & Business Media, Berlin, Germany, 2010.

[20] A. Fudholi, K. Sopian, M. H. Ruslan, M. A. Alghoul, and M. Y. Sulaiman, "Review of solar dryers for agricultural and marine products," *Renewable and Sustainable Energy Reviews*, vol. 14, no. 1, pp. 1–30, 2010.

[21] M. Davoodi, P. Vijayanand, S. Kulkarni, and K. Ramana, "Effect of different pre-treatments and dehydration methods on quality characteristics and storage stability of tomato powder," *Food Science and Technology*, vol. 40, pp. 1832–1840, 2007.

[22] R. Sharma, V. K. Joshi, and M. Kaushal, "Effect of pre-treatments and drying methods on quality attributes of sweet bell-pepper (Capsicum annum) powder," *Journal of Food Science and Technology*, vol. 52, no. 6, pp. 3433–3439, 2014.

[23] N. S. Thakur, M. M. Bhat, N. Rana, and V. K. Joshi, "Standardization of pre-treatments for the preparation of dried arils from wild pomegranate," *Journal of Food Science and Technology*, vol. 47, no. 6, pp. 620–625, 2010.

[24] S. S. Sablani, "Drying of fruits and vegetables: retention of nutritional/functional quality," *Drying Technology*, vol. 24, no. 2, pp. 123–135, 2006.

[25] L. A. Howard, A. D. Wong, A. K. Perry, and B. P. Klein, "β-Carotene and ascorbic acid retention in fresh and processed vegetables," *Journal of Food Science*, vol. 64, no. 5, pp. 929–936, 1999.

[26] J. D. Selman, "Vitamin retention during blanching of vegetables," *Food Chemistry*, vol. 49, no. 2, pp. 137–147, 1994.

[27] E. Sikora, E. Cieślik, T. Leszczyńska, A. Filipiak-Florkiewicz, and P. M. Pisulewski, "The antioxidant activity of selected cruciferous vegetables subjected to aquathermal processing," *Food Chemistry*, vol. 107, no. 1, pp. 55–59, 2008.

[28] M. N. Ramesh, W. Wolf, D. Tevini, and G. Jung, "Influence of processing parameters on the drying of spice paprika," *Journal of Food Engineering*, vol. 49, no. 1, pp. 63–72, 2001.

[29] C. Rice-Evans and N. Miller, *Antioxidant Activities of Flavonoids as Bioactive Components of Food*, Portland Press Limited, London, UK, 1996.

[30] M. C. Nicoli, B. E. Elizalde, A. Pitotti, and C. R. Lerici, "Effect of sugars and maillard reaction products on polyphenol oxidase and peroxidase activity in food," *Journal of Food Biochemistry*, vol. 15, no. 3, pp. 169–184, 1991.

14

Effect of Chitosan Edible Coating on the Biochemical and Physical Characteristics of Carp Fillet (*Cyprinus carpio*) Stored at −18°C

Ana Gabriela Morachis-Valdez,[1,2] Leobardo Manuel Gómez-Oliván,[1] Imelda García-Argueta,[3] María Dolores Hernández-Navarro,[1] Daniel Díaz-Bandera,[2] and Octavio Dublán-García[2]

[1]*Departamento de Toxicología Ambiental, Facultad de Química, Universidad Autónoma del Estado de México, Toluca, MEX, Mexico*
[2]*Departamento de Alimentos, Facultad de Química, Universidad Autónoma del Estado de México, Toluca, MEX, Mexico*
[3]*Departamento de Nutrición, Facultad de Medicina, Universidad Autónoma del Estado de México, Toluca, MEX, Mexico*

Correspondence should be addressed to Octavio Dublán-García; octavio_dublan@yahoo.com.mx

Academic Editor: Alejandro Castillo

The effect of an edible coating (EC) with 1.5% chitosan as an additive, on common carp (*Cyprinus carpio*) fillet, was determined evaluating the biochemical, physicochemical, textural, microbiological, and nutritional characteristics periodically during its storage in the freezer (−18°C), observing a decrease in the rate of biochemical reactions related to degradation ($p < 0.05$), hydroperoxides content (HPC) (0.8324 nM hydroperoxides/mg of protein versus 0.5540 nM/mg with regard to the EC sample), as well as protein carbonyl content (PCC) (0.5860 nM versus 0.4743 nM of reactive carbonyl groups/mg of protein of noncoated material), keeping properties for a longer period of time, and a lower protein solubility (7.8 mg of supernatant protein/mg of total protein versus 6.8 mg/mg) and less loss of moisture (8% less, with regard to EC); for the nutritional characteristics of the fillet, lysine is the limiting aminoacid in the sample without EC, while leucine is the limiting aminoacid for the EC sample. According to microbial growth, the count was 2.2×10^5 CFU/g of sample in mesophiles versus 4.7×10^4 in the EC sample. The results indicate that the use of EC added with chitosan maintains the quality of the product regarding lipid and protein oxidation until fourth month of storage, maintaining moisture content without variation for at least 3 months, and inhibits microbial growth up to 2 logarithmic units, during five months of frozen storage.

1. Introduction

The quality of fish is a complex concept, in which nutritional, microbiological, biochemical, and physicochemical attributes are involved. The freshness of fish decreases after its sacrifice; this is due to microbiological contamination and various biochemical reactions which produce changes in the protein fractions [1]. Some investigations have emphasized how the lipid compounds are altered due to oxidative deterioration [2]. Proteins including the sarcoplasmic, myofibrillar, and stromal proteins are susceptible to oxidative damage by intermediates of lipid oxidation [4-hydroxy-trans-2-nonenal (HNE), acrolein, malondialdehyde (MDA), glyoxal, and 4-oxo-trans-2-nonenal (ONE)], isoketals and metallic ions (such as the iron in the heme group or the copper and zinc found in enzymes and metalloenzymes) present in the muscles of animals, and those originated through processing (exposure of meat to oxygen, light, and temperature, cooling, use of additives, irradiation, and vacuum-packaging) that initiate oxidative damage, generating changes in flavor, color, texture/structure, and nutritional value [1–5]. The use of low temperatures such as freezing is a general method used for the control and decrease of biochemical changes that can occur during storage time; however, this does not completely inhibit the microbiological and chemical reactions that result in the deterioration of the quality of the fish, for which the use of edible coatings (EC) as adjuvants of preservation have demonstrated to provide an increase in shelf-life, due to their function as a barrier to oxygen, besides being employed as a vehicle of diverse components such as essential oils, bacteriocins, organic acids, and chitosan, which help in the control of oxidation and diminish the deterioration by microorganisms [6–8]. Chitosan (poly-b-(1-4)-D-glucosamine) is a versatile biopolymer, having a broad range of applications in the food industry. It has been reported to have a number of functional properties that make chitosan useful in food preservation; these include its antimicrobial activity [2] and antioxidant activity [9] and its ability to form protective films or coatings [10]. Although studies have been carried out concerning the use of chitosan as an antioxidant and/or an antimicrobial agent in EC [3, 9–11], none have thoroughly discussed its effect on nutritional properties. The purpose of this study was the use of an EC containing 1.5% chitosan in order to reduce the speed of deterioration caused by oxidation and/or microbial growth, evaluating physicochemical, textural, and nutritional properties during storage at commercial freezing temperatures (−18°C) in common carp.

2. Materials and Methods

2.1. Preparation and Treatment of Fish Samples

2.1.1. Chemicals. Chitosan, medium molecular weight, deacetylation value of 75–85%, and viscosity of 200–800 cP, was purchased from Aldrich Chemical Co.

Bovine serum albumin, acrylamide, N,N′-methylenebisacrylamide, trichloroacetic acid (TCA), $FeSO_4$, sulfuric acid, cumene hydroperoxide (CHP), butylhydroxytoluene, methanol, xylenol orange, di-nitrophenylhydrazine (DNPH), guanidine, ethanol, ethyl acetate, hydrochloric acid, Coomassie® Brilliant Blue R-250, thioglycolic acid, NBD-Cl, and o-phthalaldehyde (OPA) were purchased from Sigma-Aldrich (St. Louis, Missouri, USA).

Thiobarbituric acid was purchased (TBA) from Fluka (Sigma-Aldrich, Toluca, MX); sodium chloride, EDTA disodium salt, N,N′,N′- tetramethylethylenediamine (TEMED), Tris (base), urea, β-mercaptoethanol, glycine, acetic acid glacial, sodium phosphate monobasic, sodium phosphate dibasic, and copper sulfate pentahydrate were purchased from J.T. Baker (Pennsylvania, USA); sodium carbonate and lactic acid were purchased from Fermont (Monterrey, MX); sodium dodecyl sulfate (SDS) and bromophenol blue were obtained from Hycel (Mexico, MX); and plate count agar was purchased from Bioxon, Becton and Dickinson (Mexico, MX). All reagents used were of analytical grade.

2.1.2. Fish Sample Preparation. A total of 40 freshwater carps (Cyprinus carpio), with an average weight of 550–650 g, were purchased at Tiacaque Aquacultural Center in Toluca, State of Mexico, Mexico, and were transferred to the Food Science Laboratory in the School of Chemistry, at the Universidad Autónoma del Estado de México, and were filleted by hand using knives and cutting boards sanitized in chlorine solution

and rinsed in sterile distilled water. The fish were harvested during May 2016. Two fillets were obtained from each fish after removing the head and bone and were then immersed in the coating solution.

2.1.3. Preparation of Coating Solution and Treated Fillets. Chitosan solution was prepared with 1.5% (w/v) chitosan in 1% v/v lactic acid. To achieve complete dispersion of chitosan, the solution was stirred at 40°C for 1.5 h, on a hotplate/magnetic stirrer; the final coating forming solution consisted of 13% whey, 6% gelatin, 13% glycerol, and 4% inulin, according to Garcia-Argueta et al. [12], with a final pH of 3.5. Fillet samples were randomly assigned to two treatment batches consisting of one control batch (uncoated) and one batch treated with the coating solution. For each coated batch, approximately 20 carp fillets (12–15 cm) were immersed for 15 s in the coating solution and then allowed to stand for 1 min. Then, the fish fillets were drained on a sterile metal net and air-dried for 20 min in order to form the edible coatings, placed on polyethylene containers, and then stored at −18 ± 1°C for subsequent quality assessment in a commercial freezer (Torrey, México, MX). Chemical and microbiological analyses were performed at monthly intervals to determine the overall quality of fish, for five months.

2.2. Chemical Analyses

2.2.1. Moisture and Total Protein Analysis. Moisture content was determined by difference in weight between the fresh sample of minced fillet and the dried sample after drying in an oven at 105°C until reaching constant weight. The result is expressed as a percentage of moisture. The content of total protein was determined through the Kjeldahl method and results are expressed as g of protein/100 g of fish, as described in AOAC [13].

2.2.2. pH. 10 g of fillet muscle was weighed and homogenized at high speed in mixer/blender (Osterizer 450-20) for 1 min with 90 mL of distilled water. Connective tissue was eliminated by filtering with cloth, in accordance with that described by Owen et al. [14]. pH was determined with a digital pH meter (Conductronic pH 120, New York, USA).

2.2.3. Myofibrillar Protein Extraction. Myofibrillar protein (MP) was obtained in accordance with the methodology described by Ngapo et al. [15], with slight modifications. 100 g of common carp muscle was homogenized with a blender for 10 min with a mixture of ice-cold water 1:1:1 (w/w/v) and was then placed in an ice bath with a magnetic stirrer. The myofibrillar suspension was filtered through two layers of cloth in order to remove the connective tissue; this procedure was carried out twice. The homogenized muscle was then centrifuged at 3000 ×g at 4°C for 25 min and the supernatant was discarded. The protein concentration of the myofibrillar precipitate was determined using the Biuret method [16]. 25 mg/mL of MP was stored in a glass container with a lid for the formation of the gel. Gel forming was developed in two-step heating, first, incubation at 40°C for 30 min followed by heating with a gradual increase until 90°C was reached with constant stirring and then maintaining it for 20 min. Finally glass containers were removed and stored at 4°C.

2.2.4. Solubility.
According to Pilosof [17], 2 g of MP was centrifuged at 2500 ×g at 4°C for 30 min. The protein content in the supernatant was determined, as well as the total protein content in the MP sample prior to centrifugation. Solubility was defined by following equation:

$$\text{Solubility} = \frac{\text{Protein content in supernatant}}{\text{Protein content in the sample}} \times 100. \quad (1)$$

2.2.5. Total Sulfhydryl Content.
The total content of sulfhydryls (SH) was determined according to the method described by Ellman [18]. An aliquot of 1 mL of MP solution (5 mg/mL) reacted with 9 mL of Tris-glycine buffer (10.4 g of Tris-HCl, 6.9 g of glycine, 480 g of urea, and 1.2 g of EDTA/L at pH 8.0) at room temperature for 30 min. 0.05 mL of Ellman reagent (4 mg DTNB/mL) was added to aliquots of 3 mL and was incubated in darkness for 30 min. The reaction mixture was measured at 412 nm using a TU-1800 spectrophotometer (Beijing Purkinje General Instrument Co. Ltd., Beijing, China). The concentration of SH was expressed as total μM SH/mg of protein.

2.2.6. Determination of Total Volatile Base (TVB-N) Content.
The content of TVB-N was determined according to the Conway and Byrne method [19], with slight modifications. 5 g of the homogenized sample was added to 4% TCA in a 1 : 2 (w/v) ratio. Then, it was filtered through Whatman Number 1 paper (Schleicher & Schuell, Maidstone, England). 1 mL of the filtrate obtained was placed in the outer ring of the Conway Camara, while, in the inner ring, a solution of 1% boric acid containing Shiro Tashiro indicator was added. To initiate the reaction, 2 mL of K_2CO_3 was mixed with the filtrate. The camera was incubated at 25°C for 24 hr. The solution of the inner ring was titrated using 0.01 N HCl until a change in the color to a pink tone.

2.2.7. Determination of Hydroperoxides (HPC).
The content of HPC was determined by the Jiang et al. method [20] (FOX—ferrous oxidation-xylenol orange). A 100 μL aliquot of supernatant was obtained by the deproteinization of the sample with 10% TCA. 900 μL of the reaction mixture was added, consisting of 25 mM H_2SO_4, 0.25 mM $FeSO_4$ 0.1 mM, xylenol orange, and 4 mM butyl hydroxytoluene in 90% (v/v). The mixture obtained was incubated for 60 min at room temperature and absorbance was measured at 560 nm against the reaction blank in the spectrophotometer. The results were interpolated in a normal curve previously elaborated and were expressed as nM HPC/mg protein.

2.2.8. Determination of Lipoperoxides (LPX).
For the determination of LPX, the technique described by Büege and Aust thiobarbituric acid reactive substances (TBARS) was employed, which consists of an aliquot of 100 μL of supernatant, obtained with prior deproteinization, that was added

until 1 mL of Tris-HCl buffer solution pH 7.4 is reached. The samples were incubated at 37°C for 30 min; then 2 mL of the TBA-TCA reagent (0.375% TBA in 15% TCA) was added and thoroughly mixed using a vortex. It was taken to boiling point in a hot water bath for 45 min and was left to cool, eliminating the precipitate formed by centrifugation at 3000 ×g for 10 min. Absorbance readings were carried out at a wavelength of 535 nm against a reaction blank. The content of malondialdehyde (MDA) was calculated utilizing the molar extinction coefficient (MEC) of MDA (1.56×10^5 M/cm). The results were expressed as mM MDA/mg protein.

2.2.9. Determination of Protein Carbonyl Content (PCC).
The method is described by Levine et al. [21] and modified by Parvez and Raisuddin [22] and Burcham [23]. To an aliquot of 100 μL of supernatant obtained from the deproteinized sample, 150 μL of 10 mM DNPH dissolved in 2 M HCl was added, allowing for the reaction to be carried out in the dark for an hour at room temperature, placing 500 μL of 20% TCA and placing the mixture at rest for 15 min at 4°C. The sample was centrifuged at 11,000 ×g for 5 min. The precipitate obtained was washed at least three times with a solution of ethyl acetate : ethanol (1 : 1). Using a 6 M guanidine (pH 2.3) solution, the button was dissolved and was incubated for 30 min at 37°C. Absorbance readings were obtained at 366 nm, employing the corresponding MEC of 21,000 M/cm. The results were expressed as nM reactive carbonyls formed (C=O)/mg protein.

2.2.10. SDS-PAGE.
SDS gel electrophoresis was carried out according to Laemmli [24], with slight modifications in electrophoresis equipment, which consists of Bio-Rad Mini-PROTEAN II Cell camera, employing 10% acrylamide. MP extracts were added to 10% urea and buffer sample [0.1 M Tris-HCl (pH 6.8), 0.4% SDS, 10% glycerol, and 0.004% bromophenol blue]. The gel of 140 × 140 nm was prepared at a $T = 10\%$ in 1.2 M Tris-HCl (pH 8.8) and 0.3% SDS; the concentration gel at a $T = 4\%$ was prepared with 0.25 M Tris-HCl (pH 6.8) and 0.2% SDS. The electrode buffer contained 0.025 M Tris-HCl, 0.192 M glycine, and 0.15% SDS at pH 8.16. An electrophoretic run was carried out with a current of 200 volts; once the run was concluded, the gels were stained with a solution consisting of 40% methanol, 15% acetic acid, and 0.1% Coomassie R-250 Brilliant Blue.

2.2.11. Amino Acid Composition.
3 mg of the dehydrated simple was placed into tubes to carry out hydrolysis, with 6 N HCl and thioglycolic acid as antioxidants. Posteriorly, test tubes were heated for 6 hr at 150°C. At the end of hydrolysis, the reagent was evaporated in a Buchi rotary evaporator (Buchi, Flawil, Switzerland), obtaining a concentrate which was resuspended in 2 mL of 0.2 N sodium citrate buffer, pH 2.2.

For the determination of primary aminoacids, 250 μL of hydrolyzed extract was taken and mixed with 250 μL of o-phthalaldehyde (OPA); an aliquot of 20 μL was injected into the HPLC chromatograph (Varian 9012). For the determination of secondary aminoacids (proline and hydroxyproline),

125 μL of the lyophilized extract was put in 0.5 mL of 0.4 M borate buffer, pH 10.4. From this solution, 250 μL was taken and mixed with 250 μL of the derivative solution (NBD-Cl, 2 mg/mL, in MeOH); once filtered, the solution was heated to 60°C during 5 min in a dark vial with a lid. The derivatization reaction was stopped with the addition of 50 μL of 1 M HCl and was cooled to 0°C. For the analysis, 20 mL of the final extract was taken and injected into the HPLC; a 10 cm × 4.6 mm × 3 μm Varian Microsorb C18 column was utilized for this analysis, as was HPLC-grade methanol (with 99% purity) and sodium acetate buffer (pH 7.2) as mobile phase. A 430020-02 Fluorichrom fluorescence detector was used, and the quantification was carried out using external standards as reference.

2.3. Total Viable Counts (TVC).

A 10 g sample was homogenized in 90 mL of 0.1% peptone solution. Decimal dilutions were prepared from this solution and plated using plate count agar. The inoculated plates were incubated at 35°C for 48 h for total viable counts (log 10 CFU/g), as described by Ibrahim Sallam [25].

2.4. Statistical Analysis.

All experiments were performed in triplicate and a completely randomized design was used. All data were statistically analyzed by SPSS/PC software (version 17). One-way analysis of variance, independent sampling, and paired Student's t-tests were used for comparison of the means.

3. Results and Discussion

3.1. Physicochemical Analysis

3.1.1. Moisture. During its storage under freezing, a significant ($p < 0.05$) decrease in moisture in both treatments was observed (Figure 1(a)). For the carp with the coating, the decrease was observed until the fourth month; this could be due to the chitosan in the EC, which might promote crosslinking in the gelatin, thus diminishing the free volume of the polymeric matrix, which reduces the diffusion rate of the water molecules through the coating film. The aforementioned results in a decrease in vapor permeability in the fillet, as well as with the coating itself [26]. According to Dutta et al. [8], this is a desirable characteristic in coatings and is not observed in the control sample of this study, since a loss in moisture occurs ($p < 0.05$) during all of its storage period. After five months of storage, the percentage of loss was similar in both the control sample and the sample with the treatment; this could be due to the fact that both treatments were stored in polyethylene containers, which act as protection, where the coating would be able to maintain the moisture of the product during the first four months of storage [27].

3.1.2. pH. A significant decrease ($p < 0.05$) in pH was observed (Figure 1(c)) at the end of both treatments, which could be associated with the production of lactic acid through anaerobic glycolysis and the liberation of inorganic phosphate, a product of ATP degradation. An increase in pH is observed by the second month for the control sample and by

the third month for the coated sample, which could be due to the accumulation of basic compounds such as ammonia and trimethylamine, a result of autolytic and microbial reactions [4, 28–30]. The greatest variation in percentage at the end of storage corresponds to the noncoated sample, a lower pH in the sample with the coating can bolster microbial inhibition and contribute to the preservation of the samples inhibiting the endogenous proteases, and this result suggests that, during storage, the coating diminished the decrease in pH [11, 28].

3.1.3. Total Volatile Basic Nitrogen (TVB-N). TVB-N is a group primarily composed of primary, secondary, and tertiary amines which are used as indicators of meat deterioration; the increase in these is related to the activity of endogenous enzymes and bacteria [28]. According to Connell [31] and Giménez et al. [32], 25–40 mg of N/100 g of tissue is considered as acceptable for consumption [9, 11, 28]. Although both samples have values below said limit (Figure 1(b)), the concentration of TVB-N is greater in the control sample in each stage of storage; this could be due to the fact that the presence of chitosan helps in reducing the capacity of bacteria for oxidative deamination of nonprotein nitrogenated compounds [11, 28]. The storage time was not enough to identify when the acceptable threshold is exceeded, due to the fact that enzymatic and microbial activity diminish at low temperatures. The carp utilized in this study presented adequate conditions for human consumption at the end of the storage period, coinciding with that reported by Soares et al. [27].

3.2. Microbiological Changes

3.2.1. Total Viable Count. Microbial activity is a limiting factor in the quality of the fish, and the total viable count has been used as indicator of acceptability of the same [29, 33]. The initial value for the carp without the coating was 2.3 log 10^4 CFU/g and 1.1 log 10^4 CFU/g for the carp with the coating; these values depend on the environment from which the fish is obtained as well as postmortem conditions [11]. The evolution of the aforementioned is detailed in Figure 1(d), observing significant differences ($p < 0.05$) by the second month, obtaining a time-dependent increase in both treatments; however, the limit recommended by the ICMSF, [34] of 5 × 10^5 CFU/g, for quality fish, is not exceeded [29]. The properties of the chitosan added to the coating had an inhibitory effect, thereby obtaining 4.7 log 10^4 CFU/g, while, for the control, a value of 2.2 log 10^6 CFU/g was obtained after five months of storage. The antimicrobial effect of this compound has been widely reported by Fernández-Saiz et al. [35], Jeon et al. [36], and López-Caballero et al. [37], and its mechanism of action is related to the rupture of the lipopolysaccharide layer of the external membrane of Gram-negative bacteria and its function as a barrier against the transfer of oxygen. Another mechanism of action could be the interaction with anionic groups on the cell surface, due to its polycationic nature [3, 9]; this demonstrates that the EC inhibits microbial growth up to 2 logarithmic units, decreasing reactions involved in deterioration by microorganisms during its storage in the freezer.

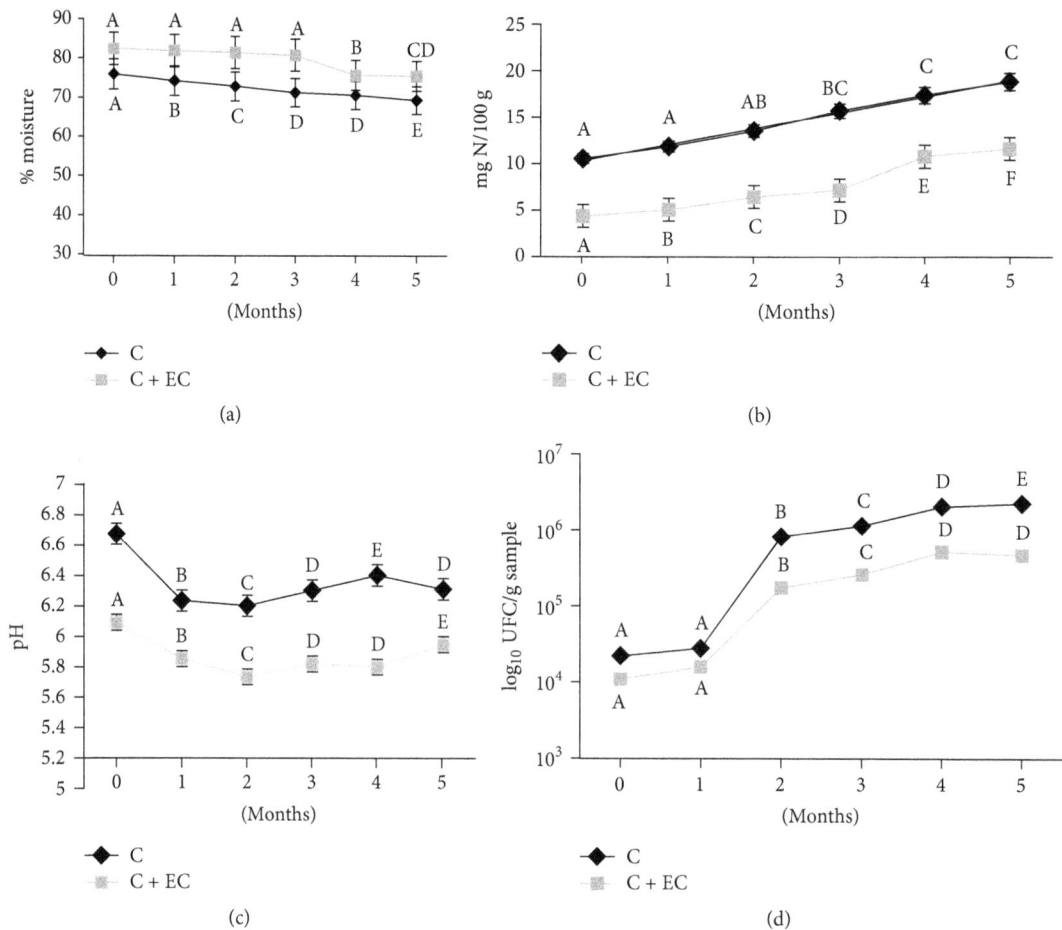

FIGURE 1: Changes in moisture (a), total volatile basic nitrogen TVB-N (b), pH (c), and total viable count (TVC) (d) values of common carp fillets stored at −18°C for 5 months. The results are the mean of three replications. C: fillet carp without coating; C + EC: fillet carp with edible coating. The different letters indicate significant differences between treatment times for the same treatment ($p < 0.05$).

3.3. Lipid Oxidation Products. The concentration of primary products of oxidation can be measured by the content of peroxides. The carp with coating shows a significant increase ($p < 0.05$) in peroxides by the fourth month and presents a final value of 0.55 nM HPOx/mg of protein (Figure 2(a)) while the control sample presents an increase by the first month, having an increase of 59% with regard to the sample with coating; this demonstrates that the coating retards lipid oxidation in the carp fillet. These results are in accordance with Ojagh et al. [9], Nowzari et al. [10], Jeon et al. [36], and Li et al. [38] and those that report that additional coatings with chitosan retard production of oxidated primary compounds in herring, trout, cod, and croaker in freezer storage as well as in ice storage. Lipid oxidation in fish is influenced by various factors like fat content, the degree of microsome associated with the oxidation system, heme group content, and the presence of ions [10, 38]. Chitosan-added coatings act as excellent barriers to the permeability of oxygen, once they are applied directly over the meat's surface, retarding the diffusion of oxygen [10].

In storage at freezing temperature (−18°C), oxidation is the most important factor in deterioration, even over

microbial activity. TBARS quantifies the compounds responsible for the loss of flavor and scent and is also important in the stages of deterioration of foods. The value of TBA is an indicator of lipid oxidation widely used, which quantifies the content of malondialdehyde (MDA), formed from hydroperoxides, which are the initial products of the oxidation of unsaturated fatty acids by oxygen [29, 39]. In the present study, the values of TBA of both treatment samples presented a significant increase ($p < 0.05$) by the second and forth months; however, by the fourth month the control sample presented a 39% increase with regard to a 6% increase in the sample with coating; this is due to the absence of chitosan in the control sample's coating. This same behavior was observed by Jeon et al. [36], in herring and cod with a chitosan coating, and Ojagh et al. [9] in rainbow trout. The antioxidative mechanism of the chitosan is due to the fact that its primary amino groups form a stable fluorosphere with volatile aldehydes such as malondialdehyde, derived from the rupture of fats during oxidation [38].

The results indicate that the chitosan employed in a 1.5% EC preserves the fish fillet through reduction of lipid oxidation.

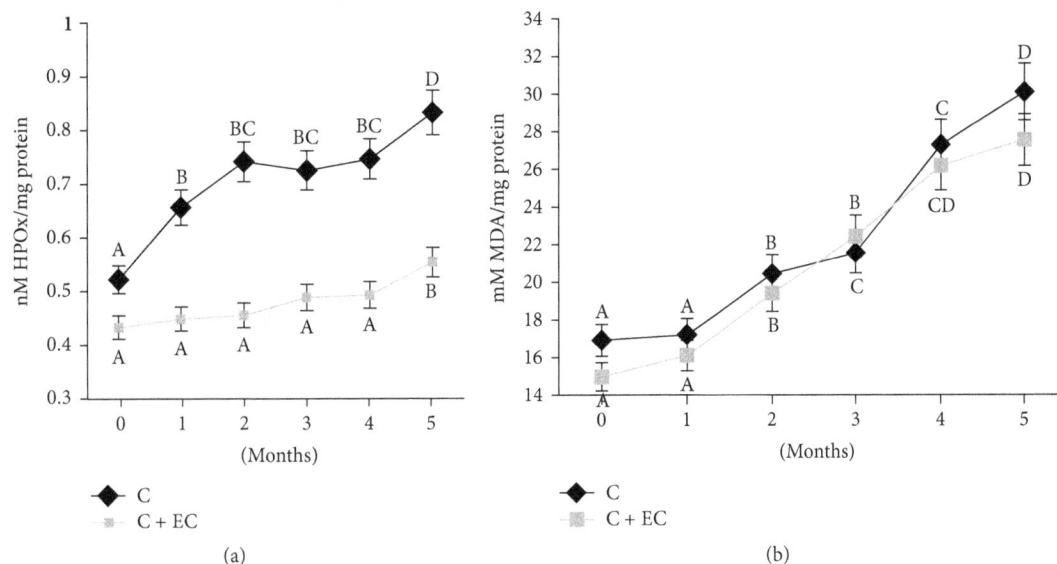

FIGURE 2: Changes in HPOx content (a) and MDA content (b) values of common carp fillets stored at −18°C for 5 months. The results are the mean of three replications. C: fillet carp without coating; C + EC: fillet carp with edible coating. The different letters indicate significant differences between treatment times for the same treatment ($p < 0.05$).

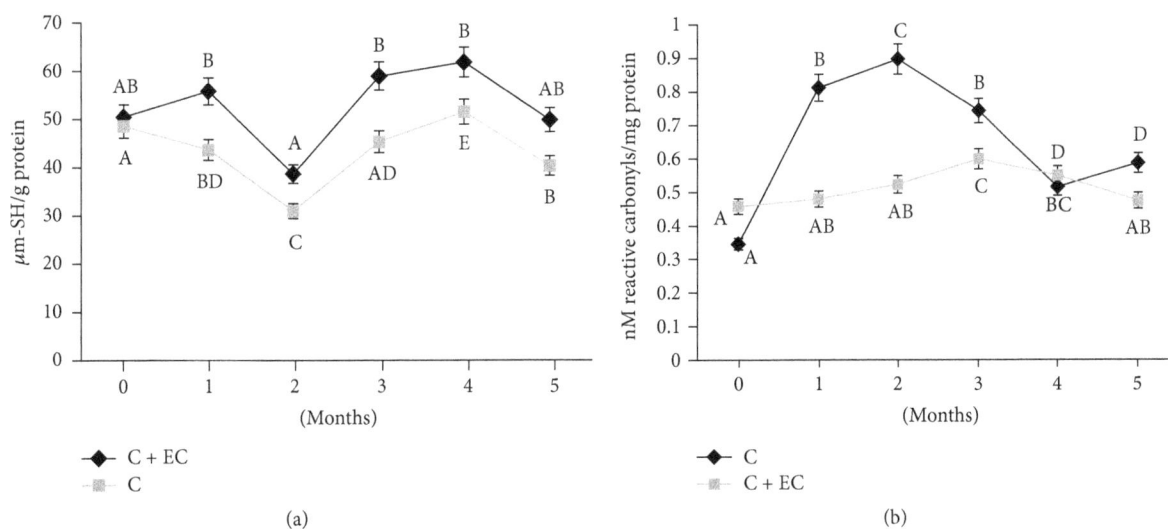

FIGURE 3: Changes in sulfhydryl content (a) and protein carbonyl content (b) values of common carp fillets stored at −18°C for 5 months. The results are the mean of three replications. C: fillet carp without coating; C + EC: fillet carp with edible coating. The different letters indicate significant differences between treatment times for the same treatment ($p < 0.05$).

3.4. Products of Protein Oxidation. Protein oxidation in carp fillet during freezer storage is shown in Figure 3; there was a significant increase ($p < 0.05$) in the formation of carbonylated proteins during storage in the control sample by the first month, while the sample with coating presented an increase until the third month and does not present differences at the end of storage. The increase was from 0.35 to 0.89 nm/mg of protein for the control sample and from 0.45 to 0.59 nm/mg of protein at maximum oxidation point. The samples of processed fillet include red and white muscle, and thus it is possible to find a high proportion of heme proteins (hemoglobin and myoglobin) and, consequently,

a concentration of iron, in addition to what can be found chelated to proteins, favoring protein and lipid oxidation. By the fourth month, both samples present a reduction in the concentration of carbonyl groups, which has been reported by other authors during a storage at −20°C, suggesting interactions between carbonyls and other cell components, for which the formation of carbonyl groups should be considered as a step in oxidation processes and not as a sole, stable marker of protein oxidation [40]. The oxidation of proteins is associated with the decrease in sulfhydryl groups, which are converted to disulfides. During freezer storage, a significant decrease ($p < 0.05$) was observed during the second month in both samples. Proteins are attacked by reactive oxygen species (ROS), where its creation in fish is due to diverse external factors, such as noise, manipulation, slaughter, or the presence of metals such as iron. The interaction can lead to the formation of carbonyl groups and the loss of sulfhydryl groups [3, 40], affecting structural, functional, and nutritional properties, for which the use of EC with chitosan could reduce mentioned effects.

3.5. Aminoacid Content.

The protein content in fresh carp is similar to that reported by FAO (21%), in frozen species, as well as frozen species but with coating (Table 1). The protein of fish is considered of the highest quality compared to standard proteins reported by FAO and although the information concerning its nutritional value is widely known, few are the studies that refer to its composition, the scarcest being in common carp [41]. It is known that the content and bioavailability of aminoacids can be affected by different operations such as drying, fermentation, extrusion, and even germination [42]. In this study, it was observed that the aminoacid content in carp suffered a significant decrease ($p < 0.05$) after 5 months of storage due to freezing, while the species with the coating only presented a decrease in the levels of Cys, His, Tyr, Thr, Met, and Lys; this could be due to protein oxidation by intermediaries of lipid oxidation and environment factors such as pH, temperature, water activity, and the presence of promoters and inhibitors like phenolic compounds [43]. The sulfur aminoacids such as Met and Cys are highly susceptible to oxidation in the presence of oxidized lipid products which lead to the formation of a variety of compounds such as sulphone, sulfoxides, and disulfide derivatives [44]. In the present study, a decrease in Met and Cys can be observed in the frozen fillet with regard to the control. Once the edible coating is added, a protective activity can be observed for Cys (0.84 g/100 g in frozen fillet and 0.91 g/100 g in coated fillet); this can be due to cross-linking, which is usually attributed to the formation of Cys (disulfide bonds) and dityrosines (from two Cys and two Tyr residues) [45]. The loss of Thr and Lys could be due to the formation of intermediary adducts such as α-amino-3-keto butyric acid and α-amino adipic semialdehyde, as a consequence of oxidation catalyzed by metals from the metalloproteins of the food matrix [43].

Common carp protein is characterized by a high content of essential aminoacids, compared to the standard protein established by the FAO, exceeding values in the case of Met + Cys and Val (Table 2) in fresh carp, having Leu as the limiting aminoacid and presenting a "limiting aminoacid" index (or chemical index) of 82.27; this index is a basic parameter used for the evaluation of the nutritional value of a food, which refers to the minor content of an essential aminoacid with regard to a given standard protein, called the "limiting aminoacid." In this study, the essential aminoacids in the sample in freezing without EC decreased, only maintaining Val as the excess aminoacid and having Lys as the limiting aminoacid. This change in the limiting aminoacid could be a consequence of enzymatic reactions and/or bacterial growth during storage [46]. On the other hand, the frozen carp with EC continued to present

TABLE 1: Aminoacid composition of common carp fillet proteins, 5-month frozen carp, and 5-month frozen carp + edible coating, stored at $-18°C$.

Aminoacids	Carp	5-month frozen carp	5-month frozen carp + edible coating
Asparagine	6.28 ± 0.18^a	5.03 ± 0.12^b	5.98 ± 0.22^a
Glutamic acid	10.75 ± 0.22^a	8.53 ± 0.18^c	9.05 ± 0.15^b
Alanine	4.98 ± 0.18^a	3.76 ± 0.15^c	4.28 ± 0.23^b
Arginine	3.74 ± 0.10^a	3.11 ± 0.12^b	3.45 ± 0.18^a
Cysteine	0.98 ± 0.07^a	0.84 ± 0.08^b	0.91 ± 0.06^a
Phenylalanine	3.51 ± 0.24^a	2.71 ± 0.28^b	3.29 ± 0.19^a
Glycine	3.34 ± 0.12^a	2.76 ± 0.19^b	$3.11 + 0.16^a$
Histidine	2.48 ± 0.08^a	1.5 ± 0.05^c	$2.01 + 0.12^b$
Isoleucine	3.89 ± 0.24^a	2.31 ± 0.22^b	$3.28 + 0.17^a$
Leucine	5.43 ± 0.32^a	4.76 ± 0.38^b	5.21 ± 0.34^a
Lysine	5.53 ± 0.28^a	4.12 ± 0.22^b	5.3 ± 0.19^c
Methionine	2.73 ± 0.08^a	1.08 ± 0.07^b	2.06 ± 0.09^c
Serine	2.72 ± 0.07^a	1.97 ± 0.08^c	2.25 ± 0.11^b
Tyrosine	2.67 ± 0.15^a	1.86 ± 0.21^b	2.22 ± 0.13^c
Threonine	3.16 ± 0.22^a	2.42 ± 0.12^b	3.08 ± 0.15^a
Valine	4.28 ± 0.19^a	3.76 ± 0.32^b	4.12 ± 0.23^a
Protein (%)	22.39 ± 1.7^a	22.65 ± 1.2^a	22.87 ± 1.4^a

a,b,c $p < 0.05$.

a higher concentration in the aforementioned aminoacids (Iso, Met + Cys, and Val), thereby presenting a protective effect in Lys, which results in great benefit since fish could be maintained as an important source of this aminoacid after 5 months of storage in the freezer [42, 47]. The chemical index diminishes 11 units in the fillet without EC and 3 units for the fillet with the EC after freezing, which suggests that EC has a protective effect over essential aminoacids.

3.6. SDS-PAGE.

The molecular weight profiles obtained through SDS-PAGE of the MPs extracted from the different treatments during storage are shown in Figure 4, in which the composition of the MPs of common carp was myosin heavy chains (MHC), actin (A), and troponin (T), observing that the sample without coating presents, after 5 months of storage, molecular weight bands lower than those of myosin, with an interval of 150, 100, and 75 kDa, approximately; this is probably due to carbonylation of MHC, coinciding with that reported by Kjærsgård et al. [48] in rainbow trout. Likewise, Passi et al. (2005) reported an increase of oxidized proteins in different species of Mediterranean fish after lipid oxidation, possibly due to the presence of the heme group in myoglobin [49]. For the samples with coating, no notable changes were observed in the SDS-PAGE profile, which could suggest that the coating is retarding the oxidation mechanism, showing that actin was the least oxidized during freezing, coinciding with that reported by Eymard et al. [40]. In spite of the high susceptibility of myofibrillary proteins to oxidation, edible coatings could help in the preservation of protein integrity

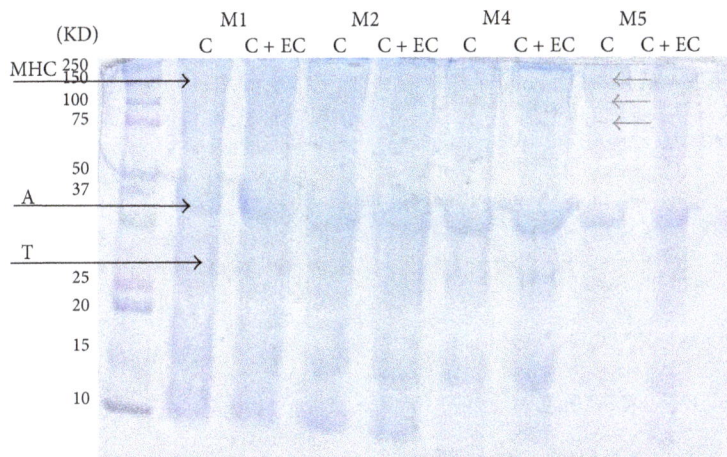

FIGURE 4: SDS-PAGE of common carp fillets stored at −18°C; control batch (C) fillet without coating and filleted carp with edible coating (C + EC); M1, M2, M4, and M5 from first to fifth month during storage.

TABLE 2: Values of the limiting aminoacid index (%).

Aminoacids	Standard FAO/WHO (1991)[c]	Carp		5-month frozen carp		5-month frozen carp + edible coating	
		g/100 g protein	%	g/100 g protein	%	g/100 g protein	%
Phe + Tyr[a]	6,30	6,18	98,10	4,57	72,54	5,51	87,46
Isoleucine	2,80	3,89	138,93	2,31	82,50	3,28	117,14
Leucine	6,60	5,43	**82,27**	4,76	72,12	5,21	**78,94**
Lysine	5,80	5,53	95,34	4,12	**71,03**	5,30	91,38
Met + Cys[b]	2,50	3,71	148,40	1,92	76,80	2,97	118,80
Threonine	3,40	3,16	92,94	2,42	71,18	3,08	90,59
Valine	3,50	4,28	122,29	3,76	107,43	4,12	117,71
Amino acid index			82,27		71,03		78,94

[a]Phenylalanine + tyrosine; [b]methionine + cysteine; [c]according to Usydus et al. [47].

of aquatic/marine species stored during freezing, keeping functional and nutritional properties for more time.

4. Conclusions

The use of EC added with chitosan allows retention of the physicochemical and nutritional characteristics of the common carp for more time during storage, diminishing the loss of moisture, bacterial growth, nutritional value, and speed of lipid and protein oxidation in oxidation products (hydroperoxides, lipoperoxidation, and carbonyl proteins), as well as an indication of the aminoacids present in the common carp fillet with EC, being an excellent alternative as an adjuvant in the conservation through freezing of aquatic species of economic importance on a global scale.

Conflicts of Interest

The authors declare that they have no conflicts of interest.

Acknowledgments

Ana Gabriela Morachis Valdez thanks the National Council of Science and Technology (Consejo Nacional de Ciencia y Tecnología, Mexico) for a graduate scholarship.

References

[1] T. Li, J. Li, W. Hu, and X. Li, "Quality enhancement in refrigerated red drum (Sciaenops ocellatus) fillets using chitosan coatings containing natural preservatives," *Food Chemistry*, vol. 138, no. 2-3, pp. 821–826, 2013.

[2] A. B. Falowo, P. O. Fayemi, and V. Muchenje, "Natural antioxidants against lipid-protein oxidative deterioration in meat and meat products: a review," *Food Research International*, vol. 64, pp. 171–181, 2014.

[3] C. O. Mohan, C. N. Ravishankar, K. V. Lalitha, and T. K. Srinivasa Gopal, "Effect of chitosan edible coating on the quality of double filleted Indian oil sardine (*Sardinella longiceps*) during chilled storage," *Food Hydrocolloids*, vol. 26, no. 1, pp. 167–174, 2012.

[4] E. Indergård, I. Tolstorebrov, H. Larsen, and T. M. Eikevik, "The influence of long-term storage, temperature and type of packaging materials on the quality characteristics of frozen farmed Atlantic Salmon (*Salmo Salar*)," *International Journal of Refrigeration*, vol. 41, pp. 27–36, 2014.

[5] R. Pamplona, "Advanced lipoxidation end-products," *Chemico-Biological Interactions*, vol. 192, no. 1-2, pp. 14–20, 2011.

[6] E. Latou, S. F. Mexis, A. V. Badeka, S. Kontakos, and M. G. Kontominas, "Combined effect of chitosan and modified atmosphere packaging for shelf life extension of chicken breast fillets," *LWT—Food Science and Technology*, vol. 55, no. 1, pp. 263–268, 2014.

[7] M. Ahmad, S. Benjakul, P. Sumpavapol, and N. P. Nirmal, "Quality changes of sea bass slices wrapped with gelatin film incorporated with lemongrass essential oil," *International Journal of Food Microbiology*, vol. 155, no. 3, pp. 171–178, 2012.

[8] P. K. Dutta, S. Tripathi, G. K. Mehrotra, and J. Dutta, "Perspectives for chitosan based antimicrobial films in food applications," *Food Chemistry*, vol. 114, no. 4, pp. 1173–1182, 2009.

[9] S. M. Ojagh, M. Rezaei, S. H. Razavi, and S. M. H. Hosseini, "Effect of chitosan coatings enriched with cinnamon oil on the quality of refrigerated rainbow trout," *Food Chemistry*, vol. 120, no. 1, pp. 193–198, 2010.

[10] F. Nowzari, B. Shábanpour, and S. M. Ojagh, "Comparison of chitosan-gelatin composite and bilayer coating and film effect on the quality of refrigerated rainbow trout," *Food Chemistry*, vol. 141, no. 3, pp. 1667–1672, 2013.

[11] J. Huang, Q. Chen, M. Qiu, and S. Li, "Chitosan-based Edible Coatings for Quality Preservation of Postharvest Whiteleg Shrimp (*Litopenaeus vannamei*)," *Journal of Food Science*, vol. 77, no. 4, pp. C491–C496, 2012.

[12] I. Garcia-Argueta, O. Dublan-Garcia, B. Quintero-Salazar, A. Dominguez-Lopez, L. M. Gomez-Olivan, and A. F. Z. M. Salem, "Effect of lactic acid bacteria on the textural properties of an edible film based on whey, inulin and gelatin," *African Journal of Biotechnology*, vol. 12, pp. 2659–2669, 2013.

[13] AOAC, *Official Methods of Analysis of the Association of Analytical Chemistry*, Association of Official Analytical Chemists, 4th edition, 1984.

[14] J. E. Owen, F. A. Nuñez, M. T. Arias et al., *Manual De Prácticas de Cursos de Tecnologías de la Carne*, Facultad de Zootecnia, Universidad de Chihuahua, Chihuahua, Mexico, 1982.

[15] T. Ngapo, B. Wilkinson, R. Chong et al., "Gelation of bovine myofibrillar protein induced by 1,5 Gluconolacone," in *Proceedings of the 38th International Congress of Meat Science and Technology*, pp. 1095–1098, Clermont Ferrand, France, 1992.

[16] A. G. Gornall, C. J. Bardawill, and M. M. David, "Determination of serum proteins by means of the biuret reaction," *The Journal of Biological Chemistry*, vol. 177, no. 2, pp. 751–766, 1949.

[17] A. M. Pilosof, "Solubilidad," in *Caracterización funcional y estructural de proteínas*, A. M. R and G. B. Barholomai, Eds., pp. 60–75, Ceudeba CYTED (Programa Iberoamericano de Ciencia y Tecnología para el Desarrollo), 2000.

[18] G. L. Ellman, "Tissue sulfhydryl groups," *Archives of Biochemistry and Biophysics*, vol. 82, no. 1, pp. 70–77, 1959.

[19] E. J. Conway and A. Byrne, "An absorption apparatus for the micro-determination of certain volatile substances I. The micro-determination of ammonia," *Journal of Biochemistry*, vol. 27, pp. 419–429, 1936.

[20] Z. Y. Jiang, J. V. Hunt, and S. P. Wolff, "Ferrous ion oxidation in the presence of xylenol orange for detection of lipid hydroperoxide in low density lipoprotein," *Analytical Biochemistry*, vol. 202, no. 2, pp. 384–389, 1992.

[21] R. L. Levine, J. A. Williams, E. R. Stadtman, and E. Shacter, "Carbonyl assays for determination of oxidatively modified proteins," *Methods in Enzymology*, vol. 233, pp. 346–357, 1994.

[22] S. Parvez and S. Raisuddin, "Protein carbonyls: Novel biomarkers of exposure to oxidative stress-inducing pesticides in freshwater fish Channa punctata (Bloch)," *Environmental Toxicology and Pharmacology*, vol. 20, no. 1, pp. 112–117, 2005.

[23] P. C. Burcham, "Modified protein carbonyl assay detects oxidised membrane proteins: A new tool for assessing drug- and chemically-induced oxidative cell injury," *Journal of Pharmacological and Toxicological Methods*, vol. 56, no. 1, pp. 18–22, 2007.

[24] U. K. Laemmli, "Cleavage of structural proteins during the assembly of the head of bacteriophage T4," *Nature*, vol. 227, no. 5259, pp. 680–685, 1970.

[25] K. Ibrahim Sallam, "Antimicrobial and antioxidant effects of sodium acetate, sodium lactate, and sodium citrate in refrigerated sliced salmon," *Food Control*, vol. 18, no. 5, pp. 566–575, 2007.

[26] S. Fakhreddin Hosseini, M. Rezaei, M. Zandi, and F. F. Ghavi, "Preparation and functional properties of fish gelatin-chitosan blend edible films," *Food Chemistry*, vol. 136, no. 3-4, pp. 1490–1495, 2013.

[27] N. M. Soares, M. S. Oliveira, and A. A. Vicente, "Effects of glazing and chitosan-based coating application on frozen salmon preservation during six-month storage in industrial freezing chambers," *LWT—Food Science and Technology*, vol. 61, no. 2, pp. 524–531, 2015.

[28] W. Fan, J. Sun, Y. Chen, J. Qiu, Y. Zhang, and Y. Chi, "Effects of chitosan coating on quality and shelf life of silver carp during frozen storage," *Food Chemistry*, vol. 115, no. 1, pp. 66–70, 2009.

[29] N. M. Soares, T. S. Mendes, and A. A. Vicente, "Effect of chitosan-based solutions applied as edible coatings and water glazing on frozen salmon preservation—a pilot-scale study," *Journal of Food Engineering*, vol. 119, no. 2, pp. 316–323, 2013.

[30] D. Liu, L. Liang, W. Xia, J. M. Regenstein, and P. Zhou, "Biochemical and physical changes of grass carp (Ctenopharyngodon idella) fillets stored at -3 and 0 °c," *Food Chemistry*, vol. 140, no. 1-2, pp. 105–114, 2013.

[31] J. J Connell, "Methods of assessing and selecting for quality," in *Control of Fish Quality*, Springer, Berlin, Germany, 3rd edition, 1990.

[32] B. Giménez, P. Roncalés, and J. A. Beltrán, "Modified atmosphere packaging of filleted rainbow trout," *Journal of the Science of Food and Agriculture*, vol. 82, no. 10, pp. 1154–1159, 2002.

[33] G. Olafsdóttir, E. Martinsdóttir, J. Oehlenschager et al., "Methods to evaluate fish freshness in research and industry," *Trends in Food Science and Technology*, vol. 8, pp. 258–265, 1997.

[34] ICMSF, *Microorganisms in Foods 2. Sampling for Microbiological Analysis: Principles and Specific Applications*, University of Toronto Press, New York, NY, USA, 2nd edition, 1986.

[35] P. Fernández-Saiz, G. Sánchez, C. Soler, J. M. Lagaron, and M. J. Ocio, "Chitosan films for the microbiological preservation of refrigerated sole and hake fillets," *Food Control*, vol. 34, no. 1, pp. 61–68, 2013.

[36] Y.-J. Jeon, J. Y. V. A. Kamil, and F. Shahidi, "Chitosan as an edible invisible film for quality preservation of herring and Atlantic cod," *Journal of Agricultural and Food Chemistry*, vol. 50, no. 18, pp. 5167–5178, 2002.

[37] M. E. López-Caballero, M. C. Gómez-Guillén, M. Pérez-Mateos, and P. Montero, "A chitosan-gelatin blend as a coating for fish patties," *Food Hydrocolloids*, vol. 19, no. 2, pp. 303–311, 2005.

[38] T. Li, W. Hu, J. Li, X. Zhang, J. Zhu, and X. Li, "Coating effects of tea polyphenol and rosemary extract combined with chitosan on the storage quality of large yellow croaker (*Pseudosciaena crocea*)," *Food Control*, vol. 25, no. 1, pp. 101–106, 2012.

[39] B. W. S. Souza, M. A. Cerqueira, H. A. Ruiz et al., "Effect of Chitosan-based coatings on the shelf life of Salmon (*Salmo salar*)," *Journal of Agricultural and Food Chemistry*, vol. 58, no. 21, pp. 11456–11462, 2010.

[40] S. Eymard, C. P. Baron, and C. Jacobsen, "Oxidation of lipid and protein in horse mackerel (Trachurus trachurus) mince and washed minces during processing and storage," *Food Chemistry*, vol. 114, no. 1, pp. 57–65, 2009.

[41] K. A. Skibniewska, J. Zakrzewski, J. Kłobukowski et al., "Nutritional value of the protein of consumer carp cyprinus carpio L.," *Czech Journal of Food Sciences*, vol. 31, no. 4, pp. 313–317, 2013.

[42] J. Boye, R. Wijesinha-Bettoni, and B. Burlingame, "Protein quality evaluation twenty years after the introduction of the protein digestibility corrected amino acid score method," *British Journal of Nutrition*, vol. 108, no. 2, pp. S183–S211, 2012.

[43] M. Estévez, "Protein carbonyls in meat systems: a review," *Meat Science*, vol. 89, no. 3, pp. 259–279, 2011.

[44] R. L. Levine, N. Wehr, J. A. Williams et al., "Determination of carbonyl groups in oxidized proteins," *Methods Molecular Biology*, vol. 99, pp. 15–24, 2000.

[45] M. N. Lund, M. Heinonen, C. P. Baron, and M. Estévez, "Protein oxidation in muscle foods: a review," *Molecular Nutrition and Food Research*, vol. 55, no. 1, pp. 83–95, 2011.

[46] A. Ciampa, G. Picone, L. Laghi, H. Nikzad, and F. Capozzi, "Changes in the amino acid composition of Bogue (Boops boops) fish during storage at different temperatures by 1H-NMR spectroscopy," *Nutrients*, vol. 4, no. 6, pp. 542–553, 2012.

[47] Z. Usydus, J. Szlinder-Richert, and M. Adamczyk, "Protein quality and amino acid profiles of fish products available in Poland," *Food Chemistry*, vol. 112, no. 1, pp. 139–145, 2009.

[48] I. V. H. Kjærsgård, M. R. Nørrelykke, C. P. Baron, and F. Jessen, "Identification of carbonylated protein in frozen rainbow trout (*Oncorhynchus mykiss*) fillets and development of protein oxidation during frozen storage," *Journal of Agricultural and Food Chemistry*, vol. 54, no. 25, pp. 9437–9446, 2006.

[49] S. Passi, S. Cataudella, L. Tiano, and G. P. Littarru, "Dynamics of lipid oxidation and antioxidant depletion in Mediterranean fish stored at different temperatures," *BioFactors*, vol. 25, no. 1-4, pp. 241–254, 2005.

15

Effect of Harvest Period on the Proximate Composition and Functional and Sensory Properties of Gari Produced from Local and Improved Cassava (*Manihot esculenta*) Varieties

Alphonse Laya,[1,2] Benoît Bargui Koubala ⓘ,[1,3] Habiba Kouninki,[1] and Elias Nchiwan Nukenine ⓘ[4]

[1]Department of Life and Earth Sciences, Higher Teachers' Training College of Maroua, University of Maroua, P.O. Box 55, Maroua, Cameroon
[2]Department of Biological Sciences, Faculty of Science, University of Maroua, P.O. Box 446, Maroua, Cameroon
[3]Department of Chemistry, Faculty of Science, University of Maroua, P.O. Box 814, Maroua, Cameroon
[4]Department of Biological Sciences, Faculty of Science, University of Ngaoundéré, P.O. Box 454, Ngaoundéré, Cameroon

Correspondence should be addressed to Benoît Bargui Koubala; bkoubala@yahoo.fr

Academic Editor: Alejandro Castillo

This study is aimed at evaluating the proximate composition and functional and sensory characteristics of gari obtained from five cassava varieties (*EN*, *AD*, *TMS92/0326*, *TMS96/1414*, and *IRAD4115*). These cassavas were harvested during the dry season 12 months after planting (12MAP) and in the rainy season (15MAP). Results showed that the characteristics of gari varied significantly ($p < 0.05$) with the variety and the harvest period. Gari from *EN* cassava harvested at 12MAP had the highest total carbohydrates (78.07% dry weight), starch (61%), and proteins content, while gari from TMS *96/1414* variety (12MAP) had high amino acids (10.25 mg/g) and phenolic compounds (9.31 mg/g) content. The gari from *IRAD4115* had the highest value of ash content (20.62 mg/g) at 12MAP. The soluble sugar content was high in the gari from cassava harvested at 12MAP while free cyanide reduced significantly in gari from cassava harvested at 12MAP. The water absorption capacity, swelling power, and bulk density were significantly ($p < 0.05$) high in the gari from *EN* cassava variety at 12MAP. Compared to commercial gari (3.30), gari from *EN* local cassava had the best overall acceptability (4.35) followed by those obtained from *TMS92/0326* and *TMS92/1414* varieties, respectively.

1. Introduction

Cassava roots yield more carbohydrates per hectare than cereal crops and can be grown at a considerably lower cost [1]. Cassava roots are a staple food that provides carbohydrates for more than 2 billion people in the tropics. However, cassava roots spoil quickly after harvest. In order to avoid this loss, they must be sold or processed into by-products after harvest. Generally, cassava and its products are poor in proteins. The deficiency in certain essential amino acids depends mostly on the varieties and geographical conditions. In order to enhance the nutritional quality of cassava, it is processed into fermented products such as gari. Gari is one of the most popular cassava products consumed in Africa, Southeast Asia, and Brazil [2]. In Africa, fermented foods and beverages are produced using fermentation. These products have been consumed for a long time because of their numerous nutritional values. In effect, lactic acid bacteria (LAB) isolated from these products have been proven to be good sources of antimicrobials and therapeutics, and are accepted as probiotics [3]. Fermented foods represent one-third of total food consumed by human beings [4]. Fermentation enhances the nutrient content of foods through the biosynthesis of vitamins, essential fatty acids, essential amino acids, and proteins and by improving protein quality and fibre digestibility [5–7]. It also enhances micronutrient bioavailability and aids in degrading antinutritional factors [8]. About 83% of the total cyanogenic glucosides (linamarin and lotaustralin) are detoxified during processing of cassava tuber into gari [9] and 98% of the cyanide is lost when gari is cooked into *eba* [10]. No detectable cyanide has been found in gari roasted with palm oil [11]. However acceptability of gari depends on the final texture and sensory attributes after processing [1]. Fermentation of cassava mash usually takes one to two days [12].

It has been reported that traditional gari contains a certain amount of residual cyanide. This is due to the tendency to shorten fermentation time in order to meet growing market demand [13]. That is why samples of gari with cyanide concentrations above 10 mg of HCN/kg are from areas where the cassava mash is fermented for less than 12 hours [14]. Halliday et al. [15] reported that the high initial moisture content and inappropriate storage container are the major factors that could encourage bacteria and fungi contamination and proliferation in gari during storage.

Gari is imported from neighbouring Nigeria or from the southern part of Cameroon to the far north region of Cameroon. In the continuous quest for solution to the problem of malnutrition in the far north region of Cameroon, improving nutritional quality and safety of local foods through better processing methods is recommended.

This work is aimed at producing gari from cassava roots of two local and three improved varieties harvested at two different growing periods.

2. Material and Methods

2.1. Trial Site and Experimental Design. The study was conducted in the far north region of Cameroon. The region is characterized by a transient equatorial climate with a long dry season (October to April) and short rainy season (May to September). The annual precipitation is 1000 mm and the mean annual temperature is 30°C. The soil is sandy and clayey.

The experimental field was a randomized complete block design with four repetitions. Each repetition measured 25 m^2 with 25 cassava plants spaced at 1 m apart. Five varieties of (local and improved) cassava (*Manihot esculenta* Crantz) were considered in this study. The improved varieties were TMS92/0326 and TMS96/1414 from IITA (International Institute of Tropical Agriculture) and IRAD4115 from IRAD (Institut de Recherche Agricole pour le Developpement) in Adamawa region. The two local varieties, *EN* (red, sweet variety) and *AD* (red, bitter variety), were highly appreciated by the population in the far north and adamawa regions, respectively. After planting, all cassava varieties were grown for 12 or 15 months. Their storage roots were harvested in May (dry season) and August (rainy season), respectively.

2.2. Cassava Processing into Gari. The preparation of the gari was done following the method described by Agbor-Egbe and Mbome [12] and Amamgbo et al. [16]. Storage roots of the local and improved cassava varieties were harvested, cleaned, peeled, washed, and grated manually with a grater (Ø = 2 mm). The resulting mash was packed in a muslin tissue which was tied with sewing cotton. Then, mash was dewatered by placing the muslin tissue between a set of thick and long wooden poles arranged beneath and on top

such that the ends were strongly fastened together with ropes. The mash was allowed to ferment for two days (48 h) under ambient conditions. The fermented mash was sieved to remove fibrous materials and then garified in a shallow pot with the addition of a small quantity of palm oil (10 ml/200 g) in order to obtain yellow gari. The different garis were obtained after 20 minutes of dry roasting at 80°C–90°C. The garis were then weighed and packed in polyethylene bags and labelled according to the cassava variety used. The gari yield was calculated as described by Sobowale et al. [17].

2.3. Determination of the Proximate Composition of Cassava Gari. The dry matter of the different cassava (*Manihot esculenta* Crantz) root and gari was determined using the standard AOAC [18] method. Slurries (10% dry matter) of all samples were made and their pH values were measured using a pH meter (HI 8424 Microcomputer Hanna instruments). The ash content of the samples was determined using standard AOAC [18] method. The titratable acidity of the gari was determined by titration with NaOH 0.01 N [18]. Values were expressed in equivalent gram acetic acid per 100 g of sample.

The total protein content of the root and in the different gari was determined using acetyl acetone/formaldehyde method proposed by Devani et al. [19]. Samples were first mineralized [20] and the nitrogen content of the mineralization was evaluated after a reaction with ammonia (NH_3) and acetyl acetone/formaldehyde. A conversion factor of 6.25 was used to determine the protein content of the samples. The Ninhydrin colorimetric method described by Michel [21] was used to evaluate the free amino acid content of the samples.

Free sugars and carbohydrates content of the roots and in the different gari were determined by Orcinol colorimetric method [22]. Free sugars were obtained after stirring dried sample in an 80% ethanol solution. As concerns the total carbohydrates, samples were first hydrolysed with 13 M H_2SO_4 (30 min, 25°C) and then heated at 100°C for two hours. Crude fibres and lipids content of the samples were determined according to the standard AOAC [18] method.

The starch content of the samples (cassava root and gari) was determined by the iodine spectrophotometric method as performed by Jarvis and Walker [23]. Results were expressed in gram per 100 g of sample. The total phenolic compounds were determined using the Folin-Ciocalteu reagent as described by Singleton et al. [24] and the results were expressed as equivalent mg of gallic acid per gram sample.

Cyanides were evaluated in the cassava roots and in the different gari sample using picrate colorimetric method proposed by de B. Baltha and Cereda [25] with some modifications. Cyanides were first extracted using a sodium phosphate buffer (0.1 M; pH 6). A standard curve was performed with KCN and results were expressed in terms of equivalent mg of HCN per gram of dry sample. All analyses were performed in triplicate. A commercial cassava gari coming from the South region of Cameroon was used as reference.

2.4. Evaluation of the Functional Properties of Cassava Gari. The bulk density of gari was determined using a measuring cylinder as performed by Adeleke and Odedeji [26].

The swelling kinetic was assessed as described by Koubala et al. [27] with slight modifications. About 500 mg of dried gari was introduced in a measuring cylinder (100 ml) where it was mixed with 50 ml of distilled water. The gari was allowed to hydrate for 60 minutes at room temperature and its volume recorded from that time till equilibrium.

The water absorption capacity (WAC) was assessed according to the method described by Koubala et al. [27]. For WAC, 200 mg of the sample was introduced in a conical flask containing 10 ml of distilled water. The sample was soaked, stirred, and left 60 minutes at room temperature (25–30°C). The slurry was put on a sintered glass filter to allow the water to leak. When water was no longer leaking for one hour, the sample was weighed, dried at 105°C (overnight), and weighed again. The WAC was expressed in terms of ml of water absorbed per gram of gari.

The swelling power (SP) was determined based on the method of Hung et al. [28] with slight modifications. Measuring cylinder (10 ml) was filled with gari to the 2 ml mark and weighed. It was later made up to 10 ml with distilled water. The top of the cylinder was tightly covered and the content was mixed by inverting the cylinder. After each two minutes the cylinder was inverted again and left to stand for eight minutes. The final volume occupied by the gari was recorded.

In a tube, suspensions were made with gari sample in 5 ml of distilled water as carried out by Koubala et al. [27]. The tubes were heated at 90°C for one hour and allowed to cool overnight. Then, the gelation capacity was determined for each sample as the least gelation concentration. That is the concentration when the sample from the inverted test tube will not slip. The concentration of gari varied from 4% to 16%. All analyses were performed in triplicate. A commercial cassava gari coming from the South region of Cameroon was used as reference.

2.5. In-House Consumer Assessment.
This assessment was conducted to determine consumer preferences and acceptability of the gari samples. Five-point hedonic scale as described by Nnanna et al. [29] was used with some modifications. A five-point hedonic scale is used (where 1 = dislike extremely, 1.5 = dislike very much, 2 = dislike moderately, 2.5 = dislike slightly, 3 = neither like nor dislike, 3.5 = like slightly, 4 = like moderately, 4.5 = like very much, and 5 = like extremely). The quality parameters assessed included the appearance or color, the taste (acidity and sweetness), the odor (aroma), the texture or mouthfeel and the overall acceptability. In-house consumer assessment was carried out within fifteen minutes of preparation. For this purpose, twenty trained panelists (students) coming from different regions of Cameroon, both males and females aged between 22 and 35, were involved. These panelists were usual consumers of gari and were chosen based on their ability to distinguish flavour, acidity, and sweetness in the gari sample. The gari samples were coded before being presented to the panelists who recorded their responses on the form provided in their slip. A commercial cassava gari coming from the South region of Cameroon was used as reference.

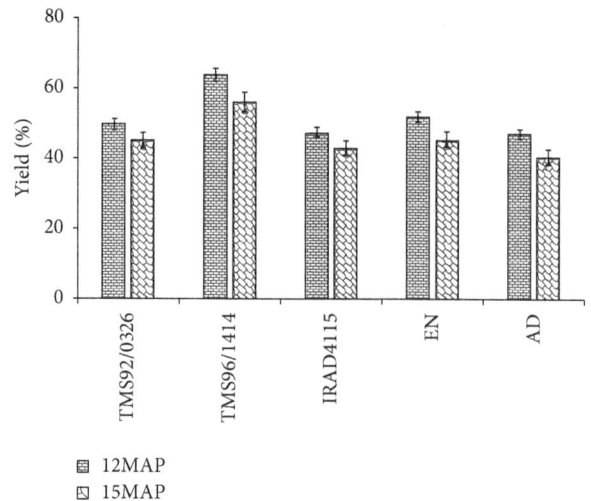

Figure 1: Effect of the harvest period (12 and 15 months after planting) on gari yield (g/100 g of wet mash). Garis are produced from local (*EN* and *AD*) and improved (*TMS92/0326*, *TMS96/1414*, and *IRAD4115*) cassava varieties harvested in dry season at 12 months and in rainy season at 15 months after planting (12MAP and 15MAP).

2.6. Statistical Analysis.
The analysis was performed with Graph Pad (version 5.0; 2007). Results were presented as means ± standard deviation. The mean values were compared using independent sample Tukey's test, and the differences between them were determined by analysis of variance. As concerns in-house consumer assessment, Chi-square test was used to identify the determining factors and the overall acceptability of the panelists for the various gari samples. Tukey's multiple comparison test was used to determine the gari sample that was different from the others [30].

3. Results

In the present work, it is found that characteristics of the different processed gari varied significantly ($p < 0.05$) according to the harvest period and from one variety to another. Differences were also observed between gari from local and improved cassava variety.

3.1. Yield and Proximate Composition of Cassava Gari.
Results in Figure 1 show that gari produced from cassava root harvested 12 month after planting (12MAP) in the dry season showed the highest yield. Whatever the harvest period (12MAP or 15MAP), gari obtained from *TMS96/1414* cassava variety significantly ($p < 0.05$) exhibited the highest yield (63.72 g/100 g of wet mash). It is also noticed that the lowest gari yield was obtained with *AD* cassava (Figure 1). Generally the best gari yield was obtained with the improved cassava varieties.

Table 1 shows that the dry matter content of the different cassava gari was significantly ($p < 0.05$) similar (96-97%); but that of the commercial gari was significantly ($p < 0.05$) low (83.48%). The pH of the samples of gari fluctuated from 4.31 to 4.72. It is noticed that the pH of gari is significantly

TABLE 1: Impact of the harvest period (12 and 15 months after planting) on the pH, the dry matter, the ash, the crude proteins, and the free amino acids content of cassava gari. Garis are produced from local (EN and AD) and improved (TMS92/0326, TMS96/1414, and IRAD4115) cassava varieties root harvested in dry season at 12 months and in rainy season at 15 months after planting (12MAP and 15MAP). These garis were compared to the commercial cassava gari (CG). Values are given on the dry weight basis.

Harvest periods and varieties		Dry matter (%)		Ash		pH	Proteins (%)		Amino acids (mg/g)	
		Gari	Root	Gari (mg/g)	Root (mg/g)	Gari	Gari	Root	Gari	Root
12MAP (May)	TMS92/0326	96.60 ± 0.37^c	25.21 ± 0.32^d	20.61 ± 0.80^a	8.55 ± 0.19^c	4.31 ± 0.14^c	4.12 ± 0.15^c	6.79 ± 0.34^d	15.51 ± 0.45^b	11.25 ± 0.84^c
	TMS96/1414	97.30 ± 0.03^a	28.39 ± 0.33^b	19.59 ± 0.71^d	8.64 ± 0.33^c	4.58 ± 0.22^{cdb}	5.03 ± 0.16^b	9.75 ± 0.41^a	14.82 ± 0.38^c	12.16 ± 0.22^a
	IRAD4115	97.28 ± 0.11^{ab}	25.55 ± 0.46^d	20.62 ± 1.19^d	8.35 ± 0.18^c	4.72 ± 0.25^d	4.17 ± 0.09^c	7.40 ± 0.21^c	16.72 ± 0.29^a	14.98 ± 0.48^b
	EN	96.91 ± 0.09^{bc}	30.25 ± 0.27^a	19.80 ± 0.36^d	8.50 ± 0.14^c	4.84 ± 0.27^{bd}	5.58 ± 0.20^a	5.86 ± 0.19^{ef}	11.97 ± 0.41^f	10.21 ± 0.33^h
	AD	96.80 ± 0.18^{bc}	25.66 ± 0.39^d	22.30 ± 1.47^{bc}	7.79 ± 0.25^d	4.50 ± 0.19^{cdb}	5.40 ± 0.18^a	5.90 ± 0.36^{ef}	14.07 ± 0.24^d	8.91 ± 0.28^g
15MAP (August)	TMS92/0326	96.60 ± 0.37^c	22.23 ± 0.09^e	24.16 ± 0.56^a	10.14 ± 0.25^b	5.28 ± 0.04^b	2.72 ± 0.14^d	5.48 ± 0.33^f	12.07 ± 0.19^f	7.82 ± 0.32^f
	TMS96/1414	96.52 ± 0.07^c	25.72 ± 0.46^{cd}	23.36 ± 0.72^{ab}	10.57 ± 0.18^b	5.29 ± 0.02^b	2.08 ± 0.08^{ef}	8.93 ± 0.25^b	14.95 ± 0.31^{bc}	7.83 ± 0.16^d
	IRAD4115	96.60 ± 0.83^c	22.26 ± 0.12^e	24.02 ± 0.47^a	11.25 ± 0.48^a	5.28 ± 0.05^b	2.28 ± 0.11^e	6.33 ± 0.35^{de}	12.59 ± 0.28^{ef}	7.21 ± 0.17^i
	EN	96.40 ± 0.06^c	26.31 ± 0.28^c	20.37 ± 0.58^d	10.80 ± 0.51^{ab}	5.30 ± 0.04^b	5.58 ± 0.18^a	6.05 ± 0.17^{de}	13.96 ± 0.45^d	9.44 ± 0.28^j
	AD	96.31 ± 0.05^c	22.38 ± 0.12^e	24.49 ± 0.74^a	10.70 ± 0.39^{ab}	5.17 ± 0.05^c	4.43 ± 0.12^c	6.61 ± 0.31^d	12.94 ± 0.33^e	6.53 ± 0.37^k
Commercial cassava gari (CG)		83.48 ± 0.28^d	ND	21.78 ± 0.35^c	ND	5.43 ± 0.03^a	2.00 ± 0.08^f	ND	6.04 ± 0.13^g	ND

Values are means ± standard deviation of triplicates ($n = 3$). Values in the same column with the different superscript are significantly different ($p < 0.05$).

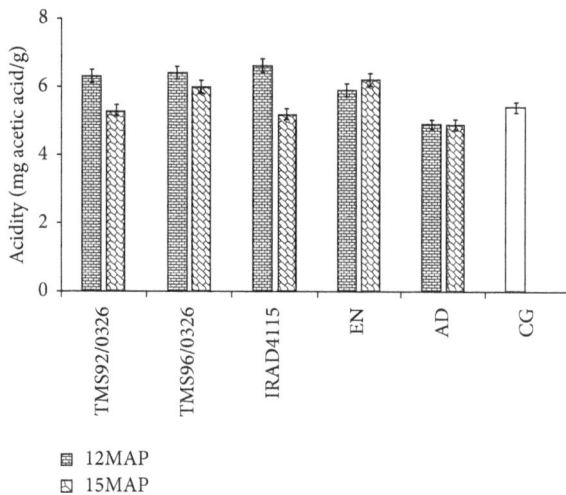

FIGURE 2: Influence of the harvest period (12 and 15 months after planting) on the titratable acidity (equivalent mg of acetic acid/g of dry weight) of cassava gari. Garis are produced from local (*EN* and *AD*) and improved (*TMS92/0326, TMS96/1414,* and *IRAD4115*) cassava varieties harvested in dry season at 12 months and in rainy season at 15 months after planting (12MAP and 15MAP). These garis were compared to the commercial cassava gari (CG).

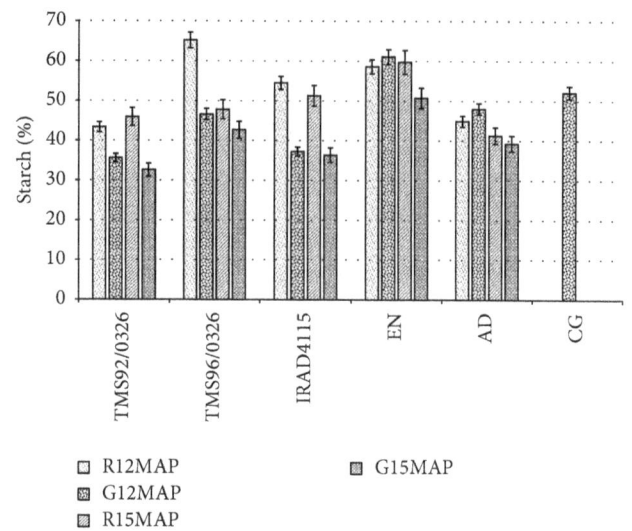

FIGURE 3: Effect of the harvest period (12 and 15 months after planting) on the starch content (g/100 g of dry weight) of cassava gari. Garis are produced from local (*EN* and *AD*) and improved (*TMS92/0326, TMS96/1414,* and *IRAD4115*) cassava varieties root harvested in dry season at 12 months and in rainy season at 15 months after planting (12MAP and 15MAP). These garis were compared to the commercial cassava gari (CG). G = gari and R = fresh root.

affected by the harvest period or age of the cassava root used for garification. Gari from cassava roots harvested at 12MAP exhibited the lowest pH values. This suggested the presence of an important quantity of organic acid in the gari. However, the commercial gari used as reference in our study recorded the highest pH value. The ash content of the cassava gari varied from 19.59 to 24.49 mg/g (dry weight). Gari from cassava roots harvested at 12MAP exhibited the lowest value of ash content (Table 1). The ash content of the commercial cassava gari was significantly ($p < 0.05$) similar to that of gari from cassava roots harvested at 12MAP in the dry season.

The acidity (equivalent mg acetic acid/g) of the different cassava gari was significantly ($p < 0.05$) affected by the harvest period. This effect was well observed with the improved cassava varieties. Whatever the harvest period, gari from local cassava varieties exhibited the lowest value of acidity (Figure 2). The titratable acidity values of the gari samples were similar to those recommended by the Codex Alimentarius Commission [31].

Table 1 shows that the protein content of gari from cassava harvested at 12 months after planting (12MAP) exhibits high value (4.12–5.58% dry weight) compared to gari from cassava harvested at 15MAP. However, the commercial gari used as reference exhibited the lowest value for protein content. According to the harvest period, gari from local cassava varieties (*EN* and *AD*) were richer in protein than those from improved cassava varieties. But these differences were not observed with the free amino acid content of the gari (Table 1), whereas the amino acid content of the commercial gari was lower than that of the others gari whatever the harvest period and the variety. However, the same differences are not observed in the roots where those harvested at 12MAP showed the highest value of proteins (Table 1).

The lipids content of gari obtained from cassava harvested at 12MAP varied between 4.17 and 8.66%. Table 2 shows that this range is higher than that of gari from cassava harvested at 15MAP. The commercial gari used as reference exhibited an intermediate value (5.43%). Gari from cassava harvested at 15MAP showed high value of fibres content (3.32–5.58%) than garis obtained from cassava harvested at 12MAP (3.05–4.74%). The phenolic compound content of the different gari samples obtained from cassava harvested at 12MAP varied significantly ($p < 0.05$) from 4.78 to 9.31 equivalent mg of gallic acid/g (dry weight) while those obtained from cassava harvested at 15MAP showed a range of 7.36 to 9.36 mg/g. It is observed in Table 2 that, from 15MAP to 12MAP, the phenolic compound content of local cassava gari (*EN* and *AD*) increases. However, these values decrease with the improved cassava gari.

Gari from cassava harvested at 15MAP exhibited the highest value (3.47–9.71% dry weight) of soluble sugars (Table 2). Among these cassava gari that was obtained from local *EN* cassava variety was richer in free sugars than those from improved cassava varieties. The commercial cassava gari showed the lowest value of free sugars. Figure 3 shows that the starch content of the cassava gari is significantly ($p < 0.05$) affected by the variety and the period of harvest. Starch content values ranged from 33 to 61% (dry weight). Whatever the variety, gari from cassava harvested at 12MAP exhibited the highest value of starch content. Gari from the *EN* local cassava variety was richer in starch than the other cassava garis. The total carbohydrates content (50 to 78%) of the cassava gari was also affected by the period of harvest and the variety. As for the starch content, it was also noticed

TABLE 2: Effect of the harvest period (12 and 15 months after planting) on the phenolic compounds (PC), the total carbohydrates, the free sugars, the fibre, and the lipids content of cassava gari. Gari are produced from local (EN and AD) and improved (TMS92/0326, TMS96/1414, and IRAD4115) cassava varieties root harvested in dry season at 12 months and in rainy season at 15 months after planting (12MAP and 15MAP). These garis were compared to the commercial cassava gari (CG). Values are given on the dry weight basis.

Harvest periods and varieties		PC (Equivalent mg gallic acid/g)	Characteristics						
			Carbohydrates (%)		Sugars (%)		Fibres (%)		Lipids (%)
			Gari	Root	Gari	Root	Gari	Root	
12MAP (May)	TMS92/0326	8.63 ± 0.16^{bc}	75.12 ± 0.57^{b}	86.14 ± 2.01^{b}	9.71 ± 0.10^{a}	25.45 ± 0.55^{b}	4.43 ± 0.10^{e}	3.70 ± 0.10^{c}	5.85 ± 0.14^{c}
	TMS96/1414	9.31 ± 0.21^{a}	65.98 ± 0.84^{e}	74.54 ± 3.26^{d}	6.18 ± 0.06^{b}	20.77 ± 0.62^{c}	3.70 ± 0.08^{g}	3.05 ± 0.18^{de}	8.66 ± 0.26^{a}
	IRAD4115	7.43 ± 0.31^{de}	71.68 ± 0.99^{c}	79.36 ± 2.48^{c}	4.57 ± 0.07^{e}	15.30 ± 0.47^{e}	3.05 ± 0.06^{i}	2.89 ± 0.11	6.66 ± 0.12^{b}
	EN	8.01 ± 0.26^{cd}	78.07 ± 0.87^{a}	90.15 ± 3.25^{a}	3.47 ± 0.06^{f}	15.68 ± 0.41^{e}	4.74 ± 0.05^{c}	3.24 ± 0.14^{d}	4.17 ± 0.33^{f}
	AD	4.78 ± 0.19^{g}	62.62 ± 1.29^{f}	74.61 ± 2.18^{d}	5.66 ± 0.09^{c}	17.17 ± 0.37^{d}	4.07 ± 0.07^{f}	2.88 ± 0.10^{e}	6.45 ± 0.25^{b}
15MAP (August)	TMS92/0326	7.36 ± 0.11^{e}	58.73 ± 1.29^{g}	82.25 ± 2.16^{bc}	5.04 ± 0.05^{d}	30.80 ± 0.48^{a}	5.58 ± 0.05^{a}	5.01 ± 0.16^{a}	3.93 ± 0.12^{fg}
	TMS96/1414	7.36 ± 0.16^{e}	52.62 ± 0.26^{i}	71.25 ± 2.04^{e}	3.21 ± 0.11^{g}	29.87 ± 0.49^{a}	4.44 ± 0.13^{e}	4.11 ± 0.13^{b}	5.22 ± 0.29^{e}
	IRAD4115	7.46 ± 0.07^{e}	49.58 ± 0.95^{j}	75.24 ± 2.43^{d}	3.44 ± 0.09^{f}	19.47 ± 0.27^{c}	3.32 ± 0.11^{h}	3.01 ± 0.11^{e}	4.18 ± 0.26^{f}
	EN	9.36 ± 0.10^{a}	68.47 ± 0.92^{d}	86.57 ± 2.37^{b}	4.56 ± 0.05^{e}	20.83 ± 0.72^{c}	5.10 ± 0.07^{b}	4.05 ± 0.14^{b}	2.66 ± 0.32^{h}
	AD	8.74 ± 0.28^{b}	54.72 ± 1.37^{hi}	72.34 ± 1.87^{de}	5.16 ± 0.08^{d}	19.71 ± 0.53^{c}	4.58 ± 0.11^{de}	4.04 ± 0.12^{b}	3.61 ± 0.24^{g}
Commercial cassava gari (CG)		6.01 ± 0.16^{f}	56.84 ± 0.71^{gh}	ND	2.45 ± 0.08^{h}	ND	4.62 ± 0.09^{cd}	ND	5.43 ± 0.03^{d}

Values are means ± standard deviation of triplicates ($n = 3$). Values in the same column with the different superscript are significantly different ($p < 0.05$).

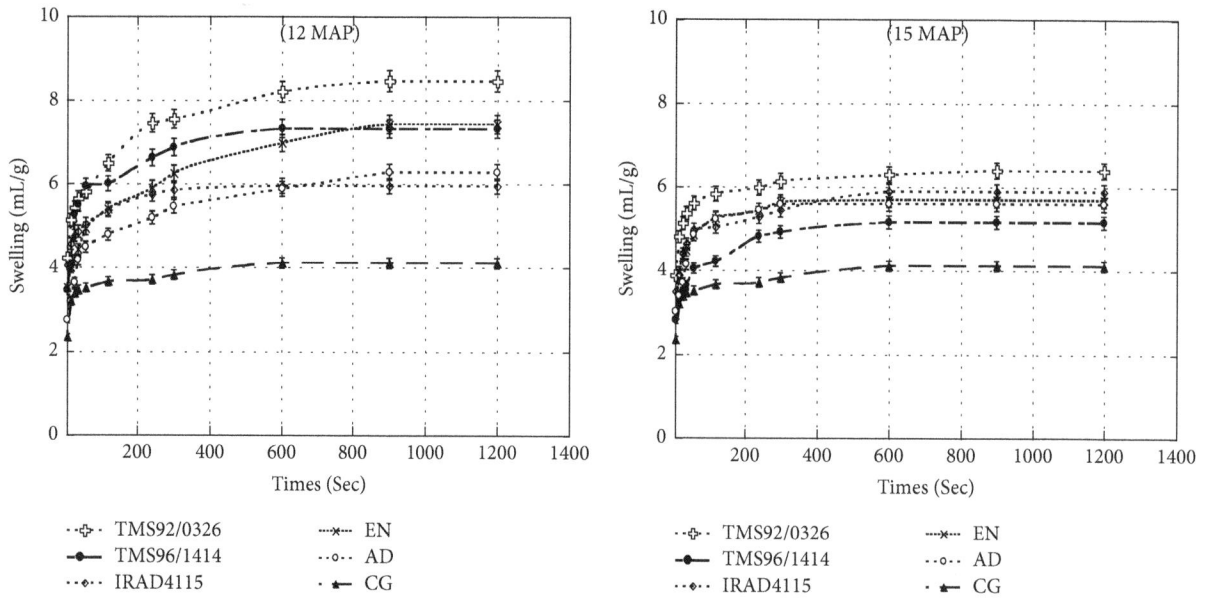

FIGURE 4: Effect of the harvest period (12 and 15 months after planting) on the swelling kinetic of cassava gari. Garis are produced from local (*EN* and *AD*) and improved (*TMS92/0326, TMS96/1414,* and *IRAD4115*) cassava varieties harvested in dry season at 12 months and in rainy season at 15 months after planting (12MAP and 15MAP). These garis were compared to the commercial cassava gari (CG).

that gari from cassava harvested at 12MAP shows the highest value (Table 2). The carbohydrates content of the commercial gari was close to those of the gari from cassava harvested at 15MAP in the rainy season. Table 2 shows that gari from cassava harvested at 15MAP exhibits the highest value of crude fibres content (3.32–5.58%).

Table 3 shows that a fermentation process of two days reduced free cyanide content of the cassava root twofold to threefold. This reduction is well marked with gari from cassava harvested at 12MAP (0.08–0.74 mg HCN/g) compared to those from cassava harvested at 15MAP (0.51–0.90 mg HCN/g). However the commercial gari used as reference exhibited high value of free cyanide (2.03) as the cassava roots were dried before fermentation.

3.2. Functional Properties of Cassava Gari. The bulk density of gari from cassava harvested at 12MAP ranged from 0.52 to 0.62 g/ml (Table 3). This value reduced for gari produced from cassava harvested at 15MAP. Commercial gari and gari from local cassava gari *EN* exhibit the highest values for bulk density.

In Table 3, it is observed that the period of cassava harvest and the varietal differences affected the water absorption capacity (WAC) of gari. The WAC of the cassava gari varied from 4.49 to 8.02 ml/g. Whatever the variety, gari from cassava harvested at 12MAP showed the highest value of WAC. The WAC of commercial gari was closed to those of gari from cassava harvested at 15MAP. As for bulk density, gari from *EN* cassava harvested at 12MAP exhibits the highest value. In Table 3, similar observations were made with the swelling power of the different cassava gari.

The least gelatinization concentration (LGC) is the concentration at which gari slurry does not break down when

tubes are inverted after a treatment at 90°C for one hour. The LGC of the different gari slurries differed significantly among varieties and for the two periods of harvest cassava root (Table 3). Gari from cassava harvested at 12MAP showed a LGC situated between 6.00 and 8.00% with the highest value recorded by the sample of TMS96/1414, IRAD4115, and *EN* cassava. These values were significantly ($p < 0.05$) lower than that of the commercial gari (14.00%). Table 4 also shows that, from 12MAP to 15MAP, the least gelatinization concentration (LGC) increased.

As Figure 4 shows, the swelling kinetics of gari from cassava at 12MAP occurred in three stages. The first 100 seconds corresponded to the phase of fast swelling. This was followed by a slow swelling between 200 and 600 seconds depending on the gari sample. After this phase, a stationary phase of swelling took place. Gari from *TMS92/0326* cassava variety swelled faster than the other garis until stabilization and recorded the highest value for swelling over time. Although gari from *IRAD4115* cassava swelled the least, it presented swelling kinetics higher than that of the commercial gari. There was a general reduction of swelling for all garis obtained from cassava harvested in the rainy season (at 15MAP). Gari from cassava harvested at 15MAP showed a similar trend; nevertheless gari from *TMS92/0326* and *EN* cassava varieties swelled faster between 10 and 250 seconds before decreasing until stabilization at 600 seconds (Figure 4).

3.3. Sensory Attributes of Cassava Gari

3.3.1. Attributes of the Soaked Cassava Gari. Garis from cassava harvested at 12 and 15 months after planting (12MAP and 15MAP) were soaked into a sucrose solution (10%) and their sensory attributes were evaluated by 20 sensory trained

TABLE 3: Effect of the harvest period (12 and 15 months after planting) on the bulk density, the water absorption capacity (WAC), the swelling power, the least gelation concentration and the free cyanide content of cassava gari. Gari are produced from local (EN and AD) and improved (TMS92/0326, TMS96/1414 and IRAD4115) cassava varieties harvested in dry season at 12 months and in rainy season at 15 months after planting (12MAP and 15MAP). These garis were compared to the commercial cassava gari (CG). Values are given on the dry weight basis.

Harvest periods	Varieties	Free Cyanide (Equivalent mg HCN/g)		Bulk density (g/mL)	WAC (g/g)	Swelling power (mL/g)	Least gelation concentration (%)
		Storage roots	Gari				
12MAP (May)	TMS92/0326	1.47 ± 0.04^e	0.08 ± 0.01^g	0.56 ± 0.02^{bc}	7.99 ± 0.06^a	8.37 ± 0.07^b	8.00
	TMS96/1414	1.56 ± 0.06^{cd}	0.23 ± 0.04^f	0.55 ± 0.02^c	6.58 ± 0.08^b	7.54 ± 0.11^c	8.00
	IRAD4115	1.20 ± 0.05^f	0.74 ± 0.03^c	0.52 ± 0.01^d	5.29 ± 0.07^e	7.24 ± 0.08^d	6.00
	EN	0.88 ± 0.03^g	0.57 ± 0.05^a	0.62 ± 0.01^a	8.02 ± 0.09^a	8.86 ± 0.11^a	6.00
	AD	1.18 ± 0.05^f	0.72 ± 0.08^{cd}	0.57 ± 0.01^b	6.79 ± 0.20^b	7.52 ± 0.05^c	6.00
15MAP (August)	TMS92/0326	2.52 ± 0.09^a	0.67 ± 0.02^d	0.47 ± 0.02^e	5.78 ± 0.04^d	6.77 ± 0.09^f	9.00
	TMS96/1414	1.54 ± 0.05^{de}	0.67 ± 0.05^d	0.50 ± 0.01^e	5.94 ± 0.09^c	6.94 ± 0.13^e	9.00
	IRAD4115	1.67 ± 0.09^c	0.51 ± 0.06^e	0.48 ± 0.03^e	5.05 ± 0.11^f	6.62 ± 0.12^f	10.00
	EN	1.49 ± 0.05^e	0.59 ± 0.05^e	0.53 ± 0.02^{cd}	4.49 ± 0.08^g	5.53 ± 0.06^h	9.00
	AD	1.85 ± 0.06^b	0.90 ± 0.08^b	0.51 ± 0.03^{de}	4.50 ± 0.06^g	5.44 ± 0.08^h	9.00
Commercial cassava gari (CG)		ND	2.03 ± 0.04^a	0.61 ± 0.02^a	5.37 ± 0.07^e	6.35 ± 0.04^g	14.00

Values are means ± standard deviation of triplicates ($n = 3$). Values in the same column with the different superscript are significantly different ($p < 0.05$).

TABLE 4: Hedonic sensory mean scores of cassava gari. Garis are produced from local (EN and AD) and improved (TMS92/0326, TMS96/1414, and IRAD4ll5) cassava varieties harvested in dry season at 12 months and in rainy season at 15 months after planting (12MAP and 15MAP). These garis were compared to the commercial cassava gari (CG).

Harvest periods	Varieties	Color	Odor	Mouthfeel	Acidity	Sweetness	Overall acceptability
12MAP (May)	TMS92/0326	3.45 ± 0.69[bc]	3.40 ± 0.82[a]	2.55 ± 0.51[e]	2.80 ± 0.77[a]	2.55 ± 0.51[bcd]	3.75 ± 0.72[b]
	TMS96/1414	2.85 ± 0.67[d]	3.20 ± 0.89[a]	2.95 ± 0.69[cde]	2.65 ± 0.75[ab]	2.65 ± 0.75[bc]	3.80 ± 0.69[b]
	IRAD4ll5	2.75 ± 0.85[d]	3.05 ± 0.89[ab]	3.30 ± 0.57[bc]	2.20 ± 0.41[bc]	2.50 ± 0.95[bcd]	2.75 ± 0.91[de]
	EN	4.50 ± 0.76[a]	3.50 ± 0.76[a]	2.75 ± 0.85[de]	1.60 ± 0.50[d]	2.60 ± 0.68[bcd]	4.30 ± 0.81[a]
	AD	3.55 ± 0.78[bc]	2.95 ± 0.94[ab]	1.90 ± 0.71[f]	1.65 ± 0.75[d]	2.30 ± 0.75[cd]	3.10 ± 0.79[cd]
15MAP (August)	TMS92/0326	2.50 ± 0.69[de]	1.80 ± 0.83[d]	4.30 ± 0.86[a]	1.90 ± 0.85[cd]	2.30 ± 0.86[cd]	2.55 ± 0.69[e]
	TMS96/1414	2.15 ± 0.81[e]	2.50 ± 0.60[bc]	3.25 ± 0.79[bc]	2.00 ± 1.03[cd]	2.25 ± 0.78[cd]	2.75 ± 0.79[de]
	IRAD4ll5	2.40 ± 0.75[de]	2.35 ± 0.49[c]	3.15 ± 0.75[bcd]	1.85 ± 0.75[cd]	3.00 ± 0.79[ab]	2.40 ± 0.50[e]
	EN	2.20 ± 0.41[e]	2.55 ± 0.83[bc]	3.60 ± 0.75[b]	1.95 ± 0.89[cd]	3.35 ± 0.99[a]	2.50 ± 0.76[e]
	AD	3.40 ± 0.68[c]	2.95 ± 0.94[ab]	1.85 ± 0.49[f]	1.65 ± 0.75[d]	2.05 ± 0.76[d]	3.00 ± 0.83[cd]
Commercial gari cassava (CG)		3.90 ± 0.55[b]	3.15 ± 0.93[a]	2.85 ± 0.81[cde]	2.50 ± 0.69[ab]	2.95 ± 0.60[ab]	3.55 ± 0.51[bc]

Values are means ± standard deviation of scores awarded by panelists ($n = 20$). Values in the same column with the different superscript are significantly different ($p < 0.05$).

panelists. 12MAP and 15MAP gari were served to the panelists in one session. Table 4 shows that there was a significant ($p < 0.05$) difference between those attributes related to the variety of cassava and the harvest period. In terms of color, the panelists significantly ($p < 0.05$) preferred gari from cassava harvested at 12MAP. Gari from *EN* cassava had the best mark (4.50) followed by the commercial gari and gari from *AD* cassava (3.55). These two garis are from the root of local cassava varieties (Table 4). The odor of gari from cassava harvested at 12MAP was also well appreciated by the panelist compared to gari from cassava harvested at 15MAP. A similar appreciation was made for the commercial gari which also exhibited a good score for odor (3.15). As concerns the mouthfeel, it was observed that gari from cassava harvested at 15MAP showed the best score when compared to those from cassava harvested at 12MAP (Table 4). The commercial gari exhibited an intermediate score for mouthfeel. According to the panelists, garis from improved cassava varieties harvested at 12MAP were more acidic than those from cassava harvested at 15MAP. The effect of harvest period on mouthfeel was found not to be significant ($p > 0.05$) for gari from local cassava varieties. Gari from EN cassava harvested at 15MAP and the commercial gari were sweeter than the other garis among which no difference of sweetness was found.

In terms of overall acceptability, the panelists preferred gari from cassava harvested at 12MAP to gari from cassava harvested at 15MAP. The commercial gari used as reference had an intermediate score of preference. Concerning all the garis processed from cassava harvested at 12MAP (in the dry season), gari from local *EN* cassava variety had the best mark (4.30) followed by those from the improved *TMS92/0326* and *TMS96/1414* cassava (3.75–3.80). The commercial gari showed an intermediate value of preference (3.55).

3.3.2. Correlation between Organoleptic Attributes and the Preference of Panelists. For the determination of the effect of preference on the sensory interest of different gari samples on the score of the overall acceptability of panelists, a Chi-square test was carried out. Results showed that the overall acceptability of panelists for gari samples was significant and positively correlated with the color ($r = +0.449$; $p < 0.01$) and the odor ($r = +0.380$; $p < 0.01$). With the acidity and sweetness, this correlation was nonsignificant. Moreover, the mouthfeel was significant and negatively correlated with the overall preference ($r = -0.278$; $p < 0.01$).

The general trend showed that gari samples processed at 12MAP were generally more acceptable for the consumers in terms of all sensory attributes while those obtained at 15MAP (rainy season) were the least preferred.

4. Discussion

4.1. Influence of Harvest Period and the Cassava Variety on the Proximate Composition of the Cassava Gari. The highest yield was observed with gari from cassava harvested during the dry season at 12 months after planting (12MAP). This may be attributed to factors such as plant age, varieties, and other environmental factors as mentioned by Oluwaniyi and Oladipo [32] and Wholey and Booth [33]. According

to Adejumo and Raji [34], Sanni [35], and Karim et al. [36] most of carbohydrate in cassava harvested in the rainy season at 15MAP has been hydrolysed into free sugars. These differences in free sugars content are shown in Table 2. These sugars are used for growth of new plant tissues. The greater yield recorded by improved variety was also reported by Oghenechavwuko et al. [37] and may be attributed to a genetic factor.

The moisture content in the gari samples was lower than that of the commercial gari (16.60%). Indeed, it is reported that the high moisture content implies that a gari sample will not have a good storage potential [38], while the values obtained from gari samples of this study were good compared to those reported by Okolie et al. [39] for good storage potential and quality as well as those recommended by Oduro et al. [40]. The gari should be properly dried to a possible very low moisture content.

The cassava variety and the period of harvest had an effect on the pH of gari. Similar results were obtained by Egebebi and Aboloma [41], Kyereh et al. [42], and Nwafor et al. [43]. The acidic pH values of the cassava gari might have been due to the fermentation activity of microorganisms which used carbohydrate to produce more organic acids. The pH values of 4.6 to 5.8 recorded during this work are similar to those reported by Nwafor et al. [43] and by Ijabo and Igbo [44] with gari from local and improved varieties. The low pH value obtained with gari from cassava harvested at 12MAP could be due to the availability of carbohydrate to be metabolised by microorganism during the fermentation process. In particular free sugars found in the different cassava can favour growth of microorganisms in the mash. The pH and titratable acidity results of gari from improved cassava variety were well correlated. However, for the local cassava varieties, no significant difference was observed between garis from cassavas 12MAP and 15MAP. The titratable acidity values of the gari were similar to those observed by Bainbridge et al. [45] and almost as the values (0.60–1.0%) ranged by the Codex Alimentarius Commission for gari [31]. Plant age is a factor affecting the ash content of cassava roots [32]. The ash contents obtained in the present study were lower than those reported by Otutu et al. [46]. But these values were close to those of the Codex Standard for gari (1.5%) [31].

The high protein content of gari from cassava harvested at 12MAP could be linked to the high level of proteins in their corresponding roots (Table 1). This could also be due to the fact that, during the fermentation, microorganisms are more active in 12MAP cassava mash than in 15MAP cassava mash. In general, the local variety exhibited the highest values of protein content. However, cassava genetic is also a factor affecting the protein content. Oluwaniyi and Oladipo [32] obviously observed that the proteins content of TME 7 cassava variety root decreases from 7 to 12 months of age. When compared to commercial gari the difference could be due to the varietal differences. Additionally, in accord with Nwafor et al. [43], the ecological conditions can favour fermentation, mixed microorganisms involved to produce more amino acids which are used to synthesize proteins. According to Kobawila et al. [47], microorganisms can increase the crude protein content from 35 to 60% during fermentation depending on the quality of microorganisms involved in mash and the locality. The amino acids content of the different cassava gari is linked to the genetic differences. Moreover, the quality of microorganisms involved in mash fermentation could significantly influence the synthesis of amino acid or the breakdown of proteins. Probably the initial quantity of amino acids in storage roots may have an influence on the final quantity of amino acid in the gari (Table 1). The high quantity of amino acids observed in gari from cassava harvested at 12MAP could be attributed to the high temperature in the dry season, whence the water stress that confirm the significant effect of harvest period. In particular, it was reported that amino acids of storage roots increase during water stress [48, 49] which could be the case in the present study.

Since all the garis were prepared using the same amount of palm oil, differences of lipids content observed could be due to the varietal difference and the period of harvest. In the dry season (12MAP), cassava root might exhibit high lipid content compared to the rainy season (15MAP). This suggests that, in the rainy season, as with carbohydrate, lipid could also be hydrolysed to provide energy needed for root growth. This could be also due to genetic factors which affect plants to synthesize pretty much lipid in different environmental conditions. The values obtained in this study are slightly higher than those found (5.71% dry weight) by Onasoga et al. [50]. On the other hand, the crude lipid increase could be due to the activities of microorganisms during mash fermentation assuming conversion of carbohydrate into lipid and lipids for cellular growth. This was also reported by Padmaja et al. [51], Oboh and Akindahunsi [52], Fagbemi and Ijah [53], Boonnop et al. [54], and Ibukun and Anyasi [55]. The age of the cassava root used may favour the fibre content of the gari. This is why gari from cassava harvested at 15MAP exhibited higher value of fibre content. Since commercial gari showed similar values as gari obtained from 15MAP cassava root, it can be suggested that gari processed with old cassava root exhibits higher fibre content. Otutu et al. [46] reported that the values of crude fibre content of cassava gari ranged from 0.38 to 7.08% which were higher than those obtained in the present study. These results might have been due to removal of some fibre during gari sieving. Nevertheless, the values obtained in the present study are close to the value of crude fibre (2%) recommended by the Codex Standard [56].

The dry season favours hydrolysis of carbohydrates into free sugars. This confirms the significant effect of harvest period of cassava roots. Free sugars content is high in gari from improved cassava varieties. This could be due to genetic characteristics that fluctuate with the season and age of the plant [32]. The low value of free sugars found in the gari from cassava harvested at 15MAP may be due to the rainy season where free sugars are used for the new regrowth of leaves. The low value of sugar of commercial gari could suggest that the cassava roots were probably harvested in the rainy season.

The starch content of gari from cassava harvested at 12 months after planting (12MAP) was higher than that of gari from cassava harvested at 15MAP. In general the dry season (12 months after planting) is the period of the optimal starch

storage in cassava than the rainy season. The highest value of starch content showed by gari from *EN* cassava variety could be due to its initial starch content in the storage roots (Figure 3). In this study, with the exception of gari from *EN* cassava, the starch contents of the other garis are similar to those of gari produced at six localities in Ghana [42]. The reduction in starch content in gari produced at 15MAP could be attributed to factors such as harvest season and plant age. In fact, gari produced from cassava harvested in the rainy season (15MAP) showed a reduction in starch content because of its mobilization for new shoot formation in the new growth cycle. This is in agreement with what was reported by Filho [57], Madore [58], Sriroth et al. [59], Sagrilo et al. [60], and Nuwamanya et al. [49]. Furthermore, cassava roots starch may decrease progressively as plant aged after optimal starch storage (Figure 3).

As shown in Table 2, the significantly high carbohydrates content showed by gari from *EN* cassava variety could be due to its initial carbohydrates content in the storage roots. The gari samples from *EN* cassava variety harvested at 12MAP showed significantly the highest value of carbohydrate than those from cassava harvested at 15 months as well as commercial gari. This might be due to varietal differences and harvest season as carbohydrate is highly concentrated during dry season compared to the rainy season. The reduction of carbohydrate for the gari samples obtained at 15 months was probably due to the rain which facilitates the hydrolysis of carbohydrate to sugar for shoots regrowth after prolonged water stress [60].

Certain cassava varieties produce more phenolic compounds when there is water stress than others justifying variation of phenolic compounds observed in this study. Gari from local *(EN)* and improved *(TMS96/1414)* cassava varieties presented significantly the same content in phenolic compounds when compared to commercial gari. This can be attributed to the processing method used. According to Etsuyankpa et al. [6] and Umezuruike et al. [61] roasting temperature and length of fermentation significantly decrease phenolic compounds.

The significant decrease in the level of residual cyanide observed in gari when compared to storage roots has been attributed to the degradation of cyanide (linamarin and lotaustralin) during cassava mash fermentation, dewatering, and roasting temperature. In fact, microorganisms involved in mash could have the ability to hydrolyse linamarin and lotaustralin. In addition, it was reported that heat significantly destroys cyanide in mash during roasting [11]. The slightly high levels of free cyanide from gari obtained at 15 months could be related to plant age and low activity of microorganisms involved in mash during the rainy season to significantly degrade cyanide into free cyanide (HCN). In fact, there is an optimum temperature for the optimum activity of any microorganisms to rapidly and completely break down the cyanogenic glucosides to cyanide acid. Gari from *TMS92/0326* improved cassava variety presented the least value of HCN. Regarding the initial storage roots cyanide content, the amount of cyanide acid in gari from local and improved cassava varieties was mostly dependent upon the processing method of garification. The highest level of free cyanide from commercial gari compared to gari produced may be due to the reduction in fermentation cassava mash to about 24 hours as well as quality of microorganisms present in mash.

4.2. Effect of Harvest Period and the Variety on the Functional Qualities of the Cassava Gari. The significant variation in bulk density (BD) may be attributed to varietal effect and climatic conditions [44, 62]. Nevertheless, the values of BD (0.47–0.49 g/ml) reported by Chika et al. [63] are lower than those observed in this study. However, bulk density obtained was lower than those reported by Olakunle et al. [64] and Oluwafemi and Udeh [7]; this could be due to fermentation length, moisture content, particles size; and processing method used. The highest value of BD recorded by *EN* gari from local variety could be correlated especially with the carbohydrates and starch content. During the dry roasting of the cassava mash, starch gelatinization occurs. This led to the formation of gari with different particles size. This suggests that gari with high particles size were from mash with high starch content.

Also, the gari samples differ in water absorption capacity (WAC). However, there was higher WAC from the gari produced at 12MAP compared to commercial gari but there was a reduction for those obtained at 15MAP. Similar results were observed by Chika et al. [63] though the values (3.58–4.17 g/g) were very low. However, these values were significantly higher when compared to some garis from three improved varieties reported by Nwancho et al. [62]. This may be attributed to size of the particles, garification length, and drying quality of gari. In fact, the WAC of gari decreases with decrease in particles size. Furthermore, a well dried gari samples should be able to absorb more water adequately when soaked compared to poorly dried gari samples.

The variation in swelling power among the gari samples when compared to commercial gari could be due to varietal factors and processing conditions used [44]. Yet we noted a general reduction in the swelling power for gari produced at 15MAP. These results can be attributed to the harvest of the storage root during the rainy season that lowered the amylose content responsible for great swelling power. In fact, all samples swelled as much as twice beyond their initial volume, which characterizes high quality gari produced from our cassava varieties [64, 65]. The variation in swelling power of gari from local and improved varieties could be explained by the amylose content of these garis.

The fluctuation of least gelation concentration (LGC) among garis when compared to commercial gari could be due to harvest season, fermentation time, and varietal differences. The highest value of LGC of commercial gari may be attributed to the size of particle and low amylose content. Furthermore, the LGC increased from 12MAP to 15MAP probably because of the decrease in roots amylase activity. In fact, the greater the percentage of amylose fraction of starch-based foods, the quicker the formation of the gel [66].

The differences linked to composition, particles size, fermentation length, and moisture contents [11, 64] could explain the significant variation of the swelling kinetics among gari samples. The *TMS96/0326* followed by

TMS96/1414 and *EN* gari samples exhibited the highest swelling ability and *IRAD4115* gari sample the lowest for the gari produced at 12 months, while, at 15 months, the *IRAD4115* gari sample exhibited the highest swelling ability and *TMS96/1414* the least. The gari sample obtained at different time exhibited the best swelling ability when compared to commercial gari. These results may be related to varietal differences, the season, and the chemical compounds as well as processing method used.

4.3. Effect of the Harvest Period and the Variety on Sensory Attribute of the Cassava Gari. A significant difference was noted between gari samples produced at 12 months and 15 months and with commercial gari. The differences observed in the color and taste could be due to the fact that different processing methods were used during production of gari samples, especially the length of fermentation which might differ from one locality to the other. Moreover, each of the mixtures of microorganism involved in mash fermentation has its effect on sensory quality as stated by Olaoye et al. [67]. It had been observed that panelists appreciated gari samples quality based on color or appearance and odor. The result is in agreement with Agbor-Egbe and Mbome [12], who reported that gari preference was attributed to the better organoleptic characteristics such as appearance, odor, particle size, swelling properties, and eating quality. This is confirmed by a significant positive correlation noted between color, odor, and preference ($r = +0.449$; $p < 0.01$ and $r = +0.380$; $p < 0.01$, resp.). The gari sample from *EN* obtained at 12 months was significantly preferred in terms of general acceptability than those produced at 15 months. This could be due to rainy season which affects the color and odor. Similarly, Agbor-Egbe and Mbome [12] observed that the gari produced using the local cassava variety was more preferred than those from the improved variety. Additionally, the choice of *EN* gari was correlated to the local sweet and red cassava used for the garification. This is in agreement with Olaoye et al. [67].

5. Conclusion

This study shows that the gari yield from *TMS96/1414* (improved variety) is 63.72% and 55.91% produced in dry season at 12 months and in rainy season at 15 months, respectively, followed by those from *EN* (local variety). The study reveals that age, season at which cassava storage roots are harvested, and variety affects some physicochemical, functional, and sensory properties of gari. The study also shows that processing cassava roots into gari significantly reduces the cyanide acid (HCN) at level considered safe. In addition, gari samples produced in the dry season are better in terms of physicochemical, functional, and sensory properties than gari produced in the rainy season. The general trends show that gari samples processed in the dry season (12MAP) were generally acceptable to the consumers in terms of all sensory attributes while those obtained in the rainy season (15MAP) and commercial gari were the least preferred. In the far north region of Cameroon, all cassava roots varieties can be processed into gari, but *EN* (red, sweet), *TMS92/0326,* and *TMS92/1414* can be processed into gari with the best physicochemical, functional, and sensory properties compared to commercial gari. Thus, quality gari can contribute to overcoming malnutrition and undernourishment in the far north region in particular and the whole of Cameroon in general.

Conflicts of Interest

The authors declare that there are no conflicts of interest concerning the publication of this paper.

References

[1] M. U. Ukwuru and S. E Egbonu, "Recent development in cassava-based products research," *Academia Journal of Food Research*, vol. 1, no. 1, pp. 001–013, 2013.

[2] L. Ola, H. Rihana, N. Siew et al., "Identification of aromatic compounds and their sensory characteristics in cassava flakes and garri (Manihot esculenta Crantz, CyTA)," *Journal of Food*, vol. 14, no. 1, pp. 154–161, 2016.

[3] P. M. Mduduzi, M. Taurai, and O. O. Ademola, "Perspectives on the probiotic potential of lactic acid bacteria from African traditional fermented foods and beverages," *Food and nutrition Research*, vol. 60, pp. 1–12, 2016.

[4] O. O. Oyeyipo, *Studies on Lafun Fortified with African breadfruit Tempeh [Msc. thesis]*, University of Port Harcourt, Nigeria, 2011.

[5] G. Oboh, "Nutrient enrichment of cassava peels using a mixed culture of Saccharomyces cerevisiae and Lactobacillus spp solid media fermentation techniques," *Electronic Journal of Biotechnology*, vol. 9, no. 1, pp. 46–49, 2006.

[6] M. B. Etsuyankpa, C. E. Gimba, E. B. Agbaji, I. Omoniyi, M. M. Ndamitso, and J. T. Mathew, "Assessment of the Effects of Microbial Fermentation on Selected Anti-Nutrients in the Products of Four Local Cassava Varieties from Niger State, Nigeria," *American Journal of Food Science and Technology*, vol. 3, no. 3, pp. 89–96, 2015.

[7] G. I. Oluwafemi and C. C. Udeh, "Effect of fermentation periods on the physicochemical and sensory properties of gari," *Journal of Environmental Science*, vol. 10, no. 1, pp. 37–42, 2016.

[8] S. C. Achinewhu, L. I. Barber, and I. O. Ijeoma, "Physicochemical properties and garification (gari yield) of selected cassava cultivars in Rivers State, Nigeria," *Plant Foods for Human Nutrition*, vol. 52, no. 2, pp. 133–140, 1998.

[9] F. A. Tetchi, O. W. Solomon, K. A. Celah, and A. N. George, "Effect of cassava variety and fermentation time on biochemical and microbiological characteristics of raw artisanal starter for Attieke production," *Innovative Romanian Food Biotechnology*, vol. 10, no. 3, pp. 40–47, 2012.

[10] N. M. Mahungwu, Y. Yamaguchi, A. M. Almazan, and S. K. Hatin, "Reduction of cyanide during processing of cyanide of cassava into some traditional African foods," *Journal of Food and Agriculture*, vol. 1, pp. 11–15, 1987.

[11] S. V. Irtwange and O. Achimba, "Effect of the Duration of Fermentation on the Quality of Gari," *Current Research Journal of Biological Science*, vol. 1, no. 3, pp. 150–154, 2009.

[12] T. Agbor-Egbe and I. L. Mbome, "The effects of processing techniques in reducing cyanogen levels during the production of some Cameroonian cassava foods," *Journal of Food Composition and Analysis*, vol. 19, no. 4, pp. 354–363, 2006.

[13] F. I. Nweke, D. S. C. Spencer, and J. K. Lynam, *The cassava transformation: Africa's best kept secret*, Michigan State University Press, Lansing, Mich, USA, 2002.

[14] O. O. Babalola, "Cyanide Content of Commercial Gari from Different Areas of Ekiti State, Nigeria," *World Journal of Nutrition and Health*, vol. 2, no. 4, pp. 58–60, 2014.

[15] D. Halliday, H. A. Qureshi, and J. A. Broadbent, "Investigation on the storage of gari Nig. Stored Prod.," *Research Institute Technical Report*, no. 16, pp. 131–141, 1967.

[16] L. E. F. Amamgbo, A. O. Akinpelu, R. Omodamiro, F. N. Nwakor, and T. O. Ekedo, "Promotion and popularization of some elite cassava varieties in Igbariam Anambra State: implication for food security and smpowerment," *Global Advanced Research Journal of Agricultural Science*, vol. 5, no. 2, pp. 061–066, 2016.

[17] S. S. Sobowale, S. O. Awonorin, T. A. Shittu, M. O. Oke, and O. A. Adebo, "Estimation of material losses and the effects of cassava at different maturity stages on garification index," *Journal of Food Processing and Technology*, vol. 07, no. 02, pp. 1–5, 2016.

[18] AOAC, "Official methods of analysis of AOAC International," in *Association of Official Analytical Chemists*, W. Herwitz, Ed., pp. 125-126, Washington DC, USA, 15 edition, 1990.

[19] M. B. Devani, C. J. Shishoo, S. A. Shal, and B. N. Suhagia, "Spectrophotometric method for microdetermination of nitrogen in Kjedahl digest," *Journal of Association Official Analytical Chemists*, vol. 72, no. 6, pp. 953–956, 1989.

[20] AFNOR (Association Française pour la Normalisation), "Produits alimentaires : directives générales pour le dosage de l'azote avec minéralisation selon la méthode de kjedahl," in *Godon et Pineau*, vol. 4, pp. 263–266, Guide Pratique des Céréales Apria, France, 1984.

[21] M. C. Michel, "Dosage des acides aminés et amines par la ninhydrine. Amélioration pratique, Annal de Biologie Animale," *Biochimie, Biophysique*, vol. 8, no. 4, pp. 557–563, 1968.

[22] M. T. Tollier and J. P. Robin, "Adaptation of the method sulfuric orcinol automatic determination of total neutral carbohydrates. Terms of adaptation to extracts of vegetable origin," *Annals of Technological Agriculture*, vol. 28, pp. 1–15, 1979.

[23] C. E. Jarvis and J. R. L. Walker, "Simultaneous, rapid; spectrophotometric determination of total starch, amylase and amylopectine," *Journal of the Science of Food and Agriculture*, vol. 63, pp. 53–57, 1993.

[24] V. L. Singleton, R. Orthofer, and R. M. Lamuela-Raventós, "Analysis of total phenols and other oxidation substrates and antioxidants by means of folin-ciocalteu reagent," *Methods in Enzymology*, vol. 299, pp. 152–178, 1999.

[25] A. D. T. de B. Baltha and M. P. Cereda, "Cassava free cyanide analysis using KCN or acetone-cyanidrin as pattern," International Meeting on Cassava Breeding, Biotechnology and Ecology. 1. Brasilia: Proceedings, Cassava improvement to enhance livelihoods in sub saharan África and Northweast, Brasil, Brasília, University of Brasília, Ministério do Meio Ambiente, p.132, 2006.

[26] R. O. Adeleke and J. O. Odedeji, "Functional properties of wheat and sweet potato flour blends," *Pakistan Journal of Nutrition*, vol. 9, no. 6, pp. 535–538, 2010.

[27] B. B. Koubala, G. Kansci, E. L. B. Enone, N. O. Dabole, N. V. Yaya, and E. M. A. Zang, "Effect of Fermentation Time on the Physicochemical and Sensorial Properties of Gari from Sweet Potato (Ipomoae batatas)," *British Journal of Applied Science and Technology*, vol. 4, no. 24, pp. 3430–3444, 2014.

[28] P. V. Hung, T. Maeda, and N. Morita, "Dough and bread qualities of flours with whole waxy wheat flour substitution," *Food Research International*, vol. 40, no. 2, pp. 273–279, 2007.

[29] H. A. Nnanna, R. D. Phillips, K. H. McWatters, and Y.-C. Hung, "Effect of germination on the physical, chemical, and sensory characteristics of cowpea products: Flour, paste, and akara," *Journal of Agricultural and Food Chemistry*, vol. 38, no. 3, pp. 812–816, 2002.

[30] AFNOR, *Collection of French Standards of General Methods of Analysis of Food Products: Chemistry, Microbiology, Sensory Analysis*, Paris, France, 1 edition, 1980.

[31] Codex Alimentarius Commission, "Codex Standard for Gari," *Codex Stan 151-1989 (Rev, 1-1995)*, vol. 7, pp. 1-6, 1989.

[32] O. Oluwaniyi and J. Oladipo, "Comparative studies on the phytochemicals, nutrients and antinutrients content of cassava varieties," *Journal of the Turkish Chemical Society, Section A: Chemistry*, vol. 4, no. 3, pp. 661–674, 2017.

[33] D. W. Wholey and R. H. Booth, "Influence of variety and planting density on starch accumulation in cassava roots," *Journal of the Science of Food and Agriculture*, vol. 30, no. 2, pp. 165–170, 1979.

[34] B. A. Adejumo and A. O. Raji, "An appraisal of gari packaging in Ogbomoso, Southwestern Nigeria," *Journal of Agricultural and Veterinary Sciences*, vol. no.2, pp. 120–127, 2010.

[35] L. O. Sanni, "Hazard analysis of critical control points in the commercial production of high quality gari," *Nigeria Journal of Science*, vol. 24, 1990.

[36] O. R. Karim, O. S. Fasasi, and S. A. Oyeyinka, "Gari yield and chemical composition of cassava roots stored using traditional methods," *Pakistan Journal of Nutrition*, vol. 8, no. 12, pp. 1830–1833, 2009.

[37] U. E. Oghenechavwuko, S. G. Olasunkanmi, T. K. Adekunbi, and A. C. Taiwo, "Effect of processing on the physico-chemical properties and yield of gari from dried chips," *Journal of Food Processing and Technology*, no. 4, pp. 1–6, 2013.

[38] E. Ukenye, U. J. Ukpabi, U. Chijoke, C. Egesi, and S. Njoku, "Physicochemical, nutritional and processing properties of promising newly bred white and yellow fleshed cassava genotypes in Nigeria," *Pakistan Journal of Nutrition*, vol. 12, no. 3, pp. 302–305, 2013.

[39] N. P. Okolie, M. N. Brai, and O. M. Atotebi, "Comparative study on some selected garri samples sold in lagos metropolis," *Journal of Food Studies*, vol. 1, no. 1, pp. 33–42, 2012.

[40] I. Oduro, W. O. Ellis, N. T. Dziedzoave, and K. Nimako-Yeboah, "Quality of gari from selected processing zones in Ghana," *Food Control*, vol. 11, no. 4, pp. 297–303, 2000.

[41] A. O. Egebebi and R. I. Aboloma, "Fungi and moisture content of gari sold in some locations in Southwestern Nigeria," *Archives of Applied Science Research*, vol. 4, no. 6, pp. 2361–2364, 2012.

[42] E. Kyereh, R. J. Bani, and D. Obeng-Ofori, "Effect of cassava processing equipment on quality of gari produce in selected processing site in Ghana," *International Journal of Agriculture Innovations and Research*, vol. 2, no. 2, pp. 1473–2319, 2013.

[43] O. E. Nwafor, O. O. Akpomie, and P. E. Erijo, "Effect of fermentation time on the physico-chemical, nutritional and sensory quality of cassava chips (Kpo-Kpo garri) a traditional nigerian food," *American Journal of BioScience*, vol. 3, no. 2, pp. 59–63, 2015.

[44] O. J. Ijabo and P. K. Igbo, "Effects of three lower benue grown cassava (Manihot esculanta) varieties and processes on seven quality indices of fresh gari," *International Journal of*

[45] Z. Bainbridge, K. Tomlins, K. Welling, and A. Westby, *Methods for Assessing Quality Characteristics of Non-Grain Starch Staple (Part 2. Filed Methods)*, Natural Resources Institute, Chatham, UK, 1996.

[46] O. L. Otutu, D. S. Ikuomola, and Q. Udom, "Comparative evaluation of quality of gari samples from six processing centres in oriade lga of osun state, Nigeria," *IJAFS*, vol. 4, no. 1, pp. 571–580, 2013.

[47] S. C. Kobawila, D. Louembe, S. Keleke, J. Hounhouigan, and C. Gamba, "Reduction of the cyanide content during fermentation of cassava roots and leaves to produce bikedi and ntoba mbodi, two food products from Congo," *African Journal of Biotechnology*, vol. 4, no. 7, pp. 689–696, 2005.

[48] B. M. Rodrigues, B. D. Souza, R. M. Nogueira, and M. G. Santos, "Tolerance to water deficit in young trees of jackfruit and sugar apple," *Revista Ciência Agronômica*, vol. 41, no. 2, pp. 245–252, 2010.

[49] E. Nuwamanya, P. R. Rubaihayo, S. Mukasa, S. Kyamanywa, J. F. Hawumba, and Y. Baguma, "Biochemical and secondary metabolites changes under moisture and temperature stress in cassava (Manihot esculenta Crantz)," *African Journal of Biotechnology*, vol. 1, no. 31, pp. 3173–3186, 2014.

[50] M. Onasoga, D. Oluwafunmilayo, Oluwafunmilayo., and O. O. Oyeyipo, "Chemical Changes during the Fortification of Cassava Meal (Gari) with African breadfruit (Treculia africana) Residue," *Journal of Applied Science and Environ Management*, vol. 18, no. 3, pp. 506–512, 2014.

[51] G. Padmaja, M. George, and S. N. Moorthy, "Detoxification of cassava during fermentation with a mixed culture inoculum," *Journal of the Science of Food and Agriculture*, vol. 63, no. 4, pp. 473–481, 1993.

[52] G. Oboh and A. A. Akindahunsi, "Biochemical changes in cassava products (flour & gari) subjected to Saccharomyces cerevisae solid media fermentation," *Food Chemistry*, vol. 82, no. 4, pp. 599–602, 2003.

[53] A. O. Fagbemi and U. J. J. Ijah, "Microbial population and biochemical changes during production of protein-enriched fufu," *World Journal of Microbiology and Biotechnology*, vol. 22, no. 6, pp. 635–640, 2006.

[54] K. Boonnop, M. Wanapat, N. Nontaso, and S. Wanapat, "Enriching nutritive value of cassava root by yeast fermentation," *Scientia Agricola*, vol. 66, no. 5, pp. 629–633, 2009.

[55] E. O. Ibukun and O. J. Anyasi, "Changes in antinutrient and nutritional values of fermented sesame (Sesanum indicum), Musk Melon (Cucumis melo) and white melon (Cucumeropsis mannii)," *International Journal of Advanced Biotechnology and Research*, vol. 4, no. 1, pp. 131–214, 2013.

[56] L. B. Sanni, J. Maziya-Dixon, C. I. Akanya, Y. Okoro, C. V. Alaya, and R. Egwuonwu, *Standards for Cassava Poducts and Gidelines for Export*, IITA, Ibadan, Nigeria, 2005.

[57] S. J. B. Filho, *Distribuição de carboidratos em plantas de mandioca (Manihot esculenta, Crantz) e o efeito do teor de reservas, na brotação e enraizamento de estacas de três posições do caule*, Tese (Mestrado em Fitotecnia) - Universidade Federal de Viçosa, Viçosa, Brazil, 1980.

[58] M. A. Madore, "Phloem transport of solutes in crop plants," in *Handbook of Plant and Crop Physiology*, M. Pessorakli, Ed., pp. 337–353, The University of Arizona, Tucson, Arizona, 1994.

[59] K. Sriroth, K. Piyachomkwan, V. Santisopasri, and C. G. Oates, "Environmental conditions during root development: Drought

constraint on cassava starch quality," *Euphytica*, vol. 120, no. 1, pp. 95–101, 2001.

[60] E. Sagrilo, P. S. Vidigal Filho, M. G. Pequeno et al., "Effect of harvest period on the quality of storage roots and protein content of the leaves in five cassava cultivars (Manihot esculenta, Crantz)," *Brazilian Archives of Biology and Technology*, vol. 46, no. 2, pp. 295–305, 2003.

[61] A. C. Umezuruike, T. U. Nwabueze, and E. N. T. Akobundu, "Anti nutrient content of food grade flour or roasted African breadfruit seeds produced under extreme condition," *Academia Journal of Biotechnology*, vol. 4, no. 5, pp. 194–198, 2016.

[62] S. O. Nwancho, F. C. Ekwu, P. O. Mgbebu, C. K. Njoku, and C. Okoro, "Effect of particle size on the functional, pasting and textural properties of gari produced from fresh cassava roots and dry chips," *The International Journal of Engineering and Science*, vol. 3, no. 3, pp. 50–55, 2014.

[63] C. Chika, C. E. Ogueke, C. I. Owuamanam, I. Ahaotu, and I. Olawuni, "Quality characteristics and HCN in gari as affected by fermentation variables," *International Journal of Life Sciences*, vol. 2, no. 1, pp. 21–28, 2013.

[64] M. M. Olakunle, S. O. Akinwale, J. O. Makanjuola, and S. O. Awonorin, "Comparative study on quality attributes of gari obtained from some processing centers in South West, Nigeria," *Advance Journal of Food Science and Technology*, vol. 4, no. 3, pp. 135–140, 2012.

[65] L. A. Sanni, O. O. Odukogbe, M. O. Faborode, and R. O. Ibrahim, "Optimization of process parameters of a conductive rotary dryer for gari production," *Journal of Emerging Trends in Engineering and Applied Sciences*, vol. 6, no. 7, pp. 255–259, 2015.

[66] L. O. Sanni, S. B. Kosoko, A. A. Adebowale, and R. J. Adeoye, "The influence of palm oil and chemical modification on the pasting and sensory properties of fufu flour," *International Journal of Food Properties*, vol. 7, no. 2, pp. 229–237, 2004.

[67] O. A. Olaoye, I. G. Lawrence, G. N. Cornelius, and M. E. Ihenetu, "Evaluation of quality attributes of cassava product (gari) produced at varying length of fermentation," *American Journal of Agricultural Science*, vol. 2, no. 1, pp. 1–7, 2015.

16

Genetics of Marbling in Wagyu Revealed by the Melting Temperature of Intramuscular and Subcutaneous Lipids

Sally S. Lloyd,[1,2] **Jose L. Valenzuela,**[1,2,3] **Edward J. Steele,**[1] **and Roger L. Dawkins**[1,2,3]

[1]*CY O'Connor ERADE Village Foundation, P.O. Box 5100, Canning Vale South, WA 6155, Australia*
[2]*CY O'Connor Centre for Innovation in Agriculture, Murdoch University, 5 Del Park Road, Box 1, North Dandalup, WA 6207, Australia*
[3]*Melaleuka Stud, 24 Genomics Rise, Piara Waters, WA 6112, Australia*

Correspondence should be addressed to Sally S. Lloyd; slloyd@cyo.edu.au

Academic Editor: Alejandro Castillo

Extreme marbling or intramuscular deposition of lipid is associated with Wagyu breeds and is therefore assumed to be largely inherited. However, even within 100% full blood Wagyu prepared under standard conditions, there is unpredictable scatter of the degree of marbling. Here, we evaluate melting temperature (T_m) of intramuscular fat as an alternative to visual scores of marbling. We show that "long fed" Wagyu generally has T_m below body temperature but with a considerable range under standardized conditions. Individual sires have a major impact indicating that the variation is genetic rather than environmental or random error. In order to measure differences of lower marbling breeds and at shorter feeding periods, we have compared T_m in subcutaneous fat samples from over the striploin. Supplementary feeding for 100 to 150 days leads to a rapid decrease in T_m of 50% Red Wagyu (Akaushi) : 50% European crosses, when compared to 100% European. This improvement indicates that the genetic effect of Wagyu is useful, predictable, and highly penetrant. Contemporaneous DNA extraction does not affect the measurement of T_m. Thus, provenance can be traced and substitution can be eliminated in a simple and cost-effective manner.

1. Introduction

Marbling (or the accumulation of intramuscular fat) is the holy grail for beef producers, chefs, and their customers, but there is still no agreed definition and therefore no universal standard of measurement [1, 2]. So as to increase commercial returns based on superior taste and health benefits, there have been countless attempts to improve the reproducibility of visual and scanning scores but with limited success [1, 3].

Lipid profiles of highly marbled samples have revealed a high content of oleic acid and therefore a reduction in melting temperature (T_m) [4–6]. A precise and high throughput method for the measurement of T_m exists [7] and is used here to interrogate the complex interplay between the genetic and environmental factors which can be optimized by the producer to the benefit of the health conscious consumer.

Because of the association of Wagyu breeds with high marbling and high oleic acid content, these traits can be assumed to be genetically determined and faithfully inherited [8, 9]. However, in spite of numerous studies [10–19], it has

not been possible, hitherto, to identify markers which allow a breeder to quantify superior genetics in individual sires and dams. Some of the explanations for the slow progress include the following:

(i) Complexity due to interactions of several metabolic processes and their regulatory mechanisms [6, 20, 21].

(ii) Contribution of many genes with small effects [22].

(iii) Uncontrolled environment factors associated with supplementary feeding [23–25].

(iv) Difficulty in quantifying marbling reproducibly [2].

(v) Unreliable tracing of meat from paddock to plate.

(vi) Perception that fat is dangerous.

Recently it has been demonstrated that that low fat diets have not improved health [26]. In fact, higher oleic acid and therefore low T_m are preferable in terms of lipid profiles [27–29]. This has led to the increasing popularity of the Wagyu brand worldwide. Not surprisingly, mislabeling is now rife resulting in the need to be able to confirm the provenance of retail samples.

Here we show that low T_m is heritable and that the same fat sample can be used for the DNA tracing without affecting the measurement of T_m.

2. Materials and Methods

Postmortem samples of meat and fat were taken from carcasses of animals harvested for routine food production. Therefore, ethics approval was not required.

2.1. Dataset 1 Full Blood Wagyu with Identified Sires. Two cohorts of Wagyu steers (n = 126) were fed for 300 ± 20 days with a proprietary ration within the same commercial feedlot. One-gram samples of meat from the longissimus dorsi were taken from between the 10th and 11th rib. AUS-MEAT marbling score (MS) was scored between the 10th and 11th rib, with an average of 7.5 and a range from 2 to 11. Steers for the comparison of sires had their paternity confirmed by DNA testing [30]. Only one progeny of each dam was included so as to focus on the effect of the sire.

2.2. Dataset 2 European and Wagyu Cross Breeds with Varied Feed Time. Melaleuka Stud, located in the Peel region of Western Australian, 100 km south of Perth, runs a variety of European breeds including Simmental, Gelbvieh, and Angus. This herd was selected to produce high quality beef on pasture, finished with 2 to 4 months of supplemental feeding. Black and Red Wagyu (full blood or pure bred) have been mated with these European breeds.

Calves stay on milk until 4 months of age when they are weaned and male calves are castrated. After weaning, they continue grazing Kikuyu and Ryegrass pasture until they reach 300 kg. Their feed is then supplemented with 9 mm EasyBeef pellets (Milne Feeds, Perth, Australia) *ad libitum*. The main ingredients of the EasyBeef pellets are lupins, barley, oats, wheat, and triticale. The nutritional composition, based on dry matter, is crude protein (min) 14.5%, metabolizable energy (est.) 11.0 MJ/kg., crude fiber (max) 20.0%, urea (max) 1.5%, and monensin 26.6 ppm.

The feeders are considered ready for slaughter when they reach a weight of 400 kg and are slaughtered to match demand. Some animals were kept on feed longer to test the effect of increased feeding on T_m and meat quality. The average live weight at slaughter for animals in this study was 461 kg, average age at slaughter was 15.4 months (range 8 to 23), and the average days on feed was 104 days (range 17 to 288). Body numbers from abattoirs were matched to farm records and pedigrees via their RFID tags, where possible identity was confirmed by in-house proprietary DNA testing [30].

Subcutaneous fat overlying the striploin (HAM number 2140) of these cattle was collected after boning and wet aging for 1 to 3 weeks.

2.3. Fat Extraction and T_m Measurement. Intramuscular fat was extracted from dataset 1 samples by digestion with proteinase K. This method allows for simultaneous extraction of intramuscular fat and DNA from 0.5 gram samples of meat if the fat content is above 20%. The samples were incubated at 56°C, digested in a proteinase K mixture for 4 hours, and centrifuged at 10,000 ×g for 2 minutes to separate the fat from the dissolved DNA and protein solution. Fat was removed for T_m measurement by pipette. DNA was extracted from the remaining mixture using a standard salting out method.

Intramuscular fat content for many of the carcasses of dataset 2 was too low to allow extraction by the above method. Instead, fat was extracted from 1-gram samples of subcutaneous fat by rendering at 90°C for at least eight hours.

Samples from 17 sirloin steaks with intramuscular fat higher than 20% were used to determine whether fat separated during a DNA extraction process could be used for T_m measurement. Fat was extracted by both digestion and rendering from the same samples and the T_m measurements compared.

T_m of all fat samples was determined in triplicate according to the thermocycler method [7], which is closely correlated to slip points, although the values are higher by 2°C for animal fat with a T_m of 40°C.

3. Results

3.1. T_m Is Affected by Sire. Samples were taken from long fed Wagyu steers differing only by sire and dam (dataset 1). The steers were fed, harvested, and tested in two cohorts two months apart. The cohorts did not differ significantly in feeding, genetics, or initial T_m (as shown in Supplemental Table 1 in Supplementary Material available online at https://doi.org/10.1155/2017/3948408) and have therefore been combined for further analysis. T_m and marble score were analyzed by sire for the three sires with more than ten progenies. As shown in Figure 1, T_m of the progeny of Sire 2 fell consistently, whereas Sire 1 had little impact. In fact, 14 progenies of Sire 1 were above 37°C, compared to only 3 of Sire 2. The cross-product ratio is 104/6 or 17, as shown in Figure 1. This difference is highly significant (p value < 0.01 by χ^2). It is noteworthy that there is more scatter with Sire 3 and all remaining sires.

By contrast with T_m, visual scores of marbling gave greater scatter, did not demonstrate a sire effect, and must be misleading in their present form.

3.2. In Wagyu, T_m Falls with Days on Feed and Proportion of Wagyu. Notwithstanding the genetic effects, there is also a major environmental effect on T_m and marbling. T_m results of dataset 2, grouped by proportion of black Wagyu, are shown in Figure 2. T_m falls with increase in Wagyu and days on feed. Separating these two variables is not yet possible but, in the meanwhile, the results suggest that increasing the content of Wagyu genes allows the benefit of long feeding. The European cattle included in this study do not show the same benefit as the Wagyu.

Importantly, the benefits are seen with only 25% Wagyu, again emphasizing the high penetrance of the Wagyu genetics.

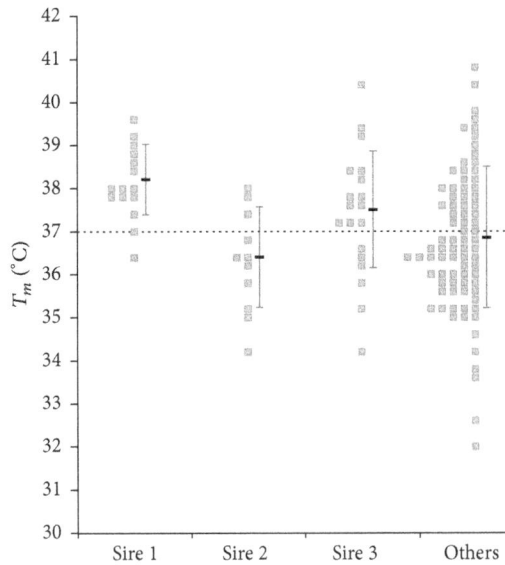

FIGURE 1: T_m distributions of Wagyu carcasses differ by sire. The melting temperature of intramuscular fat samples taken from between the 10th and 11th rib of 126 carcasses of full blood Wagyu steers. All animals were fed the same ration for 300 ± 20 days. Individual T_m measurements of carcasses are grouped by sire (mean and standard deviation). Animals with either an uncertain sire or a sire with less than 10 progeny are grouped under "other" sires. Progeny of Sire 3 shows considerable scatter, whereas 8/11 of those of Sire 2 are below 37 degrees compared with 1/15 in the case of Sire 1. The difference between Sire 1 and Sire 2 is statistically significant with a chi-square statistic of 12.2 and thus a p value < 0.01.

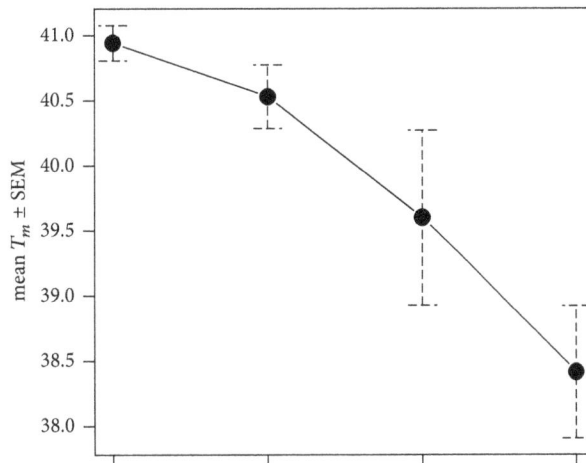

Breed	EU100	WY25	WY50	WY75+
N	176	29	11	15
DOF Mean ± SD	84 ± 41	103 ± 52	167 ± 103	225 ± 79

FIGURE 2: T_m decreases with feeding and increasing proportion of Wagyu ancestry. T_m of subcutaneous fat samples over the loin of a mix of breeds and crossbreeds including Simmental, Gelbvieh, Angus, Dexter, and Wagyu. 176 samples (EU100) came from 100% European breeds fed for an average of 81 days. WY25, WY50, and WY75+ samples had 25%, 50%, and 75–100% Wagyu ancestry, respectively. There were 29 samples of WY25, 14 samples of WY50, and 11 samples of WY75+ with average days on feed of 103, 167, and 225, respectively.

3.3. Quantitative Effect of Feeding.

So as to address the complex interaction between genetics and environment, we compared two breed groups from dataset 2: a control group of purely European cattle (EU100) and the F1 Red Wagyu, also known as Akaushi and recorded as AK50. The dams have a similar breed composition and history to the EU100 control group. So as to avoid the complexity of sampling intramuscular fat before it is visible, we have relied on T_m measurements of overlying subcutaneous fat. The effect of feeding is clear as shown in Figure 3. T_m falls progressively even with only a 50% infusion of Akaushi.

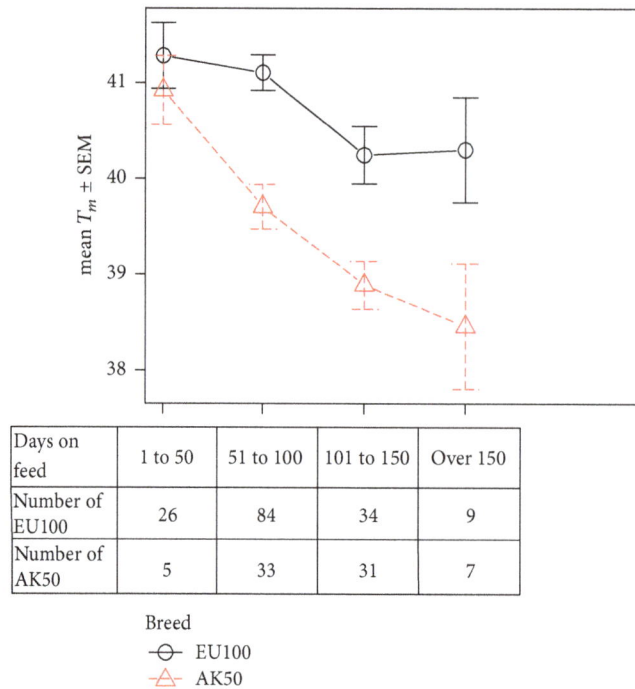

Days on feed	1 to 50	51 to 100	101 to 150	Over 150
Number of EU100	26	84	34	9
Number of AK50	5	33	31	7

Breed
- EU100
- AK50

FIGURE 3: Red Wagyu sired carcasses have lower T_m for equivalent DOF. T_m was measured for subcutaneous fat samples taken from the loins of 229 carcasses. The cattle were backgrounded on pasture and then fed on pellets until they reached a satisfactory weight and fatness. The results are grouped by days on feed and by breed of sire (European or Akaushi). The dams of all carcasses were European breeds. Breed and days on feed were both statistically significant influences on T_m, with $p < 0.01$ calculated by multiway ANOVA. The difference between the two groups was significant after only 51–100 days on feed.

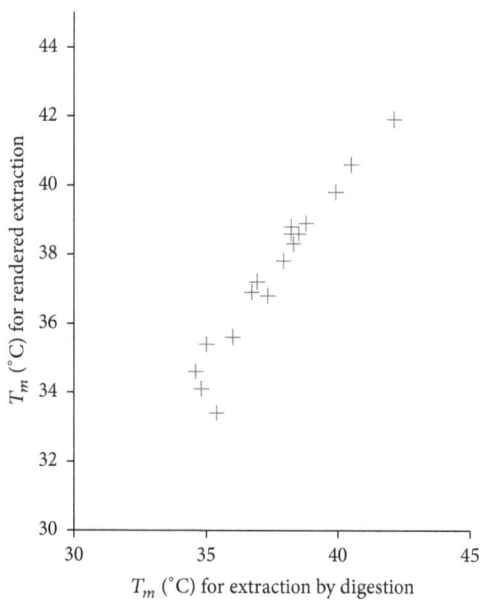

FIGURE 4: Simultaneous extraction of fat and DNA does not change T_m. There is excellent correlation between T_m measurements of fat harvested during DNA and extracted by rendering (Pearson's R = 0.97). There was no measurable bias (mean difference 0.13, SEM = 0.14). Either extraction method can be used for direct comparison without adjustment.

3.4. DNA Extraction Does Not Invalidate Measurement of T_m. In Figure 4 we show that extracting DNA with the proteinase K does not affect the measurement of T_m on the same extract. Oxidation of the polyunsaturated fatty acids in the sample that may have occurred during rendering at 90°C did not have a measurable effect on the melting point, as expected [4, 31].

4. Discussion

The intention of these studies is to resolve, in part, the manifest confusion facing producers and consumers of healthy beef.

It is clear that Wagyu beef is superior, as reflected by the commercial returns for highly marbled beef, but increasingly the brand is amenable to misuse.

A major issue is the lack of a reproducible measurement of the degree of marbling. Multiple and incompatible systems of scoring may have been retained perhaps to the advantage of some sectors. The measurement of T_m is possible at successive stages of the production line so that quality can be confirmed. At the same time, DNA can be extracted so that provenance can be confirmed.

The difficulty faced by the breeder is even more important. Nonreproducible measurement obfuscates attempt to identify breeding values and therefore confound selection of superior sires. This issue becomes particularly important in an attempt to upgrade first crosses.

The present results show that even WY25 can have reduced T_m but the scatter is substantial leading to lack of consistency. Future studies may identify those sires which are well suited to crossbreeding.

So as to reduce the number of variables we sampled AK50 at differing days on feed. All had European dams. The initial results are promising in that there was a progressive decline in T_m. Further work may define the preferred type and length of supplementary feeding. Importantly, there is also the potential to examine the controversies surrounding the use of grass versus grain. Whilst there is growing consumer demand for less intensive feeding and especially for "grass-fed," there is also the perception that corn and perhaps other grains are necessary for extreme marbling. Given reproducible measurements, it should be possible to define acceptable compromises between supplementation, on the one hand, and tastiness and healthiness, on the other hand.

A major finding of this study is the difference in T_m between the progeny of two full blood Wagyu sires. Sire 1 and Sire 2 share a paternal and a maternal grand sire and were imported from the same prefecture in Japan. Pedigree analysis alone would not predict large differences in lipid composition. It is noteworthy however that Sire 1 and Sire 2 are quite different in their C19 haplotypes, as described elsewhere [32]. A major issue remains unresolved. The degree of marbling and the lipid profile differ depending upon the site of sampling; as an approximation the intramuscular accumulation of lipid progress from the brisket backwards with the more caudal fat deposits having somewhat lower proportions of oleic acid and higher T_m [33]. Therefore, comparable samples need to be from a fixed location. Even within the same muscle group there is variation depending upon sampling [34]. We recommend further experience using subcutaneous fat so that its utility can be extended. Ultimately, it should be possible to take in vivo samples so as to monitor changes with time, genetics, and feed.

Conflicts of Interest

Collectively, the authors associated with the CY O'Connor ERADE Village Foundation have an interest in the work described in this manuscript as it forms part of the foundation's intellectual property.

Acknowledgments

The authors are grateful to the collaboration of breeders who provided samples and to Dom Bayard, Scott de Bruin, Peter Gilmour, Keith Hammond, Bruce Cheung, Alan Peggs, Bob Reed, Graham Truscott, Carel Tesling, J. R. Dawkins, and Lindsay Baker for advice and suggestions. This is publication 1504 of the CY O'Connor ERADE Village Foundation supported by funding from Pardoo Pastoral and Melaleuka Stud.

References

[1] W. Cheng, J. H. Cheng, D. W. Sun, and H. Pu, "Marbling Analysis for Evaluating Meat Quality: Methods and Techniques," *Comprehensive Reviews in Food Science and Food Safety*, vol. 14, no. 5, pp. 523–535, 2015.

[2] D. D. Johnson, J. W. Savell, D. M. Stiffler, and H. R. Cross, "Postmortem environmental factors affecting beef carcass lean maturity and marbling evaluations," *Journal of Food Quality*, vol. 8, pp. 253–264, 1986.

[3] K. Kuchida, S. Kono, K. Konishi, L. D. Van Vleck, M. Suzuki, and S. Miyoshi, "Prediction of crude fat content of longissimus muscle of beef using the ratio of fat area calculated from computer image analysis: Comparison of regression equations for prediction using different input devices at different stations," *Journal of Animal Science*, vol. 78, no. 4, pp. 799–803, 2000.

[4] K. Y. Chung, D. K. Lunt, C. B. Choi et al., "Lipid characteristics of subcutaneous adipose tissue and M. longissimus thoracis of Angus and Wagyu steers fed to US and Japanese endpoints," *Meat Science*, vol. 73, no. 3, pp. 432–441, 2006.

[5] S. De Smet, K. Raes, and D. Demeyer, "Meat fatty acid composition as affected by fatness and genetic factors: a review," *Animal Research*, vol. 53, no. 2, pp. 81–98, 2004.

[6] S. B. Smith, C. A. Gill, D. K. Lunt, and M. A. Brooks, "Regulation of fat and fatty acid composition in beef cattle," *Asian-Australasian Journal of Animal Sciences*, vol. 22, no. 9, pp. 1225–1233, 2009.

[7] S. S. Lloyd, S. T. Dawkins, and R. L. Dawkins, "A novel method of measuring the melting point of animal fats," *Journal of Animal Science*, vol. 92, no. 10, pp. 4775–4778, 2014.

[8] M. Zembayashi, K. Nishimura, D. K. Lunt, and S. B. Smith, "Effect of breed type and sex on the fatty acid composition of subcutaneous and intramuscular lipids of finishing steers and heifers.," *Journal of Animal Science*, vol. 73, no. 11, pp. 3325–3332, 1995.

[9] A. E. O. Malau-Aduli, M. A. Edriss, B. D. Siebert, C. D. K. Bottema, and W. S. Pitchford, "Breed differences and genetic parameters for melting point, marbling score and fatty acid composition of lot-fed cattle," *Journal of Animal Physiology and Animal Nutrition*, vol. 83, no. 2, pp. 95–105, 2000.

[10] L. Bartoň, D. Bureš, T. Kott, and D. Řehák, "Associations of polymorphisms in bovine DGAT1, FABP4, FASN, and PPARGC1A genes with intramuscular fat content and the fatty acid composition of muscle and subcutaneous fat in Fleckvieh bulls," *Meat Science*, vol. 114, pp. 18–23, 2016.

[11] H. S. Cheong, D. Yoon, L. H. Kim et al., "Titin-cap (TCAP) polymorphisms associated with marbling score of beef," *Meat Science*, vol. 77, no. 2, pp. 257–263, 2007.

[12] C. Han, M. Vinsky, N. Aldai, M. E. R. Dugan, T. A. McAllister, and C. Li, "Association analyses of DNA polymorphisms in bovine SREBP-1, LXRα, FADS1 genes with fatty acid composition in Canadian commercial crossbred beef steers," *Meat Science*, vol. 93, no. 3, pp. 429–436, 2013.

[13] J. Papaleo Mazzucco, D. E. Goszczynski, M. V. Ripoli et al., "Growth, carcass and meat quality traits in beef from Angus, Hereford and cross-breed grazing steers, and their association with SNPs in genes related to fat deposition metabolism," *Meat Science*, vol. 114, pp. 121–129, 2016.

[14] L. Xu, L. P. Zhang, Z. R. Yuan et al., "Polymorphism of SREBP1 is associated with beef fatty acid composition in Simmental bulls," *Genetics and Molecular Research*, vol. 12, no. 4, pp. 5802–5809, 2013.

[15] S. Zhang, T. J. Knight, J. M. Reecy, and D. C. Beitz, "DNA polymorphisms in bovine fatty acid synthase are associated with beef fatty acid composition," *Animal Genetics*, vol. 39, no. 1, pp. 62–70, 2008.

[16] M. Shibata, K. Matsumoto, K. Aikawa, T. Muramoto, S. Fujimura, and M. Kadowaki, "Gene expression of myostatin during development and regeneration of skeletal muscle in Japanese Black Cattle," *Journal of Animal Science*, vol. 84, no. 11, pp. 2983–2989, 2006.

[17] M. Saatchi, D. J. Garrick, R. G. Tait et al., "Genome-wide association and prediction of direct genomic breeding values for composition of fatty acids in Angus beef cattlea," *BMC Genomics*, vol. 14, no. 1, article no. 730, 2013.

[18] T. Matsuhashi, S. Maruyama, Y. Uemoto et al., "Effects of bovine fatty acid synthase, stearoyl-coenzyme A desaturase, sterol regulatory element-binding protein 1, and growth hormone gene polymorphisms on fatty acid composition and carcass traits in Japanese Black cattle," *Journal of Animal Science*, vol. 89, no. 1, pp. 12–22, 2011.

[19] K. Hayakawa, T. Sakamoto, A. Ishii et al., "The g.841G>C SNP of FASN gene is associated with fatty acid composition in beef cattle," *Journal of Animal Science*, pp. 737–746, 2015.

[20] P. McGilchrist, D. W. Pethick, S. P. F. Bonny, P. L. Greenwood, and G. E. Gardner, "Whole body insulin responsiveness is higher in beef steers selected for increased muscling," *Animal*, vol. 5, no. 10, pp. 1579–1586, 2011.

[21] P. McGilchrist, D. W. Pethick, S. P. F. Bonny, P. L. Greenwood, and G. E. Gardner, "Beef cattle selected for increased muscularity have a reduced muscle response and increased adipose tissue response to adrenaline," *Animal*, vol. 5, no. 6, pp. 875–884, 2011.

[22] W. G. Hill, "Applications of population genetics to animal breeding, from wright, fisher and lush to genomic prediction," *Genetics*, vol. 196, no. 1, pp. 1–16, 2014.

[23] P. R. Myer, T. P. L. Smith, J. E. Wells, L. A. Kuehn, and H. C. Freetly, "Rumen microbiome from steers differing in feed efficiency," *PLoS ONE*, vol. 10, no. 6, Article ID e0129174, 2015.

[24] Z. Durmic, P. J. Moate, R. Eckard, D. K. Revell, R. Williams, and P. E. Vercoe, "In vitro screening of selected feed additives, plant essential oils and plant extracts for rumen methane mitigation," *Journal of the Science of Food and Agriculture*, vol. 94, no. 6, pp. 1191–1196, 2014.

[25] Z. Durmic, C. S. McSweeney, G. W. Kemp, P. Hutton, R. J. Wallace, and P. E. Vercoe, "Australian plants with potential to inhibit bacteria and processes involved in ruminal biohydrogenation of fatty acids," *Animal Feed Science and Technology*, vol. 145, no. 1-4, pp. 271–284, 2008.

[26] N. Teicholz, *A Big Fat Surprise*, Simon and Schuster, New York, NY, USA, 2014.

[27] T. H. Adams, R. L. Walzem, D. R. Smith, S. Tseng, and S. B. Smith, "Hamburger high in total, saturated and trans-fatty acids decreases HDL cholesterol and LDL particle diameter, and increases TAG, in mildly hypercholesterolaemic men," *British Journal of Nutrition*, vol. 103, no. 1, pp. 91–98, 2010.

[28] L. A. Gilmore, R. L. Walzem, S. F. Crouse et al., "Consumption of high-oleic acid ground beef increases HDL-cholesterol concentration but both high- and low-oleic acid ground beef decrease HDL particle diameter in normocholesterolemic men," *Journal of Nutrition*, vol. 141, no. 6, pp. 1188–1194, 2011.

[29] R. P. Mensink, P. L. Zock, A. D. Kester, and M. B. Katan, "Effects of dietary fatty acids and carbohydrates on the ratio of serum total to HDL cholesterol and on serum lipids and apolipoproteins: a meta-analysis of 60 controlled trials," *American Journal of Clinical Nutrition*, vol. 77, no. 5, pp. 1146–1155, 2003.

[30] J. F. Williamson, E. J. Steele, S. Lester et al., "Genomic evolution in domestic cattle: Ancestral haplotypes and healthy beef," *Genomics*, vol. 97, no. 5, pp. 304–312, 2011.

[31] J. D. Wood, R. I. Richardson, G. R. Nute et al., "Effects of fatty acids on meat quality: a review," *Meat Science*, vol. 66, no. 1, pp. 21–32, 2004.

[32] S. S. Lloyd, E. J. Steele, J. L. Valenzuela, and R. L. Dawkins, "Haplotypes for type, degree, and rate of marbling in cattle are syntenic with human muscular dystrophy," *International Journal of Genomics*, vol. 2017, 14 pages, 2017.

[33] S. B. Smith and B. J. Johnson, *Marbling: Management of cattle to maximize the deposition of intramuscular adipose tissue*, National cattlemens beef association, Centennial, 2014.

[34] Y. Nakahashi, S. Maruyama, S. Seki, S. Hidaka, and K. Kuchida, "Relationships between monounsaturated fatty acids of marbling flecks and image analysis traits in longissimus muscle for Japanese Black steers," *Journal of Animal Science*, vol. 86, no. 12, pp. 3551–3556, 2008.

17

Drying Characteristics and Physical and Nutritional Properties of Shrimp Meat as Affected by Different Traditional Drying Techniques

P. T. Akonor,[1] **H. Ofori,**[1] **N. T. Dziedzoave,**[1] **and N. K. Kortei**[2]

[1]*Council for Scientific and Industrial Research-Food Research Institute, P.O. Box M 20, Accra, Ghana*
[2]*Graduate School of Nuclear and Allied Sciences, University of Ghana, P.O. Box LG 80, Legon, Ghana*

Correspondence should be addressed to P. T. Akonor; papatoah@gmail.com

Academic Editor: Fabienne Remize

The influence of different drying methods on physical and nutritional properties of shrimp meat was investigated in this study. Peeled shrimps were dried separately using an air-oven dryer and a tunnel solar dryer. The drying profile of shrimp meat was determined in the two drying systems by monitoring moisture loss over the drying period. Changes in color, proximate composition, and rehydration capacity were assessed. The rate of moisture removal during solar drying was faster than the air-oven drying. The development of red color during drying was comparable among the two methods, but solar-dried shrimps appeared darker ($L^* = 47.4$) than the air-oven-dried ($L^* = 49.0$). Chemical analysis indicated that protein and fat made up nearly 20% and 2% (wb) of the shrimp meat, respectively. Protein and ash content of shrimp meat dried under the two dryer types were comparable but fat was significantly ($p < 0.05$) higher in oven-dried meat (2.1%), compared to solar-dried meat (1.5%). Although rehydration behavior of shrimp from the two drying systems followed a similar pattern, solar-dried shrimp absorbed moisture more rapidly. The results have demonstrated that different approaches to drying may affect the physical and nutritional quality of shrimp meat differently.

1. Introduction

Shrimp belongs to a large group of crustaceans with extended abdomen and it is one of the most important commercial seafood in the world [1]. It is very popular in Ghana and fished on both commercial and artisanal scale around the Keta to Ada and Axim to Cape 3 Points areas [2]. Shrimps are estimated to contain nearly 20% protein (wb) with well-balanced amino acids and significantly high amounts of other nutrients including micronutrients such as calcium and selenium. Lipids in shrimps are largely made up of polyunsaturated fatty acids which are essential for human health [3, 4]. They have also been identified as rich in vitamin B12 and astaxanthin, a fat-soluble carotenoid which has antioxidant properties [5].

The high moisture content of shrimps, together with its high protein content, predisposes them to rapid deterioration. They begin to go bad shortly after capture, unless they are subjected to cold storage, a condition which extends its shelf life significantly. However this means of storage is expensive and may not be available in certain areas where electricity is a challenge. In Ghana and most parts of the world, drying remains one of the best options of preprocessing this seafood. It is one of the oldest means of food preservation and is applicable to a wide range of food products including shrimps. The principle behind drying is primarily reduction of moisture to levels low enough to prevent microbial growth and also slow down enzymatic and other biological reactions that may contribute to food spoilage. Dried shrimps are popular and widely acceptable. They are used (in whole or powdered form) in soups and sauces as a major protein source and for their delectable flavor.

Several drying techniques have been applied to process shrimps. Some of these methods are freeze-drying [6], superheated steam drying [7], jet-spouted bed drying [8], and heat pump drying [9], among others. Notwithstanding these improved approaches to drying shrimps, traditional sun or solar drying and hot air drying remain the most widely applicable means of processing shrimps in Ghana and most developing countries. This is because the production of dried shrimps is mostly done on artisanal and small scales which require less sophistication and relatively cheaper through-puts. Solar and hot air drying are known to affect most food products but information on their effect on the nutritional and physical quality indices of shrimps is scanty.

The effects of the aforementioned traditional drying techniques on shrimps need to be elucidated because changes that occur as a result of drying are likely to affect the quality and, consequently, market value of the final product. The aim of the present study, therefore, was to determine the effect of solar tunnel drying and air-oven drying on drying character-istics and physical and nutritional properties of shrimp meat.

2. Materials and Methods

Fresh marine shrimps (*Penaeus notialis*) were procured from a fish market in Accra and transported on ice in a cold box to the laboratory. The shrimps were sorted and cleaned by washing in potable water before the head and shells were removed and discarded. The shrimp meat was then divided into two batches and subjected to different types of drying.

2.1. Drying Experiments. Drying experiments were con-ducted in two types of dryers; a mechanical air-oven dryer (Apex B35E, London) and a solar tunnel dryer (fabricated locally by the CSIR-Food Research Institute, Ghana). Shrimp meat, weighing 200 g (in triplicates), was spread in a single layer on a wire mesh and loaded into the convective hot air dryer or solar tunnel dryer. The samples were dried for 17 hrs at $55.0 \pm 1.5°C$ in the mechanical dryer whereas solar drying was accomplished in 20 hrs at a mean temperature of $57.4 \pm 6.9°C$. Final moisture contents were $9.97 \pm 0.33\%$ and $9.68 \pm 0.26\%$ correspondingly for air-oven- and solar-dried shrimp meat.

Moisture loss during drying was determined by measur-ing the loss in weight of samples at hourly interval, with an electronic balance (Kern 510, Kern and Sohn, GMbH, Ger-many). Sampling and weighing were done until a fairly con-stant weight was attained [10]. The initial moisture content of shrimp meat, determined by standard methods [11], was $77.12 \pm 0.06\%$ (wb). Dried shrimp meat was sealed air-tight in flexible polypropylene bags and stored at room temperature.

2.2. Color Measurements. Color of the shrimp meat was measured with a Minolta tristimulus color meter (CR-310, Minolta, Japan), calibrated with a reference white porcelain tile ($L = 97.63$, $a = 0.31$, and $b = 4.63$). Measurements were done in triplicates and color described in L^*, a^*, and b^* notation, where L^* is a measure of lightness, a^* defines com-ponents on the red-green axis, and b^* defines components on the yellow-blue axis. Color determinations were done at hourly intervals over a period of 8 hrs for both mechanical and solar-dried shrimp meat.

2.3. Proximate Composition. Shrimp meat was analyzed for moisture, ash, crude fat, and crude protein using approved methods 925.10, 920.87, 920.85, and 923.03 of the Association

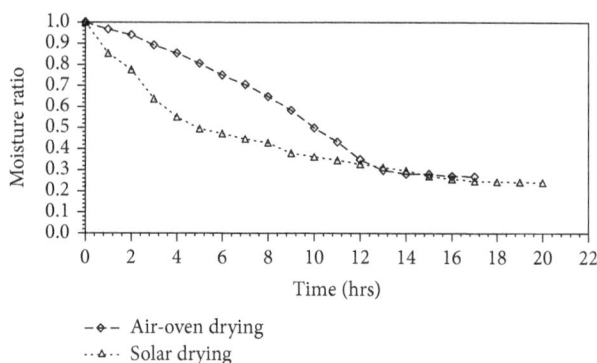

FIGURE 1: Variation of moisture ratio in solar- and air-oven-dried shrimp meat.

of Official Analytical Chemists [11]. Determinations were carried out before and after drying by the two methods.

2.4. Rehydration Studies. Dried shrimp meat was rehydrated according to the method by Doymaz and Smail [12]. Five grams of dried shrimp meat was rehydrated in distilled water at room temperature using a sample to water ratio of $1:40$. At 30 min interval, shrimp meat was removed and carefully blot-ted with tissue paper, weighed on an electronic balance, and immediately returned into the same soaking water. Dried shrimps were rehydrated over a period of 5 hrs when the weight of rehydrated samples had stabilized. Rehydration ratio was then calculated using the following relation [13]:

$$\text{Rehydration Ratio} = \frac{\text{Mass of rehydrated sample}}{\text{Mass of dried sample}}. \quad (1)$$

3. Results and Discussion

3.1. Drying Profile. The drying curves (Figure 1) show a faster rate of moisture removal during solar drying than air-oven drying, especially within the first 12 hours of drying. This outcome may be attributed to the higher average drying temperature ($57°C$) and possibly air speed, in the solar dryer compared to the air-oven ($55°C$) which is thermostat con-trolled. This, together with higher fluctuation in solar drying temperature (SD of ± 6.9), may also explain the variation in arriving at a stable weight during drying. A stable weight was reached faster in air-oven compared to solar drying.

In all two drying methods, a decline in the rate of mois-ture removal was observed after 12 hrs of drying until a fair stability in moisture content was established. This may be explained by the difficulty in moisture removal as the drying front recedes towards the innermost parts of the shrimp meat and resistance to moisture movement becomes higher [14].

3.2. Proximate Composition. Proximate composition of shrimp meat before and after drying indicates an interesting nutrient profile. As is typical of most sea foods, the fresh shrimp was made up of nearly 80% moisture. This result com-pares well with 80.5% and 77.2% reported in earlier studies for black tiger shrimp and white shrimp, respectively [4].

TABLE 1: Chemical properties of fresh and dried shrimp meat (db).

Parameter	Fresh	Air-oven-dried	Solar-dried
Protein	86.21 ± 0.08^b	85.64 ± 0.26^a	84.89 ± 0.51^a
Ash	6.93 ± 0.11^a	6.82 ± 0.10^a	6.77 ± 0.28^a
Fat	6.54 ± 0.20^c	5.98 ± 0.15^b	5.74 ± 0.11^a

Means with different superscripts along the same row are significantly different ($p < 0.05$).

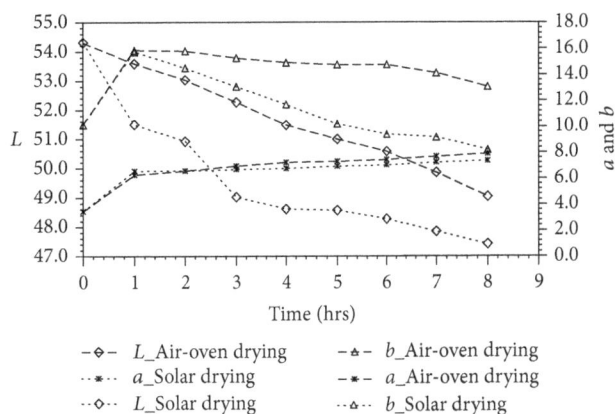

FIGURE 2: Color profile of shrimp meat during drying.

After drying, using the two methods, moisture was reduced to about 10%, which is less than the specified moisture for dried shrimps [15]. Low moisture content in dried shrimps is encouraged to safeguard the product from microbial attack and enzymatic action and therefore prevent spoilage. Moisture of shrimp from the two drying methods did not show any significant differences ($p > 0.05$), although air-dried shrimp meat appeared to have a lower moisture content.

After moisture, protein was the second most abundant component of the shrimp meat, and this made up about 86% (db) of fresh meat (Table 1). Shrimp meat from this study was slightly richer in protein compared to both black tiger and white shrimps reported by Sriket et al. [4], but lower than white leg shrimp [16] and for green tiger and speckled shrimps [3]. The discrepancies in protein content of shrimps used in these studies may be attributed to differences in species, growth stage, season, and waters in which they were shrimped [17]. The high protein content of shrimp meat makes it a good source of amino acids for human diets. Protein content of air-oven-dried shrimp meat was slightly lower but not significantly different ($p > 0.05$) from solar-dried shrimp meat. The difference in protein content before and after drying, which was remarkable ($p < 0.05$), may be related to protein denaturation and or the incidence of browning reactions, in which some amount of amino acid is used up.

Sriket et al. [4] have reported in previous studies that the fat content of some shrimp species ranges between 1.2 and 1.3% (wb). Most of this exists as membrane lipids as in the case of fish muscles [4, 18]. In the present study, shrimp meat contained about 2% fat (wb), but this reduced significantly ($p < 0.05$) to less than 2% after drying. This reduction suggests that, during drying, fat may have exuded along with moisture evaporation or oxidized into other compounds [19] since shrimp lipids are mainly made up of polyunsaturated fatty acids. Degradation of astaxanthin in the process of drying may also have contributed to the loss of lipids since this carotenoid is thought to have a high antioxidant activity [20]. The loss of fat was markedly different ($p < 0.05$) for the two drying methods, with solar-dried shrimps being the most severe. This observation is ascribed to higher drying temperature in the tunnel solar dryer. Similar results of fat loss as a result of drying were obtained by Wu and Mao [19] for dried grass carp fillets.

Ash represents the total mineral content in food and is essential in maintaining several bodily functions. Shrimp meat was found to contain appreciable amounts of ash, totaling nearly 1.6% (wb). This makes shrimp meat a good source of minerals in the diet. Due to moisture loss and concentration of chemical components, higher ash content was obtained after drying. This seemingly increased ash content was however not significantly different ($p > 0.05$) among the two drying methods. The ash content of shrimp meat in the present study is slightly higher than the ranges of 1–1.5% and 1.47% independently reported by Gunalan et al. [16] and Yanar and Çelik [3].

3.3. Color Development.
According to Yanar et al., [21], the market value of shrimp is dependent on the visual appearance of their body color, and this is attributed to the presence of the astaxanthin. This carotenoid pigment is responsible for the red-orange tissue pigmentation of shrimp meat. Figure 2 shows color development in shrimp meat as monitored over the drying period.

Lightness index (L^*) reduced with the passage of time, suggesting that shrimp meat became darker. Darkening may have occurred because of Maillard browning reactions which took place during drying. The extent of these reactions in solar drying may have been more pronounced, resulting in darker shrimp meat compared to oven-dried samples. Again, the results show that red color (a^*) was developed when shrimp meat was dried, and the two drying methods showed similar ($p > 0.05$) results. Development of redness on exposure of shrimp meat to heat is as a result of the release of astaxanthin when carotenoproteins breakdown during protein denaturation. In both solar and air-oven drying, the intensity of redness increased nearly twofold within the 1st hour of drying and only increased slightly thereafter. This is attributable to an increase in concentration of astaxanthin when water was removed from shrimp tissue [8]. Yellowness (b^*) of shrimp meat also increased over the 1st hour of drying as a result of formation of yellow pigments from browning reactions. However, a steady decline in yellowness (b^*) was noticed after 1 hour of drying. The dip in yellowness was marginal in air-oven drying but more severe in solar drying as in the case of lightness index (L^*).

3.4. Rehydration.
Rehydration refers to the process of moistening a dried product and is an indicator of quality criterion in most dried foods. It is an indicator of cellular and structural

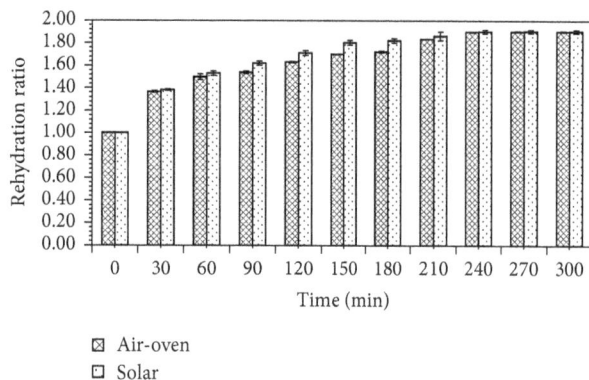

FIGURE 3: Rehydration characteristics of dried shrimp meat.

disintegration that occurs during dehydration [22]. Rehydration ratio of shrimp meat as a function of time is presented in Figure 3.

In agreement with earlier studies [10, 23, 24], there was a rapid increase in weight of shrimp meat from both air-oven and solar drying because of high rate of water absorption at the initial stages. The first 3 hours saw a rapid weight gain by the meat (from both drying methods) and this slowed down afterwards and, subsequently, flattened off between the 4th and 5th hours as the process reached equilibrium. The initial rapid uptake of moisture by shrimp meat, as posited by Sagar and Suresh Kumar [25], is the result of surface and capillary suction.

A comparison of the rehydration behavior of shrimp meat from the two drying systems indicated a clear difference, especially, between the 1st and 3rd hours of rehydration. Rehydration was faster in solar compared to mechanical drying. The extent and rate of water uptake during rehydration is largely influenced by cellular and structural arrangements in the food matrix since this provides the channels for conveying water to muscle fibers. This phenomenon has been amply demonstrated in previous studies [8, 26]. Also, as noted by Niamnuy et al. [8], higher protein contraction may have reduced the rehydration ability of mechanically dried shrimp meat.

4. Conclusion

The study showed differences in some chemical and quality properties between dried shrimps, using different techniques. Protein content generally remained unaffected by drying method, but the amount of fat was remarkably higher for oven-dried shrimp meat. Solar drying resulted in relatively darker shrimp meat with a higher rehydration rate, compared to oven-dried shrimp meat. Although drying occurred faster during solar drying, dried meat was of a lower quality compared to air-dried shrimp meat.

Conflict of Interests

The authors declare that there is no conflict of interests regarding the publication of this paper.

References

[1] P. Oosterveer, "Globalization and sustainable consumption of shrimp: consumers and governance in the global space of flows," *International Journal of Consumer Studies*, vol. 30, no. 5, pp. 465–476, 2006.

[2] M. Entsua-Mensah, K. A. A. de Graft-Johnson, M. O. Atikpo, and L. D. Abbey, "The lobster, shrimp and prawn industry in Ghana—species, ecology, fishing and landing sites, handling and export," Tech. Rep. CSIR-FRI/RE/E-MM/2002/012, Food Research Institute/AgSSIP, 2002.

[3] Y. Yanar and M. Çelik, "Seasonal amino acid profiles and mineral contents of green tiger shrimp (*Penaeus semisulcatus* De Haan, 1844) and speckled shrimp (*Metapenaeus monoceros* Fabricus, 1789) from the Eastern Mediterranean," *Food Chemistry*, vol. 94, no. 1, pp. 33–36, 2006.

[4] P. Sriket, S. Benjakul, W. Visessanguan, and K. Kijroongrojana, "Comparative studies on chemical composition and thermal properties of black tiger shrimp (*Penaeus monodon*) and white shrimp (*Penaeus vannamei*) meats," *Food Chemistry*, vol. 103, no. 4, pp. 1199–1207, 2007.

[5] V. Venugopal, *Marine Products for Healthcare: Functional and Bioactive Nutraceutical Compounds from the Ocean*, CRC Press, London, UK, 2008.

[6] G. Donsì, G. Ferrari, and P. Di Matteo, "Utilization of combined processes in freeze-drying of shrimps," *Food and Bioproducts Processing: Transactions of the Institution of of Chemical Engineers Part C*, vol. 79, no. 3, pp. 152–159, 2001.

[7] S. Prachayawarakorn, S. Soponronnarit, S. Wetchacama, and D. Jaisut, "Desorption isotherms and drying characteristics of shrimp in superheated steam and hot air," *Drying Technology*, vol. 20, no. 3, pp. 669–684, 2002.

[8] C. Niamnuy, S. Devahastin, and S. Soponronnarit, "Effects of process parameters on quality changes of shrimp during drying in a jet-spouted bed dryer," *Journal of Food Science*, vol. 72, no. 9, pp. E553–E563, 2007.

[9] G. Zhang, S. Arason, and S. V. Árnason, "Physical and sensory properties of heat pump dried shrimp (*Pandalus borealis*)," *Transactions of the Chinese Society of Agricultural Engineering*, vol. 24, no. 5, pp. 235–239, 2008.

[10] P. Akonor and C. Tortoe, "Effect of blanching and osmotic pre-treatment on drying kinetics, shrinkage and rehydration of chayote (*Sechium edule*) during convective drying," *British Journal of Applied Science & Technology*, vol. 4, no. 8, pp. 1215–1229, 2014.

[11] AOAC, *Official Methods of the Association of Analytical Chemists*, AOAC International, Washington, DC, USA, 15th edition, 1990.

[12] I. Doymaz and O. Smail, "Drying characteristics of sweet cherry," *Food and Bioproducts Processing*, vol. 89, no. 1, pp. 31–38, 2011.

[13] P. P. Lewicki, "Some remarks on rehydration of dried foods," *Journal of Food Engineering*, vol. 36, no. 1–4, pp. 81–87, 1998.

[14] M. R. Okos, O. Campanella, G. Narsimhan, R. K. Singh, and A. C. Weitnauer, "Food dehydration," in *Handbook of Food Engineering*, D. R. Heldman and D. B. Lund, Eds., Taylor and Francis Group, Boca Raton, Fla, USA, 2nd edition, 2007.

[15] R. Tapaneyasin, S. Devahastin, and A. Tansakul, "Drying methods and quality of shrimp dried in a jet-spouted bed dryer," *Journal of Food Process Engineering*, vol. 28, no. 1, pp. 35–52, 2005.

[16] B. Gunalan, N. S. Tabitha, P. Soundarapandian, and T. Anand, "Nutritive value of cultured white leg shrimp *Litopenaeus vannamei*," *International Journal of Fisheries and Aquaculture*, vol. 5, pp. 166–171, 2013.

[17] P. A. Karakoltsidis, A. Zotos, and S. M. Constantinides, "Composition of the commercially important mediterranean finfish, crustaceans, and molluscs," *Journal of Food Composition and Analysis*, vol. 8, no. 3, pp. 258–273, 1995.

[18] K. Takama, T. Suzuki, K. Yoshida, H. Arai, and T. Mitsui, "Phosphatidylcholine levels and their fatty acid compositions in teleost tissues and squid muscle," *Comparative Biochemistry and Physiology B: Biochemistry and Molecular Biology*, vol. 124, no. 1, pp. 109–116, 1999.

[19] T. Wu and L. Mao, "Influences of hot air drying and microwave drying on nutritional and odorous properties of grass carp (*Ctenopharyngodon idellus*) fillets," *Food Chemistry*, vol. 110, no. 3, pp. 647–653, 2008.

[20] Y. M. A. Naguib, "Antioxidant activities of astaxanthin and related carotenoids," *Journal of Agricultural and Food Chemistry*, vol. 48, no. 4, pp. 1150–1154, 2000.

[21] Y. Yanar, M. Çelik, and M. Yanar, "Seasonal changes in total carotenoid contents of wild marine shrimps (*Penaeus semisulcatus* and *Metapenaeus monoceros*) inhabiting the eastern Mediterranean," *Food Chemistry*, vol. 88, no. 2, pp. 267–269, 2004.

[22] N. K. Rastogi, A. Angersbach, K. Niranjan, and D. Knorr, "Rehydration kinetics of high-pressure pre-treated and osmotically dehydrated pineapple," *Journal of Food Science*, vol. 65, no. 5, pp. 838–841, 2000.

[23] K. A. Taiwo, A. Angersbach, and D. Knorr, "Rehydration studies on pretreated and osmotically dehydrated apple slices," *Journal of Food Science*, vol. 67, no. 2, pp. 842–847, 2002.

[24] M. K. Krokida and D. Marinos-Kouris, "Rehydration kinetics of dehydrated products," *Journal of Food Engineering*, vol. 57, no. 1, pp. 1–7, 2003.

[25] V. R. Sagar and P. Suresh Kumar, "Recent advances in drying and dehydration of fruits and vegetables: a review," *Journal of Food Science and Technology*, vol. 47, no. 1, pp. 15–26, 2010.

[26] Y. Namsanguan, W. Tia, S. Devahastin, and S. Soponronnarit, "Drying kinetics and quality of shrimp undergoing different two-stage drying processes," *Drying Technology*, vol. 22, no. 4, pp. 759–778, 2004.

18

Microbiological Quality Assessment of Frozen Fish and Fish Processing Materials from Bangladesh

Sohana Al Sanjee[1] and Md. Ekramul Karim[1,2]

[1]*Department of Microbiology, Faculty of Biological Sciences, University of Chittagong, Chittagong 4331, Bangladesh*
[2]*Environmental Biotechnology Division, National Institute of Biotechnology, Ganakbari, Ashulia, Dhaka 1349, Bangladesh*

Correspondence should be addressed to Md. Ekramul Karim; ekrammbio@gmail.com

Academic Editor: Thierry Thomas-Danguin

The present study aims at the microbiological analysis of export oriented frozen fishes, namely, Jew fish, Tongue Sole fish, Cuttle fish, Ribbon fish, Queen fish, and fish processing water and ice from a view of public health safety and international trade. Microbiological analysis includes the determination of total viable aerobic count by standard plate count method and enumeration of total coliforms and fecal coliforms by most probable number method. The presence of specific fish pathogens such as *Salmonella* spp. and *Vibrio cholerae* were also investigated. The TVAC of all the samples was estimated below 5×10^5 cfu/g whereas the total coliforms and fecal coliforms count were found below 100 MPN/g and 10 MPN/g, respectively, which meet the acceptable limit specified by International Commission of Microbiological Specification for Food. The microbiological analysis of water and ice also complies with the specifications having TVAC < 20 cfu/mL, and total coliforms and fecal coliforms count were below the limit detection of the MPN method. Specific fish pathogens such as *Salmonella* sp. and *V. cholerae* were found absent in all the samples under the investigation. From this study, it can be concluded that the investigated frozen fishes were eligible for export purpose and also safe for human consumption.

1. Introduction

Fish and fishery products are not only nutritionally important but also important in global trade as foreign exchange earner for a number of countries in the world [1]. Fisheries and aquaculture sectors have become the second most important contributors in export earnings of Bangladesh, providing about 3.74% in national GDP, 2.7% in export earnings, and 22.23% in agriculture sector [2]. Due to wide range of global market including USA, UK, Japan, Belgium, Netherlands, Thailand, Germany, China, France, Canada, Spain, and Italy, the export of frozen fish, dry fish, and salted and dehydrated fish is increasing day by day from Bangladesh. Though there are 129 fish processing industries in Bangladesh, only 62 plants have EU approval. So it is very important to maintain the quality of the frozen fish for its acceptance in international trade as well as avoiding the health problems of consumers.

Fish are of great concern for export earnings because of their higher nutritive value such as high protein content, with little or no carbohydrate and fat value. But fish may be contaminated at various stages of transport, handling, and processing. This contamination may be related to the raw materials, personnel, and processing tools such as forklifts through leakage, insect, and pest harborage. Additionally, seafood can become contaminated during storage and processing [3, 4]. Contamination may be caused by foodborne pathogens which are naturally present in aquatic environments, such as *Vibrio* spp., or derived from sewage contaminated water such as *Salmonella* spp. [5]. Consumption of these contaminated fish may cause infection or intoxication to the consumers.

Vibrio cholerae is responsible for the third-highest number of shellfish-related illnesses, after noncholera *Vibrio* spp. and Norwalk viruses [6]. Toxigenic Ol (epidemic biotype) infections are associated with profuse, watery diarrhea whereas nontoxigenic, non-Ol biotype (except O139) infections result in septicaemia and mild gastroenteritis. In contrast to *Vibrio* spp., the incidence of *Salmonella* infections due to seafood consumption is still low compared with salmonellosis associated with other foods. However, detection of *Salmonella* spp. in seafood cannot be skipped as it is responsible for most of the foodborne diseases or gastroenteritis characterized by diarrhea, abdominal cramp, vomiting, nausea, and fever. According to Centers for Disease Control and Prevention, *Salmonella* is the leading cause of bacterial foodborne illness causing approximately 1.4 million nontyphoidal illnesses, 15,000 hospitalizations, and 400 deaths in the USA annually [7].

Water and ice quality is also an important factor for good quality fish, because water and ice used for fish processing may contaminate the whole processing plant. EU advised Bangladesh Government to implement the Hazard Analysis Critical Control Point (HACCP) in the processing of frozen fishes [8]. So it is important to find out the quality of fish we consume as well as of the frozen fish which are exported.

Therefore, the present study was carried out to investigate the microbiological quality of the marine frozen fishes for raising food safety concern and promoting international trade. This study also investigated the microbiological quality of water and ice, as these factors were intimately correlated with the fish processing and preservation.

2. Materials and Methods

2.1. Study Area. The study was carried out in "Taj Munir Fish Preserver Ltd.," situated at the port city Chittagong of Bangladesh. The study was conducted from June 2011 to February 2014. During the study period, total viable aerobic count, total coliforms and fecal coliforms counts, and presence of pathogenic organisms (namely, *Salmonella* spp. and *Vibrio cholerae*) of public health significance from the frozen fishes (storage temperature $-20°C$) such as Jew fish (*Argyrosomus hololepidotus*), Tongue Sole fish (*Cynoglossus broadhursti*), Cuttle fish (*Sepia officinalis*), Ribbon fish (*Lepturacanthus savala*), Queen fish (*Scomberoides commersonnianus*), and water and ice which were used during the processing of samples were investigated. All the frozen fishes were gutted and organoleptically good enough to carry out further bacteriological analysis. Sampling was done each year at three months' interval, namely, June, October, and February. During study periods, triplicate samples for each fish species as well as for ice and water samples were analyzed independently.

2.2. Chemicals and Media. Pure and analytical grade chemicals purchased from BDH Chemicals Ltd., England; Merck, Germany; and Siga Chemical Co. Ltd., USA, were used throughout the study including media preparation. All the media and media ingredients such as beef extract and peptone that are used throughout the study were from Scharlau, Spain.

For the enumeration of coliforms and fecal coliforms, Lauryl Tryptose Broth (LTB) and 2% Brilliant Green Bile Broth (BGLBB) were used, respectively. Bismuth Sulfite Agar (BSA) and Xylose Lysine Deoxycholate (XLD) agar were used for the detection of *Salmonella* spp. whereas Thiosulfate Citrate Bile Salt (TCBS) agar and Cellobiose, Polymyxin, and Colistin (CPC) were used for the detection of *V. cholerae*.

2.3. Fish Samples Preparation. All glassware was sterilized ($121°C$, 15 psi, 20 minutes) before use. Triplicate fish samples (each about 25 g) of each fish type were measured separately in an analytical balance (Model: ML204/01, Mettler Toledo, Switzerland) in aseptic condition and then dissolved into about 225 mL buffered peptone water (BPW) and blended for (30–60) seconds in a sterilized blender machine [8]. Each fish sample was blended and homogenized separately.

2.4. Water and Ice Collection. The water (namely, WS_1, WS_2, and WS_3) and ice (namely, IS_1, IS_2, and IS_3) were collected in 1-liter sterilized container from different location. The collected samples were preserved in the refrigerator ($4°C$), when analysis was delayed for more than 3 hours.

2.5. Enumeration of TVAC of Fish, Water, and Ice. Total viable aerobic bacteria of fish, water, and ice were enumerated by standard plate count (SPC) method [9]. For the enumeration of TVAC, serial dilution of each sample was carried out up to 10^{-5} dilutions with 9 mL sterilized 0.1% peptone water, from which an aliquot of 1 mL of each dilution was aseptically poured into duplicate sterile Petri plate, and sterile melted (around 40–45°C) Plate Count Agar was poured over it, rotated clockwise-anticlockwise, allowed to solidify, and finally incubated at inverted position at 37°C for 24–48 hours.

After incubation, the plates having well spaced colonies (30–300) were used for counting and the colonies were counted by a colony counter (Stuart Scientific, UK). Total viable aerobic count per mL or per g was calculated by multiplying the average number of colonies per plate by reciprocal of the dilution and expressed as colony forming units (cfu) per mL or per g of sample [10].

2.6. Enumeration of Total Coliforms of Fish. Most Probable Number (MPN) method is used for the quantitative estimation for coliform [11]. Serial dilution of the samples was prepared as described earlier. Nine test tubes containing about 9 mL Lauryl Tryptose Broth (LTB) with inverted Durham's tube were sterilized. Three test tubes were inoculated with 1 mL from 10^{-1} dilution, another three test tubes were inoculated from 10^{-2} dilution, and the remaining three test tubes were inoculated from 10^{-3} dilution. The inoculated tubes were incubated at 37° for 48 hours. Test tubes showing positive results (gas production in Durham's tube) were counted and recorded as presumptive positive for coliforms.

2.7. Enumeration of Fecal Coliforms (Presumptive E. coli Test) of Fish. About one loopful from each gas positive LTB was inoculated into test tube of sterilized BGLBB and a test tube of sterilized 10 mL Tryptone Broth and then incubated at $44.5° ± 0.5°C$ for 48 hours. After incubation, gas production was recorded and 2-3 drops of Kovac's reagent were added to each of the positive tubes. A positive indole reaction in Tryptone Broth that has produced cherry red color indicates the presence of *E. coli*. The positive gas production tubes were recorded and results were compared using Most Probable Number (MPN) chart to determine the total fecal coliforms number (*E. coli*) per gram [12].

2.8. Enumeration of Total Coliforms and Fecal Coliforms in Water and Ice. About 50 mL of water was inoculated to 50 mL of sterilized LTB (double strength) in one mega test tube whereas about 10 mL of water was inoculated in five test tubes containing 5 mL of sterile LTB (double strength) and about 1 mL of water was also inoculated in five test tubes containing 5 mL of sterile LTB (single strength) in each of the test tubes. All of the test tubes contained inverted Durham's tubes. After incubation at 37°C for 24–48 hours, the positive results were recorded which indicate total coliforms count (MPN/100 mL). About one loopful from the gas positive tubes was inoculated in BGLBB with inverted Durham's tubes and sterilized Tryptone Broth and then incubated at 44.5° ± 0.5°C for 24–48 hours. After incubation, 2-3 drops of Kovac's reagent were added to the Tryptone Broth and cherry red color indicated total fecal coliforms/100 mL. The total coliforms and fecal coliforms of ice samples were also enumerated similarly.

2.9. Detection of Salmonella spp. Salmonella spp. were detected following the procedure as described by Andrews and Hammack [13]. About 25 g samples were dissolved in about 225 mL of sterilized buffered peptone water (BPW), blended, and incubated at 37°C for 16–20 hours. About 10 mL from the incubated BPW culture was selectively enriched into the 100 mL sterilized Selenite Cystine Broth and incubated again at 37°C for 24–48 hours. After incubation, 1 loopful inoculum from the selective enrichment culture was streaked onto the preincubated BSA and XLD agar plate. Typical *Salmonella* spp. produce pink colonies with or without black centers on XLD agar and brown, grey, or black colonies on BSA agar. Then the suspected colonies were identified by their cultural, morphological, and biochemical characteristics, namely, TSI (Triple Sugar Iron), Urease test, MR-VP test, Oxidase test, Citrate test, fermentation of carbohydrates (Glucose, Sucrose, Arabinose, Mannose, Mannitol, and Inositol), and Decarboxylase (Lysine, Arginine, and Ornithine) following the taxonomic guides of *Bergey's Manual of Determinative Bacteriology*, 8th ed. [14]. All cultures giving biochemical reactions were confirmed by agglutination test with *Salmonella* polyvalent (O) somatic antisera [15].

2.10. Detection of Vibrio cholera. Vibrio cholera was detected following the procedure as described by Kaysner and Angelo [16]. About 25 g samples were blended with 225 mL sterilized Alkaline Peptone Water (APW) and incubated at 37°C for 16–18 hours. Then 1 loopful inoculum from the APW culture was streaked on the preincubated TCBS and CPC agar plate and incubated at 37° for 24 hours. Typical colonies of *V. cholerae* on TCBS agar are large, yellow, and smooth whereas on CPC agar they are small, smooth, opaque, and green to purple in color. Then the suspected colonies were identified by their cultural, morphological, and biochemical characteristics, namely, Oxidase test, fermentation of carbohydrates (Glucose, Sucrose, Arabinose, Mannose, Mannitol, and Inositol), and Decarboxylase test (Lysine, Arginine, and Ornithine) according to the taxonomic guides of *Bergey's Manual of Determinative Bacteriology*, 8th ed. [14]. Finally

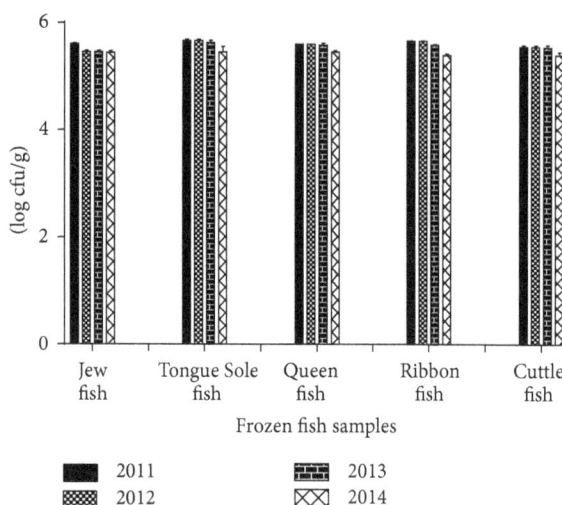

FIGURE 1: Total viable aerobic count (log cfu/g) of the frozen fish samples. The data were representatives of the three independent experiments using triplicate samples and mean ± SD values were expressed.

the *V. cholerae* were confirmed by agglutination test using polyvalent *V. cholerae* (O) antiserum [15].

3. Results and Discussion

Fish and seafoods hold an important position as a food component for a large section of world population [17]. In Bangladesh, the export of fish and fishery products has gained a remarkable position in the earnings of foreign currencies in the last few years. So, maintenance of appropriate quality of the products is regarded as vital for achieving desired success in the global trade of this product.

Jew fish (*A. hololepidotus*), Queen fish (*S. commersonnianus*), Tongue Sole fish (*C. broadhursti*), Ribbon fish (*L. savala*), and Cuttle fish (*S. officinalis*) are the most commonly exported marine fishes from Bangladesh. The maximum microbiological limit for the TVAC which separates the good quality products from bad quality is 5×10^5 cfu/g [18]. The TVAC of the studied samples ranged from 2.8×10^5 to 4.9×10^5 cfu/g which was below the maximum acceptable limit. So all the samples of each type of the fish meet the acceptable limit specified by ICMSF which points out the good quality of the frozen fishes.

Figure 1 showed the TVAC of all frozen fish samples from which it was observed that the density of total aerobic bacteria detected in the Tongue Sole fish samples was comparatively higher than all of the fishes whereas the lowest bacterial count was observed in the samples of Jew fish. Loads of bacteria in fish samples decreased gradually over time in all of the fishes. This may be due to the aseptic processing and handling, proper sampling, trained personnel, improved storage conditions, and increased awareness for preservation.

The acceptable limits of total coliforms (TC) and fecal coliforms (FC) for fresh and frozen fish are <100 MPN/g and <10 MPN/g, respectively [18]. The presence of TC is indicator of sewage contamination which may also occur during

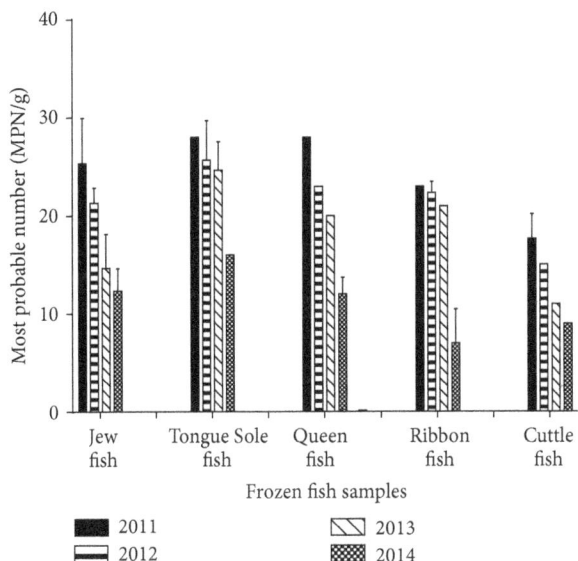

FIGURE 2: Total coliforms count (MPN/g) of frozen fish samples. The data were representatives of the three independent experiments using triplicate samples and mean ± SD values were expressed.

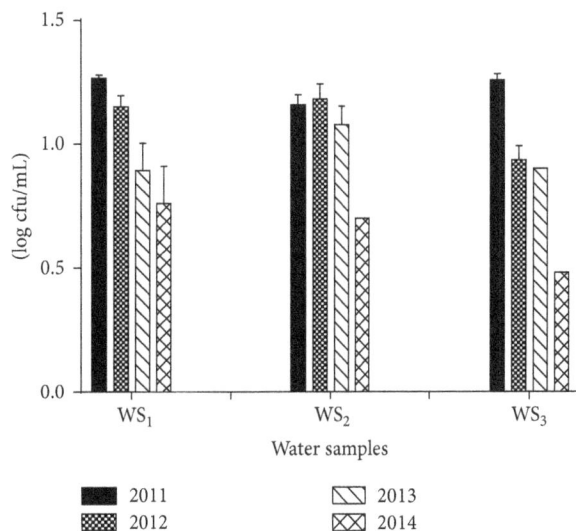

FIGURE 3: Total fecal coliforms count (MPN/g) of frozen fish samples. The data were representatives of the three independent experiments using triplicate samples and mean ± SD values were expressed.

different processing steps such as transport and handling. Moreover, the contamination may also be caused by the water used for washing or icing [19]. The more accurate indicator of fecal contamination is fecal coliforms that is *E. coli* [20]. The lower number of coliforms can be beneficial for pointing out the effectiveness of safety procedures during processing and handling [21]. In the present study, the total coliforms count ranged from 5 MPN/g to 28 MPN/g and fecal coliforms count was from 3 MPN/g to 8.3 MPN/g. Figures 2 and 3 showed the highest number of coliforms and fecal coliforms bacteria in Tongue Sole fish whereas the lowest count was observed in the samples of Jew fish, respectively. Our study revealed that all the samples were within the recommended limits which indicated that the samples were collected from pollution-free water and also the food processors and handlers maintained aseptic conditions throughout the processing.

Water and ice are the most important factor for the processing of exported fish. These two factors contribute to determining and maintaining of the standard quality of the frozen fishes. Figures 4 and 5 showed the TVAC of water and ice samples, respectively, over the study period. It was found that TVAC of water samples and ice samples ranged from 3 to 18 cfu/mL. Significant reduction of TVAC for both water and ice samples was observed over the time period. The total coliforms and fecal coliforms count were found absent for both samples. Hence, our study revealed that all the tested samples complied with the recommended limit specified by ICMSF, that is, TVAC having <20 cfu/mL, and coliforms and fecal coliforms count were below the limit detection of the MPN method. This may be due to the advanced and improved facilities for water and ice purification, treatment, and handling.

Seafood infections are caused by variety of bacteria, viruses, and parasites. According to Centers for Disease Prevention and Control (CDC), during 1973 to 2006, 188

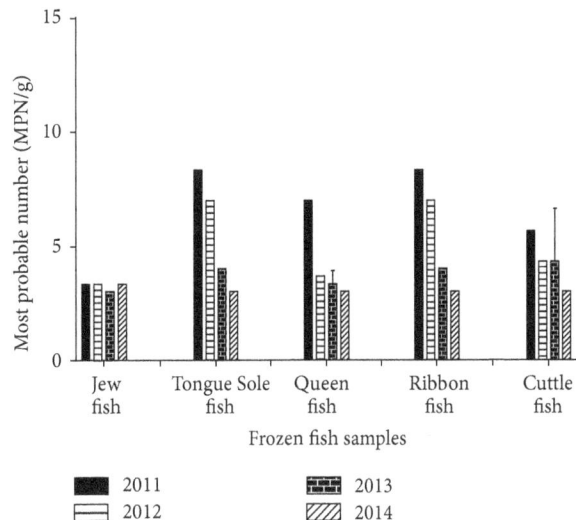

FIGURE 4: Total viable aerobic count (log cfu/mL) of water samples. The data were representatives of the three independent experiments using triplicate samples and mean ± SD values were expressed.

outbreaks of seafood-associated infections, causing 4,020 illnesses, 161 hospitalizations, and 11 deaths, were reported to the Foodborne Disease Outbreak Surveillance System. Most of these seafood-associated outbreaks (143 (76.1%)) were due to a bacterial agent; 40 (21.3%) outbreaks had a viral etiology; and 5 (2.6%) had a parasitic cause. According to the report, *Vibrio* were the most commonly reported cause of seafood-associated outbreaks where toxigenic *V. cholerae* caused 3 outbreaks and 10 illnesses without deaths and non-O1, non-O139 *V. cholerae* caused 4 outbreaks and 12 illnesses without deaths, whereas *Salmonella* was responsible for 18 outbreaks, 374 illnesses, and 28 hospitalizations during the study period [22].

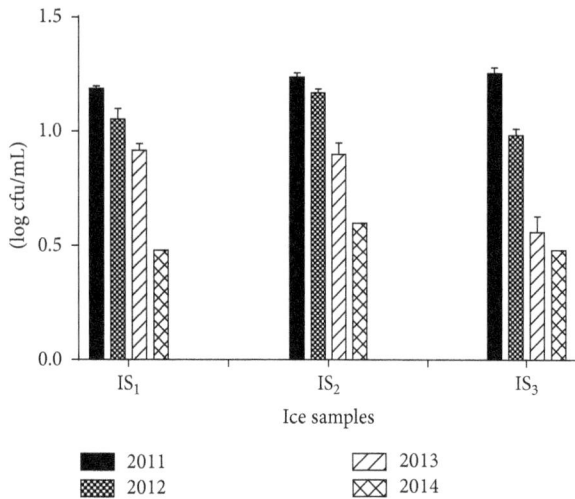

FIGURE 5: Total viable aerobic count (log cfu/mL) of ice samples. The data were representatives of the three independent experiments using triplicate samples and mean ± SD values were expressed.

Recently CDC reported that about 62 people were infected with *Salmonella* Paratyphi B variant L (+) tartrate (+) (formerly known as *Salmonella* Java) from 11 states of USA related to the consumption of frozen raw tuna. The infection was characterized by diarrhea, fever, and abdominal cramps after 12–72 hours' exposure without paratyphoid fever, enteric fever, or typhoid fever [23].

Although in Bangladesh, foodborne illness related to fresh or frozen seafood consumption has not been traced yet or data on this issue is still lacking. In this context, microbiological analysis of frozen fish and fishery products seems to be an important issue. This type of study generates scientific information which would help in preventing and controlling future outbreaks related to seafood consumption. According to the rule of International Association of Microbiology Society, fresh and frozen fish should possess neither *Vibrio* spp. nor *Salmonella* spp. The investigated frozen samples were of good quality as all the samples were free from these pathogenic microorganisms.

Bacterial growth in the frozen fishes is one of the main causes of food spoilage or contamination of fish. Hence, the microbiological analysis of the frozen fish samples and fish processing materials (water and ice) acts as the indicator of fish quality determination. The present study reported that all of the fish samples along with the materials meet the standard levels suggested by ICMSF which indicated that aseptic and proper hygienic conditions were maintained properly throughout all steps such as catching, landing, transportation, processing, handling, and preservation.

4. Conclusion

Although seafood is part of a healthful diet, its consumption is not out of risk. Worldwide continued outbreaks of seafood-associated infections have rendered the existing control strategies questionable. An understanding of the etiologic agents, seafood commodities associated with illness, and mechanisms of contamination that are amenable to control is thus necessary for the prevention of seafood-associated infection outbreaks. Coordinated efforts from government sector and private industry together with federal agencies are urgently needed in this context. There is a need for routine surveillance systems using pathogen-specific techniques to avoid any future outbreaks. However, the current study revealed that microbiological quality of the investigated frozen fishes and fish processing materials (ice and water) was within the specified limit of ICMSF. So it can be concluded that these fishes were processed with properly treated pathogen-free water and ice and, finally, maintained at good storage condition. Hence, the investigated frozen fishes were qualified enough for export as well as human consumption from bacteriological point of view. The presence of viruses, parasites, viable but nonculturable (VBNC) state of the pathogenic bacteria, and biochemical parameters such as histamine risk might be a problem in frozen fish products which is the limitations of this study. Beyond ICMSF, in order to comply with more stringent indigenous quality standards of the exporting countries, these quality parameters must be taken into consideration.

Conflict of Interests

The authors declare that there is no conflict of interests regarding the publication of this paper.

Acknowledgment

The authors are grateful to the authorities of "Taj Munir Fish Preserver Ltd.," Chittagong, Bangladesh, for providing their facilities and also for their guidance.

References

[1] S. O. Yagoub and T. M. Ahmed, "Pathogenic Microorganisms in fresh water samples collected from Khartoum central market," *Sudan Journal of Veterinary Science and Animal Husbandry*, vol. 43, no. 1-2, pp. 32–37, 2003.

[2] DoF (Department of Fisheries), *Fisheries Statistical Yearbook of Bangladesh 2008-09. Fisheries Resources Survey System*, Department of Fisheries (Dof), Ministry of Fisheries and Livestock, Government of the People's Republic of Bangladesh, Dhaka, Bangladesh, 2011.

[3] F. L. Bryan, "Epidemiology of foodborne diseases transmitted by fish, shellfish and marine crustaceans in the United States, 1970-1978," *Journal of Food Protection*, vol. 43, pp. 859–876, 1980.

[4] E. J. Gangarosa, A. L. Bisno, E. R. Eichner et al., "Epidemic of febrile gastroenteritis due to Salmonella java traced to smoked whitefish," *American Journal of Public Health*, vol. 58, no. 1, pp. 114–121, 1968.

[5] K. Gnanambal and J. Patterson, "Biochemical and microbiological quality of frozen fishes available in Tuticorin supermarkets," *Fishery Technology*, vol. 42, no. 1, pp. 83–84, 2005.

[6] R. J. Wittman and G. J. Flick, "Microbial contamination of shellfish: prevalence, risk to human health, and control strategies," *Annual Review of Public Health*, vol. 16, pp. 123–140, 1995.

[7] Centers for Disease Control and Prevention (CDC), "Preliminary foodnet data on the incidence of infection with pathogens

transmitted commonly through food," *Morbidity and Mortality Weekly Report*, vol. 59, no. 14, pp. 418–422, 2010, http://www.cdc.gov/mmwr/preview/mmwrhtml/mm5914a2.htm.

[8] W. H. Andrews and T. S. Hammack, "Food sampling and preparation of sample homogenate," in *United States Food and Drug Administration (US FDA) Bacteriological Analytical Manual*, chapter 1, United States Food and Drug Administration, Silver Spring, Md, USA, 2001, http://www.fda.gov/Food/FoodScienceResearch/LaboratoryMethods/ucm063335.htm.

[9] L. J. Maturin and J. T. Peeler, "Aerobic plate count," in *Bacteriological Analytical Manual*, chapter 3, United States Food and Drug Administration (US FDA), 2001, http://www.fda.gov/Food/FoodScienceResearch/LaboratoryMethods/ucm063346.htm.

[10] C. H. Collins and M. P. Lyne, *Microbiological Methods*, Butterworth, London, UK, 5th edition, 1984.

[11] P. Feng, D. W. Stephen, and A. G. Michael, "Enumeration of *Escherichia coli* and the coliform bacteria," in *Bacteriological Analytical Manual*, chapter 4, United States Food and Drug Administration (US FDA), 2002, http://www.fda.gov/Food/FoodScienceResearch/LaboratoryMethods/ucm064948.htm.

[12] J. L. Oblinger and J. A. Koburger, "Understanding and teaching the most probable number technique," *Journal of Milk and Food Technology*, vol. 38, pp. 540–545, 1975.

[13] W. H. Andrews and T. S. Hammack, "*Salmonella*," in *United States Food and Drug Administration (US FDA) Bacteriological Analytical Manual*, chapter 5, United States Food and Drug Administration, Silver Spring, Md, USA, 2007, http://www.fda.gov/downloads/Food/FoodScienceResearch/UCM244774.pdf.

[14] R. E. Buchanan and N. E. Gibbons, *Bergeys Manual of Determinative Bacteriology*, The Williams and Wilkins Company, Baltimore, Md, USA, 8th edition, 1974.

[15] P. K. Surendran, N. Thampuran, and K. Gopakumar, "Microbial profile of cultured fishes 1 and prawn viz a viz their spoilage and contamination," in *Proceedings of the 9th Session of the Indo-Pacific 2 Fishery Commission Working Party on Fish Technology and Marketing*, D. James, Ed., vol. 3, pp. 1–12, FAO, Rome, Italy, 1995.

[16] C. A. Kaysner and D. J. Angelo, "*Vibrio*," in *United States Food and Drug Administration (US FDA) Bacteriological Analytical Manual*, chapter 9, United States Food and Drug Administration, Silver Spring, Md, USA, 2004, http://www.fda.gov/Food/FoodScienceResearch/LaboratoryMethods/ucm070830.htm.

[17] W. M. K. Bakr, W. A. Hazzah, and A. F. Abaza, "Detection of *Salmonella* and *Vibrio* species in some seafood in Alexandria," *Journal of American Science*, vol. 7, no. 9, pp. 663–668, 2011.

[18] ICMSF (International Commission of Microbiological Specification for Food), *Microorganisms in Food 2. Sampling for Microbiological Analysis: Principles and Specific Applications*, University of Toronto Press, Toronto, Canada, 2nd edition, 1986.

[19] C. E. Boyd, *Water Quality in Ponds for Aquaculture*, Alabama Agricultural Experiment Station, Auburn University, Auburn, Ala, USA, 1990.

[20] V. Suvanich, D. L. Marshall, and M. L. Jahncke, "Microbiological and color quality changes of channel catfish frame mince during chilled and frozen storage," *Journal of Food Science*, vol. 65, no. 1, pp. 151–154, 2000.

[21] N. Elhadi, S. Radu, C.-H. Chen, and M. Nishibuchi, "Prevalence of potentially pathogenic *Vibrio* species in the seafood marketed in Malaysia," *Journal of Food Protection*, vol. 67, no. 7, pp. 1469–1475, 2004.

[22] M. Iwamoto, T. Ayers, B. E. Mahon, and D. L. Swerdlow, "Epidemiology of seafood-associated infections in the United States," *Clinical Microbiology Reviews*, vol. 23, no. 2, pp. 399–411, 2010.

[23] Centers for Disease Prevention and Control (CDC), "Multistate outbreak of *Salmonella* paratyphi B variant L(+) tartrate(+) infections linked to frozen raw tuna," July 2015, http://www.cdc.gov/salmonella/paratyphi-b-05-15/index.html.

19

Effects of Acidification and Preservatives on Microbial Growth during Storage of Orange Fleshed Sweet Potato Puree

Joyce Ndunge Musyoka ⓘ,[1] George Ooko Abong',[1] Daniel Mahuga Mbogo,[2] Richard Fuchs,[3] Jan Low,[2] Simon Heck,[4] and Tawanda Muzhingi ⓘ[2]

[1]Department of Food Science, Nutrition and Technology, University of Nairobi, P.O. Box 29053-00625, Kangemi, Kenya
[2]International Potato Center (CIP), Sub-Saharan Africa (SSA) Regional Office, Old Naivasha Road, P.O. Box 25171-00603, Nairobi, Kenya
[3]Food and Markets Department, Natural Resources Institute of University of Greenwich, Central Avenue, Chatham Maritime, Chatham, Kent ME4 4TB, UK
[4]International Potato Center (CIP), Regional Office, Plot 106, Katalima Road, Naguru, P.O. Box 22274, Kampala, Uganda

Correspondence should be addressed to Tawanda Muzhingi; T.Muzhingi@cgiar.org

Academic Editor: Alejandro Castillo

Orange Fleshed Sweet Potato (OFSP) puree, a versatile food ingredient, is highly perishable limiting its use in resource constrained environments. It is therefore important to develop shelf-stable puree. A challenge test study was carried out to determine the effect of combinations of chemical preservatives and acidification on microbial growth in stored puree. Puree was prepared and treated as follows: control (**A**); 0.05% potassium sorbate+0.05% sodium benzoate+1% citric acid (**B**); 0.1% potassium sorbate+0.1% sodium benzoate+1% citric acid (**C**); 0.2% potassium sorbate+0.2% sodium benzoate+1% citric acid (**D**); 1% citric acid (**E**). Samples were inoculated with *Escherichia coli* and *Staphyloccocus aureus* at levels of 5.2 x 10^9 cfu/100g and 1.5 x 10^9 cfu/100g, respectively, before being evaluated during storage for 10 weeks at prevailing ambient temperature (15-25°C) and refrigeration temperature (4°C). Total aerobic counts, yeasts, and molds were also evaluated. *E. coli* and *S. aureus* counts declined significantly ($p<0.05$) by 4 log cycles in all puree treatments except for control and puree with only citric acid. Total viable count, yeasts, and molds were completely inhibited except for puree with only citric acid. Combination of chemical preservatives and acidification is effective in inhibiting pathogens and spoilage microorganisms in sweet potato puree.

1. Introduction

Sweet potato (*Ipomoea batatas*) is an important crop for food security and income generation in Sub-Saharan Africa (SSA) [1]. Sweet potato roots occur in various colors ranging from white, yellow, purple, and orange that is rich in β-carotene, an important pro-vitamin A carotenoid [2]. Biofortified orange fleshed sweet potato (OFSP) has been promoted in Kenya as an effective and sustainable source of vitamin A [3] that can be used to mitigate vitamin A deficiency (VAD), a major public health concern in the Western and Nyanza parts of Kenya [4].

In east Africa, OFSP roots are processed into puree (boiled and mashed) that is used at household level to make food products such as fried doughnuts (*mandazi*), *chapatti*,

and porridges for children. OFSP puree is being used as a partial substitute for wheat flour in bakery products in Rwanda and Kenya. There are several economic advantages of using OFSP puree compared to OFSP flour. According to research, 1.25 kg of OFSP fresh roots makes 1 kg of puree while 4-5 kg of OFSP fresh roots is needed to make 1 kg of flour [5]. The major challenge in the use of OFSP puree is that it is highly perishable and requires refrigeration during storage and distribution. An advanced technology of processing sweet potato puree in the USA is done through sterilization and aseptic packaging using a continuous flow microwave system [6] giving a shelf-life of up to 36 months. However, this system is expensive in low income countries such as Kenya where OFSP puree is processed on small-scale.

Therefore, there is an urgent need for developing shelf-stable OFSP puree which does not require refrigeration to ensure a continuous supply for bakeries throughout the year. Shelf-stable OFSP puree will also help ensure all year supply of OFSP puree in countries with one sweet potato growing season. One of the approaches in producing shelf-stable OFSP puree is by controlling the growth of spoilage and pathogenic microorganisms through the use of 'hurdles'. Hurdle technology is a method that ensures microbial stability and safety of foods as well as nutritional and sensory quality based on the application of several preservation factors [7]. The hurdles mainly used in food preservation include the use of preservatives, the type of packaging, temperature, water activity, and pH, among others [8].

Natural preservatives which give the much desired "clean label" have been shown to be effective against microbial growth in food products [9, 10]. However, natural preservatives such as nisin and natamycin are not cost-effective for small-scale processors in SSA. Chemical preservatives such as sodium benzoate (E211) and potassium sorbate (E202) are commonly used to retard or stop the growth of pathogenic microorganisms in food. Sodium benzoate and potassium sorbate are weak acids and their antimicrobial action is known to be due to the accumulation of protons and anions inside the microbial cell that disrupts normal metabolism [11]. Potassium sorbate and sodium benzoate are permitted in food products in levels of 0.1 % [12]. These two compounds have been used before in inhibiting microbial growth in food products. For instance, Jin [9] recorded the antimicrobial effect of sodium benzoate and potassium sorbate against *Escherichia coli* in strawberry puree. High acid foods such as fruit formulations require less heat treatment for their stability and therefore citric acid is added to maintain their pH at < 4.6 [13].

Citric acid (E330) also increases the acidity of the food product thus increasing the effectiveness of sodium benzoate and potassium sorbate since a greater proportion of the acids is in undissociated form [14]. Therefore, developing a shelf-stable OFSP puree using sodium benzoate, potassium sorbate, and citric acid may be effective as well as cost-effective. The quality and shelf-life of OFSP puree can be improved by application of alternative technologies for foodstuff packaging, distribution, and storage, such as modified atmosphere packaging (MAP) or vacuum packaging. MAP involves the replacement of the atmosphere surrounding food products with controlled mixtures of oxygen, nitrogen, and carbon dioxide before sealing in barrier materials [15]. During vacuum packaging the air is removed from within the packaging material and the product is enclosed with an airtight seal which prevents the return of air [16]. In SSA vacuum packing will be more cost-effective in the development of shelf-stable OFSP puree.

The potential growth of spoilage and pathogenic microorganisms in OFSP puree could affect its quality and shelf-life, as well as its safety. In order to determine the growth of pathogens in OFSP puree, a microbial challenge test (MCT) is important. This involves the inoculation of a food product with pathogenic microorganisms, storing the food under controlled conditions while evaluating the growth potential of the pathogens in the food product over time [17]. MCT is important as it helps in assessing whether a specific pathogen is able to grow in a food product if the properties of the food product such as pH are not able to control the particular pathogen [18]. MCT has been used to examine the growth potential of pathogens in foods in case of contamination. For instance, the growth and survival of pathogens such as *S. aureus* and *E. coli* in salad vegetables have been examined [17]. One important consideration in conducting a challenge test is selection of the appropriate pathogens. Contamination of OFSP puree with *S. aureus* could take place during postprocess handling of the puree by operatives [19]. This pathogen is commonly found on human skin and can be transferred to foods if good hygienic practices are not followed. OFSP puree could also be contaminated with *E. coli* through the use of nonpotable water for cleaning surfaces and equipment [20]. *E. coli* has been known to survive and persist on surfaces for months [21]. These are real scenarios during OFSP puree processing at small-scale level in rural settings.

Therefore, the current study was designed to examine survival and growth potential of *S. aureus* and *E. coli* in stored OFSP puree with preservatives. Conducting the challenge test in OFSP puree will help in establishing the ability of potassium sorbate, sodium benzoate, and citric acid to inhibit the growth of *S. aureus*, *E.coli*, total counts, yeasts, and molds in OFSP puree held under ambient (15-25°C) and refrigeration (4°C) conditions and therefore facilitate more extensive use of the puree. Results from this study will determine the feasibility of developing shelf-stable OFSP puree which can be stored without refrigeration. Shelf-stable OFSP puree will bridge the OFSP puree shortage during off seasons and will also help reduce postharvest losses for OFSP farmers.

2. Materials and Methods

2.1. Design and Methodology. The current study employed an experimental study design by preparing orange fleshed sweet potato (OFSP) puree and carrying out microbial evaluation. The experimental study design included two independent variables: use of different combinations of preservatives and two different storage temperatures. The dependent variables were *E. coli* counts, *S. aureus* counts, total aerobic counts, and yeast and mold counts.

2.2. Preparation of OFSP Puree. About 25 kg of root samples was randomly collected from a pool of five batches from farmers in Homabay-Kenya. Kabode variety which is the most adopted and widely grown OFSP variety was used for the current study. The processed puree was packaged in 5-kilogram polyethylene bags of 300 micron and vacuum sealed before being transported overnight to the Department of Food Science, Nutrition and Technology (University of Nairobi) laboratory for evaluation. Figure 1 shows a flow diagram of processing sweet potato puree. Sweet potato roots were sorted to remove the diseased and badly damaged roots before being washed to remove soil and loose dirt. The roots were then peeled manually using knives before being

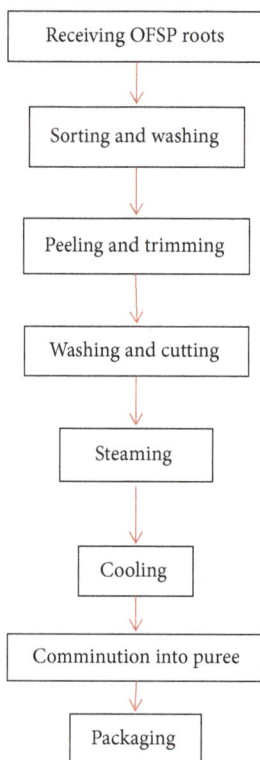

FIGURE 1: Preparation of OFSP puree.

trimmed to remove the fibrous ends, surface blemishes, and the diseased ends of the roots. The washed roots were cut into approximately 0.5-0.75 cm slices in thickness and steamed at 100°C for 30 minutes in a steaming pot and then cooled for 30 minutes at ambient temperature. Cooked and cooled roots were then comminuted into puree using a pureeing machine (OMAS Food Machinery, AEE1T0, Euro ingredients limited, Italy), packaged in 5-kilogram polyethylene bags, and vacuum sealed using vacuum packaging machine (MINIPACK-TORRE S.p.A MVS 45X, ANNO-2015-Euro ingredients, Italy). This activity was carried out in the current puree processing plant in Kenya and with the help of workers in the plant in order to ensure that the study utilized similar product usually produced in the set up.

2.3. Bacterial Load in OFSP Puree. Before conducting the challenge test, OFSP puree was assessed microbiologically for *E. coli* and *S. aureus* and then microwaved and assessed again. The initial enumeration of *E. coli* and *S. aureus* in the puree was carried out as described in previous studies [22]. Puree sample (25 g) was homogenized with 0.85 % NaCl and serial dilutions were prepared up to 10^{-6}. A volume of 0.1 mL from each dilution was spread in triplicate onto Brilliance *E. coli*/Coliform agar (Oxoid, Hampshire, England) and incubated at 37°C for 24 hours for enumeration of *E. coli*. Similarly, 0.1 mL of each dilution was spread in triplicate onto Baird parker agar (Oxoid, Hampshire, England) and incubated at 37°C for 48 hours for the enumeration of *S. aureus*. Enumeration was done for plates with 30-300

TABLE 1: Bacteria counts in OFSP puree (log cfu/g).

	Raw OFSP puree (log cfu/g)	OFSP puree after microwaving (Log cfu/g)
S. aureus	6.9 ± 0.04	nd∗
E. coli	5.7 ± 0.05	nd∗

Each value is mean ± standard deviation for triplicate experiments. nd∗: not detected.

colonies. All microbial counts were expressed as mean base-10 logarithms of colony forming units per gram (log cfu/g). Data points were expressed as means from the triplicate analysis. The results indicated high levels of *E. coli* and *S. aureus* in puree before treatment with preservatives. *E. coli* and *S. aureus* were not detected after microwaving OFSP puree as shown in Table 1. This formed the basis for the level of inoculation of *E. coli* and *S. aureus* into the puree.

2.4. Preparation of Bacterial Inoculum. *Escherichia coli* ATCC 8739 and *Staphylococcus aureus* ATCC 6538 pellets were obtained from the American Type Culture Collection (Microbiologics, MN 56303-USA). The pellets were activated by suspending a single pellet of each microorganism in phosphate buffer (0.1 M) and incubating it at 38°C for 30 minutes. From the buffer, 1mL was transferred to nutrient broth and incubated at 35°C for 24 hours to allow for growth of the bacteria. The levels of inoculum obtained after plating

Sample ID	Treatments
A	Puree without preservatives
B	Puree with 0.05 % sodium benzoate + 0.05 % potassium sorbate + 1 % citric acid
C	Puree with 0.1 % sodium benzoate + 0.1 % potassium sorbate + 1 % citric acid
D	Puree with 0.2 % sodium benzoate + 0.2 % potassium sorbate + 1 % citric acid
E	Puree with 1 % citric acid

were 5.2×10^9 cfu/mL for *E. coli* and 1.5×10^9 cfu/mL of *S. aureus*. The inoculum was then stored at -80°C to avoid changes that may affect growth [23].

2.5. Treatment of OFSP Puree Samples with Combination of Chemical Preservatives. OFSP puree was first sterilized for 3 minutes in a microwave. OFSP puree samples (100 g) were then dosed with combinations of selected chemical preservatives as shown in Table 2.

2.6. Inoculation Strategy and Growth Assessment. Puree samples (100 g) treated with preservative combinations and that without preservatives were inoculated with 1000 μL of bacterial suspension containing 5.2×10^9 cfu of *E. coli* and 1.5×10^9 cfu of *S. aureus* resulting in a load of 5.2×10^7 cfu (7.8 log cfu) of *E. coli*/g of puree and 1.5×10^7 cfu (7.2 log cfu) of *S. aureus*/g of puree and vacuum sealed. Some inoculated samples were incubated at ambient (15-25°C) and others at refrigeration temperatures (4°C) with enumeration of bacterial load at weekly intervals. Serial dilutions of all samples were prepared up to 10^{-6} and each dilution was plated in triplicate for the different tested microorganisms. The experiments were performed independently three times.

2.7. Determination of Total Viable Count (TVC) and Yeast and Molds in OFSP Puree during Storage. OFSP puree was prepared as shown in Figure 1 with the addition of selected chemical preservatives (as in Table 2) before packaging 100 g in polyethylene bags. Some of the puree was packaged without preservatives and analyzed for TVC and yeasts and molds. The puree samples (100 g) were then stored at ambient temperature of (15-25°C) and refrigeration temperature (4°C) with TVC and yeast and mold evaluation weekly for a period of 10 weeks of puree storage. A sample of puree (25 g) was placed into 225 mL of sterile saline solution (0.85 % NaCl), vortexed for 1 minute to homogenize, and serially diluted to a dilution of 10^{-7}. TVC was determined by transferring 1 mL of each sample dilution to sterile Petri dishes in triplicate to which approximately 20 mL of Plate Count Agar (PCA, LAB, UK) was added. The plates were swirled and allowed to solidify before being incubated at 30°C for 72 hours [24]. Yeasts and molds were determined by spread plating 0.1 mL of each sample dilution in triplicate onto Dichloran-Rose Bengal Chloramphenicol (DRBC) agar (Oxoid, Hampshire). The plates were incubated at 25°C for 5 days [25]. Enumeration was done for plates with 30-300 colonies. All microbial counts were expressed as mean base-10 logarithms of colony forming units per gram (log cfu/g). Data points

were expressed as mean from the triplicate experiments and results were expressed as logarithm of colony forming units per gram (log cfu/g). The experiment was performed three times independently.

2.8. Determination of PH in OFSP Puree during Storage. One gram of OFSP puree sample was homogenized in 1mL of distilled water in a test tube. The pH values of the samples were measured using pH meter (model HI 98107, USA) by immersing the electrode directly into the sample in the test tube. Before the measurements, pH meter was calibrated using pH 4.0 and 7.0 buffers.

2.9. Statistical Analysis. Experiments were carried out in triplicate and quality control measures were taken into account. Data was analyzed by analysis of variance (ANOVA) using SPSS software (Version 20.0 SPSS Inc). Tukey test was used to determine the significant difference of mean values. The significance level was expressed at 5 % level. Microsoft Excel was used to plot line graphs.

3. Results

3.1. Changes in PH in OFSP Puree during Storage. The initial pH of OFSP puree was 5.23 before treatment with preservatives. Addition of different combinations of sodium benzoate and potassium sorbate and citric acid led to a decline in pH of the puree kept at ambient (15-25°C) and refrigeration (4°C) conditions as shown in Figure 2.

Immediately after treatment of puree with preservatives, the highest pH values of 5.19 and 5.18 were obtained in the control sample at ambient and refrigeration temperatures, respectively, while least values of 4.60 and 4.61 were recorded in the sample with 1 % citric acid at ambient and refrigeration temperatures, respectively. At the end of the storage period (10 weeks), the highest pH values of 4.99 and 4.63 were recorded in the control sample, while least values of 3.94 and 3.98 were obtained in samples with 1 % citric acid at ambient and refrigeration temperatures, respectively.

3.2. Growth and Survival of Escherichia coli in Stored OFSP Puree. Figure 3 shows the growth of *E. coli* in OFSP puree treated with a combination of selected chemical preservatives and that without preservatives and stored at ambient (15-25°C) and refrigeration temperatures (4°C) for 10 weeks.

In nonsupplemented puree, *E. coli* counts increased significantly after inoculation by 2 logs with subsequent increase in storage from 7 log cfu/g to 9 log cfu/g. All combinations of potassium sorbate, sodium benzoate, and 1 % citric acid led

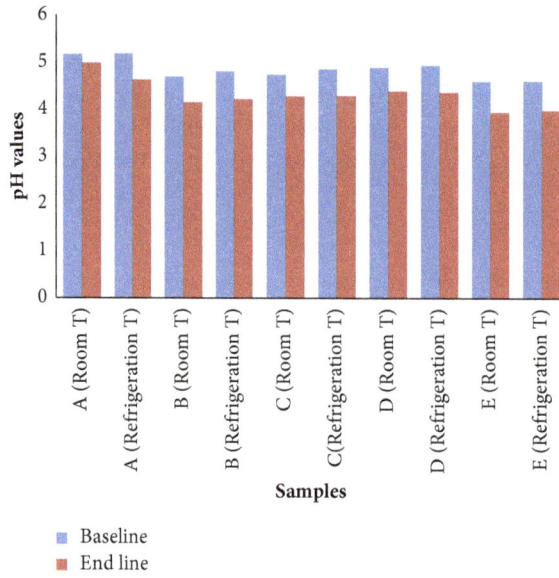

FIGURE 2: Changes in pH in OFSP puree during storage at ambient (15-25°C) and refrigeration temperature (4°C). **A**: puree without preservatives, **B**: puree with 0.05 % potassium sorbate+0.05 % sodium benzoate+1% citric acid, **C**: puree with 0.1 % potassium sorbate+0.1 % sodium benzoate+1% citric acid, **D**: puree with 0.2 % potassium sorbate+0.2 % sodium benzoate+1% citric acid, and **E**: puree with 1 % citric acid.

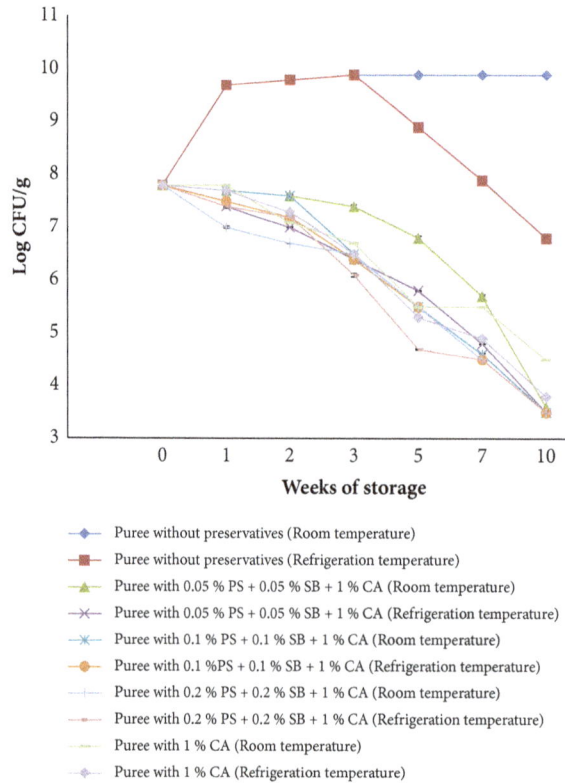

FIGURE 3: Growth of *E. coli* in OFSP puree with preservatives during storage at ambient temperatures (15-25°C) and refrigeration temperature (4°C). **PS**: potassium sorbate, **SB**: sodium benzoate, and **CA**: citric acid.

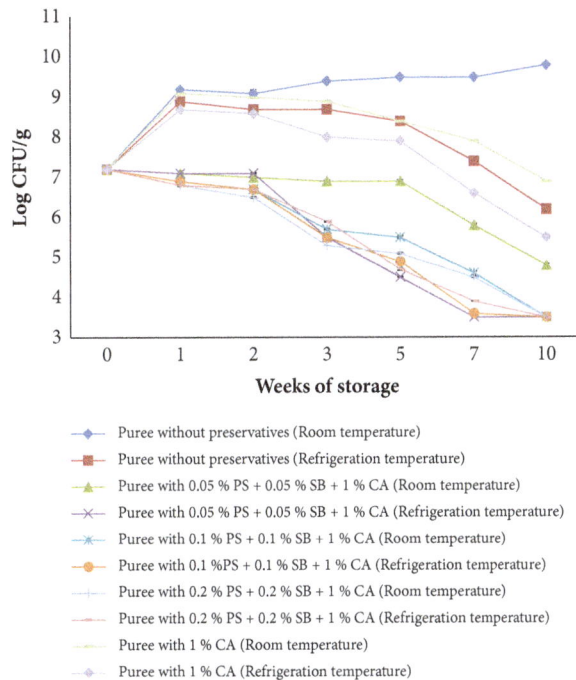

FIGURE 4: Growth of *S. aureus* in OFSP puree with preservatives during storage at ambient temperatures (15-25°C) and refrigeration temperature (4°C). **PS**: potassium sorbate, **SB**: sodium benzoate, and **CA**: citric acid.

to a significant (p<0.05) 4-log reduction in *E. coli* counts in puree kept at ambient temperature from 7 log cfu/g to 3 log cfu/g while 1 % citric acid gave a 3-log reduction from 7 log cfu/g to 4 log cfu/g.

Similarly, preservative treatment with all combinations of potassium sorbate, sodium benzoate, and citric acid led to a significant (p<0.05) reduction of *E. coli* by 4 log cycles (from 7 log cfu/g to 3 log cfu/g) in OFSP puree stored at refrigeration temperature. 1 % citric acid also resulted in a 4-log reduction in the numbers of *E. coli*. *E. coli* counts in nonsupplemented puree increased significantly by 2 logs immediately after inoculation but declined after 3 weeks storage from 7 log cfu/g to 6 log cfu/g. There was no significant difference (p>0.05) in *E. coli* populations in OFSP puree with different combinations of sodium benzoate, potassium sorbate, and citric acid in the two storage conditions.

3.3. Growth and Survival of Staphylococcus aureus in Stored OFSP Puree. Figure 4 shows the growth of *S. aureus* in OFSP puree treated with a combination of selected chemical preservatives and that without preservatives and stored at ambient (15-25°C) and refrigeration temperatures (4°C) for a period of 10 weeks.

Combined use of sodium benzoate and potassium sorbate at different concentrations with 1 % citric acid led to a 4-log reduction in *S. aureus* counts in OFSP puree kept at ambient and refrigeration temperatures from 7 log cfu/g to 3 log cfu/g. *S. aureus* counts in nonsupplemented puree kept at ambient temperature recorded a 2-log increase immediately after inoculation with subsequent increase in counts during storage from 7 log cfu/g to 9 log cfu/g while that at

refrigeration recorded 2-log increase after inoculation with subsequent decline in counts during storage from 7 log cfu/g to 6 log cfu/g.

Treatment with 1 % citric acid recorded a reduction in *S. aureus* population by 1 log cycle at ambient temperature from 7 log cfu/g to 6 log cfu/g and by 2 log cycles at refrigeration temperature from 7 log cfu/g to 5 log cfu/g. Treatment of OFSP puree with potassium sorbate and sodium benzoate at different concentrations together with 1 % citric acid had a slightly greater effect on *S. aureus* growth compared to 1 % citric acid when used alone.

3.4. Total Viable Count (TVC) in Stored OFSP Puree. Figure 5 shows the growth of aerobic microorganisms in OFSP puree with and without preservatives and stored at ambient (15-25°C) and refrigeration temperatures (4°C) for 10 weeks.

OFSP puree was found to contain high levels of bacteria (9.0 log cfu/g) immediately after preparation. The counts declined significantly (p<0.05) from 9 log cfu/g to nondetectable levels at the end of the storage period in OFSP puree treated with different combinations of potassium sorbate and sodium benzoate together with 1 % citric acid both at ambient and refrigeration conditions. However, there was a decline in aerobic microorganisms from 9 log cfu/g to 3 log cfu/g in puree with 1 % citric acid at ambient temperature at week 10 of puree storage.

3.5. Levels of Yeasts and Molds in OFSP Puree during Storage. Figure 6 shows the growth of yeast and molds in OFSP puree with and without preservatives and stored at ambient (15-25°C) and refrigeration temperatures (4°C) for 10 weeks.

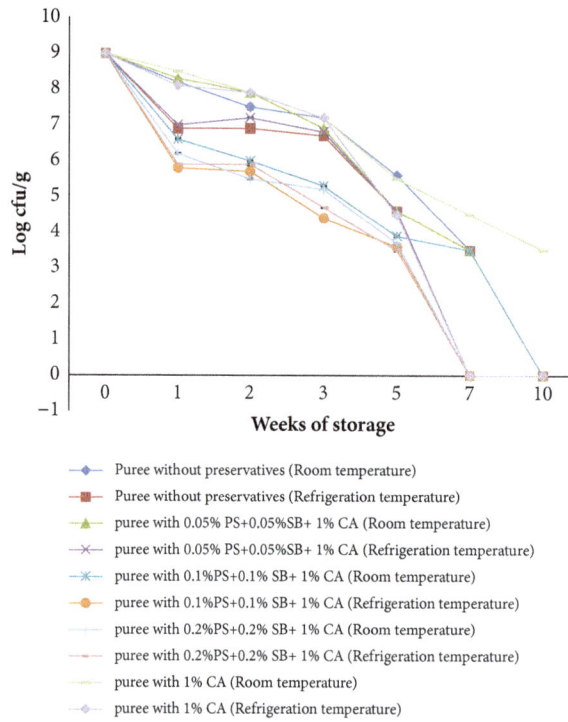

FIGURE 5: Total aerobic microorganisms in OFSP puree with preservatives during storage at ambient temperatures (15-25°C) and refrigeration temperature (4°C). **PS**: potassium sorbate, **SB**: sodium benzoate, and **CA**: citric acid.

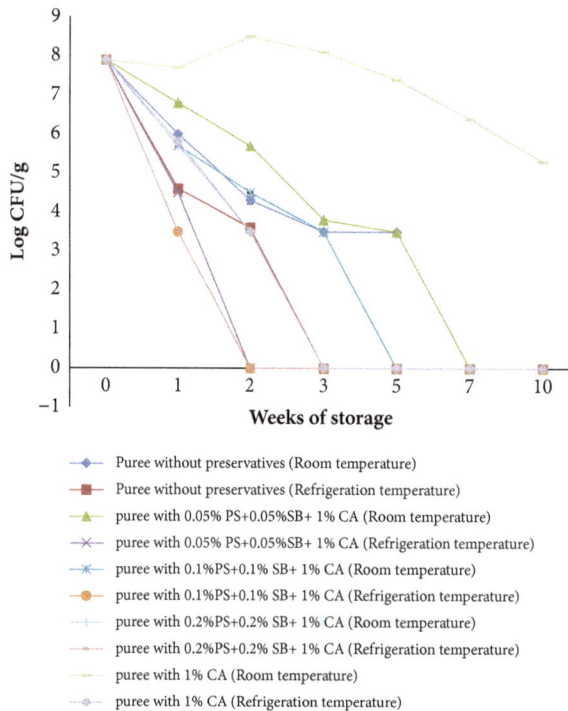

FIGURE 6: Microbial counts of yeast and molds in OFSP puree with and without preservatives during storage at ambient temperatures (15-25°C) and refrigeration temperature (4°C). **PS**: potassium sorbate, **SB**: sodium benzoate, and **CA**: citric acid.

The level of yeast and molds counts in OFSP puree after preparation was 7.92 log cfu/g. The counts declined significantly ($p<0.05$) with storage from 7.92 log cfu/g to nondetectable levels at week 7 of puree storage in all the treatments kept at ambient temperature except for the puree with 1 % citric acid in which yeast and mold counts declined from 7.92 log cfu/g to 5 log cfu/g at week 10 of puree storage. For the puree kept at refrigeration temperature, yeast and mold counts declined significantly ($p<0.05$) from 7.92 log cfu/g to nondetectable levels at week 3 of puree storage. Pack distention and alcoholic odors were noted in OFSP puree treated with citric acid only after one week of storage.

4. Discussion

4.1. Effect of PH on Microbial Growth in Orange Fleshed Sweet Potato (OFSP) Puree. Effectiveness of preservative is dependent on pH of the product [26] and pH is also one of the factors that determine the growth and survival of microorganisms during processing and storage [27]. The interest of food processors is to determine the pH of a food product and maintain that pH at a certain level in order to control microbial growth thus preventing product spoilage [28]. The reduction in pH of puree treated with preservatives was mainly because of addition of citric acid. pH is important to the antimicrobial effect of potassium sorbate and sodium benzoate because their effect is due to the undissociated form of their molecule which is dependent on pH [29]. Preservatives such as sodium benzoate and potassium sorbate have also been shown to have an effect on the pH of a food product and therefore the decrease in pH of puree during storage would also be attributed to the presence of preservatives [27].

4.2. Growth and Survival of Escherichia coli and Staphylococcus aureus in OFSP Puree. Information on the survival potential of pathogenic microorganisms in orange fleshed sweet potato (OFSP) puree is limited and such knowledge would be of significance in OFSP puree storage period by the consumer further contributing to food safety and quality. If postprocessing contamination of OFSP puree by *E. coli* and *S. aureus* would occur, the data presented indicate that these pathogens would grow extensively in the puree under ambient temperature assuming a 10-week storage period. The increase in *E. coli* and *S. aureus* counts in nonsupplemented puree (control) can be attributed to nutrient availability and favorable environment for their growth [30] such as pH and water activity.

The decline in *E. coli* and *S. aureus* populations in stored OFSP puree with preservatives can be attributed to various factors. Sodium benzoate and potassium sorbate activity is largely dependent on the pH of a food product. The optimum inhibitory activity of these preservatives takes place at low pH which favors the undissociated form of the molecule that freely moves across the plasma membrane into the cytoplasm [11]. The low pH of the puree was achieved through the addition of citric acid in combination with the preservatives. Due to the neutral pH of the cytoplasm, the acid dissociates into anions and protons. These molecules are not able to

diffuse back across the cell membrane and hence accumulate in the cytoplasm. Acidification of the cytoplasm and the energy depletion lead to physiological malfunction finally inhibiting microbial growth [31, 32]. In addition to the preservatives, the storage conditions and vacuum packaging of the puree also contributed towards inactivation of *E. coli* and *S. aureus* in the puree. The results are similar to those reported by Chikthimmah [33], who demonstrated that the use of chemical preservatives was critical for a significant reduction in *E. coli* counts in apple cider stored under ambient and refrigeration conditions.

The antimicrobial activity of citric acid was due to reduction of pH of OFSP puree below the optimal range of pH values for *E. coli* and *S. aureus* growth which is 6-7 or due to the disruption of the pathogens' membrane permeability thus preventing entry of essential nutrients for its growth. Other researchers have demonstrated the effectiveness of citric acid in inhibiting the growth of *S. aureus*. For instance, Seo [34] found that 2 % citric acid was effective in reducing counts of *S. aureus* in chicken meat in 5 days. Abu-ghazaleh [35] showed that 0.03 % citric acid significantly inhibited *S. aureus* growth in growth medium after 24 hours of incubation. Similarly, the use of 2 % citric acid alone on chicken meat led to the decline in *E. coli* counts by 4 log cycles in 12 hours [34].

There were no significant differences ($p>0.05$) in *E. coli* and *S. aureus* populations in OFSP puree with different combinations of sodium benzoate, potassium sorbate, and citric acid in the two storage conditions. This suggests that even the lowest concentration of preservatives used in combinations was effective in inhibiting the growth of *E. coli* and *S. aureus* in OFSP puree during the storage period. There was a slightly better inhibition of *E. coli* and *S. aureus* in treatments with different combinations of potassium sorbate, sodium benzoate, and citric acid as compared to citric acid alone at ambient and refrigeration temperatures. This suggests that combination of a number of hurdles (preservative factors) gives higher or multiple inhibitory effects against microorganisms compared to a single hurdle. According to Lotte Dock [29], the effect of combined treatments with preservatives in apple cider was significantly greater than that of a single preservative used alone. For instance, antimicrobial activity against *E. coli* was enhanced through the combined use of 0.1 % potassium sorbate and 0.1 % sodium benzoate at 8°C with survival time being reduced by 50 % compared with the one with 0.1 % sodium benzoate alone [29].

Temperature is also known to be one of the significant factors affecting microbial growth in food products [36]. The enzyme activity of microorganisms is optimum at a certain temperature range beyond which the enzyme undergoes denaturation; thus microbial growth was inhibited. As expected, growth of *S. aureus* and *E. coli* was more rapid at 25°C compared to 4°C. This is because, at low temperatures, the fluidity of the cytoplasmic membrane of microorganisms is reduced, thus interfering with transport mechanisms [37]. Therefore, microbial growth rate increases with increasing temperature until the maximum temperature for growth is attained [38]. *E. coli* is able to grow at a temperature range of 4-45°C with an optimum of 37°C but can survive refrigeration and freezing temperatures. A study carried out on the effect of temperature on the growth of *E. coli* revealed that it can grow and survive well on a range of temperatures but can grow well at 37°C compared to other temperatures [39]. *S. aureus* on the other hand grows on a temperature range of 7-48°C with an optimum of 37°C [40].

4.3. Growth of Aerobic Microorganisms, Yeast, and Molds in OFSP Puree. The high levels of aerobic microorganisms in OFSP puree before treatment with preservatives would be attributed to poor handling during preparation. The reduction in levels of aerobic counts in OFSP puree samples treated with different combinations of potassium sorbate, sodium benzoate, and citric acid indicates the benefits of the combination of antimicrobial chemicals having multiple effects against bacterial growth in OFSP puree. Vacuum packaging of the puree eliminates oxygen which is essential for the growth of aerobic microorganisms thus eliminating them even in the nonsupplemented puree. Other researchers showed the effect of preservatives on the growth of aerobic microorganisms. For instance, Ogiehor & Ikenebomeh [41] recorded a decline in aerobic microorganisms in Garri product treated with 0.2 % sodium benzoate stored for 6 months at 30°C. Results obtained by Momoh [42] showed that 0.1 % sodium benzoate along with refrigeration was able to inhibit aerobic microorganism multiplication of up to 13 days of storage while at room temperature the inhibition lasted for only 4 days.

The complete inhibition of yeast and molds in OFSP puree with and without preservatives would be attributed to the vacuum packaging of the puree that eliminates oxygen and therefore prevents the growth of molds and oxidative yeasts since they do not grow in the absence of oxygen [43]. However, yeast and molds counts were still detected in puree with 1 % citric acid and kept at ambient temperature even at week 10 of puree storage. Pack distention and alcoholic odors were noted in OFSP puree treated with citric acid only after one week of storage. This could be attributed to the growth of fermentative yeasts and/or lactic acid bacteria in the puree metabolizing simple sugars into ethanol and carbon dioxide. According to Rawat [44], yeasts can grow at very low pH values. Other researchers have demonstrated the effect of preservatives on fungal growth in food products. For instance, Omojowo [45] reported that 3-5 % potassium sorbate led to a decline in levels of yeast and molds in smoked fish stored for 8 weeks. A study by Guynot [46] demonstrated that potassium sorbate at concentrations of 0.15-0.30 % was effective in preventing fungal growth.

5. Conclusion

The study sought to investigate the effect of different combinations of preservatives on microbial growth in OFSP puree at different storage conditions and time. Different combinations of sodium benzoate, potassium sorbate, and citric acid improved microbial keeping quality and inhibited the growth of *Escherichia coli* and *Staphylococcus aureus* in OFSP puree as indicated by the microbial challenge test. Use of citric acid alone was less effective in controlling the growth of these pathogens.

Disclosure

The manuscript has been presented before in a conference as shown in the following link: http://www.sweetpotatoknowledge.org/files/presentation-4-effects-acidification-preservatives-microbial-growth-storage-orange-flesh-sweetpotato-puree/.

Conflicts of Interest

The authors declare no conflicts of interest.

Acknowledgments

The authors would like to acknowledge International Potato Centre (CIP-ILRI), Kenya, for funding this research.

References

[1] C. Wheatley and C. Loechl, "Acritical review of sweetpotato processing research conducted by CIP and partners in Sub-Saharan Africa," *Social Sciences Working Paper Series*, vol. 2008, p. 48, 2008.

[2] G. O. Fetuga, K. Tomlins, A. Bechoff, F. O. Henshaw, M. A. Idowu, and A. Westby, "A survey of traditional processing of sweet potato flour for amala, consumption pattern of sweet potato amala and awareness of orange-fleshed sweet potato (OFSP) in South West Nigeria," *Journal of Food, Agriculture and Environment (JFAE)*, vol. 11, no. 3-4, pp. 67–71, 2013.

[3] A. Saltzman, E. Birol, H. E. Bouis et al., "Biofortification: Progress toward a more nourishing future," *Global Food Security*, vol. 2, no. 1, pp. 9–17, 2013.

[4] World Health Organization, *Global prevalence of vitamin A deficiency in populations at risk 1995-2005*, WHO global database on vitamin A deficiency, 2009.

[5] J. Low, A. Ball, P. J. van Jaarsveld, A. Namutebi, M. Faber, and F. K. Grant, *Assessing Nutritional Value And Changing Behaviours regarding Orange-Fleshed Sweetpotato Use in Sub-Saharan Africa*, Africa Transforming value Chains for food and Nutrition Security, 2015.

[6] L. E. Steed, V.-D. Truong, J. Simunovic et al., "Continuous flow microwave-assisted processing and aseptic packaging of purple-fleshed sweetpotato purees," *Journal of Food Science*, vol. 73, no. 9, pp. E455–E462, 2008.

[7] L. Leistner, "Basic aspects of food preservation by hurdle technology," *International Journal of Food Microbiology*, vol. 55, no. 1-3, pp. 181–186, 2000.

[8] S.-Y. Lee, "Microbial Safety of Pickled Fruits and Vegetables and Hurdle Technology," *Internet Journal of Food Safety*, vol. 4, pp. 21–32, 2004.

[9] T. Jin, H. Zhang, and G. Boyd, "Incorporation of preservatives in poiyiactic acid films for inactivating escherichia coli O157:H7 and extending microbiological shelf life of strawberry puree," *Journal of Food Protection*, vol. 73, no. 5, pp. 812–818, 2010.

[10] S. Theivendran, N. S. Hettiarachchy, and M. G. Johnson, "Inhibition of Listeria monocytogens by nisin combined with grape seed extract or green tea extract in soy protein film coated on turkey frankfurters," *Journal of Food Science*, vol. 71, no. 2, pp. M39–M44, 2006.

[11] A. López-Malo, J. Barreto-Valdivieso, E. Palou, and F. S. Martín, "Aspergillus flavus growth response to cinnamon extract and

sodium benzoate mixtures," *Food Control*, vol. 18, no. 11, pp. 1358–1362, 2007.

[12] A. C. Gören, G. Bilsel, A. Şimşek et al., "HPLC and LC-MS/MS methods for determination of sodium benzoate and potassium sorbate in food and beverages: Performances of local accredited laboratories via proficiency tests in Turkey," *Food Chemistry*, vol. 175, pp. 273–279, 2015.

[13] D. M. Barrett and B. Lloyd, "Advanced preservation methods and nutrient retention in fruits and vegetables," *Journal of the Science of Food and Agriculture*, vol. 92, no. 1, pp. 7–22, 2012.

[14] E. Mani-López, H. S. García, and A. López-Malo, "Organic acids as antimicrobials to control Salmonella in meat and poultry products," *Food Research International*, vol. 45, no. 2, pp. 713–721, 2012.

[15] C. Rowswell, "Food Standards Agency guidance on the safety and shelf-life of vacuum and modified atmosphere packed chilled foods with respect to non-proteolytic Clostridium June," 2016.

[16] K. W. McMillin, "Where is MAP Going? A review and future potential of modified atmosphere packaging for meat," *Meat Science*, vol. 80, no. 1, pp. 43–65, 2008.

[17] F. Feroz, J. D. Senjuti, and R. Noor, "Determination of Microbial Growth and Survival in Salad Vegetables through in Vitro Challenge Test," *International Journal of Nutrition and Food Sciences*, vol. 2, no. 6, p. 312, 2013.

[18] National Advisory Committee on Microbiological Criteria for Foods, "Parameters for Determining Inoculated Pack/Challenge," *Journal of Food Protection*, vol. 73, no. 1, pp. 140–202, 2010.

[19] S. P. Chawla and R. Chander, "Microbiological safety of shelf-stable meat products prepared by employing hurdle technology," *Food Control*, vol. 15, no. 7, pp. 559–563, 2004.

[20] S. G. D. N. L. Reddi, R. N. Kumar, N. Balakrishna, and V. S. Rao, "Microbiological quality of street vended fruit juices in Hyderabad, India and their association between food safety knowledge and practices of fruit juice vendors," *International Journal of Current Microbiology and Applied Sciences*, vol. 4, no. 1, pp. 970–982, 2015.

[21] S. A. Wilks, H. Michels, and C. W. Keevil, "The survival of Escherichia coli O157 on a range of metal surfaces," *International Journal of Food Microbiology*, vol. 105, no. 3, pp. 445–454, 2005.

[22] S. A. Baluka, R. Miller, and J. B. Kaneene, "Hygiene practices and food contamination in managed food service facilities in Uganda," *African Journal of Food Science*, vol. 9, no. 1, pp. 31–42, 2015.

[23] J. E. L. Corry, B. Jarvis, and A. J. Hedges, "Minimising the between-sample variance in colony counts on foods," *Food Microbiology*, vol. 27, no. 5, pp. 598–603, 2010.

[24] I. M. Pérez-Díaz, V.-D. Truong, A. Webber, and R. F. McFeeters, "Microbial growth and the effects of mild acidification and preservatives in refrigerated sweet potato puree," *Journal of Food Protection*, vol. 71, no. 3, pp. 639–642, 2008.

[25] A. Landl, M. Abadias, C. Sárraga, I. Viñas, and P. A. Picouet, "Effect of high pressure processing on the quality of acidified Granny Smith apple purée product," *Innovative Food Science and Emerging Technologies*, vol. 11, no. 4, pp. 557–564, 2010.

[26] D. Stanojevic, L. Comic, O. Stefanovic, and S. Solujic-Sukdolak, "Antimicrobial effects of sodium benzoate, sodium nitrite and potassium sorbate and their synergistic action in vitro," *Bulgarian Journal of Agricultural Science*, vol. 15, no. 4, pp. 307–311, 2009.

[27] N. Beales, "Adaptation of microorganisms to cold temperatures weak acid preservatives low pH and osmotic stress: a review," *Comprehensive Reviews in Food Science and Food Safety*, vol. 3, no. 1, pp. 1–20, 2004.

[28] S. C. Ricke, "Perspectives on the use of organic acids and short chain fatty acids as antimicrobials," *Poultry Science*, vol. 82, no. 4, pp. 632–639, 2003.

[29] L. L. Dock, J. D. Floros, and R. H. Linton, "Heat inactivation of Escherichia coli O157:H7 in apple cider containing malic acid, sodium benzoate, and potassium sorbate," *Journal of Food Protection*, vol. 63, no. 8, pp. 1026–1031, 2000.

[30] A. N. Olaimat and R. A. Holley, "Factors influencing the microbial safety of fresh produce: a review," *Food Microbiology*, vol. 32, no. 1, pp. 1–19, 2012.

[31] A. E. Ghaly, D. Dave, S. Budge, and M. S. Brooks, "Fish spoilage mechanisms and preservation techniques: Review," *American Journal of Applied Sciences*, vol. 7, no. 7, pp. 846–864, 2010.

[32] R. Hazan, A. Levine, and H. Abeliovich, "Benzoic acid, a weak organic acid food preservative, exerts specific effects on intracellular membrane trafficking pathways in Saccharomyces cerevisiae," *Applied and Environmental Microbiology*, vol. 70, no. 8, pp. 4449–4457, 2004.

[33] N. Chikthimmah, L. F. LaBorde, and R. B. Beelman, "Critical factors affecting the destruction of Escherichia coli O157:H7 in apple cider treated with fumaric acid and sodium benzoate," *Journal of Food Science*, vol. 68, no. 4, pp. 1438–1442, 2003.

[34] S. Seo, D. Jung, X. Wang et al., "Combined effect of lactic acid bacteria and citric acid on Escherichia coli O157:H7 and Salmonella Typhimurium," *Food Science and Biotechnology*, vol. 22, no. 4, pp. 1171–1174, 2013.

[35] M. A. Bayan, "Effects of ascorbic acid, citric acid, lactic acid, NaCl, potassium sorbate and Thymus vulgaris extract on Staphylococcus aureus and Escherichia coli," *African Journal of Microbiology Research*, vol. 7, no. 1, pp. 7–12, 2013.

[36] M. Islam, J. Chen, M. P. Doyle, and M. Chinnan, "Control of Listeria monocytogenes on Turkey frankfurters by generally-recognized-as-safe preservatives," *Journal of Food Protection*, vol. 65, no. 9, pp. 1411–1416, 2002.

[37] L. Valík, A. Medved'ová, B. Bajúsová, and D. Liptáková, "Variability of growth parameters of Staphylococcus aureus in milk," *Journal of Food and Nutrition Research*, vol. 47, no. 1, pp. 18–22, 2008.

[38] H. Fujikawa, A. Kai, and S. Morozumi, "A new logistic model for Escherichia coli growth at constant and dynamic temperatures," *Food Microbiology*, vol. 21, no. 5, pp. 501–509, 2004.

[39] M. T. Nguyen, "The effect of temperature on the growth of the bacteria Escherichia coli DH5a," *Biology Journal*, vol. 1, no. may, pp. 87–94, 2006.

[40] R. Lindqvist, S. Sylvén, and I. Vagsholm, "Quantitative microbial risk assessment exemplified by Staphylococcus aureus in unripened cheese made from raw milk," *International Journal of Food Microbiology*, vol. 78, no. 1-2, pp. 155–170, 2002.

[41] I. S. Ogiehor and M. J. Ikenebomeh, "Extension of shelf life of garri by hygienic handling and sodium benzoate treatment," *African Journal of Biotechnology*, vol. 4, no. 7, pp. 744–748, 2005.

[42] J. E. Momoh, C. E. Udobi, and A. A. Orukotan, "Improving the microbial keeping quality of home made soymilk using a combination of preservatives, pasteurization and refrigeration," *British Journal of Dairy Science*, vol. 2, no. 1, pp. 1–4, 2011.

[43] M. Barth, T. R. Hankinson, H. Zhuang, and F. Breidt, *Compendium of the Microbiological Spoilage of Foods and Beverages*,

Food Microbiology and Food Safety, Springer Science Business Media, New York, NY, USA, 2009.

[44] S. Rawat, "Food Spoilage?: Microorganisms and their prevention," *Asian Journal of Plant Science and Research*, vol. 5, no. 4, pp. 47–56, 2015.

[45] F. S. Omojowo, I. G. Libata, and I. J. Adoga, "Comparative assessment of potassium sorbate and sodium metabisulphite on the safety and shelf life of smoked catfish," *Nature and Science*, vol. 7, no. 10, pp. 10–17, 2009.

[46] M. E. Guynot, A. J. Ramos, V. Sanchis, and S. Marín, "Study of benzoate, propionate, and sorbate salts as mould spoilage inhibitors on intermediate moisture bakery products of low pH (4.5-5.5)," *International Journal of Food Microbiology*, vol. 101, no. 1-2, pp. 161–168, 2005.

20

Sensory Profile of Chihuahua Cheese Manufactured from Raw Milk

Sarai Villalobos-Chaparro, Erika Salas-Muñóz, Néstor Gutiérrez-Méndez, and Guadalupe Virginia Nevárez-Moorillón ⓘD

Facultad de Ciencias Químicas, Universidad Autónoma de Chihuahua, Chihuahua, 31125, Mexico

Correspondence should be addressed to Guadalupe Virginia Nevárez-Moorillón; vnevare@uach.mx

Academic Editor: Thierry Thomas-Danguin

Chihuahua cheese is a local artisanal cheese traditionally produced from raw milk. When this cheese is produced with pasteurized milk, cheesemakers complain that there are differences in taste and aroma as compared with traditional manufacturing. This work aimed to obtain a descriptive sensory analysis of Chihuahua cheese manufactured with raw milk under traditional conditions. Samples were collected in five cheese dairies at two different seasons (summer and autumn), and a Quantitative Descriptive Sensorial Analysis was done by a panel of trained judges. For aroma descriptors, cooked descriptor showed differences between dairies, and whey was different among dairies and sampling seasons ($P<0.01$); diacetyl, fruity ($P<0.01$), as well as free fatty acids, nutty and sulphur ($P<0.05$) descriptors varied between seasons. For flavour descriptors, bitter perception was different between dairies and seasons ($P<0.01$). Salty and creamy cheese was also different among dairies ($P<0.01$). A Principal Component Analysis for differences among dairies and sampling season demonstrated that the first three components accounted for 90% of the variance; variables were more affected by the sampling seasons than by the geographical location or if the dairy was operated by Mennonites. Chihuahua cheese sensorial profile can be described as a semi-matured cheese with a bitter flavour, slightly salted, and with a cream flavour, with aroma notes associated with whey and sour milk. Principal Component Analysis demonstrated season influence on flavour and aroma characteristics.

1. Introduction

Cheese has been produced for centuries, using the milk of many domestic animals, including cow, sheep, and buffalo. There are many types of cheese, all of them have been initially manufactured within a particular human community where a specific cheesemaking process was developed. Traditional cheese products are characterized by diversity in their manufacturing processes, and when raw milk is used, a rich and diverse microbiota is associated with its production. Traditional cheeses need to be studied considering the geographic location, history, and other social variables of the human communities where they are produced. Also, climate, milk characteristics, microbiota associated with cheese factories, and other environmental conditions contribute to specific and particular flavour and aroma of each product [1, 2]. In order to protect and preserve the production of traditional food products, certifications such as Protected Designation of Origin (PDO) and Protected Geographical Indication (PGI) have been awarded to food products that demonstrate their differences and particular social and cultural conditions of production. In Mexico, many traditional types of cheese have been elaborated for centuries with raw milk, but foodborne outbreaks associated with dairy products have led to mandatory use of pasteurized milk in the production of most cheese types, except some PDO aged cheeses (minimum six months of ripening period) [3].

Chihuahua cheese manufactured with unpasteurized milk has been produced for almost a century in the northern state of Chihuahua, Mexico, and is the result of the interaction between the Mennonite community and local farmers [4]. Chihuahua cheese manufacturing is similar to Cheddar cheese, but it is usually consumed without maturation; it is then considered as a young semi-hard cheese elaborated

TABLE 1: Sensory evaluation terms used for evaluation of Chihuahua cheese [13, 14].

Term	Definition	Reference
Cooked	Aromatics associated with cooked milk	Skim milk heated to 85°C for 30 min
Whey	Aromatics associated with fresh whey	Fresh whey
Diacetyl (butyric fat)	Aromatics associated with diacetyl	Diacetyl, 20 ppm
Fruity	Aromatics associated with different fruits	Fresh pineapple, Ethyl hexanoate, 20 ppm
Sulphur	Aromatics associated with sulphurous compounds	Boiled mashed egg
Free fatty acid	Aromatics associated with short-chain fatty acids	Butyric acid 20 ppm
Nutty	The nut-like aromatic associated with different nuts	lightly toasted unsalted nuts, wheat germ, unsalted wheat thins, roasted peanut oil extract
Bitter	Fundamental taste sensation elicited by caffeine, quinine	Caffeine 0.08% in water
Salty	Fundamental taste sensation elicited by salts	Sodium chloride 0.5% in water
Sweet	Fundamental taste sensation elicited by sugars	Sucrose 5% in water
Sour	Fundamental taste sensation elicited by acids	Citric acid 0.08% in water
Umami	Chemical feeling factor elicited by certain peptides and nucleotides	Monosodium glutamate 1% in water
Creamy	Characteristic flavour of cream milk	Cream milk

Concentrated chemical references were prepared in 95% ethanol.

with cow milk [5]. It is usually consumed within weeks of manufacturing and is usually not accepted by local consumers after a few months of maturation. Nowadays, different products are identified as Chihuahua cheese, including traditional cheese manufactured by the Mennonite community using either pasteurized or unpasteurized milk; traditional cheese manufactured by non-Mennonite farmers, as well as industrialized cheese, was elaborated outside Chihuahua state or even outside Mexico [6]. Chihuahua cheese manufacturing process has been reported previously, including time and temperature conditions throughout processing. A considerable variation in manufacturing conditions has been observed, due to nonstandardization of the process [7].

There have been few efforts to characterize and define the particular flavour and aroma characteristics of Chihuahua cheese, including cheese produced in Chihuahua state [5, 8, 9] as well as Chihuahua-type cheese [10]. Almanza-Rubio et al. [6] demonstrated in a study with consumers from the state of Chihuahua that there is a preference to consume cheese elaborated in Chihuahua State as compared with Chihuahua-type cheese elaborated elsewhere.

The traditional manufacturing process used raw milk as starting material, but because of government regulations [11], pasteurization has been included in the process, although cheesemakers claim that there are differences on flavour and aroma after milk pasteurization. In order to propose the use of pasteurised milk for the artisanal manufacturing of Chihuahua cheese, State government has an intensive training program in good manufacturing practices, as well as support to improve infrastructure in small and medium-size cheesemaking facilities. Therefore, it is essential to describe the sensorial profile of Chihuahua cheese manufactured with raw milk, to serve as a reference for cheese produced with pasteurised milk and with the addition of starter cultures.

Therefore, this work aimed at providing information for the development of a sensorial profile of Chihuahua cheese

manufactured with raw milk, as well as to describe the effect of season and area of food production.

2. Materials and Methods

2.1. Chihuahua Cheese Samples. In order to obtain a complete description of the sensorial characteristics of Chihuahua cheese elaborated by Mennonites and non-Mennonites farmers, five dairies located in the state of Chihuahua, Mexico, were selected for this study. Two locations were owned and operated by Mennonites (cheese dairies B and C) and three by non-Mennonites (A, D, and E). Dairies elaborated artisanal Chihuahua cheese using raw milk and without the addition of a starter culture; samples were taken during summer and autumn 2013, according to the procedure described in Sánchez-Gamboa et al. [7]. For each season, one block of cheese (1-2 kg) was obtained from the dairies and transported to the laboratory within 24 h of manufacture. In order to complete the initial maturation process used by cheesemakers, cheese blocks were stored at 18°C for three days. Cheese portions (250 g) were vacuum-packed using a commercial system, including their trademarked vacuum bags (Food Saver V2830, Sunbeam Products, Oklahoma City, USA) and stored at 6.5°C in a container under dark controlled conditions for three months before sensory analysis.

2.2. Panel Selection and Quantitative Descriptive Analysis of Chihuahua Cheese. After an initial selection from 30 university members [12], eight were included in the panel. Panel members (2 male, 6 female, ages 24-32) were further trained (minimum 75 h) using the SpectrumTM 15-point intensity scale [12] for cheese flavour and aroma, based on descriptors suggested by Drake et al. [13, 14] for Cheddar cheese. During training, panelists have presented references, as indicated in Table 1 as well as Chihuahua cheese samples for the identification of descriptive flavour and aroma terms.

TABLE 2: Effect of dairy and sampling season on the sensory scores of Chihuahua cheese flavour.

	Cooked	Whey	Diacetyl	Fruity	Free fatty acid	Sulphur	Nutty	Bitter	Salty	Sweet	Sour	Umami	Creamy
Dairy						**Summer**							
A	1.5[b]	5.2[a]	2.8[a]	1.4[ab]	3.0[a]	1.2[a]	1.1[a]	6.1[a]	3.5[a]	1.5[a]	3.8[a]	3.1[a]	3.4[abc]
B	3.3[a]	3.9[ab]	1.4[a]	0.9[ab]	0.8[b]	0.8[a]	0.5[a]	4.2[abc]	1.4[bc]	0.6[a]	3.8[a]	1.6[a]	2.8[bcde]
C	3.1[ab]	3.7[ab]	2.3[a]	0.6[b]	1.8[ab]	1.1[a]	0.4[a]	4.3[ab]	1.5[bc]	0.6[a]	2.9[a]	2.2[a]	3.5[ab]
D	2.7[ab]	3.3[b]	2.4[a]	0.6[b]	1.5[ab]	0.9[a]	1.3[a]	2.5[bc]	2.9[abc]	0.8[a]	2.0[a]	2.1[a]	2.0[bcde]
E	1.9[ab]	2.4[b]	2.1[a]	0.4[b]	1.1[ab]	1.1[a]	0.6[a]	2.1[c]	2.8[abc]	0.9[a]	2.4[a]	2.6[a]	1.1[e]
						Autumn							
A	2.1[ab]	3.3[b]	1.4[a]	1.1[ab]	0.8[ab]	0.2[a]	0.8[a]	2.9[bc]	1.9[abc]	0.3[a]	2.0[a]	2.3[a]	1.4[de]
B	2.5[ab]	2.8[b]	1.6[a]	1.3[ab]	1.1[ab]	0.7[a]	0.4[a]	3.8[bc]	1.2[c]	0.4[a]	2.1[a]	1.6[a]	2.5[bcde]
C	2.4[ab]	2.6[b]	2.3[a]	1.5[ab]	1.0[b]	0.4[a]	0.3[a]	2.4[bc]	1.9[abc]	1.0[a]	1.9[a]	2.1[a]	5.1[a]
D	2.1[ab]	2.9[b]	1.6[a]	2.0[a]	0.9[ab]	0.3[a]	0.3[a]	2.5[bc]	3.4[ab]	1.2[a]	2.3[a]	2.1[a]	3.1[bcd]
E	1.5[b]	3.0[b]	1.5[a]	1.1[ab]	1.0[ab]	1.2[a]	0.5[a]	2.2[bc]	2.1[abc]	0.8[a]	2.0[a]	2.5[a]	1.5[cde]
					F ratio from ANOVA								
Dairy	4.25**	4.81**	1.04	1.05	1.60	0.77	1.86	875**	6.08**	0.97	1.36	1.79	13.02**
Season	2.68	11.69**	4.50*	12.64**	5.84*	5.01*	4.98*	13.08**	1.60	0.65	9.31**	0.47	0.46
Dairy*Season	1.25	3.37*	1.28	2.72*	2.60*	0.83	1.27	4.57**	2.09	2.71*	1.60	0.33	5.61**

Values with the same letters in a column indicate that samples do not differ significantly at a significance level of 5%. F values identified with an asterisk show statistical differences at $P<0.05$, while those identified with two asterisks show differences at $P<0.01$.

Descriptors related to aroma terms were evaluated by nasal perception.

Cheese samples were cut into 2 cm by side, tempered for 30 minutes to 20 ($\pm$2) °C before evaluation, and presented in plastic cups with a randomly assigned code. For training purposes, panelists received information on Chihuahua cheese manufacturing, and the team evaluated and discussed their evaluation of cheese samples used for training, in order to minimize variability. Sensory evaluations were done in a sensory lab, with environmentally controlled booths and white lights. Sensory evaluations were carried out when cheese samples were stored for three months; therefore, evaluations of summer and autumn samples were done at different times. In each session, panel members evaluated three cheese samples; salt crackers and coffee beans were provided for the panelists to eliminate residual flavour and odour between each sample analysed. Data were collected on a paper ballot, with a nine-point scale.

2.3. Statistical Analysis. Data collected were subjected to analysis of variance (Block Full Factorial ANOVA design) for each sensorial parameter, using production area and sampling season as independent variables, and panelists were considered as blocks for a Complete Block Design. Tukey means analysis was used to differentiate groups, using a 5% significance level. Mean values obtained for each sensorial parameter were used for Quantitative Descriptive Analysis to generate radar charts for each dairy. A Principal Component Analysis (PCA) and a K-means clustering were also carried out to evaluate if there was a relationship between individual cheese samples and dairy or sampling season. In the graphic presented, samples are observed as dots [15]. Statistical

analysis was done using the statistical software Minitab 17 [16].

3. Results

The descriptive analysis of Chihuahua cheese was based on descriptors used by Drake et al. [13] (Table 1) that were used to train a sensorial panel composed of eight members. Mean score values for each descriptor are summarized in Table 2; statistical analysis was done to determine differences between dairies, sampling period, and the interaction of those two variables; the table also includes a summary of the results of analysis of variance for each descriptor.

Based on the statistical analysis (Table 2), the aroma descriptor "cooked" showed difference between dairies ($P<0.01$) being dairy B the one with the highest mean score (2.9). The aroma descriptor "whey" showed a highly significant difference for both, the dairy where the cheese was produced and the sampling season ($P<0.01$). Cheese sampled during summer had the highest value for this parameter (3.7 versus 2.9 for autumn), and dairy A obtained the highest value (4.2). The diacetyl aroma perceived by the sensory panel was different between seasons ($P<0.05$), but not among dairies; the same pattern was observed in the fruity ($P<0.01$), free fatty acids ($P<0.05$), nutty ($P<0.05$), and sulphur ($P<0.05$) descriptors of aroma.

For the flavour descriptors, bitter perception was statistically different between dairies (($P<0.01$) (A, B > C > D, E) as well as between seasons ($P<0.01$) (Summer 3.8, Autumn 2.7). Regarding the perception of salty in cheese, differences were observed between dairies (($P<0.01$). The creamy descriptor, which can be related to the highest concentration of milk

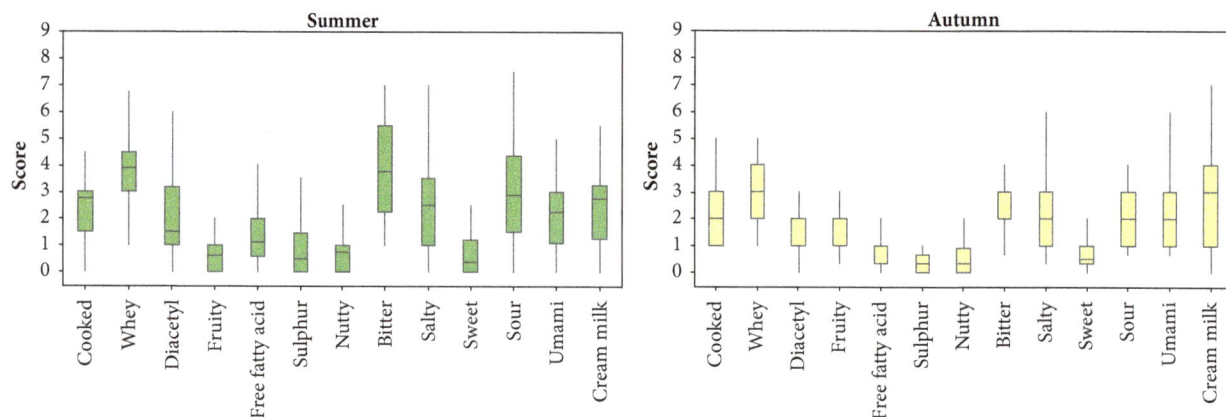

FIGURE 1: Boxplot of Quantitative Descriptive Analysis of Chihuahua cheese manufactured with raw milk and sampled during summer and autumn. Median, 1st and 3rd quartile, minimum, and maximum scores are presented for each descriptor.

fat in this dairy product, was perceived different among the dairies analysed (P<0.01) (C > B, D > A > E) but not among sampling seasons. No differences were presented for the sweet and umami descriptors among dairies or sampling season.

The descriptors that presented a significant interaction between dairy and sampling season are indicated with asterisks in Table 2. Cheese from dairy A sampled during summer had the highest scores for whey, diacetyl, free fatty acids, bitter, and salty terms. In general, the highest scores were presented for Chihuahua cheese manufactured during summer. The terms with the highest scores were bitter, creamy, and whey.

The effect of season on the flavour and aroma of Chihuahua cheese can be better observed in Figure 1, where the median, quartiles, and minimum and maximum score values for each descriptor are incorporated in a boxplot graphic, separated by season. Although in both seasons the descriptors with the highest scores are the same, score values for the cheese manufactured during autumn have lower values than those manufactured in summer, except for the fruity descriptor.

In order to relate all descriptors used in the sensorial analysis of Chihuahua cheese with seasonality and dairy analysed, a PCA was done using dairies and seasons as independent variables. The first three components explained 90.9% of data variation; PC1 accounted for 73.8% of the variation, while PC2 explained 10.3% and PC3 6.8% of variance. Figure 2 shows the position of each of the dairies in each season related to the three first components of the analysis. PC1 has an almost homogeneous positive load in all cheese dairies and sampling seasons (coefficient values 0.25-0.34), while PC2 has a positive correlation with cheese manufactured in dairy C (non-Mennonite owner) during autumn (0.539), and a negative correlation with cheese from dairy E, also from non-Mennonite owner produced in summer (-0.562). PC3 coefficients were positive for Mennonite owned dairies and were negative correlated to cheese produced in autumn from dairies C and D (-0.50 and -0.60, respectively).

K-means analysis identified three groups that are more related to the season of sampling than to dairy or even

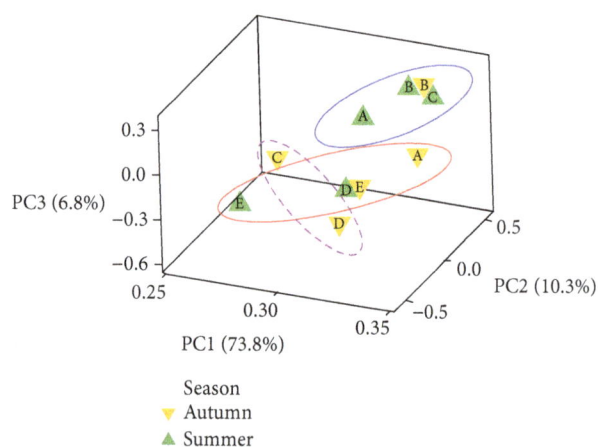

FIGURE 2: Principal Component Analysis for dairies and sampling period of Chihuahua cheese for sensory analysis. Scores of each sample for the three first Principal Component Analyses (PCA) are shown.

geographical location of each farm. One of the groups included dairies A and E sampled during autumn, as well as summer dairies D and E. Another group included cheese samples from dairy B autumn, and dairies A, B, and C summer. The third group identified included autumn samples from dairies C and D.

4. Discussion

4.1. Aroma and Flavour Descriptors for Chihuahua Cheese. In order to generate a sensorial profile of Chihuahua cheese manufactured with raw milk, the aroma and flavour descriptors proposed by Drake et al. [13] for Cheddar cheese were used in this study. The term creamy was also added, based on previous analysis (data not published).

Aroma descriptors included the terms cooked, whey, diacetyl, fruity, free fatty acid, sulphur, and nutty. The last two terms obtained low scores by the sensory panel; therefore, they are not related to specific aroma notes in Chihuahua cheese. "Sulphur" aroma has been associated with Cheddar cheese [13], related to the release of sulphur compounds from amino acid breakdown, and is commonly found in riped cheeses [17]. The term "nutty" is given by volatile compounds also derived from amino acid metabolism that have been reported in Emmental or Gruyere cheese [18]. Such cheese types are also consumed after long ripening.

The aroma terms of "diacetyl", "free fatty acids", and "fruity" are related to the metabolism of fatty acids, either by auto-oxidation to give the aroma related to butter, the release by lipolysis of volatile fatty acids such as butanoic, hexanoic, or octanoic acid, or fruity notes derived from esterification reactions of alcohol and fatty acids [19]. The notes related to fruits have been associated to Italian cheese, where they balance the presence of free fatty acids [20] and have also been associated with Cheddar, as well as with the Italian cheeses, Grana Padano, and Ragusano [21, 22]. These three terms were identified as different between sampling seasons, but not between dairies, which can be related to the difference in milk composition due to variation in the composition of the cow's milk through the year.

The term "cooked" is related to the presence of compounds derived from the Maillard reaction and is expected to be higher in cheese manufactured with pasteurized milk; although all cheese samples studied in this work were elaborated with raw milk, differences among dairies can be related to physicochemical characteristics of the milk used or to differences in processing conditions [7]. The term "whey" relates to the smell of fresh whey developed during cheese production that contains sugar, proteins, and fatty acids not retained in the curd [23]. Mean scores for this descriptor had a high mean score in the samples analysed; therefore, this descriptor can be associated with Chihuahua cheese. The term was also different among dairies and seasons, so is one of the most important aroma descriptors related to the differences in Chihuahua cheese production. The differences can be related to the milk composition and the manufacturing processes.

Flavour descriptors evaluated included the five basic tastes and creamy. The term "bitter" can have different sources related to their perception including ripening time, microorganisms present and their metabolism, or milk composition. This term had a high mean score in Chihuahua cheese, as well as differences among dairies and seasons. Chihuahua cheese was also identified as "salty" by the sensory panel; differences among dairies can be related to differences in traditional manufacturing of Chihuahua cheese [7]. "Cream-like" flavour is related to fresh cheese types rather than to ripped ones. Chihuahua cheese has a high score for creamy that is influenced by the dairy and by the season of manufacturing.

"Sweet" and "umami" descriptors did not show statistical differences among dairies or between sampling seasons; "sour" descriptor was different among seasons (Table 2), and Chihuahua cheese had relatively high mean scores for sour and umami terms. The umami descriptor is related to glutamate presence, and the concentration of this aminoacid increases during ripening. Aged Cheddar cheese has been reported to increase tenfold the concentration of glutamate after eight months of ripening [24].

4.2. Quantitative Descriptive Analysis (QDA). Differences for the descriptors between seasons can be observed in Figure 1, which summarizes the QDA analysis of Chihuahua cheese included in this study. Chihuahua cheese can be identified with a whey aroma, and flavour associated with bitter, sour, and creamy descriptors. It is important to consider the ripening time of three months applied to samples analysed taking into account that Chihuahua cheese is usually consumed within four months of production. As opposed to Cheddar cheese that is consumed after 8-12 months of maturation, Chihuahua cheese is preferred as a semi-matured cheese. The main characteristics appreciated by consumers are its stretchability and melting properties [4]. The rheological profile of Chihuahua cheese has been more related to fresh Colby than to Cheddar or Chester [25].

Previous reports have described Chihuahua cheese with salty, sour, and bitter flavour and diacetyl, cooked, whey as main descriptors for aroma in cheese manufactured with pasteurized milk and with intense notes of sour and bitter in cheese manufactured with raw milk; the authors reported differences among manufacturers [8, 9]. In another report, Chihuahua cheese manufactured with raw milk was described as having characteristics of young, basic cheese with slight bitter notes [5]. On the other hand, Hernández-Morales et al. [2] included the descriptors bitter and sour (mean scores of 2.2 and 3.4, respectively) for Añejo (aged) Mexican cheese; these values are similar to the scores presented in samples of Chihuahua cheese manufactured during summer.

In a recent report by Lopez-Díaz and Martinez-Ruiz [26], the authors describe a sensorial profile from commercial samples of Chihuahua cheese by trained judges and consumer preference tests. They used cheese manufactured in the state of Chihuahua that was obtained from retail supermarkets, but there was no information on their manufacture conditions. Based on the results from consumer preferences, cheese was described by their trained panelists as intense aroma, cooked, butter, and fresh milk aroma descriptors. On the other hand, consumers did not appreciate high acidity, salty taste, weak odour, high-fat content, or bitter taste in Chihuahua cheese. In contrast to the report by Lopez-Díaz and Martínez-Ruiz [26], we tested Chihuahua cheese produced with raw milk, and we have previously reported their microbiological and physicochemical profile [7, 27].

Principal Component Analysis can help on the identification of association not distinguished at first sight of the analysed variables. We used PCA to determine if sampling season or if the fact that Mennonite or non-Mennonite owners operated the dairies can affect the sensorial profile of Chihuahua cheese. The result of plotting the response for the first three components (Figure 2) demonstrated that PC1, PC2, and PC3 explained 90.9% of variance, and three groups were identified (K-means analysis) with higher influence of sampling season on the group formation, rather than if the cheesemaker was either Mennonite or non-Mennonite, or the dairy geographical location. Contrary to what we report here, Olson et al. [28] described that season was not important in the functional properties of Chihuahua cheese; instead, aging or the use of pasteurized or raw milk had an effect on cheese functional properties, although the effect of seasonality was previously reported for rheological characteristics [29].

Mexican sanitary authorities are requesting cheese producers to use pasteurized milk as starter material; therefore, it was important to describe the sensory profile of the few farmers that still manufactured Chihuahua cheese using raw milk and with traditional techniques [1]. The composition of the microbiota responsible for cheese production and ripening has an essential role in the development of a cheese's sensory profile cheese. Although commercial starter cultures are mixed to provide similar cheese characteristics than those presented in traditional manufacturing, cheese manufactured with indigenous bacteria presented stronger flavour descriptors in buffalo and cow milk as compared to commercial cultures [30].

The complex mixture of geographical, seasonal, and environmental characteristics associated with cheese production can influence the particular sensory profile of cheese. In a report on semi-matured and matured Cheddar cheese, consumers were able to differentiate cheeses elaborated in different farms, even when they were only 80 km apart, regardless of the use of pasteurized or raw milk for Cheddar cheese manufacturing [31]. Almanza et al. [6] carried out a consumer preference analysis and demonstrated that the average consumer from Chihuahua City preferred the cheese manufactured in the state of Chihuahua as opposed to cheese elaborated in other Mexican states or even in other countries. All the information related to Chihuahua cheese characteristics, rheological, microbiological, and sensory properties can help on the search for a certificate of origin of denomination, which will help on the preservation of the traditional manufacturing process [3].

5. Conclusions

The sensorial analysis of Chihuahua cheese produced in traditional cheese factories, using raw milk as starting material, can be used to create a sensorial profile of the product, which can be used to describe the distinctive characteristics of Chihuahua cheese better. Based on the Quantitative Descriptive Analysis, the sensory profile of Chihuahua cheese can be described as a semi-matured cheese with a bitter flavor, slightly salted and with a cream flavor, with aroma notes associated with whey and sour; all this related to its short ripening time.

The season of production of Chihuahua cheese had a considerable influence on flavour and aroma characteristics, more than dairies' geographical location.

Conflicts of Interest

The authors declare that there are no conflicts of interest regarding the publication of this paper.

Acknowledgments

The authors want to thank Comisión Estatal para la Protección contra Riesgos Sanitarios (COESPRIS) of Chihuahua State, Mexico, for the facilities provided for the visit and sampling in the dairies studied in this work. This work had financial support provided by CONACYT [Grant CB-2011-168960]. SVC also had a CONACYT scholarship for her graduate studies [Fellowship No. 269448].

References

[1] G. Licitra, "World wide traditional cheeses: Banned for business?" *Dairy Science & Technology*, vol. 90, no. 4, pp. 357–374, 2010.

[2] C. Hernández-Morales, A. Hernández-Montes, E. Aguirre-Mandujano, and A. V. De Gante, "Physicochemical, microbiological, textural and sensory characterisation of Mexican Añejo cheese," *International Journal of Dairy Technology*, vol. 63, no. 4, pp. 552–560, 2010.

[3] K. Koppel and D. H. Chambers, "Flavor Comparison of Natural Cheeses Manufactured in Different Countries," *Journal of Food Science*, vol. 77, no. 5, pp. S177–S187, 2012.

[4] N. Gutiérrez-Méndez and GV. Nevárez-Moorillón, "Chihuahua cheese: the history of a Mexican cheese," *Carnilac Industrial*, vol. 24, pp. 27–34, 2009.

[5] M. Paul, A. Nuñez, D. L. Van Hekken, and J. A. Renye, "Sensory and protein profiles of Mexican Chihuahua cheese," *Journal of Food Science and Technology*, vol. 51, no. 11, pp. 3432–3438, 2014.

[6] J. L. Almanza-Rubio, R. E. Orozco-Mena, and N. Gutiérrez-Méndez, "Assessing consumer preference toward Chihuahua cheese and Chihuahua-type cheese," *Tecnociencia Chihuahua*, vol. 7, pp. 123–131, 2013.

[7] C. Sánchez-Gamboa, L. Hicks-Pérez, N. Gutiérrez-Méndez, N. Heredia, S. García, and G. V. Nevárez-Moorillón, "Seasonal influence on the microbial profile of Chihuahua cheese manufactured from raw milk," *International Journal of Dairy Technology*, vol. 71, pp. 81–89, 2018.

[8] D. Van Hekken, M. Drake, F. M. Corral, V. G. Prieto, and A. Gardea, "Mexican Chihuahua Cheese: Sensory Profiles of Young Cheese," *Journal of Dairy Science*, vol. 89, no. 10, pp. 3729–3738, 2006.

[9] D. L. Van Hekken, M. A. Drake, M. H. Tunick, V. M. Guerrero, F. J. Molina-Corral, and A. A. Gardea, "Effect of pasteurization and season on the sensorial and rheological traits of Mexican Chihuahua cheese," *Dairy Science & Technology*, vol. 88, no. 4-5, pp. 525–536, 2008.

[10] J. Alvaradobr, D. Almarazbr, and L. Rivera, "Determination of the quality of cheese 'Chihuahua' Type: Sensory and physicochemical approaches," *Emirates Journal of Food and Agriculture*, vol. 25, no. 6, p. 409, 2013.

[11] Diario Oficial de la Federación (2010). NOM-243-SSA1-2010, Productos y servicios. Leche, fórmula láctea, producto lácteo combinado y derivados lácteos. Disposiciones y especificaciones sanitarias. Métodos de prueba.

[12] M. C. Meilgaard, B. T. Carr, and G. V. Civille, *Sensory Evaluation Techniques*, CRC Press, Boca Raton, FL, USA, 2006.

[13] M. Drake, S. Mcingvale, P. Gerard, K. Cadwallader, and G. Civille, "Development of a Descriptive Language for Cheddar

Cheese," *Journal of Food Science*, vol. 66, no. 9, pp. 1422–1427, 2001.

[14] M. A. Drake, P. D. Gerard, J. P. Kleinhenz, and W. J. Harper, "Application of an electronic nose to correlate with descriptive sensory analysis of aged Cheddar cheese," *LWT- Food Science and Technology*, vol. 36, no. 1, pp. 13–20, 2003.

[15] D. S. Kemp, D. T. Hollowood, and D. J. Hort, *Sensory Evaluation*, Wiley-Blackwell, Oxford, UK, 2009.

[16] Minitab 17 Statistical Software. 2016. [Computer software]. State College, PA: Minitab, Inc. (www.minitab.com).

[17] W. Bockelmann, "Secondary Cheese Starter Cultures," in *In Law BA and Tamime AY Technology of Cheesemaking*, B. A. Law and A. Y. Tamime, Eds., pp. 193–230, Wiley-Blackwell, Oxford, UK, 2nd edition, 2010.

[18] J. B. Lawlor, C. M. Delahunty, M. Wilkinson, and G. Sheehan, "Relationships between the odor and flavor attributes and the volatile composition and gross composition of cheese," in *Food Flavors and Chemistry*, Special Publications, pp. 108–117, Royal Society of Chemistry, Cambridge, 2001.

[19] M. Kim, S. Drake, and M. Drake, "Evaluation of Key Flavor Compounds in Reduced- And Full-Fat Cheddar Cheeses Using Sensory Studies on Model Systems," *Journal of Sensory Studies*, vol. 26, no. 4, pp. 278–290, 2011.

[20] C. Vialloninsta, B. Martin, I. Verdier-Metz et al., "Transfer of monoterpenes and sesquiterpenes from forages into milk fat," *Le Lait*, vol. 80, no. 6, pp. 635–641, 2000.

[21] F. Bellesia, A. Pinetti, U. M. Pagnoni et al., "Volatile components of Grana Parmigiano-Reggiano type hard cheese," *Food Chemistry*, vol. 83, no. 1, pp. 55–61, 2003.

[22] L. Moio and F. Addeo, "Grana Padano cheese aroma," *Journal of Dairy Research*, vol. 65, no. 2, pp. 317–333, 1998.

[23] T. Janhøj and K. B. Qvist, "The Formation of Cheese Curd," in *Technology of Cheesemaking*, B. A. Law and A. Y. Tamime, Eds., pp. 130–165, Wiley-Blackwell, Oxford, UK, 2nd edition, 2010.

[24] O. G. Mouritsen and K. Styrb, "Aged, Dried and Hard Cheeses," in *Umami: Unlocking secrets of the Fifth Taste*, Columbia University Press, New York, NY, USA, 2014.

[25] D. L. Van Hekken, M. H. Tunick, P. M. Tomasula, F. J. Corral, and A. A. Gardea, "Mexican Queso Chihuahua: rheology of fresh cheese," *International Journal of Dairy Technology*, vol. 60, no. 1, pp. 5–12, 2007.

[26] J. A. Lopez-Diaz and N. D. R. Martinez-Ruiz, "Sensorial and physicochemical profile of Chihuahua cheese considering consumer preferences," *Agrociencia*, vol. 52, no. 3, pp. 361–378, 2018.

[27] C. Sánchez-Gamboa, L. Hicks-Pérez, N. Gutiérrez-Méndez, N. Heredia, S. García, and G. Nevárez-Moorillón, "Microbiological Changes during Ripening of Chihuahua Cheese Manufactured with Raw Milk and Its Seasonal Variations," *Foods*, vol. 7, no. 9, p. 153, 2018.

[28] D. Olson, D. Van Hekken, M. Tunick, P. Tomasula, F. Molina-Corral, and A. Gardea, "Mexican Queso Chihuahua: Functional properties of aging cheese1," *Journal of Dairy Science*, vol. 94, no. 9, pp. 4292–4299, 2011.

[29] M. H. Tunick, D. L. Van Hekken, J. Call, F. J. Molina-Corral, and A. A. Gardea, "Queso Chihuahua: effects of seasonality of cheesemilk on rheology," *International Journal of Dairy Technology*, vol. 60, no. 1, pp. 13–21, 2007.

[30] M. Murtaza, S. Rehman, F. Anjum, and N. Huma, "Descriptive sensory profile of cow and buffalo milk Cheddar cheese prepared using indigenous cultures," *Journal of Dairy Science*, vol. 96, no. 3, pp. 1380–1386, 2013.

[31] G. Turbes, T. D. Linscott, E. Tomasino, J. Waite-Cusic, J. Lim, and L. Meunier-Goddik, "Evidence of terroir in milk sourcing and its influence on Cheddar cheese," *Journal of Dairy Science*, vol. 99, no. 7, pp. 5093–5103, 2016.

21

Viability of Molds and Bacteria in Tempeh Processed with Supercritical Carbon Dioxides during Storage

Maria Erna Kustyawati (ID),[1] Filli Pratama (ID),[2] Daniel Saputra (ID),[2] and Agus Wijaya[2]

[1]*Department of Agriculture Product Technology, University of Lampung, Bandar Lampung 34145, Indonesia*
[2]*Department of Agriculture Technology, University of Sriwijaya, Palembang 30662, Indonesia*

Correspondence should be addressed to Maria Erna Kustyawati; maria.erna@fp.unila.ac.id

Academic Editor: Vita Di Stefano

Application of supercritical carbon dioxide for processing of food products has an impact on microbial inactivation and food quality. This technique is used to preserve tempeh due to no heat involved. The quality of tempeh is highly influenced by mold growth because of its role in forming a compact texture, white color, and functional properties as well as consumer acceptance. This study aims to observe viability of molds and bacteria in tempeh after processed with supercritical CO_2 and to determine the best processing conditions which can maintain mold growth and reduce the number of bacteria in tempeh. For that purpose, tempeh was treated using high pressure CO_2 at 7.6 MPa (supercritical CO_2) and at 6.3 MPa (sub/near supercritical CO_2) with incubation period of 5, 10, 15, and 20 min. The best treatment obtained was used to process tempeh for storage study. The results showed that there was a significant interaction between pressure and incubation period for bacterial and mold viability at $\rho > 0.05$. Reduction of bacteria and molds increased with longer incubation period. Molds were undetectable after treatment for 20 min with either supercritical CO_2 or sub-supercritical, and bacteria significantly reduced up to 2.40 log CFU/g. On the other hand, sub-supercritical CO_2 for 10 min was the best processing method because molds survived 4.3×10^4 CFU/gram after treatment and were able to grow during storage at 30°C, producing white mycelium as indicated by increasing the $L*$ color value and tempeh acceptability. The inactivation of mold was reversible causing it to grow back during storage under suitable conditions. Tempeh matrix composition can provide protection against the destructive effects of supercritical CO_2. Gram-positive bacteria were more resistant than Gram-negative. In conclusion, sub-supercritical CO_2 can act as a method of cold pasteurization of tempeh and can be used as an alternative method to preserve tempeh.

1. Introduction

Consumer needs for food are not only in terms of health and food safety but also food with minimal processing that can maintain the quality of freshness and taste for a certain length of storage. Thermal food preservation is an effective technique for reducing microbial count of foods. However, for heat sensitive food products it can give undesirable sensorial changes and destroy the nutritional quality of the food. High pressure carbon dioxide technology is a nonthermal alternative processing to improve the microbial safety of the product while preserving nutritional and sensorial characteristics. It is known that carbon dioxide under the supercritical phase (7.4 M.Pa and 31.06°C) has unique properties. Carbon dioxide has dual characteristics where it is like a gas with high diffusivity and a liquid with high solubility which enable it to easily diffuse through complex matrices and extract substances "Liao [1]". The supercritical CO_2 ($scCO_2$) characteristic has expanded its use for the inactivation of various vegetative microorganisms in food as a nonthermal technology without loss of taste, color, and nutrients "Calvo and Torres [2]". Sub-supercritical carbon dioxides (sub-$scCO_2$) treatment causes microbial inactivation and can avoid changes in sensory attributes of food quality. In relation to microbial growth and food processing, "Garcia-Gonzales [3]" and "Guo [4]" found that carbon dioxide can stimulate and inhibit cellular development, where inhibitory measures have been used to improve the hygiene of liquid and solid food by inactivating bacterial growth. The study conducted by "Kustyawati [5]" found that processing with supercritical

CO_2 and sub-supercritical CO_2 retained the texture, vitamin B, Ca, and protein content, but reduced fat, water content, and some volatile compounds in tempeh.

Tempeh is generally sold in fresh form, even though it is not consumed in raw state, but needs to be processed further before consuming. Tempeh is a fermented soybean product by *Rhizopus oligosporus*, but bacteria and yeasts are also involved during the fermentation and contributed significantly to the production of functional metabolites. The microbial community structure in tempeh is a very important feature in maintaining not only the sensory appearance but also the functional nature of the tempeh. Supercritical carbon dioxide technology can be an alternative process for tempeh which is expected to reduce the number of bacteria and at the same time maintain high mold growth. The high number of bacteria in tempeh can interfere with mold growth and consequently the tempeh will spoil more quickly. Molds growth is needed to produce tempeh with a compact texture, white gray color, and being palatable. Previous research has shown that sub-supercritical CO_2 at 6.3 MPa for 10 min did not significantly affect tempeh color and the tempeh was acceptable "Kustyawati [5]". However, the survival of microorganisms in tempeh processed with supercritical CO_2 has not been revealed. Minimal processing technology without involving heat that can maintain the growth of mold in tempeh is needed in an effort to increase the shelf-life and maintain the freshness of tempeh, nutritional value, and consumer preferences. The aim of this study was to observe viability of molds and bacteria in tempeh after processed with supercritical CO_2 and determine best processing which can reduce bacteria but maintain mold life and to observe the ability of the mold to grow during storage.

2. Materials and Methods

2.1. Processing of Tempeh. The high pressure CO_2 installation used for experimental treatments consists of a CO_2 gas cylinder, a cylindrical pressure chamber, pressure gauges, and a water bath at constant temperature "Saputra [6]" (see Figure 1). Tempeh, in the form of cylinder with 3.5cm in diameter and 10 cm in length, was obtained from the Center of Home Industry Tempeh Making Palembang, Indonesia. Fresh tempeh was placed in a pressure chamber and then closed tightly. When the designated temperature in water bath was reached and all pipe connections were secured, commercially available CO_2 (PERTAMINA, Jakarta, Indonesia) was injected through the gas inlet valve from the gas cylinder into the pressure chamber until it reached the desired pressures of 6.3 and 7.6 MPa (showed in pressure gauge) within 1 min. After being subjected to high pressure CO_2 treatment for the specified incubation period, the pressure was lowered to atmospheric pressure within 3 minutes by slowly opening the gas outlet valve. Then the tempeh was aseptically removed from the pressure chamber using a sterilized tong, placed in the sterilized container, and stored in a refrigerator before conducting the analysis such as SEM, but the samples were directly analyzed for microorganism analysis.

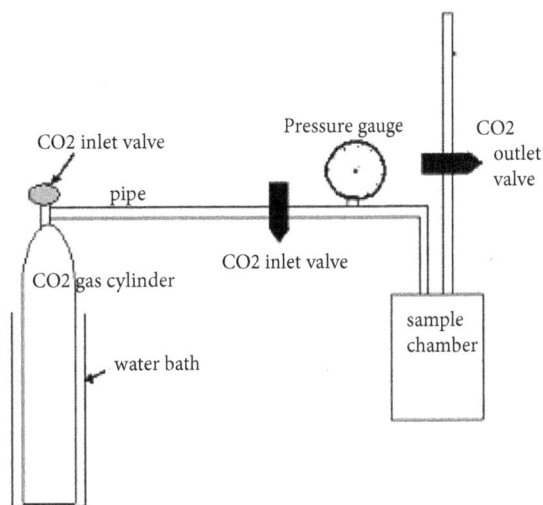

FIGURE 1: The diagram of the experimental apparatus.

The experiment was conducted in a full factorial design with the factors as follows: supercritical CO_2 (scCO$_2$) treatment at 7.6 MPa for 5, 10, 15, and 20 min and sub-supercritical CO_2 (sub-scCO$_2$) treatment at 6.3 MPa for 5, 10, 15, and 20 min. Each treatment was replicated three times.

2.2. Enumeration of Bacteria and Molds. The tempeh was plated no later than 1 hour after the processing. Samples (5 g) were homogenized (1:3) with phosphate buffer solution (BPS) in a Stomacher 400 for 1 min, and appropriate dilutions of the homogenate were made. The enumeration of the bacteria in the Nutrient agar (NA Difco, USA) and molds in the potatoes dextrose agar (PDA, Difco USA) plates was done after incubation period 32°C for 24 h for bacterial and of 27°C for 4 days for molds counts. Oxytetracycline 0.05% and chloramphenicol 0.05% were added to the media to inhibit the growth of bacteria, and cyclohexemide 0.05% was added to inhibit the growth of yeasts. Results were reported as log CFU/g for each treatment (CFU, colony forming units). The degree of inactivation was determined by evaluating the log (N/N0) versus time, where N0 (CFU/g) was the number of microorganisms initially present in the unprocessed sample and N (CFU/g) was the number of survivors after the processing.

The surviving bacteria after supercritical CO_2 processing were isolated and identified by the PCR sequencing analysis. The isolated strains were identified based on morphological characteristics, the biochemical profile according to the manufacturer's instructions (API system, Biomerieux, France), and sequencing 16S (bacteria) rDNA as described below.

2.3. Analysis of DNA. The work of bacterial identification was done according to the method developed by "Parton [7]" as followed. To each of the isolates three to four colonies were picked up and suspended into 100 μL of sterile ddH$_2$O (double-distilled water). Extraction was done by heating at 100°C for 5 min to lyse the cells and centrifuged at 13.000

g for 15 min at 4°C. The supernatant which is containing DNA was transferred to an Eppendorf tube. Two μL of each DNA sample was used as a template in the Polymerase Chain Reaction assay. The primers 355F (5'-CCT ACG GGA GGC AGC AG-3') and 910r (5' –CCC GTC AAT TCC GAG TT– 3') were used for bacterial cells. A final 50 μL volume was used containing 5 μL of forward primer 10 μm (Sigma), 5 μL of reverse primer 10 μm (Sigma), 2 μL of template DNA, 5 μL of 10x Taq DNA polymerase buffer (Sigma), 4 μL $MgCl_2$ (25 mm), and 0.5 μL Taq DNA polymerase (5 u/μL, Sigma). The PCR conditions applied for bacteria were 95°C for 1min, followed by 30 cycles at 95°C for 30 sec, 50°C for 1 min, and 72°C for 1 min, followed by one final extension at 72°C for 6 min. The Microcon PCR columns (Millipore, CA, USA) were used to purify the amplicons, and the purified products were eluted with 35 μL of Milli-Q sterile water. PCR sequencing reaction, which was of eight *ng* of the DNA, was performed as follows: 3.2 μL of forward primer (1μM), 6 μL of Big Dye Buffer (Applied Bio Systems), 2 μL of Big Dye Mix (Terminator RR Mix, Applied Bio Systems), and Milli-Q sterile water up to 20 μL. The PCR conditions were followed as the one previously described. The extracted DNA was treated with 45 μL of pure ethanol of 4°C and 3.75 μL of EDTA of 125 μM, incubated in the dark for 15 min, and centrifuged at 15,000 g for 15 min at 4°C. The pellet obtained was washed with 150 μL of 70% ethanol and again centrifuged at 15,000 g for 15min at 4°C. The pellets were dried for 15 min at 37°C and supernatants were discharged. The samples obtained were then suspended into the 15 μL formamide and sequencing was carried out at an ABI PRISM 310 Genetic Analyzer (Perkin Elmer). For the identification, the sequences obtained were searched against and compared to those present in the National Center for Biotechnology Information (NCBI) genome bank. The similarity was determined by the percentage of similarity greater than 98.5%.

2.4. Storage Study. The optimal condition of the process that was found in the experiment was used to treat the tempeh that would be used for storage study. Tempeh processed at the optimal process condition was stored at 20 and 30°C for 5 days, together with the unprocessed tempeh. During the storage, the total of molds and $L*$ color were analyzed daily. A storage time of 5 days was chosen considering that the shelf-life of fresh tempeh is normally around 1-2 days at room temperatures (30±2°C), while processing of sub-scCO$_2$ was expected to extend the shelf-life of the tempeh.

2.5. Color Measurement. The surface color analysis of processed and unprocessed tempeh was evaluated as CIE $L*a*b*$ value and LCH color scale using color difference meter (TC-1500, Tokyo, Japan). Results were expressed as $L*$ (Lightness), $a*$ (redness), and $b*$ (yellowness). The $L*$, $a*$, and $b*$ values represent the means of the three measurements for each sample. The total color difference ($\Delta E*$) between the control and the treated tempeh was obtained using the following equation: $\Delta E^* = \sqrt{\Delta L*^2 + \Delta a*^2 + \Delta b*^2}$ where the $\Delta L*$, $\Delta a*$, and $\Delta b*$ values meant the difference between

the $L*$, $a*$, and $b*$ values after the treatment and the $L*$, $a*$, and $b*$ values of the standard color. The standard color used in this experiment was the $L*$, $a*$, and $b*$ values of the tempeh control.

2.6. Scanning Electron Microscope (SEM). Analysis microstructure of tempeh mycelium microstructure was conducted by scanning electron microscope (SEM JEOL JSM 5310 LV) following "Hong and Pyun [8]" procedures adjusted to tempeh sample. Sample preparation procedures before being observed with SEM were as follows: (1) tempeh was cut according to the *stub* size and affixed to the top of the *stub*, (2) then tempeh was coated with gold by using *IB2* ion coater tool for 5 minutes with ions current of 6-8 miliAmpere, and (3) finally tempeh was observed with ACC 20kV-voltage devices at 2000x, 3000x, and 10000x magnification.

2.7. Statistical Analysis. Statistically significant differences ($\rho < 0.05$) between the two types of treatment were determined using analysis of variance (ANOVA) and Duncan's multiple range tests "Gomez and Gomez [9]".

3. Results and Discussion

3.1. Bacterial and Mold Inactivation. The study showed that there was a significant interaction with pressure and holding time for bacteria and molds inactivation at $\rho > 0.05$ (see Table 1). The initial bacterial and the mold counts were 2.3×10^7 CFU/g and 6.1×10^6 CFU/g, respectively. The number of bacteria and molds decreased with increasing pressure and time period applied. Molds were more affected than bacteria (Figure 2). Bacterial number decreased to about 1.7 logs at supercritical CO_2 and 1.08 logs at sub-supercritical CO_2 while molds decreased to about 4.88 logs at supercritical CO_2 and 3.73 logs at sub-supercritical CO_2. It is suggestive that the inactivation process was not caused by the pressure of pressurized CO_2. Inactivation process occurred at either sub/near supercritical CO_2 or supercritical CO_2. In some studies such as Kimchi processing, the inactivation of microorganisms was achieved at very high CO_2 pressures, namely, 600 MPa, compared to the inactivation process only at the pressure of 6.3 and 7.6 MPa in this study. The decrease in the number of microorganisms at elevated pressurized CO_2 can be explained as high solvating power of supercritical CO_2 extract vital constituents from the cells or cell membranes, resulting in death of the cell. In this case, pressurized CO_2 penetrates into the cells to build up the density within the cells and expand the cell wall, then it removes intracellular constituent including phospholipids and hydrophobic compounds when the pressure is suddenly released. The removals of constituents alter the structure of the membranes and the balance of the biological system promoting inactivation.

Figure 2 showed that bacteria decreased up to 2.4 logs at supercritical (7.6 MPa) CO_2 for 20 min, whereas they decreased to 1.5 logs at sub-supercritical (6.3 MPa) CO_2 for 20 min. In contrast to our study, "Ferrentino [10]" was reported that the optimal conditions to obtain about 3.0, 1.6, and 2.5 log (CFU/g) reductions of mesophilic aerobic,

TABLE 1: Interaction of high pressure CO_2 processing and time period on the inactivation of bacteria and molds in tempeh.

Treatments: pressures/time periods	Bacterial reduction log N/No	Mold reduction log N/No
scCO₂ /5min	0.54 ± 0.52^b	$2,33\pm0.15^c$
scCO₂ /10min	1.54 ± 0.22^e	$4,23\pm0.02^d$
scCO₂ /15min	2.34 ± 0.12^f	$6,47\pm0.11^f$
scCO₂ /20min	2.40 ± 0.13^f	$6,53\pm0.52^f$
sub scCO₂ /5min	0.30 ± 0.12^a	$1,17\pm0.12^a$
sub scCO₂ /10min	1.10 ± 0.13^c	$1,93\pm0.20^b$
sub scCO₂ /15min	1.40 ± 0.11^d	$5,27\pm0.07^e$
sub scCO₂/20min	1.50 ± 0.23^e	$6,47\pm0.41^f$

Note: the numbers in the column followed by the same letter were not significantly different under p 0.05.

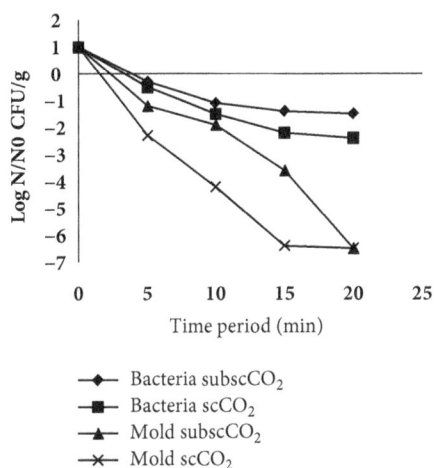

FIGURE 2: The effect of supercritical (7.6 MPa) and sub-supercritical CO_2 (6.3 MPa) processing on bacterial and mold inactivation.

psychrophilic, and lactic acid bacteria in cubed cooked ham were scCO₂ processing at 12 MPa, 50°C, 5 min. "Gunes [11]" demonstrated that supercritical CO_2 at 8 MPa can be an effective nonthermal alternative process for pasteurization of grape juice and tomato paste. The inactivation process of microorganisms in tempeh increased in relation to the increase of incubation period of pressurized CO_2 (see Table 1). Increase in incubation period from 5 to 20 min showed a significant increase in the inactivation values of bacteria and molds where bacterial population decreased from 0.42 to 1.95 logs and molds decreased from 1.75 to 6.5 logs. The explanation to this was that mass transfer rate of CO_2 was greater with the longer incubation period of the pressure of pressurized CO_2. At longer incubation period the amount of CO_2 increases, accumulates into the lipophilic inner layer, and dissolves into and forms hydrogen bond with phospholipid "Mulakhudair [12]", resulting in destruction to cell structure and function due to breakdown of lipid chains. This will further increase the permeability of cell membrane, making it easier for CO_2 to enter the cytoplasm cell. In the cytoplasm CO_2 binds to water and forms HCO_3^- ions, lowering the cytosolic pH which interferes with cell metabolic processes and results in cell death "Garcia-Gonzales [13]".

At the increased pressure and time, molds inactivation increased (see Figure 2). Molds decreased to an undetectable number (log 6) for 20 min at sub-supercritical CO_2 (6.3 MPa) and 15 min at supercritical CO_2 (7.6 MPa). However, it was found that the countable numbers of molds were $4.3x10^4$ CFU/g at the tempeh processed with sub/near supercritical CO_2 for 10 min, indicating processing under these conditions is optimal to be applied to tempeh because there are still a number of fungal growths on the surface of tempeh at 10^4 CFU / g. This is supported by the fact that the mycelium was inflated (see Figure 3(b)). Hyphae mycelium (tempeh control) without inactivation process showed an elastic-rigid texture while mycelium after the inactivation process at 6.3 MPa for 10 min showed that hyphae was inflated (see Figure 3(a)). The minimum number of molds is 10^3 CFU / g for tempeh to have a compact structure, produce grayish white color, have a functional role, and be accepted by consumers "Kustyawati".

This finding was in contrast with published data by "Shon and Lee [14]" where molds remained relatively constant in *Kimchi* after treatment up to 600 MPa. Other finding showed that yeasts and molds were undetected in herbs dried with supercritical CO_2 at 10 MPa for 150 min "Zambon [15]" which was in agreement with our results. The type and chemical contents of products processed with the supercritical CO_2 could be the reason for differences in findings. During high pressure processing in this experiment, CO_2 diffused easily into the tempeh matrix because the tempeh contains soluble proteins, fat, carbohydrates, and other polar compounds. The interaction between CO_2 and matrix macromolecules caused changes in the matrix structure, providing protection for microorganisms in tempeh. In addition, the penetration of CO_2 into the hyphae caused the cell to be inflated which was reversible due to its elastically rigid texture, comprising a double layer of glycoprotein, glucan, chitin, and melanin "Madigan [16]".

The number and type of microorganisms present in tempeh depend on the inoculums used in fermentation and fermentation process conditions. Tempeh is a fresh food with water content ranging from 65 to 65.7% (dry weight) "Kustyawati [5]". When tempeh is processed with supercritical CO_2, water in contact with pressurized CO_2 becomes acidic due to the formation and dissociation of H_2CO_3 which liberates H^+ ions, resulting in lowering pH in the tempeh (pH of extracellular). Even though this low pH may diminish

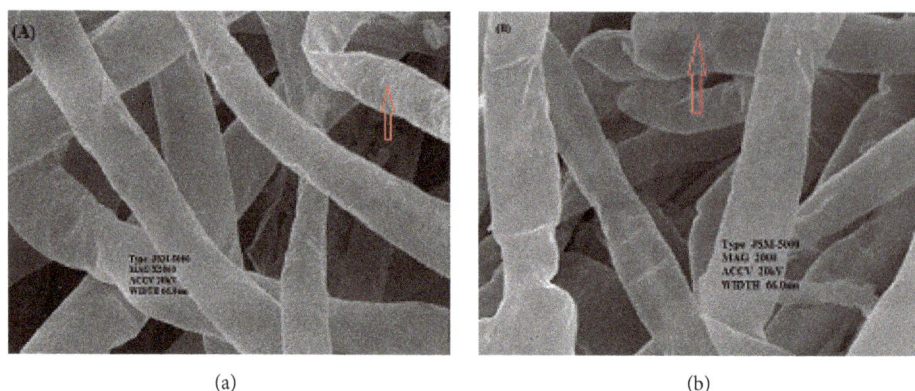

FIGURE 3: The changes of hyphae mycelium. (a) Hyphae were elastically rigid before the inactivation process at 6.3MPa for 10 min. (b) Hyphae were inflated after the inactivation process at 6.3Mpa for 10 min.

resistance to inactivation of microorganisms, the reduction in pH is not enough to cause the lethal effect of CO_2 on some bacteria in the tempeh. Therefore in this experiment, *Bacillus subtilis, Lactobacillus* sp., *Pediococcus* sp., and *Streptococcus* sp. were found and isolated from tempeh after processed with pressurized CO_2 with the pressure of 7.8 MPa for 20 min. In addition to these bacteria, *Klebsiella pneumonia, Citrobacter freundii*, and *Enterobacter cloacae* were also found from tempeh without processing. This finding was in agreement with the published data "Mathias [17]" that supercritical CO_2 exposure to *Bacillus subtillis* ranging from 2 to 25 MPa did not influence its inactivation. Another study reported that application of high pressure CO_2 ranging from 200 to 600 MPa resulted in more than 99.99% of cells which were sublethally injured "Ulmer [18]". Lowered pH of the tempeh may contribute to an increase in cell permeability which facilitates penetration of CO_2 into microbial cell and accumulates in the cytoplasmic interior of bacterial cell. Cell walls of Gram-negative bacteria are composed of lipopolysaccharide on the outside and a thin layer of peptidoglycan in the inside. Supercritical CO_2 has hydrophobic properties, which can penetrate the cell wall, and dissolves lipopolysaccharide layer. If too much amount of dissolved CO_2 enters the cytoplasm, the cell may be unable to maintain the pH homeostasis and pH of the internal cell will begin to decrease to coincide with disruption of cellular activity and result in the cell death "Dillows [19]". This likely is the reasons of the lethal effect on Gram-negative bacteria in tempeh, whereas it is possible that the resistance of Gram-positive bacteria is due to high impermeability of their cell membrane owing to the thick layers rich of peptidoglycan and basic protein, and thin layer of phospholipid-content, resulting in the limited penetration of CO_2.

3.2. *Storage Study*. It was found that the best processing condition was sub/near supercritical CO_2 for 10 min. For the storage study, tempeh was processed with sub-supercritical CO_2 for 10 min and then stored at temperature of 20°C and 30°C. The initial number of molds was 2.5×10^6 CFU/g before storage and slightly increased during 3 days of storage at

30°C. Relationship between countable mold and storage time showed that the molds increased at a storage temperature of 30°C but decreased at a temperature of 20°C (see Figure 4). High CO_2 concentration increases the acidity of the medium because CO_2 reacts with water in the tempeh matrix and produces carbonic acid. Carbonic acid is a weak acid which dissociates to produce H^+ ions so that the acidity of tempeh (pH of the tempeh in this study was 5.9-6.1) is favored by molds for their growth, beside optimal growth of mold is at 30°C. This may explain why mold can grow during the storage process.

Fresh tempeh has bright white color produced by the growth of mold, *Rhizopus oligosporus*. The brightness of color in tempeh is measured using $L*$ value. Tempeh which has a brownish yellow color indicates that the tempeh has been spoilage, and tempeh that has a dark color produced by spores shows that the tempeh undergoes overfermentation. The color changes in tempeh are caused by, for example, damage to the mycelium of *R. oligosporus*, increased concentration of soy color in a particular area, occurrence of other reactions in tempeh, and spore formations. The $L*$ color kinetic value was showed in Figure 5. High $L*$ (lightness) values showed the whitest bright color of fresh tempeh, while low $L*$ value showed the dark brown color of spoilage tempeh. Compared with unprocessed tempeh, $L*$ color of the processed tempeh showed a slight increase during storage. The $L*$ value of processed tempeh increased after day 1 at the level of 0.94 while it slightly decreased on storage of 20°C at the -3.42 level. Meanwhile, the $L*$ value of unprocessed tempeh showed a rapid decline from the first day of storage at the level of -10.3 to -14.0. An increase of $L*$ value indicated that there was a growing mold. Similar results were reported by "Ferrentino [20]" who observed a significant reduction in color lightness and redness for untreated samples of cloudy apple juice while samples treated with supercritical CO_2 appeared to have a smaller change when compared with untreated ones. "Kincal [21]" provides that orange juice treated with a continuous high pressure carbon dioxides (HPCD) system has higher lightness and yellowness when compared with untreated samples during storage.

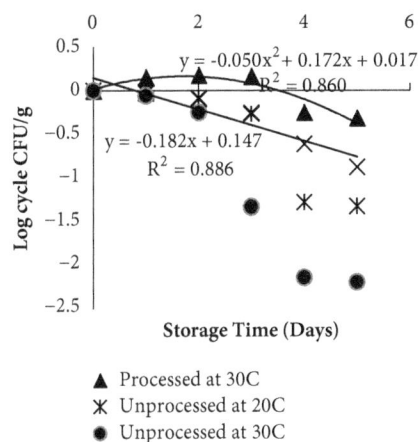

FIGURE 4: The relationship between mold abundance and storage time at 20 and 30°C of processed tempeh.

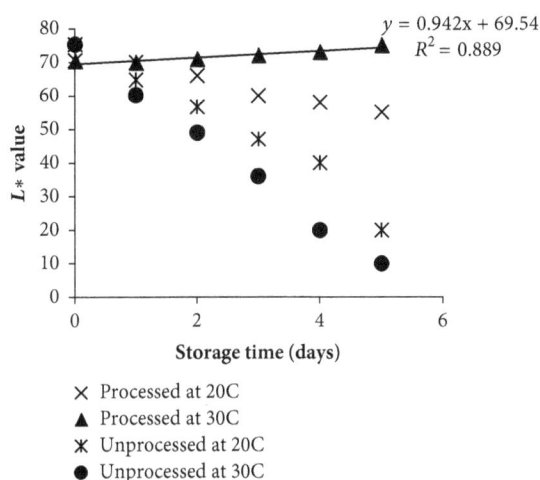

FIGURE 5: Effect of sub-supercritical CO_2 on the change of lightness ($L*$) during storage at 20 and 30°C.

4. Conclusions

Tempeh was used as a model for food product processed with supercritical CO_2. Ratio of survivor microorganisms (bacteria and molds) in tempeh after each treatment was calculated after spread plating the bacteria and molds in nutrient agar and potatoes dextrose agar plates, respectively. There was a significant interaction with the pressure and incubation period for bacterial and molds reduction at $\rho > 0.05$. Reduction of bacteria and molds increased with longer incubation time. Reduction of bacteria 1.5 log was achieved after treatment with supercritical CO_2 for 10 min and sub-supercritical CO_2 for 20 min, while mold reduced 6.0 logs after treatment for 20 min with either supercritical or sub-supercritical CO_2. The longer incubation period may influence microbial reduction in tempeh. Composition of tempeh matrix may give protection against destructive effect of supercritical CO_2. Gram-negative bacteria in tempeh were dying but Gram-positive bacteria were more resistant to supercritical CO_2. The inactivation of mold was reversible causing it to grow back during storage under suitable conditions. Therefore, processing with sub/near supercritical CO_2 for 10 min was the best method to apply to tempeh because molds survived up to 4.3×10^4 CFU/g and bacteria reduced 1.1 logs, and tempeh is still acceptable to consumer. The treatment can act as a method of cold pasteurization of tempeh and can be an alternative method to preserve tempeh.

Disclosure

This work referred to the abstract published in https://waset.org/abstracts/nutrition-and-food-engineering/50233.

Conflicts of Interest

The authors declare that there are no conflicts of interest regarding the publication of this work.

Acknowledgments

The author would like to acknowledge the Ministry of Research, Technology and Higher Education of The Republic of Indonesia for partly funding this research.

References

[1] H. Liao, L. Zhang, X. Hu, and X. Liao, "Effect of high pressure CO2 and mild heat processing on natural microorganisms in apple juice," *International Journal of Food Microbiology*, vol. 137, no. 1, pp. 81–87, 2010.

[2] L. Calvo and E. Torres, "Microbial inactivation of paprika using high-pressure CO2," *The Journal of Supercritical Fluids*, vol. 52, no. 1, pp. 134–141, 2010.

[3] L. Garcia-Gonzalez, A. H. Geeraerd, S. Spilimbergo et al., "High pressure carbon dioxide inactivation of microorganisms in foods: The past, the present and the future," *International Journal of Food Microbiology*, vol. 117, no. 1, pp. 1–28, 2007.

[4] M. Guo, J. Wu, Y. Xu, G. Xiao, M. Zhang, and Y. Chen, "Effects on microbial inactivation and quality attributes in frozen lychee juice treated by supercritical carbon dioxide," *European Food Research and Technology*, vol. 232, no. 5, pp. 803–811, 2011.

[5] M. Erna Kustyawati, F. Pratama, D. Saputra, and A. Wijaya, "Modification of texture, color and aroma of tempeh after being process with supercritical CO2," *Jurnal Teknologi dan Industri Pangan*, vol. 25, no. 2, pp. 168–175, 2014.

[6] D. Saputra, F. A Payne, and P. L. Cornelius, "Puffing degydrated green bell pappers with CO2," *American Society of Agricultural and Biological Engineers*, 1991.

[7] T. Parton, A. Bertucco, and G. Bertoloni, "Pasteurisation of grape must and tomato paste by dense-phase CO2," *Italian Journal of Food Science*, vol. 19, no. 4, pp. 425–437, 2007.

[8] S. Hong and Y. Pyun, "Inactivation Kinetics of Lactobacillus plantarum by High Pressure Carbon Dioxide," *Journal of Food Science*, vol. 64, no. 4, pp. 728–733, 1999.

[9] G. H. Freeman, K. A. Gomez, and A. A. Gomez, "Statistical Procedures for Agricultural Research with Emphasis on Rice.," *Biometrics*, vol. 34, no. 4, p. 721, 1978.

[10] G. Ferrentino, S. Balzan, and S. Spilimbergo, "Optimization of supercritical carbon dioxide treatment for the inactivation of the natural microbial flora in cubed cooked ham," *International Journal of Food Microbiology*, vol. 161, no. 3, pp. 189–196, 2013.

[11] G. Gunes, L. K. Blum, and J. H. Hotchkiss, "Inactivation of yeasts in grape juice using a continuous dense phase carbon dioxide processing system," *Journal of the Science of Food and Agriculture*, vol. 85, no. 14, pp. 2362–2368, 2005.

[12] A. R. Mulakhudair, M. Al-Mashhadani, J. Hanotu, and W. Zimmerman, "Inactivation combined with cell lysis of Pseudomonas putida using a low pressure carbon dioxide microbubble technology," *Journal of Chemical Technology and Biotechnology*, vol. 92, no. 8, pp. 1961–1969, 2017.

[13] L. Garcia-Gonzalez, A. H. Geeraerd, J. Mast et al., "Membrane permeabilization and cellular death of Escherichia coli, Listeria monocytogenes and Saccharomyces cerevisiae as induced by high pressure carbon dioxide treatment," *Food Microbiology*, vol. 27, no. 4, pp. 541–549, 2010.

[14] S. Kyung-Hyun and L. H-Joo, *Effect of high pressure treat*, no. 359-365, 1998.

[15] A. Zambon, F. Michelino, S. Bourdoux et al., "Microbial inactivation efficiency of supercritical CO2 drying process," *Drying Technology*, pp. 1–6, 2018.

[16] M. T. Madigan, J. M. Martinko, and D. A. Stahl, *Brock Biology of Microorganisms*, Pearsons Education Inc, San Fransisco, 2012.

[17] O. Mathias, T. Kablan, and A. Joseph, "Inactivation of Bacillus Subtilis spores with pressurized CO2and influence of O2, N2O and CH2CH2OH on its sporicidal activity," *European Journal of Scientific Research*, vol. 40, no. 1, pp. 6–14, 2010.

[18] H. M. Ulmer, M. G. Ganzle, and R. F. Vogel, "Effects of high pressure on survival and metabolic activity of Lactobacillus plantarum TMW1.460," *Applied and Environmental Microbiology*, vol. 66, no. 9, pp. 3966–3973, 2000.

[19] A. Dillow, F. Dehghani, and J. S. Hrkah, "Bacterial inactivation by using near-and supercritical CO2," *Proceedings of the National Academy of Sciences of the United States of America*, vol. 96, no. 18, pp. 10344–10348, 2008.

[20] G. Ferrentino, M. Bruno, G. Ferrari, M. Poletto, and M. O. Balaban, "Microbial inactivation and shelf life of apple juice treated with high pressure carbon dioxide," *Journal of Biological Engineering*, vol. 3, 2009.

[21] D. Kincal, W. Hill, M. Balaban et al., "A Continuous High-Pressure Carbon Dioxide System for Cloud and Quality Retention in Orange Juice," *Journal of Food Science*, vol. 71, no. 6, pp. C338–C344, 2006.

22

Effectiveness of a Commercial Lactic Acid Bacteria Intervention Applied to Inhibit Shiga Toxin-Producing *Escherichia coli* on Refrigerated Vacuum-Aged Beef

Katie R. Kirsch,[1] **Tamra N. Tolen,**[2] **Jessica C. Hudson,**[1] **Alejandro Castillo,**[2] **Davey Griffin,**[3] **and T. Matthew Taylor**[2]

[1]*Department of Nutrition and Food Science, Texas A&M University, College Station, TX, USA*
[2]*Department of Animal Science, Texas A&M University, College Station, TX, USA*
[3]*Department of Animal Science, Texas A&M AgriLife Extension, College Station, TX, USA*

Correspondence should be addressed to T. Matthew Taylor; matt_taylor@tamu.edu

Academic Editor: Marie Walsh

Because of their antagonistic activity towards pathogenic and spoilage bacteria, some members of the lactic acid bacteria (LAB) have been evaluated for use as food biopreservatives. The objectives of this study were to assess the antimicrobial utility of a commercial LAB intervention against O157 and non-O157 Shiga-toxigenic *E. coli* (STEC) on intact beef strip loins during refrigerated vacuum aging and determine intervention efficacy as a function of mode of intervention application. Prerigor strip loins were inoculated with a cocktail ($8.9 \pm 0.1 \log_{10}$ CFU/ml) of rifampicin-resistant ($100.0\,\mu$g/ml; Rif^R) O157 and non-O157 STEC. Inoculated loins were chilled to $\leq 4°$C and treated with $8.7 \pm 0.1 \log_{10}$ CFU/ml LAB intervention using either a pressurized tank air sprayer (conventional application) or air-assisted electrostatic sprayer (ESS). Surviving STEC were enumerated on tryptic soy agar supplemented with $100.0\,\mu$g/ml rifampicin (TSAR) to determine STEC inhibition as a function of intervention application method (conventional, ESS) and refrigerated aging period (14, 28 days). Intervention application reduced STEC by $0.4 \log_{10}$ CFU/cm^2 ($p < 0.05$), although application method did not impact STEC reductions ($p > 0.05$). Data indicate that the LAB biopreservative may assist beef safety protection when utilized within a multi-intervention beef harvest, fabrication, and aging process.

1. Introduction

An estimated 175,905 Shiga-toxigenic *Escherichia coli* (STEC) foodborne disease cases occur in the United States each year, with non-O157 STEC being reportedly the causative agents in 64.1% of cases [1]. The US Department of Agriculture Food Safety and Inspection Service (USDA-FSIS) declared raw nonintact beef, as well as intact beef intended to be processed into nonintact beef, adulterated if found positive for *E. coli* belonging to serogroups O26, O45, O103, O111, O121, O145, and/or O157 [2]. Cattle serve as a reservoir of STEC [3–5]; eradication of these pathogens from the beef supply chain remains a challenge. Chemical food safety interventions such as lactic acid, peroxyacetic acid, and chlorine are commonly applied to reduce and/or eliminate spoilage and pathogenic

organisms from beef surfaces [6–8]. However, consumer demand for natural or minimally processed foods [9, 10] suggests a need for alternative beef safety interventions [11].

The lactic acid bacteria (LAB), as a type of biopreservative, are reported to be useful for preventing the growth of pathogenic microbes on meat products [12, 13] and are in some instances classified as generally recognized as safe (GRAS) for use in nonintact, whole muscle cuts or carcasses, and ready-to-eat meats [14, 15]. These organisms antagonize other bacteria, including human pathogens, through competition for nutrients and/or attachment sites, production of antimicrobial metabolites (e.g., reuterin, diacetyl, and fatty acids), bacteriocins (e.g., nisin, pediocin), and weak organic acids (e.g., lactic, acetic acid) [16, 17]. Previous studies have explored the inhibitory mechanisms of specific protective

TABLE 1: Shiga-toxigenic *Escherichia coli* isolate identification and sources.

STEC serotype	Isolate ID[a]	Source
O104:H4	TY-2482 ATCC BAA-178	Human stool
O157:H7	USDA-FSIS 380-94	Salami isolate
O26:H11	H30	Infant with diarrhea
O103:H2	CDC 90-3128	Human stool
O45:H2	CDC 96-3285	Human stool
O145:NM	83-75	Human stool
O111:H-	JB1-95	Clinical isolate
O121:H19	CDC 97-3068	Human stool

[a]Isolates were provided by Luchansky, Ph.D. (USDA-Agricultural Research Service, Wyndmoor, PA).

cultures from members of the LAB for inhibition of *E. coli* O157:H7 in meats [13, 18–25]. Nevertheless, to date little data are published detailing the antimicrobial efficacy of LAB food safety interventions for inhibiting members of the non-O157 STEC on fresh beef during storage and handling prior to retail. Therefore, the objectives of this study were to (i) assess the efficacy of a commercial LAB biopreservative (LactiGuard™, Guardian Food Technologies, LLC, Overland Park, KS) for the inhibition of eight STEC serogroups on beef subprimals during refrigerated vacuum aging and (ii) determine whether mode of intervention application (conventional spray, electrostatic spray [ESS]) impacts antimicrobial efficacy.

2. Materials and Methods

2.1. Bacterial Culture Maintenance. Rifampicin-resistant (Rif^R) STEC strains encompassing O157 and the six O-groups of non-O157 STEC named as adulterants in raw nonintact beef [2], and *E. coli* O104:H4 (2011 European sprout outbreak) (STEC8), were provided by J. B. Luchansky, Ph.D. (US Department of Agriculture-Agricultural Research Service, Wyndmoor, PA, USA) (Table 1). Culture revival and maintenance procedures were completed according to previous methods [26]. Working cultures of isolates were prepared by transferring a loopful of culture from tryptic soy agar (TSA; Becton, Dickinson and Co., Sparks, MD, USA) slants into 10 ml sterile tryptic soy broth (TSB) and incubating statically at 35°C for 18–24 h. Each isolate was individually subcultured by inoculating a 50 ml volume of sterile TSB supplemented with 0.1% (w/v) rifampicin (Sigma-Aldrich, St. Louis, MO, USA) with one loopful of fresh culture and incubating statically at 35°C for 18–24 h. Immediately prior to use, a cocktail of STEC isolates was prepared by transferring 50 ml of each culture into a calibrated misting bottle (previously sanitized by immersion in 70% ethanol for 5 min followed by triplicate flushing with sterile distilled water). The targeted STEC concentration in the inoculum was 9.0 log$_{10}$ CFU/ml and was determined by serially diluting in 0.1% peptone water (Becton, Dickinson and Co.) and spreading on TSA supplemented with 100 µg/ml rifampicin (TSAR). Colonies were enumerated following incubation at 35°C for 18–24 h. Inoculum preparation procedures were completed to provide

nondiffering counts of each of the eight *E. coli* isolates by mixing equivalent volumes into the inoculation bottle and thoroughly mixing isolates together prior to application. Isolates were prepared for inoculation according to methods previously published by Kirsch et al. [26], who reported that STEC isolates identical to those used in the current study were able to achieve non-statistically differing counts following 24 h incubation at 35°C in a nutritious medium. Rif^R organisms used in the current study were previously compared to antibiotic-sensitive parents for tolerance to common food safety interventions including heating, lactic acid exposure, and high pressure, with no statistically significant differences in survivors between parents and mutants [27].

The commercial biopreservative LactiGuard consisted of lyophilized powders containing a manufacturer-described set of *Lactobacillus*, *Lactococcus*, and *Pediococcus* spp. Immediately prior to use, a LactiGuard suspension was prepared by combining 7.5 g (11.0 log$_{10}$ CFU/g) of each organism (4 total) into 3.0 liters of sterile water to a concentration of 8.4 log$_{10}$ CFU/ml. LAB numbers in the antimicrobial suspension prior to application to beef surfaces were verified by enumerating on de Man, Rogosa, and Sharpe (MRS) lactobacilli agar (Becton, Dickinson and Co.) supplemented with streptomycin sulfate (40 µg/ml, Amresco, Solon, OH, USA), sodium oxacillin (0.4 µg/ml, Chem-Impex Int., Inc., Wood Dale, IL, USA), and gentamycin sulfate (5 µg/ml, Amresco), as per intervention supplier instructions. Colonies were enumerated following anaerobic incubation for 48 h at 35°C.

2.2. Meat Preparation and Inoculation. Prerigor beef strip loins were procured from a federally inspected establishment in Texas and harvested within 2 h of animal death. Once collected, each strip loin was transferred into a polyethylene bag and swathed in a thermal blanket (EverReady First Aid, Brooklyn, NY, USA) to minimize heat loss. Beef pieces were then immediately transported in insulated coolers containing activated instant hot packs (Dynarex, Orangeburg, NY, USA) to a Texas A&M AgriLife Research facility maintaining a BSL-2 Laboratory located within a 30 min drive of the beef slaughter establishment to inoculate strip loins as quickly as possible after collection (simulating prerigor carcass cross-contamination during beef harvest). At the facility, a prepared inoculum application spray bottle was primed and held 25 to 31 cm above the meat for application of three pumps of inoculum (1.0–1.5 ml per pump) onto the lean side of the strip loin surface. The bag containing the inoculated strip loin was closed with a zip tie and hand-tumbled for 1 min to distribute inoculum over beef surfaces. Beef pieces were held at 25°C for 30 min for inoculum attachment, loaded into a second polypropylene bag, and placed in insulated coolers containing frozen ice packs to initiate chilling. Coolers were transported to the Food Microbiology Laboratory (Texas A&M University) within 8 h of inoculation and bacterial attachment. Upon return, bagged strip loins were removed from coolers and placed on shelves in a single layer in a 4°C walk-in cooler and held until a total chilling period of 24 h had elapsed.

TABLE 2: Least squares means of bacterial populations on beef strip loins ($\log_{10}$ CFU/cm^2) treated with lab and vacuum aged at 4°C.

Target organisms	Experimental process stage[a]					p value
	Postinoculation	Postchilling	Posttreatment	14 days of aging	28 days of aging	
Control[b]	6.7A	6.6A	—	6.6A	6.3A	0.4047
STEC[c]	—	7.2A	6.8B	7.1AB	6.7B	0.0151
Lactic acid bacteria (LAB[d])	—	—	6.5A	6.2B	6.1B	0.0124

[a]Values are least square means from two replications with triplicate samples in each replication ($n = 6$). Means within a row lacking the same capitalized letter (A, B) differ at $p = 0.05$ by Tukey's Honestly Significant Differences (HSD) multiple comparisons test. [b]Control indicates STEC counts from STEC-inoculated, nontreated beef strip loins. STEC were enumerated on tryptic soy agar supplemented with 100.0 μg/ml (TSAR) following 48 h incubation at 35°C. [c]STEC denotes STEC means from strip loins treated with the LAB intervention by pressurized spray or ESS. STEC were enumerated on tryptic soy agar supplemented with 100.0 μg/ml (TSAR) following 48 h incubation at 35°C. Significant differences in STEC counts were not detected as a function of intervention application by pressurized spray versus ESS; counts of organisms are therefore compiled for both application methods; [d]LAB denotes numbers of LAB enumerated from intervention-treated strip loins (pressurized spray, ESS). As significant differences in LAB counts were not detected as a function of intervention application by pressurized spray versus ESS, counts of organisms were compiled for both application methods. LAB from the biopreservative LactiGuard LAB were enumerated on de Man, Rogosa, and Sharpe (MRS) agar supplemented with streptomycin sulfate (40 μg/ml), sodium oxacillin (0.4 μg/ml), and gentamycin sulfate (5 μg/ml), as per manufacturer guidance.

2.3. Intervention Application to Beef. Following chilling, inoculated beef strip loins were readied for intervention application (under biosafety level 2 [BSL2] containment) and sample analysis. Prior to treatment application, a sterile meat hook was inserted into the distal end of the strip loin and strip loins were hung lean side facing outward in a model spray cabinet (Birko Corp., Centennial, CO, USA). Each piece was randomly assigned to a treatment: (a) conventionally spray-applied LactiGuard [25°C for 100 s at 310 kPa, 1.7 ml/min flow rate]; (b) ESS-applied LactiGuard [25°C for 120 s at 207 kPa]; or (c) STEC-inoculated, untreated control. ESS application was performed using a XT-3 air-assisted ESS sprayer (Electrostatic Spraying Solutions, Inc., Watkinsville, GA, USA) charged to ≤ -10 amps at a flow rate of 2.1 ml/s, while the conventional spray application was achieved using a hand-held, pressurized tank air sprayer (Roundup, Marysville, OH, USA) at a flow rate of 1.7 ml/s. Interventions were sprayed approximately 90 cm from the strip loin surface in a sweeping horizontal zig-zag motion. Strip loins were then inserted into commercial-grade vacuum bags (oxygen transmission rate: ≤ 50 cm^3/m^2·24 h·0.1 MPa; Weston, Strongville, OH, USA) and packaged in a vacuum sealer. Strip loins were arranged in a single layer and stored at 4°C for 14 or 28 days prior to postaging sampling.

2.4. Sampling and Microbiological Analysis. In order to track changes in meat pH as a function of intervention application, external pH of individual strip loins was measured in triplicate using an ExStik® pH and temperature meter (Extech Instruments Corp., Nashua, NH, USA) before and after STEC inoculation, before and after LAB intervention treatment, and again after 14 and 28 days of refrigerated vacuum aging. For microbiological sampling, three 10 cm^2 outlines were marked on the lean tissue surface using a flame-sterilized stainless steel borer, excised to a depth of 1-2 mm using flame-sterilized scalpels and forceps, and composited into a sterile stomacher bag. Samples were then sealed and transported in insulated coolers packed with ice to the Food Microbiology Laboratory for analysis. Beef samples were assayed by adding 99 ml phosphate buffered saline (PBS; Sigma-Aldrich Co.) to each sample pouch, pummeling for 1 min in a stomacher,

serially diluting in 0.1% peptone water, and spreading on an appropriate medium. LactiGuard LAB were enumerated on MRS agar supplemented with antibiotics described above and incubated anaerobically 48 h at 35°C prior to colony counting. Rif^R STEC were spread on TSAR and colonies enumerated following incubation at 35°C for 24 h.

2.5. Statistical Analysis. The experiment was completed via triplicate identical samples conducted over two replications ($N = 6$). All statistical analyses were performed using JMP Pro v11.0 (SAS Institute, Inc., Cary, NC, USA). Colony counts were transformed to $\log_{10}$ CFU/cm^2; the limit of detection for the plating assays was 0.5 $\log_{10}$ CFU/cm^2. Differences between main effects (vacuum aging duration, and biopreservative intervention application method [control, conventional spray, ESS]) and their interaction were identified by analysis of variance (ANOVA) at a $p = 0.05$. Statistical differences among means were separated with Tukey's Honestly Significant Differences (HSD) multiple comparisons test ($p < 0.05$).

3. Results and Discussion

To simulate cross-contamination during commercial animal slaughter prior to intervention application, beef strip loins were collected within 2 h of animal slaughter and inoculated as rapidly as possible. STEC inoculum fluid contained 8.9 ± 0.1 $\log_{10}$ CFU/ml prior to application to beef surfaces; STEC populations on inoculated strip loins remained unchanged during chilled transportation to the food microbiology laboratory prior to treatment (Table 2). Although control strip loins (STEC-inoculated, untreated) were handled identically to those subjected to LAB intervention treatment, STEC numbers enumerated from nontreated controls after chilling (6.6 ± 0.1 $\log_{10}$ CFU/cm^2) were statistically lower than those from intervention-treated samples (7.2 ± 0.1 $\log_{10}$ CFU/cm^2) ($p < 0.05$) (Table 2). This was unexpected; authors are uncertain as to the cause(s) behind this observed difference in STEC counts between controls and other samples eventually treated with LAB.

Table 2 presents surviving populations of STEC and LAB on strip loins immediately following treatment during vacuum aging at 4°C. Analysis of microbiological data determined method of intervention application (conventional spray, ESS) did not influence LAB numbers recovered from treated beef strip loins ($6.5 \pm 0.1 \log_{10}$ CFU/cm^2). Likewise, STEC numbers on treated strip loins did not differ as a function of the mode of LAB application method ($p \geq 0.05$). Nevertheless, STEC numbers on biopreservative-treated strip loins differed by intervention application with respect to STEC counts before and after treatment on strip loins and by posttreatment aging period ($p = 0.015$) (Table 2). Applying LAB onto beef strip loins resulted in a reduction in STEC of $0.4 \pm 0.1 \log_{10}$ CFU/cm^2 ($p < 0.05$). Once treated, subprimals were individually vacuum packaged and aged at 4°C for 14 or 28 days. On day 14 posttreatment, STEC numbers increased to $7.1 \log_{10}$ CFU/cm^2 and were not different from STEC counts obtained immediately after intervention applications. Conversely, at 28 days of vacuum refrigerated aging, STEC counts ($6.7 \pm 0.1 \log_{10}$ CFU/cm^2) were significantly lower than pretreatment means ($p < 0.05$). At 14 days' vacuum refrigerated aging, LAB numbers on treated strip loin surfaces declined from posttreatment application counts to $6.2 \log_{10}$ CFU/cm^2, though no further changes in LAB counts were detected at day 28 ($p \geq 0.05$) (Table 2). Overall, reductions in STEC counts following intervention application and refrigerated aging were modest and not likely of great antimicrobial significance.

Outcomes from previous studies evaluating the antimicrobial activity of LAB biopreservatives against pathogens on beef products are mixed with respect to pathogen reduction results. Smith et al. [22] inoculated ground beef with $5.0 \log_{10}$ CFU/g of E. coli O157:H7 and $7.0 \log_{10}$ CFU/g of LAB (Lactobacillus acidophilus strains NP 51, NP 25, NP 7, and NP 3) and then stored the product at 5°C under vacuum for up to 5 days prior to analysis of pathogen survival. Compared to untreated controls, E. coli O157:H7 in treated beef was reduced by $>3.0 \log_{10}$ CFU/g following refrigeration for 5 days ($p < 0.05$). Echeverry et al. [28], conversely, reported that E. coli O157:H7 inoculated on refrigerated strip loins ($5.0 \log_{10}$ CFU/cm^2) treated with a spray consisting of $7.7 \log_{10}$ CFU/ml LAB remained unchanged during refrigerated vacuum aging over 21 days, similar to findings reported herein. Finally, previous research has reported that reductions in E. coli O157:H7 on beef following treatment with LAB intervention were independent of LAB numbers applied, with no differences in E. coli O157:H7 numbers observed following incubation [29]. In the current study, numbers of LAB and STEC were nearly equivalent throughout the refrigerated aging period, yet LAB did not exert strong pathogen inhibition activity (Table 2). This may have resulted from insufficient levels of fermentable sugars for production of organic acids, lack of available oxygen for production of peroxides, or possibly other unanticipated factors. Finally, no effort to differentially count surviving STEC from intervention-treated or control strip loins during the experiment was attempted, disallowing authors from determining whether some members of the STEC inoculum

were inhibited to greater or lesser extents than other inoculum members. While Kirsch et al. [26] reported differing morphological characteristics of STEC strains identical to those used herein on selective/differential media, it was noted that STEC isolates not bearing good isolation were frequently subject to misidentification on selective/differential plating medium surfaces. The enumeration of total surviving STEC from the inoculum would be expected to primarily yield counts of STEC isolates most tolerant to the antimicrobial intervention, though quantification of any differences in inhibition on individual STEC isolates was therefore precluded.

Mean surface pH of strip loins immediately prior to inoculation was 6.2 ± 0.1, while pH following meat inoculation and chilling of inoculated product declined to 5.8 ± 0.1 ($p < 0.05$) (data not shown). Whereas application of the LAB intervention by conventional pressurized sprayer versus ESS did not produce statistically significant differences in meat surface pH, surface pH significantly decreased from the point of treatment application (LAB, control) to 14 days of refrigerated vacuum storage (5.4 ± 0.1) ($p < 0.05$). A small, nonstatistically significant decline in meat surface pH was observed to occur between days 14 and 28 for refrigerated vacuum-aged strip loins (5.3 ± 0.1) ($p > 0.05$) (data not shown). These pH declines mirror those for similarly chilled and vacuum-aged refrigerated beef in other studies where researchers determined LAB-driven declines in beef meat pH during chilled aging [30–32]. The inhibition of STEC by LAB is thought to result largely from production of organic acids [33]. However, exposure to sublethal stress (cold storage, carbohydrate limitation) may have limited acid fermentation output of LAB [34, 35]. Data indicate that acid fermentation of endogenous carbohydrate by applied LAB was insufficient to exert a greater degree of pathogen inhibition than that observed. This study investigated the efficacy of a commercial LAB food safety biopreservative intervention for inhibiting growth of eight O157 and non-O157 STEC isolates on beef strip loins during refrigerated vacuum aging. STEC numbers were reduced following treatment, and STEC populations on beef strip loins after 28 days of aging were lower than at pretreatment. However, pathogen reductions were small ($<1.0 \log_{10}$-cycle). While not likely effective as a sole antimicrobial intervention for beef safety protection, biopreservative food safety interventions such as LactiGuard can be utilized to gain useful reductions in STEC when integrated into a multi-intervention process for beef safety protection during beef harvest and fresh beef products manufacture.

Disclosure

Katie R. Kirsch current affiliation is as follows: Department of Epidemiology and Biostatistics, Texas A&M Health Science Center School of Public Health. The funding source did not have any role in the design of the experiment, experiment completion or data analysis, and manuscript preparation or submission. Guardian Food Technologies, LLC, did not have any role in experimental design, research completion, or manuscript preparation/submission.

Conflicts of Interest

The authors declare that they have no conflicts of interest.

Acknowledgments

This material is based upon work that is supported by the National Institute of Food and Agriculture, US Department of Agriculture, Agriculture and Food Research Initiative Competitive Grants Program, under Award no. 2012-68003-30155. Authors thank John Luchansky, Ph.D., US Department of Agriculture, Agricultural Research Service, for provision of STEC isolates. Authors express thanks to Forrest Mitchell, Ph.D., Department of Entomology, Texas A&M AgriLife Research, for access to BSL-2 containment facilities to complete beef inoculation. Authors thank Gary Acuff, Ph.D., and the Texas A&M Center for Food Safety for access to BSL-2 containment facilities to complete beef intervention application. LactiGuard was donated by Guardian Food Technologies, LLC (Overland Park, KS, USA). Authors thank Jason Sawyer, Ph.D., Tarleton State University, Stephenville, TX, for technical assistance provided.

References

[1] E. Scallan, R. M. Hoekstra, F. J. Angulo et al., "Foodborne illness acquired in the United States—major pathogens," *Emerging Infectious Diseases*, vol. 17, no. 1, pp. 7–15, 2011.

[2] USDA-FSIS, "Shiga toxin-producing *Escherichia coli* in certain raw beef products," *Federal Register*, vol. 76, pp. 58157–58165, 2011.

[3] J. Isiko, M. Khaitsa, and T. M. Bergholz, "Novel sequence types of non-O157 Shiga toxin-producing *Escherichia coli* isolated from cattle," *Letters in Applied Microbiology*, vol. 60, no. 6, pp. 552–557, 2015.

[4] H. S. Hussein, "Prevalence and pathogenicity of Shiga toxin-producing *Escherichia coli* in beef cattle and their products," *Journal of Animal Science*, vol. 85, no. 13, supplement, pp. E63-E72, 2007.

[5] A. B. Ekiri, D. Landblom, D. Doetkott, S. Olet, W. L. Shelver, and M. L. Khaitsa, "Isolation and characterization of shiga toxin-producing *Escherichia coli* serogroups O26, O45, O103, O111, O113, O121, O145, and O157 shed from range and feedlot cattle from postweaning to slaughter," *Journal of Food Protection*, vol. 77, no. 7, pp. 1052–1061, 2014.

[6] J. H. Chen, Y. Ren, J. Seow, T. Liu, W. S. Bang, and H. G. Yuk, "Intervention technologies for ensuring microbiological safety of meat: current and future trends," *Comprehensive Reviews in Food Science and Food Safety*, vol. 11, no. 2, pp. 119–132, 2012.

[7] M. Koohmaraie, T. M. Arthur, J. M. Bosilevac, M. Guerini, S. D. Shackelford, and T. L. Wheeler, "Post-harvest interventions to reduce/eliminate pathogens in beef," *Meat Science*, vol. 71, no. 1, pp. 79–91, 2005.

[8] T. L. Wheeler, N. Kalchayanand, and J. M. Bosilevac, "Pre- and post-harvest interventions to reduce pathogen contamination in the U.S. beef industry," *Meat Science*, vol. 98, no. 3, pp. 372–382, 2014.

[9] A. E. J. McGill, "The potential effects of demands for natural and safe foods on global food security," *Trends in Food Science and Technology*, vol. 20, no. 9, pp. 402–406, 2009.

[10] M. Dickson-Spillmann, M. Siegrist, and C. Keller, "Attitudes toward chemicals are associated with preference for natural food," *Food Quality and Preference*, vol. 22, no. 1, pp. 149–156, 2011.

[11] M. Singh, S. M. Simpson, H. R. Mullins, and J. S. Dickson, "Thermal tolerance of acid-adapted and non-adapted *Escherichia coli* O157:H7 and *Salmonella* in ground beef during storage," *Foodborne Pathogens and Disease*, vol. 3, no. 4, pp. 439–446, 2006.

[12] A. Amézquita and M. M. Brashears, "Competitive inhibition of *Listeria monocytogenes* in ready-to-eat meat products by lactic acid bacteria," *Journal of Food Protection*, vol. 65, no. 2, pp. 316–325, 2002.

[13] A. Echeverry, J. C. Brooks, M. F. Miller, J. A. Collins, G. H. Loneragan, and M. M. Brashears, "Validation of lactic acid bacteria, lactic acid, and acidified sodium chlorite as decontaminating interventions to control *Escherichia coli* O157:H7 and *Salmonella* typhimurium DT 104 in mechanically tenderized and brine-enhanced (nonintact) beef at the purveyor," *Journal of Food Protection*, vol. 73, no. 12, pp. 2169–2179, 2010.

[14] FDA, "GRAS notification for the use of lactic acid bacteria to control pathogenic bacteria in meat and poultry products," 2012, https://www.fda.gov/downloads/Food/IngredientsPackagingLabeling/GRAS/NoticeInventory/ucm349355.pdf.

[15] USDA-FSIS, "Safe and suitable ingredients used in the production of meat, poultry, and egg products. Directive 7120.1, Rev. 40," 2017, http://www.fsis.usda.gov/wps/wcm/connect/bab10e09-aefa-483b-8be8-809a1f051d4c/7120.1.pdf?MOD=AJPERES.

[16] M. Ogawa, K. Shimizu, K. Nomoto et al., "Inhibition of in vitro growth of Shiga toxin-producing *Escherichia coli* O157:H7 by probiotic *Lactobacillus* strains due to production of lactic acid," *International Journal of Food Microbiology*, vol. 68, no. 1-2, pp. 135–140, 2001.

[17] L. Vold, A. Holck, Y. Wasteson, and H. Nissen, "High levels of background flora inhibits growth of *Escherichia coli* O157:H7 in ground beef," *International Journal of Food Microbiology*, vol. 56, no. 2-3, pp. 219–225, 2000.

[18] M. M. Brashears, S. S. Reilly, and S. E. Gilliland, "Antagonistic action of cells of *Lactobacillus lactis* toward *Escherichia coli* O157:H7 on refrigerated raw chicken meat," *Journal of Food Protection*, vol. 61, no. 2, pp. 166–170, 1998.

[19] S. Bredholt, T. Nesbakken, and A. Holck, "Protective cultures inhibit growth of *Listeria monocytogenes* and *Escherichia coli* O157:H7 in cooked, sliced, vacuum- and gas-packaged meat," *International Journal of Food Microbiology*, vol. 53, no. 1, pp. 43–52, 1999.

[20] S. Chaillou, S. Christieans, M. Rivollier, I. Lucquin, M. C. Champomier-Vergès, and M. Zagorec, "Quantification and efficiency of Lactobacillus sakei strain mixtures used as protective cultures in ground beef," *Meat Science*, vol. 97, no. 3, pp. 332–338, 2014.

[21] J. R. Ruby and S. C. Ingham, "Evaluation of potential for inhibition of growth of *Escherichia coli* O157:H7 and multidrug-resistant *Salmonella* serovars in raw beef by addition of a presumptive *Lactobacillus sakei* ground beef isolate," *Journal of Food Protection*, vol. 72, no. 2, pp. 251–259, 2009.

[22] L. Smith, J. E. Mann, K. Harris, M. F. Miller, and M. M. Brashears, "Reduction of *Escherichia coli* O157:H7 and *Salmonella* in ground beef using lactic acid bacteria and the impact on sensory properties," *Journal of Food Protection*, vol. 68, no. 8, pp. 1587–1592, 2005.

[23] P. Muthukumarasamy, J. H. Han, and R. A. Holley, "Bactericidal effects of *Lactobacillus reuteri* and allyl isothiocyanate on *Escherichia coli* O157:H7 in refrigerated ground beef," *Journal of Food Protection*, vol. 66, no. 11, pp. 2038–2044, 2003.

[24] P. Muthukumarasamy and R. A. Holley, "Survival of *Escherichia coli* O157:H7 in dry fermented sausages containing microencapsulated probiotic lactic acid bacteria," *Food Microbiology*, vol. 24, no. 1, pp. 82–88, 2007.

[25] S. Erkkilä, M. Venäläinen, S. Hielm, E. Petäjä, E. Puolanne, and T. Mattila-Sandholm, "Survival of *Escherichia coli* O157:H7 in dry sausage fermented by probiotic lactic acid bacteria," *Journal of the Science of Food and Agriculture*, vol. 80, no. 14, pp. 2101–2104, 2000.

[26] K. R. Kirsch, T. M. Taylor, D. Griffin, A. Castillo, D. B. Marx, and L. Smith, "Growth of Shiga toxin-producing *Escherichia coli* (STEC) and impacts of chilling and post-inoculation storage on STEC attachment to beef surfaces," *Food Microbiology*, vol. 44, pp. 236–242, 2014.

[27] J. B. Luchansky, "Personal communication," 2017.

[28] A. Echeverry, J. C. Brooks, F. M. Markus, A. C. Jesse, H. L. Guy, and M. B. Mindy, "Validation of intervention strategies to control *Escherichia coli* O157:H7 and *Salmonella* typhimurium DT 104 in mechanically tenderized and brine-enhanced beef," *Journal of Food Protection*, vol. 72, no. 8, pp. 1616–1623, 2009.

[29] A. R. Hoyle, J. C. Brooks, L. D. Thompson, W. Palmore, T. P. Stephens, and M. M. Brashears, "Spoilage and safety characteristics of ground beef treated with lactic acid bacteria," *Journal of Food Protection*, vol. 72, no. 11, pp. 2278–2283, 2009.

[30] K. M. Wójciak and E. Solska, "Evolution of free amino acids, biogenic amines and N-nitrosoamines throughout ageing in organic fermented beef," *Acta Scientiarum Polonorum, Technologia Alimentaria*, vol. 15, no. 2, pp. 191–200, 2016.

[31] R. Reid, S. Fanning, P. Whyte et al., "The microbiology of beef carcasses and primals during chilling and commercial storage," *Food Microbiology*, vol. 61, pp. 50–57, 2017.

[32] D. A. Mohrhauser, S. M. Lonergan, E. Huff-Lonergan, K. R. Underwood, and A. D. Weaver, "Calpain-1 activity in bovine muscle is primarily influenced by temperature, not pH decline," *Journal of Animal Science*, vol. 92, no. 3, pp. 1261–1270, 2014.

[33] M. M. Brashears and W. A. Durre, "Antagonistic action of *Lactobacillus lactis* toward *Salmonella* spp. and *Escherichia coli* O157:H7 during growth and refrigerated storage," *Journal of Food Protection*, vol. 62, no. 11, pp. 1336–1340, 1999.

[34] R. Cárcoba and A. Rodríguez, "Influence of cryoprotectants on the viability and acidifying activity of frozen and freeze-dried cells of the novel starter strain *Lactococcus lactis* ssp. *lactis* CECT 5180," *European Food Research and Technology*, vol. 211, no. 6, pp. 433–437, 2000.

[35] H. Velly, F. Fonseca, S. Passot, A. Delacroix-Buchet, and M. Bouix, "Cell growth and resistance of *Lactococcus lactis subsp. lactis* TOMSC161 following freezing, drying and freeze-dried storage are differentially affected by fermentation conditions," *Journal of Applied Microbiology*, vol. 117, no. 3, pp. 729–740, 2014.

23

Processing Challenges and Opportunities of Camel Dairy Products

Tesfemariam Berhe,[1,2] **Eyassu Seifu,**[3] **Richard Ipsen,**[2]
Mohamed Y. Kurtu,[1] **and Egon Bech Hansen**[4]

[1]*School of Animal and Range Sciences, Haramaya University, P.O. Box 138, Dire Dawa, Ethiopia*
[2]*Department of Food Science, University of Copenhagen, Rolighedsvej 30, 1958 Frederiksberg C, Denmark*
[3]*Department of Food Science and Technology, Botswana University of Agriculture and Natural Resources,*
 Private Bag 0027, Gaborone, Botswana
[4]*Division for Diet, Disease Prevention and Toxicology, National Food Institute,*
 Technical University of Denmark, 2860 Søborg, Denmark

Correspondence should be addressed to Tesfemariam Berhe; lucyselam@gmail.com

Academic Editor: Salam A. Ibrahim

A review on the challenges and opportunities of processing camel milk into dairy products is provided with an objective of exploring the challenges of processing and assessing the opportunities for developing functional products from camel milk. The gross composition of camel milk is similar to bovine milk. Nonetheless, the relative composition, distribution, and the molecular structure of the milk components are reported to be different. Consequently, manufacturing of camel dairy products such as cheese, yoghurt, or butter using the same technology as for dairy products from bovine milk can result in processing difficulties and products of inferior quality. However, scientific evidence points to the possibility of transforming camel milk into products by optimization of the processing parameters. Additionally, camel milk has traditionally been used for its medicinal values and recent scientific studies confirm that it is a rich source of bioactive, antimicrobial, and antioxidant substances. The current literature concerning product design and functional potential of camel milk is fragmented in terms of time, place, and depth of the research. Therefore, it is essential to understand the fundamental features of camel milk and initiate detailed multidisciplinary research to fully explore and utilize its functional and technological properties.

1. Introduction

In many countries, especially the dry zones of Sub-Saharan Africa, camels (Camelus dromedarius) play a significant role in the lifestyle of many communities owing to their adaptation to the hostile climatic conditions by providing milk, meat, and transportation. Phenotypic characterization indicated that camels are physiologically, anatomically, and behaviorally adapted to the desert environment. Recent genomic study on the camelid family revealed that the animals carry genetically adapted and evolved genes for desert adaptation such as fat metabolism, osmoregulation, heat stress response, ultraviolet radiation, and choking dust [1]. The genus Camelus consists of two different species that live in vast pastoral areas: Camelus dromedarius, the one-humped camel, which mainly lives in the desert areas of Africa/Middle East, and Camelus bactrianus, the two-humped camel that mainly lives in the cooler dry areas of Asia. More than half of the world's 28 million camel population (old and new world camel) are found in the East African countries, namely, Somalia, Sudan, Ethiopia, and Kenya [2]. Nowadays, the camel is becoming the subject of increasing scientific and commercial interest since climate change is already influencing traditional cattle productivity and the camel is an exceptional animal capable of surviving in the hostile climatic conditions.

Camel milk is composed of lactose, fat, and protein in roughly the same proportion as bovine milk (Table 1).

TABLE 1: Proximate composition of camel milk compared to milk of other species.

Species	Total solids (%)	Fat (%)	Protein (%)	Lactose (%)	Ash (%)
[a]Camel	12.0	3.5	3.1	4.4	0.8
[b]Bovine	12.7	3.7	3.4	4.8	0.7
[b]Caprine	12.3	4.5	2.9	4.1	0.8
[b]Ovine	19.3	7.4	4.5	4.8	1.0
[b]Human	12.2	3.8	1.0	7.0	0.2

Sources. [a]Al haj and Al Kanhal [17]; [b]Fox et al. [18].

However, their relative composition, distribution, and the molecular structures of the milk components are different. Proteomic analysis of whey proteins of camel, cow, buffalo, goat, and yak milk has revealed that camel milk whey proteins are hierarchically clustered differently than those from the other species [3].

It is commonly claimed that camel milk is difficult to process into products and is only suitable for drinking as fresh or sour milk. But, currently, the possibility of producing various products from camel milk including soft cheese [4, 5], yoghurt [6], and butter [7, 8] has been reported. Also, camel milk has traditionally been recognized for its medicinal values and experimental results indicate that camel milk has antiallergic, antimicrobial, and antidiabetic properties [9, 10]. The objective of this review is to summarize the processing technologies and functional potential of camel milk. Comparison is made with bovine milk, unless otherwise mentioned.

2. The Challenges of Processing Camel Milk

The mean gross chemical composition of camel milk is 3.5% fat, 3.1% protein, 4.4% lactose, 0.8% ash, and 12% total solids, which is comparable to bovine milk (Table 1). However, their relative composition, distribution, and the molecular structures of the milk proteins (Tables 2 and 3) and fatty acids (Table 4) are reported to be different.

2.1. Processing of Cheese. Processing camel milk into cheese is difficult and has even been considered as impossible [11]. The relative distribution and amino-acid composition of camel milk caseins are different from bovine milk. Camel milk casein has high beta casein (β-CN) (65% versus 39%), low alpha S_1-casein (α_{s1}-CN) (22% versus 38%), and low kappa casein (κ-CN) (3.5% versus 13%) as compared to bovine milk caseins (Table 2). Moreover, the camel milk caseins have low homology to bovine milk caseins, being 39% for α_{s1}-CN, 64% for β-CN, 56% for α_{s2}-CN, and 56% for κ-CN [12]. The chymosin cleavage site of camel milk k-CN was found at the Phe^{97}–Ile^{98} amino-acid sequence site, whereas the hydrolysis site in bovine milk is Phe^{105}–Met^{106} [12]. Thus, the amount of κ-CN in camel milk is relatively small and coagulation of milk in cheese making is typically achieved by enzymatic hydrolysis of κ-CN at the surface of casein micelles.

Alpha-lactalbumin (α-LA) is the major protein in the camel and human milk whey protein group and β-lactoglobulin (β-LG) is absent from camel and human milk

(Table 3). Camel milk has been reported to contain higher whey protein to casein ration compared to bovine milk which is responsible for a soft and easily digestible curd in the gastrointestinal tract [13]. Camel milk casein has large micelle size with an average diameter of 380 versus 150, 260, and 180 nm compared to bovine, caprine, and ovine milk, respectively [14]. Smaller casein micelles have been reported to improve the gelation properties of bovine milk [15]. Thus, the lower amount of k-CN, the high ratio of whey protein to casein, and the larger micelle size in camel milk are reported reasons for the difficulty of cheese making. These properties result in formation of a less firm coagulum and lower yield during cheese processing. Such low efficiency of cheese processing trials is reported by Bornaz et al. [14] and Konuspayeva et al. [16].

2.2. Processing of Butter. The fat content of camel milk ranges from 1.2 to 6.4% [31], which is comparable to that of bovine milk. Nevertheless, butter is not a traditional product made from camel milk and is difficult to produce by using the same technology of production as for butter from bovine milk. The somewhat higher melting point [7, 8] of camel milk fat (41–43°C) makes it difficult to churn the cream at temperatures 10–14°C, which is the optimum churning temperature for bovine milk. Processing of camel milk into butter is also difficult because camel milk shows little tendency to cream up due to deficiency of the protein agglutinin, small fat globule size, and a thicker fat globular membrane [32]. Camel milk is reported to have a higher proportion of long chain fatty acids and a lower amount of short chain fatty acids (Table 4). The high melting point of camel milk butter can be attributed to the high proportion of long chain fatty acids in the fatty acid profile.

However, butter can be made from camel milk under optimum conditions of churning temperature and agitation method. Berhe et al. [7] reported that vigorous shaking of fermented camel milk in a vertical direction instead of the traditional back- and fro-agitation method at a relatively high churning temperature (22-23°C) was able to extract butter from camel milk with a fat recovery efficiency of 80%. This method exerts more force to rupture the fat globule membrane and allow the globules to adhere to one another. Farah et al. [8] also reported that camel milk butter was made at churning temperatures between 15 and 36°C. According to Farah et al. [8], the highest butter fat recovery of 85% was obtained at a churning temperature of 25°C. Camel milk butter is prominently white with a more viscous consistency

TABLE 2: Casein protein distribution of camel, bovine, and human milk.

	Caseins (% of total caseins)			Amino acid residues		
	Camel	Bovine	Human	Camel	Bovine	Human
α_{s1}-casein	22.0[a]	38.0[b]	11.8[c]	217[a]	199[b]	170[d]
α_{s2}-casein	9.5[a]	10.0[b]	*Absent[d]	178[a]	207[b]	*Absent[d]
β-casein	65.0[a]	39.0[b]	64.0[e]	217[a]	209[b]	212[d]
κ-casein	3.5[a]	13.0[b]	24.0[e]	162[a]	169[b]	162[d]
Total caseins (g/100 ml milk)	2.4[c]	2.5[c]	0.4[e]			

*Absent indicates corresponding coding sequence is absent in genome. *Sources.* *Kappeler et al. [12], [b]Eigel et al. [19], [c]El-Agamy [20], [d]Martin et al. [21], and [e]Malacarne et al. [22].

TABLE 3: Whey protein distribution of camel, bovine, and human milk.

	Whey proteins (g/L) in milk			Amino acid residues		
	Camel	Bovine	Human	Camel	Bovine	Human
β-Lactoglobulin	*Absent[f]	1.3[b]	*Absent[c]	*Absent[f]	162[b]	*Absent[cg]
α-Lactalbumin	5.0[f]	1.2[bc]	1.8[cg]	123[f]	123[b]	123[d]
Serum albumin	2.4[h]	0.4[bg]	0.5[gc]	—	582[b]	585[e]
Whey acidic protein	0.16[f]	—	—	117[f]	—	—
Lactoferrin	0.22[a]	0.14[b]	1.5[g]	689[a]	700[b]	700[c]
Immunoglobulins	0.73[fh]	0.7[b]	1.2[g]	—	—	—
Total whey protein	9.3[f]	7.3[f]	7.6[g]			

*Absent indicates that corresponding coding sequence is absent in genome. *Sources.* [a]Kappeler et al. [23], [b]Madureira et al. [24] [c]Inglingstad et al. [25], [d]Findlay and Brew [26], [e]Meloun et al. [27], [f]El-Agamy [20], [g]Malacarne et al. [22], and [h]El-Hatmi et al. [28].

TABLE 4: Fatty acid profile of camel milk compared to bovine and human milk (% fatty acid).

Carbon number	Fatty acid	Camel milk[ab]	Bovine milk[abc]	Human milk[c]
4:0	Butyric (%)	0.8	1.4	0.1
6:0	Caproic (%)	0.4	2.1	0.2
8:0	Caprylic (%)	0.3	1.7	0.3
10:0	Capric (%)	0.4	3.5	2.0
12:0	Lauric (%)	0.7	3.9	6.8
14:0	Myristic (%)	11.0	12.6	10.4
16:0	Palmitic (%)	29.1	29.5	28.1
18:0	Stearic (%)	12.4	13.3	6.9
Monounsaturated				
14:1		0.5	—	—
16:1	Palmitoleic (%)	10.1	1.7	3.5
18:1	Oleic (%)	24.5	26.3	33.6
Polyunsaturated				
18:2	Linoleic (%)	3.1	2.9	6.4
18:3	Linolenic (%)	1.4	1.1	1.7
Unsaturated/saturated		0.7	0.47	0.82
Short chain (C4–C14)		14.6	25.2	19.8
Long chain (C16–C20)		84.5	72.18	80.2

[a]El-Agamy [10], [b]El-Agamy [20], and [c]Malacarne et al. [22].

compared to bovine milk butter. It is reported that pastoralists in the Sahara region produce small amounts of butter from camel milk and they usually use it for medicinal purposes [11].

2.3. Processing of Yoghurt. Manufacturing of yoghurt or other fermented products from camel milk is reported to be difficult. Dromedary milk coagulum does not have a desirable curd formation and firmness and the curd is instead fragile and heterogeneous and consists of dispersed flakes [33]. The problem with camel milk yoghurt is thus the thin consistency and weak texture of the product. Yoghurt texture is a very important parameter that affects the appearance, mouth feel, and overall acceptability. Camel milk has been reported to be not easily fermentable because of its antibacterial properties mainly due to the presence of protective proteins. However, growth of commercial starter cultures in camel milk has been found to be possible. The acidification rate in camel milk was, however, lower than in bovine milk [34, 35].

Nevertheless, there are reports that indicate the possibility of yoghurt production from camel milk [36, 37]. Hashim et al. [38] have reported that the firmness of camel milk yoghurt could be improved by supplementation of the milk with gelatin, alginate, and calcium. On the other hand, Al-Zoreky and Al-Otaibi [39] reported that supplementation of stabilizers to camel milk could not improve the consistency of camel milk yoghurt. Ibrahim [40] suggested that the use of exopolysaccharide producing starter cultures could improve the texture of camel milk yoghurt better than additives. There are also reports that indicate the weak texture of camel milk yoghurt can be improved by mixing of camel milk with milk of other livestock species such as ovine milk [41]. The weak texture and thin consistency of camel milk yoghurt can be attributed to the compositional properties of the milk such as lack of β-LG and lower amount of k-CN [12], high whey protein to casein ratio [13].

3. Functional Potential of Camel Dairy Products

The unique characteristics of camel milk such as its therapeutic potential and absence β-LG have made it a focus area of research in the fields of health science and nutrition as an antimicrobial, antidiabetic and antihypertensive supplement. Camel milk is showing encouraging results in the treatment of autism, cancer, diabetes, and hepatitis and it is safe for children with bovine milk allergy [42–44].

Camel milk has high vitamin C and high mineral contents (sodium, potassium, iron, copper, zinc, and magnesium) and can be good nutritional source for the people living in the arid zones [17, 45].

3.1. Therapeutic Properties of Camel Milk. Camel milk has been indicated as safe and efficient in improving long-term glycemic control with a significant reduction in the doses of insulin in type 1 diabetic patients [46, 47]. Agrawal et al. [48] reported that the prevalence of diabetes in the camel milk consuming society of Indian Raica communities was zero. Camel milk is reported to contain insulin-like proteins which is characterized by resistance to gastrointestinal proteolysis, mimics insulin interaction with its receptor, and is easy to absorb into calcium and be encapsulated into lipid nanocapsules [49]. Agrawal et al. [50] have reported that the half cysteine rich protein in camel milk [51] is similar to the insulin family proteins and may be attributed for the glycemic control of camel milk for diabetic patients. Abdulrahman et al. [52] demonstrated that an allosteric effect of camel milk on insulin receptor conformation and activation on the intracellular signaling process may be responsible for the antidiabetic properties of camel milk.

Camel milk was reported to have an antimicrobial effect against Gram-positive and Gram-negative bacteria including *Escherichia coli, Listeria monocytogenes, Staphylococcus aureus,* and *Salmonella typhimurium* [53]. This inhibitory activity was reported to be due to presence of higher amounts of protective proteins in camel milk including lysozyme, lactoferrin, lactoperoxidase, and immunoglobulins [23, 53, 54].

Camel milk is reported to have antiviral properties. Immunoglobulin and lactoferrin isolated from camel milk could inhibit the hepatitis C virus and demonstrated strong signal against its synthetic peptides, while human counterpart failed to do so [55, 56]. Similarly, camel lactoferrin and lactoperoxidase demonstrated higher inhibition against the entry and direct interaction of hepatitis C virus to Huh7.5 (hepatocyte-derived carcinoma) and HepG2 (human hepatoma) cells [55, 57, 58]. Korashy et al. [59] investigated that camel milk significantly inhibited HepG2 and MCF7 (human breast) cells proliferation and showed induction of death receptors in both cell lines and oxidative stress mediated mechanisms. It is reported that camel milk could play an important role in decreasing oxidative stress by alteration of enzymatic and nonenzymatic antioxidant molecules as well as thymus activation-regulated chemokine levels which resulted in the improvement of child autism in children [60, 61].

Bioactive peptides derived from milk proteins are of great scientific interests due to their nutritional, technological, and potential health benefits. Bioactive peptides can be enriched or released from milk proteins by the use of selected starter cultures and enzymes and by manipulation of the manufacturing processes such as nanofiltration and encapsulation [62–64]. Bioactive peptides with various biological activities such as antihypertensive, antimicrobial, immunomodulatory, antioxidative, antithrombotic, and antiulcerogenic activities have been reported from milk of camels and other livestock species [9]. Many diseases such as Alzheimer, diabetes, atherosclerosis, rheumatoid arthritis, and cancer result from uncontrolled oxidative stresses by excess of free radicals and other reactive oxygen species present in cellular organism [65–67]. Thus, bioactive peptides derived from milk proteins used as an antioxidants play significant role by preventing the formation of radicals or by scavenging the radicals [68, 69].

Camel milk is reported to be a rich source of proteins with potential antimicrobial, angiotensin-converting enzyme (ACE) inhibitory, and antioxidative activities [70–72]. Casein peptides derived from camel milk showed higher antioxidant and ACE-inhibitory activities after enzymatic digestion [73–75]. Most ACE-inhibitory and opioid peptide fractions of the fermented milk mainly contained β-CN derived peptides as precursor molecules [62]. The relatively higher content of β-CN in camel milk may be advantageous from this point of view. Relatively higher digestibility, antimicrobial activity, and antioxidant activity from camel milk α-LA and whey protein was reported [76, 77]. Moslehishad et al. [72] reported that higher ACE-inhibitory and antioxidant activity was observed in cultured camel milk than bovine milk as a result of structural differences and the presence of higher proline content in camel milk caseins. Homayouni-Tabrizi et al. [71]

identified two novel antioxidant peptides from camel milk proteins using digestive proteases. Thus, the bioactive properties of camel milk may be responsible for the therapeutic properties of the milk.

3.2. The Potential of Camel Milk in Infant Formulations.

Infants who are allergic to bovine milk proteins suffer a severe immune response when they ingest nonhuman milk and thus many studies have been done to reduce the allergenicity of bovine milk or to find milks that can substitute bovine milk without producing an allergenic response [78–80]. Human milk is naturally designed to be optimal nutrition for the neonate of the same species. Bovine milk is frequently used to supplement or substitute the human milk for children. Unfortunately, hypersensitivity to bovine milk proteins is a major source of food allergy, which affects primarily infants. The majority of children suffering from bovine milk protein allergy synthesize antibodies primarily against the β-LG and α_{s1}-CN [81, 82].

In vitro and *in vivo* experiments showed that camel milk is hypoallergenic and a promising substitute for children who are allergic to bovine milk [83]. Variations in the amino-acid compositions between bovine milk and human milk are reported to be the problems in feeding bovine milk-based infant formulation (Table 5). Human milk casein pattern revealed that the dominance of β-CN and α_{s1}-CN is found in small proportion. Kappeler et al. [12] reported high β-CN, lower α_{s1}-CN, and absence of β-LG from camel milk. Thus, this tendency indicates the similarity between camel and human milk. Reports indicated that there was no indicator of immunoglobulin E recognition site at the epitopes of camel milk casein and whey proteins indicating the antigenic difference of camel against bovine, goat, and buffalo milk proteins [80, 84]. Restani et al. [84] indicated that the homologies in amino-acid composition or the phylogenetic differences could justify for difference observed in cross-reactivity between camel and the other livestock species. α-LA is a major whey protein in camel milk and camel α-LA hydrolysates showed higher digestibility and more antioxidative activity than bovine α-LA [76]. These results showed that the chemical composition of camel milk varies considerably from bovine milk. However, chemical composition of camel milk shows remarkable similarity to human milk showing its potential to substitute bovine milk in infant formulations.

3.3. Storage Stability of Fermented Camel Milk.

Fermented milk products are probably developed from the need to extend the shelf life of milk in the absence of cooling facility, their high nutrient contents, and potential health benefits. Traditionally fermented camel milk is the commonly available camel dairy product unlike camel milk cheese, butter, and yoghurt. Fermented camel milk has different names in different parts of the world; *Shubat* in Turkey, Kazakhstan, and Turkmenistan, *Suusac/susa* in Kenya and Somalia, *Gariss* in Sudan, and *Dhanaan* in Ethiopia. Fermented camel milk is reported to remain relatively stable for a longer time at ambient temperature as a result of the antimicrobial properties of the milk.

TABLE 5: Amino acid composition of camel milk proteins compared with bovine and human milk proteins (g/100 g protein).

	Camel[a]	Bovine[b]	Human[c]
Essential			
Arginine	4.0	3.7	3.3
Histidine	2.7	3.3	2.8
Isoleucine	5.1	4.9	3.7
Leucine	9.7	9.3	9.5
Lysine	7.2	8.1	10.1
Methionine	3.2	2.5	1.7
Phenylalanine	5.0	4.2	3.9
Threonine	5.7	7.3	8.3
Tryptophan	1.2	1.4	0.5
Valine	6.7	7.6	8.2
Nonessential			
Alanine	3.0	4.0	4.2
Aspartic	7.0	7.0	6.7
Cysteine	1.2	0.9	1.0
Glycine	1.5	2.5	2.1
Glutamic	21.7	18.6	16.8
Proline	12.0	9.9	10.6
Serine	5.2	6.2	4.1
Tyrosine	4.6	4.6	2.9

Sources. [a]Mehaia and Al-Kahnal [29], [bc]El-Agamy et al. [30].

Dhanaan is reported to have higher storage stability. Pastoralists in Eastern Ethiopia responded that the storage stability of *Dhanaan* is long and it can stay several months when using continuous back slopping, that is, inoculating a new batch of milk with part of a previous batch [85]. Similarly, El-Hadi Sulieman et al. [86] reported that the process of removal of the accumulated *garris* and replacement with fresh camel milk continues for months of time. This might be speculated to the inherent antimicrobial properties of the milk. Potential probiotic lactic acid bacteria strains have been isolated from traditional fermented camel milk and their bacteriocin activities showed inhibitory potential against pathogenic microorganisms [87, 88]. In addition to lactic acid bacteria, yeasts play significant role in the fermentation of traditional camel dairy products since they have strong proteolytic and lipolytic enzymes [87].

4. Conclusion

The main reason for the difficulty of producing products from camel milk is due to the unique structural and functional properties of the milk components. Hence, manufacturing of traditional dairy products using the same technology as for dairy products from bovine milk resulted in processing difficulties. Compositional analyses showed that camel milk is similar to human milk. This indicates the potential of using camel milk in infant formulations to alleviate bovine milk allergy in children. The rich source of bioactive components of camel milk and its compositional properties could be attributed to the therapeutic potential of the milk. Information about the processing technologies and functional properties of camel milk is limited. Hence, more detailed study and holistic approach are needed to fully utilize its technological and functional potentials.

Conflicts of Interest

The authors declare that there are no conflicts of interest regarding the publication of this article.

Acknowledgments

The authors want to express their great thanks to Danish International Development Agency (Danida) for funding "Haramaya Camel Dairy Project." The partners of the project are the Technical University of Denmark, University of Copenhagen (Denmark), Chr. Hansen A/S (Denmark), and Haramaya University (Ethiopia).

References

[1] H. Wu, X. Guang, M. B. Al-Fageeh et al., "Camelid genomes reveal evolution and adaptation to desert environments," *Nature Communications*, vol. 5, article no. 5188, 2014.

[2] FAO stat, 2014. Availlable: http://faostat3.fao.org/, (2017).

[3] Y. Yang, D. Bu, X. Zhao, P. Sun, J. Wang, and L. Zhou, "Proteomic analysis of cow, yak, buffalo, goat and camel milk whey proteins: Quantitative differential expression patterns," *Journal of Proteome Research*, vol. 12, no. 4, pp. 1660–1667, 2013.

[4] T. Ahmed and R. Kanwal, "Biochemical characteristics of lactic acid producing bacteria and preparation of camel milk cheese by using starter culture," *Pakistan Veterinary Journal*, vol. 24, no. 2, pp. 87–91, 2004.

[5] M. A. Mehaia, "Manufacture of fresh soft white cheese (Domiati-type) from ultrafiltered goats' milk," *Food Chemistry*, vol. 79, no. 4, pp. 445–452, 2002.

[6] A. A. Al-Saleh, A. A. M. Metwalli, and E. A. Ismail, "Physicochemical properties of probiotic frozen yoghurt made from camel milk," *International Journal of Dairy Technology*, vol. 64, no. 4, pp. 557–562, 2011.

[7] T. Berhe, E. Seifu, and M. Y. Kurtu, "Physicochemical properties of butter made from camel milk," *International Dairy Journal*, vol. 31, no. 2, pp. 51–54, 2013.

[8] Z. Farah, T. Streiff, and M. R. Bachmann, "Manufacture and characterization of camel milk butter," *Milchwissenschaft*, vol. 44, no. 7, pp. 412–414, 1989.

[9] M. H. A. El-Salam and S. El-Shibiny, "Bioactive peptides of buffalo, camel, goat, sheep, mare, and yak milks and milk products," *Food Reviews International*, vol. 29, no. 1, pp. 1–23, 2013.

[10] E. I. El-Agamy, "Bioactive Components in Camel Milk," *Bioactive Components in Milk and Dairy Products*, pp. 159–194, 2009.

[11] R. Yagil, "Camels and camel milk," *FAO Animal Production and Health Paper*, vol. 26, 69 pages, 1982.

[12] S. Kappeler, Z. Farah, and Z. Puhan, "Sequence analysis of Camelus dromedarius milk caseins," *Journal of Dairy Research*, vol. 65, no. 2, pp. 209–222, 1998.

[13] S. M. Shamsia, "Nutritional and therapeutic properties of camel and human milks," *International Journal of Genetics and Molecular Biology*, vol. 1, no. 2, pp. 52–58, 2009.

[14] S. Bornaz, A. Sahli, A. Attalah, and H. Attia, "Physicochemical characteristics and renneting properties of camels' milk: a comparison with goats', ewes' and cows' milks," *International Journal of Dairy Technology*, vol. 62, no. 4, pp. 505–513, 2009.

[15] M. Glantz, T. G. Devold, G. E. Vegarud, H. Lindmark Månsson, H. Stålhammar, and M. Paulsson, "Importance of casein micelle size and milk composition for milk gelation," *Journal of Dairy Science*, vol. 93, no. 4, pp. 1444–1451, 2010.

[16] G. Konuspayeva, B. Camier, F. Gaucheron, and B. Faye, "Some parameters to process camel milk into cheese," *Emirates Journal of Food and Agriculture*, vol. 26, no. 4, pp. 354–358, 2014.

[17] O. A. Al haj and H. A. Al Kanhal, "Compositional, technological and nutritional aspects of dromedary camel milk," *International Dairy Journal*, vol. 20, no. 12, pp. 811–821, 2010.

[18] P. F. Fox, T. Uniacke-Lowe, P. L. H. McSweeney, and J. A. O'Mahony, "Dairy chemistry and biochemistry, second edition," *Dairy Chemistry and Biochemistry, Second Edition*, pp. 1–584, 2015.

[19] W. Eigel, J. Butler, C. Ernstrom et al., "Nomenclature of proteins of cow's milk: fifth revision," *Journal of Dairy Science*, vol. 67, no. 8, pp. 1599–1631, 1984.

[20] I. E. El-Agamy, "Camel milk," in *Handbook of Milk of Non-Bovine Mammals*, Y. W. Park and G. F. W. Haenlein, Eds., p. 297, 1st edition, 2008.

[21] P. Martin, C. Cebo, and G. Miranda, "Interspecies comparison of milk proteins: quantitative variability and molecular diversity," in *Encyclopedia of Dairy Sciences*, J. W. Fuquay, P. E. Fox, and., and P. L. H. McSweeney, Eds., pp. 3–821, Elsevier Ltd, 1st edition.

[22] M. Malacarne, F. Martuzzi, A. Summer, and P. Mariani, "Protein and fat composition of mare's milk: Some nutritional remarks with reference to human and cow's milk," *International Dairy Journal*, vol. 12, no. 11, pp. 869–877, 2002.

[23] S. R. Kappeler, M. Ackermann, Z. Farah, and Z. Puhan, "Sequence analysis of camel (Camelus dromedarius) lactoferrin," *International Dairy Journal*, vol. 9, no. 7, pp. 481–486, 1999.

[24] A. R. Madureira, C. I. Pereira, A. M. P. Gomes, M. E. Pintado, and F. Xavier Malcata, "Bovine whey proteins - Overview on their main biological properties," *Food Research International*, vol. 40, no. 10, pp. 1197–1211, 2007.

[25] R. A. Inglingstad, T. G. Devold, E. K. Eriksen et al., "Comparison of the digestion of caseins and whey proteins in equine, bovine, caprine and human milks by human gastrointestinal enzymes," *Dairy Science and Technology*, vol. 90, no. 5, pp. 549–563, 2010.

[26] J. B. C. Findlay and K. Brew, "The Complete Amino-Acid Sequence of Human α-Lactalbumin," *European Journal of Biochemistry*, vol. 27, no. 1, pp. 65–86, 1972.

[27] B. Meloun, L. Morávek, and V. Kostka, "Complete amino acid sequence of human serum albumin," *FEBS Letters*, vol. 58, no. 1-2, pp. 134–137, 1975.

[28] H. El-Hatmi, A. Levieux, and D. Levieux, "Camel (Camelus dromedarius) immunoglobulin G, α-lactalbumin, serum albumin and lactoferrin in colostrum and milk during the early post partum period," *Journal of Dairy Research*, vol. 73, no. 3, pp. 288–293, 2006.

[29] M. A. Mehaia and M. A. Al-Kahnal, "Studies on camel and goat milk proteins: nitrogen distribution and amino acid composition," *Nutrition Reports International*, vol. 39, no. 2, pp. 351–357, 1989.

[30] E. I. El-Agamy, I. A.-S. Zeinab, and Y. Abdel-Kaader, "A comparative study of milk proteins from different species II. Electrophoretic patterns, molecular characterization, amino acid

composition and immunological relationships," in *Proceedings of the 3rd Alexandria Conference of Food Science and Technology*, pp. 67–87, 1997.

[31] G. Konuspayeva, B. Faye, and G. Loiseau, "The composition of camel milk: a meta-analysis of the literature data," *Journal of Food Composition and Analysis*, vol. 22, no. 2, pp. 95–101, 2009.

[32] Z. Farah, camel milk properties, 1st ed. SKAT, Sweden, 1996.

[33] H. Attia, N. Kherouatou, and A. Dhouib, "Dromedary milk lactic acid fermentation: Microbiological and rheological characteristics," *Journal of Industrial Microbiology and Biotechnology*, vol. 26, no. 5, pp. 263–270, 2001.

[34] H. M. Abu-Tarboush, "Comparison of Associative Growth and Proteolytic Activity of Yogurt Starters in Whole Milk from Camels and Cows," *Journal of Dairy Science*, vol. 79, no. 3, pp. 366–371, 1996.

[35] T. Berhe, R. Ipsen, E. Seifu, M. Y. Kurtu, M. Eshetu, and E. B. Hansen, "Comparison of the acidification activities of commercial starter cultures in camel and bovine milk," *LWT-Food Sceince Technol*, 2017.

[36] S. K. Ahmed, R. Haroun, and M. O. Eisa, "Banana frozen yoghurt from camel milk," *Pakistan Journal of Nutrition*, vol. 9, no. 10, pp. 955-956, 2010.

[37] E. A. Eissa, A. E. A. Yagoub, E. E. Babiker, and I. A. Mohamed Ahmed, "Physicochemical, microbiological and sensory characteristics of yoghurt produced from camel milk during storage," *Electronic Journal of Environmental, Agricultural and Food Chemistry*, vol. 10, no. 6, pp. 2305–2313, 2011.

[38] I. B. Hashim, A. H. Khalil, and H. Habib, "Quality and acceptability of a set-type yogurt made from camel milk," *Journal of Dairy Science*, vol. 92, no. 3, pp. 857–862, 2009.

[39] N. S. Al-Zoreky and M. M. Al-Otaibi, "Suitability of camel milk for making yogurt," *Food Science and Biotechnology*, vol. 24, no. 2, pp. 601–606, 2015.

[40] A. H. Ibrahim, "Effects of exopolysaccharide-producing starter cultures on physicochemical, rheological and sensory properties of fermented camel's milk," *Emirates Journal of Food and Agriculture*, vol. 27, no. 4, pp. 374–383, 2015.

[41] S. A. Ibrahem and I. E. M. El Zubeir, "Processing, composition and sensory characteristic of yoghurt made from camel milk and camel-sheep milk mixtures," *Small Ruminant Research*, vol. 136, pp. 109–112, 2016.

[42] U. S. Dubey, M. Lal, A. Mittal, and S. Kapur, "Therapeutic potential of camel milk," *Emirates Journal of Food and Agriculture*, vol. 28, no. 3, pp. 164–176, 2016.

[43] A. G. M. Abdel Gader and A. A. Alhaider, "The unique medicinal properties of camel products: A review of the scientific evidence," *Journal of Taibah University Medical Sciences*, vol. 11, no. 2, pp. 98–103, 2016.

[44] T. Mihic, D. Rainkie, K. J. Wilby, and S. A. Pawluk, "The Therapeutic Effects of Camel Milk: A Systematic Review of Animal and Human Trials," *Journal of Evidence-Based Complementary and Alternative Medicine*, vol. 21, no. 4, pp. NP110–NP126, 2016.

[45] A. K. Yadav, R. Kumar, L. Priyadarshini, and J. Singh, "Composition and medicinal properties of camel milk: a review," *Asian Journal of Dairy and Food Research*, vol. 34, no. 2, p. 83, 2015.

[46] R. P. Agrawal, S. Jain, S. Shah, A. Chopra, and V. Agarwal, "Effect of camel milk on glycemic control and insulin requirement in patients with type 1 diabetes: 2-years randomized controlled trial," *European Journal of Clinical Nutrition*, vol. 65, no. 9, pp. 1048–1052, 2011.

[47] R. H. Mohamad, Z. K. Zekry, and H. A. Al-Mehdar, "Camel milk as an adjuvant therapy for the treatment of type 1 diabetes: verification of a traditional ethnomedical practice," *Journal of Medicinal Food*, vol. 12, no. 2, pp. 461–465, 2009.

[48] R. P. Agrawal, S. Budania, P. Sharma et al., "Zero prevalence of diabetes in camel milk consuming Raica community of northwest Rajasthan, India," *Diabetes Research and Clinical Practice*, vol. 76, no. 2, pp. 290–296, 2007.

[49] A. B. Shori, "Camel milk as a potential therapy for controlling diabetes and its complications: A review of in vivo studies," *Journal of Food and Drug Analysis*, vol. 23, no. 4, pp. 609–618, 2015.

[50] R. P. Agrawal, R. Beniwal, D. K. Kochar et al., "Camel milk as an adjunct to insulin therapy improves long-term glycemic control and reduction in doses of insulin in patients with type-1 diabetes: A 1 year randomized controlled trial [1]," *Diabetes Research and Clinical Practice*, vol. 68, no. 2, pp. 176-177, 2005.

[51] O. U. Beg, H. Von Bahr-Lindstrom, Z. H. Zaidi, and H. Jornvall, "A camel milk whey protein rich in half-cystine. Primary structure, assessment of variations, internal repeat patterns, and relationships with neurophysin and other active polypeptides," *European Journal of Biochemistry*, vol. 159, no. 1, pp. 195–201, 1986.

[52] A. O. Abdulrahman, M. A. Ismael, K. Al-Hosaini et al., "Differential effects of camel milk on insulin receptor signaling - Toward understanding the insulin-like properties of camel milk," *Frontiers in Endocrinology*, vol. 7, article no. 4, 2016.

[53] E. I. Elagamy, "Effect of heat treatment on camel milk proteins with respect to antimicrobial factors: A comparison with cows' and buffalo milk proteins," *Food Chemistry*, vol. 68, no. 2, pp. 227–232, 2000.

[54] G. Konuspayeva, B. Faye, G. Loiseau, and D. Levieux, "Lactoferrin and immunoglobulin contents in camel's milk (Camelus bactrianus, Campus dromedarius, and Hybrids) from Kazakhstan," *Journal of Dairy Science*, vol. 90, no. 1, pp. 38–46, 2007.

[55] E. M. El-Fakharany, L. Sánchez, H. A. Al-Mehdar, and E. M. Redwan, "Effectiveness of human, camel, bovine and sheep lactoferrin on the hepatitis C virus cellular infectivity: Comparison study," *Virology Journal*, vol. 10, article no. 199, 2013.

[56] E.-R. M. Redwan and A. Tabll, "Camel lactoferrin markedly inhibits hepatitis C virus genotype 4 infection of human peripheral blood leukocytes," *Journal of Immunoassay and Immunochemistry*, vol. 28, no. 3, pp. 267–277, 2007.

[57] E. M. Redwan, E. M. EL-Fakharany, V. N. Uversky, and M. H. Linjawi, "Screening the anti infectivity potentials of native N- and C-lobes derived from the camel lactoferrin against hepatitis C virus," *BMC Complementary and Alternative Medicine*, vol. 14, no. 1, article no. 219, 2014.

[58] E. M. Redwan, H. A. Almehdar, E. M. El-Fakharany, A.-W. K. Baig, and V. N. Uversky, "Potential antiviral activities of camel, bovine, and human lactoperoxidases against hepatitis C virus genotype 4," *RSC Advances*, vol. 5, no. 74, pp. 60441–60452, 2015.

[59] H. M. Korashy, Z. H. Maayah, A. R. Abd-Allah, A. O. S. El-Kadi, and A. A. Alhaider, "Camel milk triggers apoptotic signaling pathways in human hepatoma HepG2 and breast cancer MCF7 cell lines through transcriptional mechanism," *Journal of Biomedicine and Biotechnology*, vol. 2012, Article ID 593195, 9 pages, 2012.

[60] L. Y. Al-Ayadhi and N. E. Elamin, "Camel milk as a potential therapy as an antioxidant in autism spectrum disorder (ASD)," *Evidence-Based Complementary and Alternative Medicine*, vol. 2013, Article ID 602834, 8 pages, 2013.

[61] S. Bashir and L. Y. Al-Ayadhi, "Effect of camel milk on thymus and activation-regulated chemokine in autistic children: double-blind study," *Pediatric Research*, vol. 75, no. 4, pp. 559–563, 2014.

[62] H. Korhonen and A. Pihlanto, "Bioactive peptides: production and functionality," *International Dairy Journal*, vol. 16, no. 9, pp. 945–960, 2006.

[63] S. Le Maux, A. B. Nongonierma, B. Murray, P. M. Kelly, and R. J. FitzGerald, "Identification of short peptide sequences in the nanofiltration permeate of a bioactive whey protein hydrolysate," *Food Research International*, vol. 77, pp. 534–539, 2015.

[64] V. Đorđević, B. Balanč, A. Belščak-Cvitanović et al., "Trends in encapsulation technologies for delivery of food bioactive compounds," *Food Engineering Reviews*, vol. 7, no. 4, pp. 452–490, 2015.

[65] Q. Liu, A. K. Raina, M. A. Smith, L. M. Sayre, and G. Perry, "Hydroxynonenal, toxic carbonyls, and Alzheimer disease," *Molecular Aspects of Medicine*, vol. 24, no. 4-5, pp. 305–313, 2003.

[66] S. R. J. Maxwell and G. Y. H. Lip, "Free radicals and antioxidants in cardiovascular disease," *British Journal of Clinical Pharmacology*, vol. 44, no. 4, pp. 307–317, 1997.

[67] F. K. Salawu, J. T. Umar, and A. B. Olokoba, "Alzheimers disease: A review of recent developments," *Annals of African Medicine*, vol. 10, no. 2, pp. 73–79, 2011.

[68] B. Halliwell, "Role of free radicals in the neurodegenerative diseases: therapeutic implications for antioxidant treatment," *Drugs and Aging*, vol. 18, no. 9, pp. 685–716, 2001.

[69] A. Pihlanto, "Antioxidative peptides derived from milk proteins," *International Dairy Journal*, vol. 16, no. 11, pp. 1306–1314, 2006.

[70] H. El Hatmi, Z. Jrad, T. Khorchani et al., "Identification of bioactive peptides derived from caseins, glycosylation-dependent cell adhesion molecule-1 (GlyCAM-1), and peptidoglycan recognition protein-1 (PGRP-1) in fermented camel milk," *International Dairy Journal*, vol. 56, pp. 159–168, 2016.

[71] M. Homayouni-Tabrizi, H. Shabestarin, A. Asoodeh, and M. Soltani, "Identification of Two Novel Antioxidant Peptides from Camel Milk Using Digestive Proteases: Impact on Expression Gene of Superoxide Dismutase (SOD) in Hepatocellular Carcinoma Cell Line," *International Journal of Peptide Research and Therapeutics*, vol. 22, no. 2, pp. 187–195, 2016.

[72] M. Moslehishad, M. R. Ehsani, M. Salami et al., "The comparative assessment of ACE-inhibitory and antioxidant activities of peptide fractions obtained from fermented camel and bovine milk by Lactobacillus rhamnosus PTCC 1637," *International Dairy Journal*, vol. 29, no. 2, pp. 82–87, 2013.

[73] Z. Jrad, J.-M. Girardet, I. Adt et al., "Antioxidant activity of camel milk casein before and after in vitro simulated enzymatic digestion," *Mljekarstvo*, vol. 64, no. 4, pp. 287–294, 2014.

[74] M. Rahimi, S. M. Ghaffari, M. Salami et al., "ACE- inhibitory and radical scavenging activities of bioactive peptides obtained from camel milk casein hydrolysis with proteinase K," *Dairy Science and Technology*, vol. 96, no. 4, pp. 489–499, 2016.

[75] M. Salami, A. A. Moosavi-Movahedi, F. Moosavi-Movahedi et al., "Biological activity of camel milk casein following enzymatic digestion," *Journal of Dairy Research*, vol. 78, no. 4, pp. 471–478, 2011.

[76] M. Salami, R. Yousefi, M. R. Ehsani et al., "Enzymatic digestion and antioxidant activity of the native and molten globule states of camel α-lactalbumin: Possible significance for use in infant formula," *International Dairy Journal*, vol. 19, no. 9, pp. 518–523, 2009.

[77] M. Salami, A. A. Moosavi-Movahedi, M. R. Ehsani et al., "Improvement of the antimicrobial and antioxidant activities of camel and bovine whey proteins by limited proteolysis," *Journal of Agricultural and Food Chemistry*, vol. 58, no. 6, pp. 3297–3302, 2010.

[78] J. Rytkönen, T. J. Karttunen, R. Karttunen et al., "Effect of heat denaturation on beta-lactoglobulin-induced gastrointestinal sensitization in rats: Denatured βLG induces a more intensive local immunologic response than native βLG," *Pediatric Allergy and Immunology*, vol. 13, no. 4, pp. 269–277, 2002.

[79] I. Sélo, G. Clément, H. Bernard et al., "Allergy to bovine β-lactoglobulin: Specificity of human IgE to tryptic peptides," *Clinical and Experimental Allergy*, vol. 29, no. 8, pp. 1055–1063, 1999.

[80] E. I. El-Agamy, M. Nawar, S. M. Shamsia, S. Awad, and G. F. W. Haenlein, "Are camel milk proteins convenient to the nutrition of cow milk allergic children?" *Small Ruminant Research*, vol. 82, no. 1, pp. 1–6, 2009.

[81] A. Ametani, T. Sakurai, Y. Katakura et al., "Amino acid residue substitution at t-cell determinant-flanking sites in β-lactoglobulin modulates antigen presentation to t cells through subtle conformational change," *Bioscience, Biotechnology and Biochemistry*, vol. 67, no. 7, pp. 1507–1514, 2003.

[82] C. Bevilacqua, P. Martin, C. Candalh et al., "Goats' milk of defective αs1-casein genotype decreases intestinal and systemic sensitization to β-lactoglobulin in guinea pigs," *Journal of Dairy Research*, vol. 68, no. 2, pp. 217–227, 2001.

[83] E. I. El-Agamy, "The challenge of cow milk protein allergy," *Small Ruminant Research*, vol. 68, no. 1-2, pp. 64–72, 2007.

[84] P. Restani, A. Gaiaschi, A. Plebani et al., "Cross-reactivity between milk proteins from different animal species," *Clinical and Experimental Allergy*, vol. 29, no. 7, pp. 997–1004, 1999.

[85] E. Seifu, "Handling, preservation and utilization of camel milk and camel milk products in Shinile and Jijiga Zones, eastern Ethiopia," *Livestock Research for Rural Development*, vol. 19, no. 6, 2007.

[86] A. M. El-Hadi Sulieman, A. A. Ilayan, and A. El-Awad El Faki, "Chemical and microbiological quality of Garris, Sudanese fermented camel's milk product," *International Journal of Food Science and Technology*, vol. 41, no. 3, pp. 321–328, 2006.

[87] K. Maurad and K.-H. Meriem, "Probiotic characteristics of Lactobacillus plantarum strains from traditional butter made from camel milk in arid regions (Sahara) of Algeria," *Grasas y Aceites*, vol. 59, no. 3, pp. 218–224, 2008.

[88] S. Takeda, K. Yamasaki, M. Takeshita et al., "The investigation of probiotic potential of lactic acid bacteria isolated from traditional Mongolian dairy products," *Animal Science Journal*, vol. 82, no. 4, pp. 571–579, 2011.

24

Isolation and Molecular Identification of Lactic Acid Bacteria using 16s rRNA Genes from Fermented *Teff* (*Eragrostis tef* (Zucc.)) Dough

Belay Tilahun ⓘ,[1] Anteneh Tesfaye ⓘ,[2] Diriba Muleta ⓘ,[2] Andualem Bahiru,[3] Zewdu Terefework,[4] and Gary Wessel[5]

[1]*Department of Biotechnology, College of Natural Sciences, Wolkite University, P.O. Box 07, Wolkite, Ethiopia*
[2]*Institute of Biotechnology, Addis Ababa University, Addis Ababa, Ethiopia*
[3]*Addis Ababa Institute of Technology, Addis Ababa University, Addis Ababa, Ethiopia*
[4]*MRC-ET Molecular Diagnostics Laboratory, Addis Ababa, Ethiopia*
[5]*Department of Molecular Biology, Cellular Biology and Biochemistry, Brown University, Providence, RI 02912, USA*

Correspondence should be addressed to Belay Tilahun; belayt21@gmail.com

Academic Editor: Salam A. Ibrahim

Injera is soft fermented baked product, which is commonly prepared from teff (*Eragrostis tef* (Zucc.)) flour and believed to be consumed on daily basis by two-thirds of Ethiopians. As it is a product of naturally fermented dough, the course of fermentation is done by consortia of microorganisms. The study was aimed at isolating and identifying some dominant bacteria from fermenting *teff* (*Eragrostis tef*) dough. A total of 97 dough samples were collected from households, microenterprises, and hotels with different fermentation stage from Addis Ababa. The bacterial isolates obtained from the fermenting *teff* dough samples were selected on the basis of their acid production potentials. A total of 24 purified bacterial isolates were found to be Gram-positive (they are coccus and rod under microscope) and were good acid producers. Genomic DNA of bacterial isolates were extracted using Invisorb® Spin DNA Extraction kit. 16S rRNA of bacterial isolates were amplified using the bacteria universal primers (rD1 and fD1). The amplified product was sequenced at Genewiz, USA. Sequence analysis and comparison with the resources at the database were conducted to identify the isolated microbes into species and strain levels. The bacterial isolates were identified as *Lactobacillus paracasei*, *Lactobacillus brevis*, *Enterococcus durans*, *Enterococcus hirae*, *Enterococcus avium*, and *Enterococcus faecium*. All identified lactic acid bacteria were able to produce acid at 12 h time of incubation. This study has confirmed the presence of different bacterial species in the fermenting *teff* dough and also supports the involvement of various groups of bacterial species in the course of the fermentation.

1. Introduction

A wide variety of fermented foods and beverages are consumed in Ethiopia being prepared from a wide range of raw materials using traditional techniques. These include *injera, kocho, tella, awaze, borde,* and *tejj* [1]. *Injera* is one of the fermented foods that is made from different cereals, including sorghum, *teff*, corn, wheat, barley, or a combination of some of these cereals [2]. *Injera* from *teff* (*Eragrostis tef*) is much more relished by most Ethiopians than that from any other source. It is a thin soft fermented baked food usually obtained after the flour of cereals has been subjected to 24 to 96 h of fermentation depending on the ambient temperature [1, 3].

The fermentation process uses natural inoculants from different sources in a mixed form [4]. *Teff injera* is getting popularity in the developed world because of its gluten free nature and being a whole grain product [5]. *Teff* is a cereal crop which is mainly cultivated in Ethiopia for the purpose of making *injera* [2, 3]. For injera making, *teff* grain is considered by many as superior when compared to other cereal grains used in the country [6].

Research activities that investigate the microbial diversity and their roles during the course of fermentation of locally fermented products including *injera* have been started in the early 1980s in Ethiopia [7]. However, researches on such regard are not recent and/or limited to only [4] investigating microbial ecology of *injera* employing only phenotypic characterization of some of the fermenting microbial flora [7, 8] and the use of phenotypic characterization of fermenting microbial flora of fermenting dough during the preparation of *injera* employing only biochemical identification [9]. Only a very limited research activity was conducted recently using the present state-of-the-art technology for identifying of the fermenting microbial flora of local products [10, 11].

So, it is important to identify microorganisms with molecular approaches which have been developed to provide more rapid and accurate identification of bacteria using 16S rRNA gene sequences [12].

Therefore, the study was aimed at conducting molecular identification of microorganisms found in fermented *teff* dough hoping to give better understanding of microbial community found in fermented *teff* dough. The general objective of the study was to isolate, identify, and characterize lactic acid bacteria from fermented *teff* (*Eragrostis tef*) dough.

2. Materials and Methods

2.1. Isolates Designation. Microbial isolates were designated as follows: AAUBT for bacterial isolates followed by numbers and capital letters A, B, and C which represent the range of time of fermentation. A represents 48 h, B represents 60 h, and C represents 72 h of fermentation.

2.2. Sample Collection and Description of the Study Sites. A total of 97 teff dough samples (two hundred grams each) were collected from each of 14 sampling sites of fermenting dough samples at 48 h, 60 h, and 72 h time of fermentation [12] from hotels, households, and *injera* baking microenterprises in Addis Ababa, the capital of Ethiopia located at latitude of 8°58′N, longitude of 38°47′E, and altitude/elevation of 2324 m (7625 ft.). The samples were transported aseptically to Holeta Biotechnology Institute, Microbial Biotechnology Laboratory, for processing and microbial isolation. Sample processing, laboratory isolation, and identification of bacteria were carried out at Holeta Biotechnology Institute, Microbial Biotechnology laboratory. Molecular characterization was carried out at MRC-ET Molecular Diagnostics Laboratory, Addis Ababa, Ethiopia, and DNA sequencing was done at Genewiz in collaboration with Brown University, Boston, USA.

2.3. Isolation and Selection of Bacteria. Ten-gram dough from each sample was transferred aseptically into separate flask with 90mL sterile 0.1% peptone water and homogenized. Aliquots of 0.1 mL from appropriate dilutions were spread plated on presolidified de Man, Rogosa, and Sharpe (MRS) agar and incubated at 30°C for 48-72 h. Representative 10-20 colonies of lactic acid bacteria were randomly picked from countable MRS agar plates. Each bacterial isolate was purified by repeated streak-plating on MRS agar for three times. The pure isolates were maintained on MRS agar slants at 4°C and subcultured every four weeks until required for

characterization [1]. The isolates were further examined for cellular morphology and staining characteristics using Gram stain and acid production test [13]. Cell morphology and colonial characteristics were observed on MRS agar [14].

2.3.1. Acid Production Test of Bacteria. Bacterial isolates were refreshed on MRS broth, and then the broth was incubated at 37°C for 24 h. Turbidity of bacterial suspension was adjusted to 0.1 to 0.5 McFarland standard using spectrophotometer at the absorbance of 600 nm [13]. Each well of ELIZA plates was filled with 990 μl of MRS broth with Bromocresol Green (0.04 gm/1000 ml) with pH range of 3.8-5.4. The first row of wells of the ELIZA plates served as a negative control without being inoculated with bacterial isolates. A total of 149 isolates of lactic acid bacteria were inoculated separately into each well with duplication. Finally, the plates were incubated at 37°C for 12 to 48 h. The formation of yellow color on the well indicated a positive result for fermentation or acidification, whereas the absence of color change was considered as a negative result.

2.4. Molecular Characterization of Bacteria

2.4.1. Genomic DNA Extraction. Bacterial isolates were subcultured on MRS medium and incubated at 30°C for 48-72 h; i.e., 30 were selected on the basis of their morphological and acid production characteristics. The DNA of LAB isolates were extracted and purified using an Invisorb Spin DNA Extraction kit, according to the instructions of the manufacturer [15].

2.4.2. Amplification of 16S rDNA of Bacterial Isolates. Fragments of the 16S rRNA genes of each bacterial isolate were separately amplified using the eubacteria universal primers rD1 (5′-AGA GTT TGA TCC TGG CT C AG-3′) and fD1 (5′-AAG GAG GTG ATC CAG CC-3′) [16]. For amplification of 16S rDNA genes of each bacterial isolate, PCR reaction mixtures (50 μl) contained 1μl of the extracted DNA, 5 μl dNTPs, 1 μl of each of the primers rD1 and fD1, 1 μl of *Taq* DNA polymerase (Fermentas, St. Leon-Rot, Germany), and 5 μl PCR buffer. To this content reverse osmosis purified water up to volume of 50 μl was added. The temperature program and the cycle of reactions were as initial denaturation step at 95°C for 60 sec, followed by 35 cycles of denaturation at 94°C for 60 sec, primer annealing at 51°C for 30 sec, and primer extension at 72°C for 60 sec with a final extension at 72°C for 60 sec [17]. After running the PCR, the amplicons of LABs were separated by gel electrophoresis using 3% Agarose gel and 1 μL loading dye with 5 μL PCR products and stained with ethidium bromide for gel documentation.

2.5. Cloning and Sequencing of 16S rDNA Bacteria. The PCR amplified genomic region of interest (i.e., 16S rDNA) of each isolate was ligated into pGEM®-T easy vector (Figure 1) according to manufacturer's instructions, transformed into XL1-blue cells, and inoculated into medium containing 100μg/ml ampicillin for selection. Isolates were grown for 16 h at 37°C with vigorous shaking. Plasmids were isolated as described in QIA quick Gel Extraction Kit. DNA was then

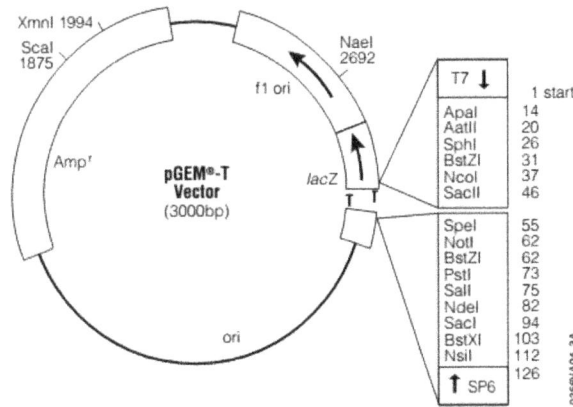

FIGURE 1: pGEM-T or pGEM-T easy vector (Genewiz, 2018).

TABLE 1: Characteristics of the colonial morphology of the isolated LAB and yeasts.

Isolates	Species (16S rRNA gene analysis)	Colony Morphology				Gram stain
		Pigmentation	Shape	Elevation	Size	
AAUBT1C	Unidentified	White	Circular	Flat	Small	+
AAUBT3A	Unidentified	White	Circular	Flat	Small	+
AAUBT4C	Unidentified	White	Circular	Flat	Small	+
AAUBT5C	Unidentified	White	Circular	Flat	Small	+
AAUBT6B	Unidentified	White	Circular	Flat	Small	+
AAUBT8B	Unidentified	White	Circular	Flat	Small	+
AAUBT9A	*Bacillus subtilis*	White	Circular	Flat	Small	+
AAUBT10B	*Bacillus subtilis*	White	Irregular	Flat	Small	+
AAUBT11B	Unidentified	White	Circular	Flat	Small	+
AAUBT12C	*Enterococcus faecium*	White	Circular	Flat	Small	+
AAUBT13B	*Enterococcus avium*	White	Circular	Flat	Small	+
AAUBT14B	*Enterococcus hirae*	White	Circular	Flat	Small	+
AAUBT15C	*Enterococcus hirae*	White	Circular	Flat	Small	+
AAUBT16C	Unidentified	White	Irregular	Flat	Small	+
AAUBT18B	Unidentified	White	Circular	Flat	Small	+
AAUBT19A	*Enterococcus avium*	White	Irregular	Flat	Small	+
AAUBT21B	*Enterococcus durans*	White	Circular	Flat	Small	+
AAUBT22A	*Lactobacillus paracasei*	White	Circular	Flat	Small	+
AAUBT23A	Unidentified	White	Circular	Flat	Small	+
AAUBT24B	*Lactobacillus brevis*	White	Irregular	Flat	Small	+

sequenced by automated DNA sequencer (ABI model 377; Applied Biosystems) at Genewiz.

2.6. Phylogenetic Analysis. DNA sequences were edited, and consensus sequences were obtained using the Bioedit software package. Final sequences were then aligned using CLUSTAL (version: 1.2.4) [18] for each of the sequences. The sequences of bacterial isolates of this study were then compared to those in GenBank (National Centre for Biotechnology Information; http://www.ncbi.nih.gov/) using the Basic Local Alignment Search Tool [19] for nucleotide sequences (*blastn*). Phylogenetic tree construction was performed using the Maximum Likelihood method based on the Tamura-Nei model with MEGA 6.06 [20, 21].

3. Results and Discussion

3.1. Isolation of Bacterial Isolates. From 97 samples collected and processed a total of 249 bacterial isolates were recovered and purified. Purified isolates that are found to be Gram-positive (they are coccus and rod under microscope, Table 1) were tested for the bacterial acid production potential.

3.2. Acidification Test for Bacterial Isolates. Out of 249 bacterial isolates, 24 (9.64%) were found to be good acid producers by changing the color of the medium completely from green to yellow within 48h. Of these 24 isolates, 11 (45.83%) were found to change the color within 12h of incubation and the remaining 13 (54.17%) LAB isolates completed the color

TABLE 2: Acid production capacity of the isolates after incubation.

Species	Frequency of bacterial isolate (no./%)		Total
	12h	48h	
Bacillus subtilis	-	2(10%)	2(10%)
Enterococcus avium	2(10%)	-	2(10%)
Enterococcus durans	1(5%)	-	1(5%)
Enterococcus faecium	1(5%)	-	1(5%)
Enterococcus hirae	2(10%)	-	2(10%)
Lactobacillus brevis	1(5%)	-	1(5%)
Lactobacillus paracasei	1(5%)	-	1(5%)
Unidentified LAB	2(10%)	8(40%)	10(50%)
Total	10(50%)	10(50%)	20(100%)

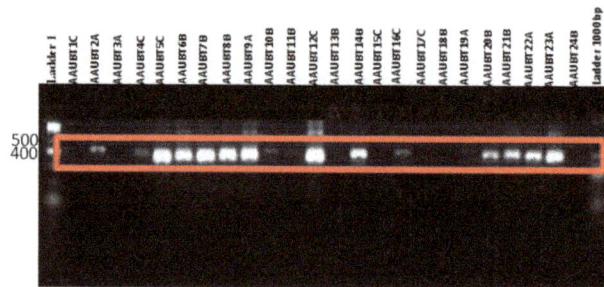

FIGURE 2: PCR amplification 16s rRNA using rD1 and fD1 bacterial universal primers.

change after 48h of incubation. Based on these results the isolates were examined for their acid production potential in relation to time of incubation. *Bacillus subtilis* and other unidentified bacterial species produced acid after 48h time of incubation. As indicated in Table 2 all the identified lactic acid bacteria (i.e., *Lactobacillus paracasei, Enterococcus durans, Enterococcus hirae, Enterococcus avium, Lactobacillus brevis,* and *Enterococcus faecium*) were able to produce acid at 12 h time of incubation.

From 249 isolates 24 isolates were selected on the basis of their acid production potential. As Ashenafi [8], indicated, lactic acid bacteria were responsible for the acidic characteristics of the dough and reduced the pH to about 4.7 and were selected for molecular identification. Lactic acid bacteria produced acid at 12h of incubation. The color change is due to the production of lactic acid, and the pH was reduced so the color of bromophenol blue was changed from blue to yellow color.

Ayele et al. [22] have demonstrated that strong acid producing lactic acid bacteria belong to genera *Pediococcus, Lactobacillus, Streptococcus, Leuconostoc,* and *Bacillus* species. In addition, Brhanu, [23] has also showed that different genera of lactic acid bacteria were responsible for the acidic characteristics of the dough and these included *Pediococcus cerevisiae, Lactobacillus brevis, Lactobacillus plantarum,* and *Lactobacillus fermentum*.

3.3. Molecular Characterization of the Isolates

3.3.1. Amplification and Sequencing of 16S rRNA Region of Bacterial Isolates. All the isolates were shown to have PCR

amplified fragments with around 500bp DNA. And the DNA of some isolates were not amplified; and no band was found; this may be due to the concentration sample DNA used for PCR amplification (Figure 2).

3.3.2. 16S rRNA Sequence Analyses. After all the 16S rRNA sequences of 20 lactic acid bacteria were edited using the Bioedit software package, consensus sequences obtained were blasted in GenBank of NCBI and only 10 samples showed significant similarity of 92-98% in the GenBank. Accordingly, the 10 bacterial isolates were identified and belonging to genera *Lactobacillus* and *Enterococcus* and one genus *Bacillus* as presented in Table 3.

The two LAB genera contained six different species, namely, *Lactobacillus paracasei* (one isolate, AAUBT24B), *Lactobacillus brevis* (one isolate, AAUBT21B), *Enterococcus durans* (one isolate, AAUBT21B), *Enterococcus hirae* (two isolates, AAUBT14B and AAUBT15C), *Enterococcus avium* (two isolates, AAUBT13B and AAUBT19A), and *Enterococcus faecium* (one isolate, AAUBT12C). The genus *Bacillus* only was represented with one species and was named as *Bacillus subtilis* (two isolates AAUBT9A and AAUBT10B).

In this study *Lactobacillus paracasei, Enterococcus durans, Enterococcus hirae, Enterococcus avium, Lactobacillus brevis, Enterococcus faecium,* and *Bacillus subtilis* were also identified from fermenting teff dough. Different workers indicated that microbial flora of fermenting teff dough were complex and shown to include Enterobacteriaceae and aerobic mesophilic bacteria [8], aerobic spore formers, and lactic acid bacteria [7–9]. Brhanu et al. [9] have indicated that the microbes were involved in *teff* flour, indicating that there is the involvement

TABLE 3: Phylogenetic neighbors of bacteria on the basis of similarity to the partial 16S rDNA sequence.

Sequence ID	E-value	Identity	Species (16S rRNA gene analysis)	Accession
AAUBT9A	0.0	98%	*Bacillus subtilis* TAT1-8	HQ_236066.1
AAUBT10B	0.0	98%	*Bacillus subtilis* TAT1-8	HQ_236066.1
AAUBT19A	0.0	94%	*Enterococcus avium* ATCC 14025	NR_114777.2
AAUBT13B	0.0	92%	*Enterococcus avium* ATCC 14025	NR_114777.2
AAUBT21B	0.0	97%	*Enterococcus durans* KLDS 6.0318	DQ_340072.1
AAUBT12C	0.0	98%	*Enterococcus faecium* AT15	KP_137385.1
AAUBT15C	0.0	94%	*Enterococcus hirae* ATCC 9790	NR_075022.1
AAUBT14B	0.0	97%	*Enterococcus hirae* ATCC 9790	NR_075022.1
AAUBT24B	0.0	92%	*Lactobacillus brevis* ATCC 14687	EF_120367.1
AAUBT22A	0.0	97%	*Lactobacillus paracasei* ATCC 25302	NR_117987.1

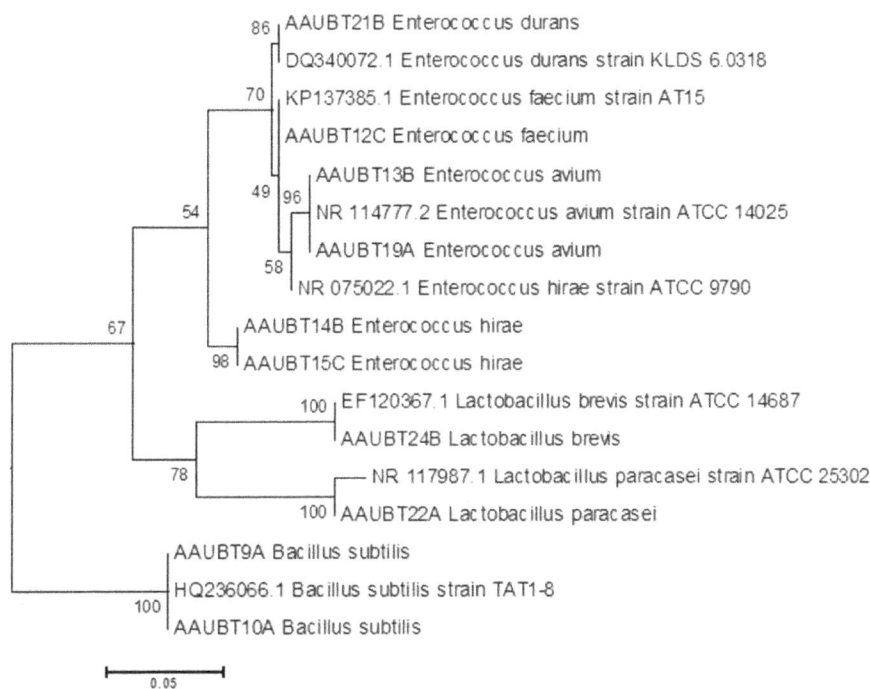

FIGURE 3: Phylogenetic tree of nucleotide sequence of 10 bacteria isolates from *teff* dough, i.e., AAUBT15C, AAUBT9A, AAUBT10A, AAUBT22A, AAUBT21B, AUBT24B, AAUBT13B, AAUBT1A9, AAUBT14B, and AAUBT12C. And 7 from GenBank were HQ236066.1, DQ340072.1, NR075022.1, NR117987.1, EF120367.1, NR 114777.2, and KP137385.1.

of mold, Enterobacteriaceae, aerobic mesophilic bacteria, yeasts, fermentative aerogenic, Gram negative bacteria rods, lactic acid bacteria, and *Bacillus* spp.

Bacillus subtilis were also identified which is supported by Ayele et al. [22], who isolated and identified different *Bacillus* species from teff dough that included *Bacillus subtilis*, *Bacillus licheniformis*, *Bacillus circulans*, *Bacillus laterosporus*, *Bacillus firmus*, *Bacillus alvei*, and *Bacillus larvae*.

Phylogenetic tree made from sequenced 16S rRNA region of seven bacterial isolates of those identified from fermented *teff* dough (Table 3) and evolutionary analyses were conducted in MEGA6.06. The phylogenetic grouping indicated that strains having similar sequences were clustered in the same group and presumably were considered as close relatives. Maximum Likelihood phylogenetic trees based on the 16S rRNA genes of the isolated strains and their closest related species Bootstrap values calculated for 1000 replications are indicated, bar, 5 nt substitution per 100 nt indicated in Figure 3 with the sum of branch length of 0.567189932. The scale length of the tree was 0.05, with branch lengths in the same units as those of the evolutionary distances used to infer the phylogenetic tree. The evolutionary history was inferred by using the Maximum Likelihood method based on the Tamura-Nei model with the highest log likelihood value -1981.7596 and cutoff point 50. The tree was drawn to scale, with branch lengths measured in the number of substitutions per site. The analysis involved 24 nucleotide sequences. All positions containing gaps and missing data were eliminated. There were a total of 522 positions in the final dataset.

4. Conclusion

The study has indicated that different bacterial groups were involved in the fermentation process of *teff* dough. The result of this study may contribute to the future effort of the formulation of starter culture for *injera* dough fermentation.

5. Recommendation

It is important to examine the fermentation potential of each identified bacterial species in order to facilitate the formulation and development of starter cultures.

Conflicts of Interest

The authors declare that they have no conflicts of interest.

Acknowledgments

The authors would like to thank Dr. Abiy Zegeye who has kindly assisted in sequence analysis and bioinformatics part of the research. They would also like to acknowledge the following institutions: Holeta National Agricultural Biotechnology Research Institute, Federal Ministry of Science and Technology, and MRC-ET Molecular Diagnostics Laboratory, for support to accomplish all molecular experiments. This study was financially supported by the Institute of Biotechnology, Addis Ababa University, Ministry of Science and Technology, Ethiopia, USA NIH grant NIGMS 5R01GM125071-24 (GMW), and The Wally Foundation [https://www.brown.edu/research/labs/wessel/wally-foundation-educations].

References

[1] D. Askal and A. Kebede, "Isolation, characterization and identification of lactic acid bacteria and yeast involved in fermentation of Teff (*Eragrostist eff*)," *International Journal*, vol. 1, pp. 36–44, 2013.

[2] A. Mogessie, "A Review on the Microbiology of Indigenous Fermented Foods and Beverages of Ethiopia," *Ethiopian Journal of Biological Sciences*, vol. 5, pp. 189–245, 2006.

[3] G. Bultosa, B. R. Hamaker, and J. N. BeMiller, "An SEC–MALLS Study of Molecular Features of Water-soluble Amylopectin and Amylose of Tef [Eragrostis tef (Zucc.) Trotter] Starches," *Starch - Stärke*, vol. 60, no. 1, pp. 8–22, 2008.

[4] B. R. Stewart and A. Getachew, "Investigation of the nature of Injera," *Econ. Bota*, vol. 16, pp. 127–130, 1962.

[5] T. A. Hiwot, Z. W. Ashore, and D. H. Gulelat, "Preparation of Injera From Pre-Fermented Flour: Nutritional and Sensory Quality," *International Journal of Science, Technology and Society*, vol. 3, pp. 165–175, 2013.

[6] S. Yetneberk, L. W. Rooney, and J. R. N. Taylor, "Improving the quality of sorghum injeraby decortication and compositing with teff," *Journal of the Science of Food and Agriculture*, vol. 85, pp. 1252–1258, 2005.

[7] C. Gifawesen and B. Abraham, "Yeast flora of fermenting *teff* (*Eragrostis teff*) dough," *SINET: Ethiopian Journal of Science*, vol. 5, pp. 21–25, 1982.

[8] M. Ashenafi, "Microbial flora and some chemical properties of ersho, a starter for teff (Eragrostis tef) fermentation," *World Journal of Microbiology & Biotechnology*, vol. 10, no. 1, pp. 69–73, 1994.

[9] A. G. Brhanu, G. Meaza, and B. Abraham, "Teff Fermentation, The role of microorganisms in fermentation and their effect on the nitrogen content of teff," *SINET: Ethiopian Journal of Science*, vol. 5, pp. 69–76, 1982.

[10] K. Bacha, H. Jonsson, and M. Ashenafi, "Microbial dynamics during the fermentation of wakalim, a traditional Ethiopian fermented sausage," *Journal of Food Quality*, vol. 33, no. 3, pp. 370–390, 2010.

[11] T. Anteneh, M. Tetemke, and A. Mogessie, "Antagonism of lactic acid bacteria against foodborne pathogens during fermentation and storage of borde and shamita, traditional Ethiopian fermented beverages," *International Food Research Journal*, vol. 18, no. 3, 2011.

[12] J. M. Janda and S. L. Abbott, "16S rRNA gene sequencing for bacterial identification in the diagnostic laboratory: pluses, perils, and pitfalls," *Journal of Clinical Microbiology*, vol. 45, no. 9, pp. 2761–2764, 2007.

[13] A. Menconi, G. Kallapura, J. D. Latorre et al., "Identification and Characterization of Lactic Acid Bacteria in a Commercial Probiotic Culture," *Bioscience of Microbiota, Food and Health*, vol. 33, no. 1, pp. 25–30, 2014.

[14] M. Haque, F. Akter, K. Hossain, M. Rahman, M. Billah, and K. Islam, "Isolation, identification and analysis of probiotic properties of lactobacillus spp. from selected regional yoghurts," *World Journal of Dairy & Food Sciences*, vol. 5, pp. 39–46, 2010.

[15] D. Gevers, G. Huys, and J. Swings, "Applicability of rep-PCR fingerprinting for identification of Lactobacillus species," *FEMS Microbiology Letters*, vol. 205, no. 1, pp. 31–36, 2001.

[16] D. R. Benson, D. W. Stephens, M. L. Clawson, and W. B. Silvester, "Amplification of 16S rRNA genes from Frankia strains in root nodules of Ceanothus griseus, Coriaria arborea, Coriaria plumosa, Discaria toumatou, and Purshia tridentata," *Applied and Environmental Microbiology*, vol. 62, no. 8, pp. 2904–2909, 1996.

[17] M. Evelyn, "Molecular Characterization and Population Dynamics of Lactic Acid Bacteria during the Fermentation of Sorghum," *University of Pretoria*, pp. 1–54, 2009.

[18] J. D. Thompson, D. G. Higgins, and T. J. Gibson, "CLUSTAL W: improving the sensitivity of progressive multiple sequence alignment through sequence weighting, position-specific gap penalties and weight matrix choice," *Nucleic Acids Research*, vol. 22, no. 22, pp. 4673–4680, 1994.

[19] S. F. Altschul, W. Gish, W. Miller, E. W. Myers, and D. J. Lipman, "Basic local alignment search tool," *Journal of Molecular Biology*, vol. 215, no. 3, pp. 403–410, 1990.

[20] K. Tamura, G. Stecher, D. Peterson, A. Filipski, and S. Kumar, "MEGA6: Molecular Evolutionary Genetics Analysis version 6.0," *Molecular Biology and Evolution*, vol. 30, no. 12, pp. 2725–2729, 2013.

[21] N. Sunil and J. D. Vora, "Licensed Under Creative Commons Attribution CC BY Analysis of 16S rRNA," *Gene of Lactic Acid Bacteria Isolated from Curd*, vol. 5, pp. 2319–7064, 2013.

[22] N. Ayele, A. G. Berhanu, and A. Tarekegn, "Bacilius spp. from Fermented Teff Dough and Kocho: Identify and Role in the

two Ethiopian Fermented Foods," *SINET: Ethiopian Journal of Science*, vol. 20, pp. 101–114, 1997.

[23] A. G. Brhanu, "Involvement of lactic acid bacteria in the fermentation of teff (Eragrostisteff, an Ethiopian fermented food," *Journal of Food Science*, vol. 50, pp. 800-801, 1985.

25

Heat Pump Drying of Fruits and Vegetables: Principles and Potentials for Sub-Saharan Africa

Folasayo Fayose[1] and Zhongjie Huan[2]

[1]*Agricultural and Bio-Resources Engineering, Federal University Oye-Ekiti, PMB 373, Oye-Ekiti 371010, Nigeria*
[2]*Tshwane University of Technology, Pretoria, South Africa*

Correspondence should be addressed to Folasayo Fayose; folasayo.fayose@fuoye.edu.ng

Academic Editor: Alejandro Castillo

Heat pump technology has been used for heating, ventilation, and air-conditioning in domestic and industrial sectors in most developed countries of the world including South Africa. However, heat pump drying (HPD) of fruits and vegetables has been largely unexploited in South Africa and by extension to the sub-Saharan African region. Although studies on heat pump drying started in South Africa several years ago, not much progress has been recorded to date. Many potential users view heat pump drying technology as fragile, slow, and high capital intensive when compared with conventional dryer. This paper tried to divulge the principles and potentials of heat pump drying technology and the conditions for its optimum use. Also, various methods of quantifying performances during heat pump drying as well as the quality of the dried products are highlighted. Necessary factors for maximizing the capacity and efficiency of a heat pump dryer were identified. Finally, the erroneous view that heat pump drying is not feasible economically in sub-Saharan Africa was clarified.

1. Introduction

Consumers, in a bid to have healthier and more natural foodstuffs, have been encouraged to increase their daily intake of fruits and vegetables because their nutritional values as suppliers of vitamins, minerals, fiber, and low fat are well recognized. However, the water content of most fruits and vegetables is higher than 80%, which limits their shelf-life and makes them more susceptible to storage and transport problems. Vegetables and fruits can be made more acceptable to consumers by drying [1]. In addition, there is market for dehydrated fruits and vegetables which increases the importance of drying for most of the countries worldwide [2]. In sub-Saharan Africa, a lot of losses of fruit and vegetables are usually experienced during the peak seasons and only a few cold storage of fruits and vegetables is practiced. Although drying is an energy intensive operation, it is highly very indispensable.

According to Bonazzi and Dumoulin [3], drying is needed to extend the shelf-life of foods without the need for refrigerated storage; to reduce weight and bulk volumes, for saving in the cost of transportation and storage; to convert perishable products (surplus) to stable forms (e.g., milk powder); to produce ingredients and additives for industrial transformation (so-called intermediate food products (IFPs), like vegetables for soups, onions for cooked meats, fruits for cakes, binding agents, aroma, food coloring agents, gel-forming and emulsifying proteins, etc.); and to obtain particular convenience foods (potato flakes, instant drinks, breakfast cereals, dried fruits for use as snacks, etc.), with rapid reconstitution characteristics and good sensorial qualities, for special use, such as in vending machines, or directly for consumers. Also, the loss of product moisture content during drying results in an increasing concentration of nutrients in the remaining mass making proteins, fats, and carbohydrates present in larger amounts per unit weight in the dried food than in the fresh.

In the process of drying, heat is required to evaporate moisture from the product and a flow of air to carry away the evaporated moisture, making drying a high energy consuming operation [4]. There are different heat sources available for drying and these have been well discussed in many articles [5]. However, due to the increasing prices of fossils and electricity and the emission of CO_2 in conventional drying methods, green energy saving and other heat recovery methods for processing and drying of produce become very important. Heat pump technology has been successfully used for drying agricultural products as well as for other domestic dehumidification/heating applications. It has been used for heating, ventilation, and air-conditioning in domestic and industrial sectors in most developed countries of the world including South Africa. However, heat pump drying (HPD) of fruits and vegetables has been largely unexploited in South Africa and by extension to the sub-Saharan African region.

Although studies on heat pump drying started in South Africa several years ago, not much progress has been recorded to date in sub-Saharan Africa. Many potential users view heat pump drying technology as fragile, slow, and high capital intensive when compared with the conventional dryer. However, heat pump drying has been found to be more effective in drying of material with higher amount of free moisture such as tomato [6]. In view of the relevance of heat pump drying, this paper tried to divulge the principles and potentials of heat pump drying technology and the conditions for its optimum use. The paper attempts to bring together the basic information on the effects of heat pump drying, which are inconveniently scattered in several journals and texts in order to justify the need to carry out cutting-edge research on heat pump drying in sub-Saharan Africa.

2. Case for Heat Pump Drying Application in Sub-Saharan Africa

Drying to produce high quality agricultural produce especially fruits and vegetables is yet a bottleneck in most sub-Saharan countries, especially Nigeria. Up till now, a lot of food losses are being experienced due to inadequate storage and processing techniques. Heat pump applications are highly required in the sub-Saharan regions of Africa being typical of the warmer regions of the world [7]. Despite the fact that energy failure is a common experience in most sub-Saharan Africa, the use of air conditioning and refrigeration has been on the increase both in the industries as well as in domestic uses because of the prevalent hot conditions of climate in the region. As the use of air conditioning is increasing in this region, so also must the use of heat pump drying be explored. In fact, harnessing of the recoverable heat from these processes which otherwise would have been wasted to some useful purpose would be a worthwhile exercise.

Heat pump drying has also recorded less drying time than other drying methods and it is simple to design [8, 9] making it suitable for low technology countries in the sub-Saharan region. For development of sustainable energy, three important technological changes have been required: energy economies on the demand side, efficiency improvements in the energy production, and renewing of fossil fuels by various sources of renewable energy. In this regard, HPD systems improve energy efficiency and cause less fossil fuel consumption. Since heat pump drying is a low temperature drying process, it will give a double advantage over the conventional, common, and unreliable sun drying in the region.

In addition, the fact that the major source of electricity in some sub-Saharan regions including Nigeria is hydropower gives confidence that there is safety to the environment as well as reduced energy costs [10, 11].

Even though heat pump attracts more initial cost, when placed vis-a-vis, the reduced variable electricity cost of running it substantiates its preference to other conventional drying methods. Heat pump dryers are known to be cost effective in many drying applications because it can extract and utilize the latent energy of the air and water vapor for product drying [12]. It has been established that heat pump drying consumes only about half or one-third of the electricity of conventional condenser dryers [7, 13]. Earlier published works in the area of heat pump assisted grain drying found the concept to be mechanically feasible but not attractive economically due to the low fuel prices prevailing at the time [8, 14]. However, Prasertsan and Saen-saby [15] showed that HPD had the lowest operating cost when compared to electrically heated convective dryers and direct-fired dryers.

For heat pump dryers, the total cost of removing a liter of water from a product was observed to be considerably lower at long hours than at short hours of operation. Also Sosle et al. [16] confirmed that HPD is useful for materials with high initial moisture content and in regions with high humidity of ambient air. Therefore, HPD is preferable where high value or quality retention outweighs other considerations. In addition, even though the initial cost of acquiring a heat pump drying set up may be high, yet because of its importance and benefits, the use of heat pump drying technology can be enhanced in developing countries through the assistance of government through incentives, supporting policies and advertising [7]. Moreover, the economic value of purchasing a heat pump depends on the relative costs of the energy types that are consumed and saved.

Jangam and Mujumdar [5] observed that the capital and running costs of heat pumps can be reduced by using heat pumps only over the initial drying period, beyond which the dehumidified drying air does not enhance the drying rate any longer. Also, Mujumdar [17] suggested ways of making heat pump drying technology more cost effective including reducing the capital costs by selecting smaller heat pumps and reducing operational costs by reducing the running time in order to decrease cost of electricity utilization or supplementary use of renewable energy such as solar energy where possible. In addition, heat pumps with multiple modes of heat input and intermittent operation allow the use of smaller heat pumps to service more than a single drying chamber for simultaneous drying of different products.

Also, using mathematical models, concurrent and sequential application of heat by radiation, conduction, and convection can enhance the drying kinetics while improving quality at reduced capital and operating costs. There may be need to switch between different modes of heat input to get the optimum and energy efficient drying condition [18]. Moreover, among the many other observations about heat pump drying that worth future study is the use of the clean water which is gained by condensation. According to *current concerns* [19], the water might be used as a side

TABLE 1: Comparison of HPD with other commonly used dryers.

Item	HPD drying	Hot air drying	Vacuum drying	Freeze drying
SMER (kg/kWh)	1.0–4.0	0.1–1.3	0.7–1.2	0.4 and lower
Operating temperature range (°C)	−10 to 80	40 to very high	30–60	−35 to >50
Operating % RH range	10–80	Varies depending on temperature	Low	Low
Drying efficiency (%)	Up to 95	35–40	Up to 70	Very low
Drying rate	Faster	Average	Very slow	Very slow
Capital cost	Moderate	Low	High	Very high
Running cost	Low	High	Very high	Very high
Control	Very good	Moderate	Good	Good

Source: Mujumdar and Jangam [20].

product. Finally, the heat pump can also be used as cooling plants, which is a basis for further developments towards the cooling and storing of fruit.

3. Comparison of Conventional and Heat Pump Drying Processes

Heat pump drying has the ability to recover the latent and sensible heat by condensing moisture from the drying air which may other drying methods cannot do [5]. The recovered heat is recycled back to the dryer through heating of the dehumidified drying air; hence the energy efficiency is increased substantially as a result of heat recovery which otherwise is lost in the atmosphere in conventional dryers [19]. This enables drying at lower temperatures, lower cost, and operation even under humid ambient conditions. A comparison of efficiencies and advantages of heat pump dryers over vacuum and hot air dryers is shown in Table 1. Many articles have been written on the available different types of heat pump dryers [20].

According to FAO [4], the capacity of air to remove moisture depends on its initial temperature and humidity. The relative humidity of the air ought to be controlled so that it does not depend on the absolute humidity of the ambient and the drying temperature [15]. The changes in air conditions when air is heated and passed through a bed of moist product are shown in Figure 1. The heating of air from temperature T_A to T_B is represented by the line AB. During heating, the absolute humidity remains constant at H_A whereas the relative humidity falls from h_A to h_B. This low relative humidity removes moisture from the materials. As air moves through the material bed it absorbs moisture.

Under adiabatic drying, sensible heat in the air is converted to latent heat and the change in air conditions is represented along a line of constant enthalpy, BC. The air will have increase in both absolute humidity, H_C, and relative humidity, h_C, but fall in temperature, T_C. The absorption of moisture by the air would be the difference between the absolute humidities at C and B, that is, $H_C - H_A$. If unheated air was passed through the bed, the drying process would be represented along the line AD. Assuming that the air at D was

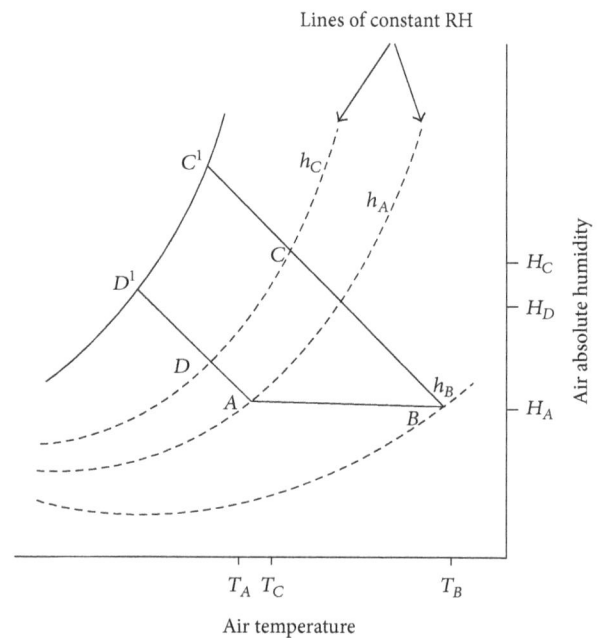

FIGURE 1: Psychrometric representation of the conventional air drying system. Adapted from FAO [4].

at the same relative humidity, h_C, as the heated air at C, then the absorbed moisture would be $(H_D - H_A)$, considerably less than that absorbed by the heated air $(H_C - H_A)$.

At the final stage of drying, there will be little difference of the moisture ratios at the inlet and outlet of the drying chamber. The corresponding temperature difference will also be minimal and these will result in ineffective drying and low thermal efficiency. However, with heat pump drying, there is control of the moisture and temperature of the air as well as heat recovery. In this way, heat pump dryer can improve the product quality while using less energy. The components' arrangement of a typical heat pump drying process is shown in Figure 2. The figure shows that the drying air is dehumidified in the evaporator and reheated to the desired temperature in the condenser before its further passage through the material, thereby offering an advantage

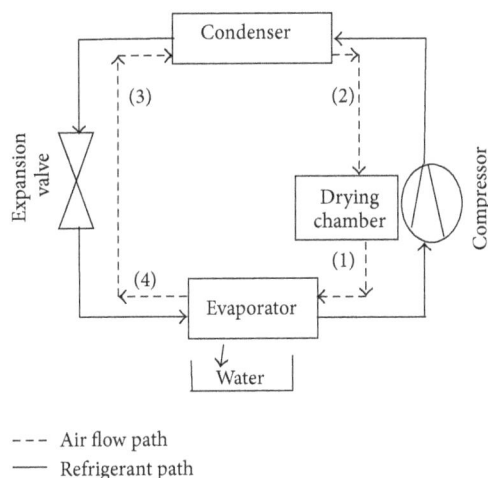

--- Air flow path
——— Refrigerant path

FIGURE 2: Component arrangement of a heat pump dryer.

of better drying rate and product quality over conventional drying.

4. Different Ways of Application of Heat Pump Drying

There are many possible ways of applying heat pump drying. According to Mujumdar and Jangam [20] these possibilities include varying the following: mode of operation, HPD cycle, drying media, supplementary heating, heat pump dryer operation, number of heat pump stages, and temperature for drying. One improvement that heat pump has over other heat sources for drying is that it can be applied to any kind of dryers. Any dryer that uses convection as the primary mode of heat input can be fitted with a suitably designed heat pump, but dryers that require large amounts of drying air, for example, flash or spray dryers, are not suited for HP operation [5, 21]. Heat pump drying technology has been combined with other drying techniques to overcome some problems encountered in those techniques and to achieve improved product quality, reduced energy consumption, high coefficient of performance, and high thermal efficiency [22].

Examples of heat pump assisted drying include heat pump assisted solar drying, microwave drying, infrared drying, fluidized bed drying, atmospheric freeze drying, radiofrequency drying, and chemical heat pump assisted drying [21, 23]. This is in particular with heat sensitive materials like fruits and vegetables that need only low temperature. For example, combining HPD with solar drying enhances the drying and reduces cost. A heat pump is attractive because it can deliver more energy as heat than the electrical energy it consumes. Also it can use modified atmospheres to dry sensitive materials like fruits and vegetables. Moreover, the number of stages of heat pump in the dryer and other arrangements can be varied to improve the performance of the dryer [21]. In addition, chemical heat pump dyer has the advantage of being designed for continuous operation [24] which allows for stable optimum operating conditions [25].

5. Quantification of Performances during Heat Pump Drying

According to Yagcioglu et al. [26], the goals of drying in the food industry can be classified into three groups, as follows: (a) economic considerations, (b) environmental concerns, and (c) product quality aspects. Heat pump drying technology achieves all these goals, by drying in environmentally friendly conditions since gases and fumes are not given off into the atmosphere and operates independently outside ambient weather conditions. In addition, energy consumption is drastically reduced (energy cost savings of between 60% and 80%) and quality of products is safely maintained [27]. Also, the condensate can be recovered and disposed off in an appropriate manner, and a potential also exists to recover valuable volatiles from the condensate [13, 23, 28].

The quality attributes of heat pump dried products are presented below.

5.1. Quality. The three features of heat pump drying technology that help in controlling quality characteristics include the ability to operate at an absolute humidity less than that of the environment, the ability to select a drying temperature less or higher than the environmental temperature, and the ability to dry in a nonvented chamber by using a modified drying atmosphere [29].

The quality of dried products, as enhanced by heat pump drying, can be comprised by a number of physical, chemical, and sensory characteristics, some of which are discussed below.

5.1.1. Microbial Safety. Quality deterioration caused by microorganisms is undesirable commercially because they limit shelf-life or lead to quality deterioration. Drying helps in reducing or overcoming potential microbial damages. With heat pump drying, microbial safety is minimized by ensuring that all raw materials conform to recognized standards of preparation [29]. Heat pump dryers are able to enhance microbial safety in the foodstuffs by maintaining the relative humidity data at acceptable low level. Also, the operating temperature of heat pump dryers is not limited by the environmental humidity. Moreover, Britnell et al. [30] found that air recirculation in heat pump dryers was not a significant problem for commercial fruit and meat products in Australia, the total bacterial count being typically less than 10^3 per gram of product. This is because heat pump dryer does not support a large microbial population on the coils or any other site throughout the dryer. However, to guide against microbial activity build-up, sterilization must be done as and when due.

5.1.2. Colour. Colour degradation is a major cause of loss in food drying. The colour of foods is important to their acceptability. Although sulfating agents inhibit enzymatic and nonenzymatic browning (NEB) reactions, their use is surrounded by health and safety concerns. However, enzymatic browning in the drying of peach halves can be reduced by reducing the relative humidity (20%) without the use of sulfites when the moisture content is high (2 kg/kg dry matter) by increasing the air velocity. Mujumdar [17] observed the need to reduce the drying temperature towards the end of the drying cycle in order to avoid NEB. This strategy is relevant to heat pump dryers because the humidity can be controlled independently of the environment.

Also drying of fruits under nitrogen has been found to be effective in inhibiting browning during the initial critical drying period when the moisture content is high. This condition is possible by using modified atmospheric heat pump drying, thereby producing high quality fruits and vegetables. Another way to achieve nonenzymatic browning in banana and other fruits is by using heat pump dryer to produce specific temperature-humidity schedules [29].

5.1.3. Ascorbic Acid Content (AA), Volatile Compound, and Active Ingredients Retention.

The impact of constant temperature drying on product quality is well recorded. Most of the product quality parameters, such as NEB and AA content, are often manifested by a progressive loss with increasing temperature. Carrington [29] reported that with proper selection of the temperature schedule, the AA content of the guava pieces can be increased to 20% higher than that in the isothermal drying without significant enhancement in drying time. However, results from Perera and Rahman [28] indicated that using reduced air temperatures at the onset of drying as in the case of heat pump drying, followed by temperature elevation as drying proceeds, yields a better quality product.

Also, the concentration of volatile compound is usually increased by drying, particularly at lower temperatures, typical of heat pump drying. Sunthonvit et al. [31] evaluated the effects of different dryer types, namely, cabinet dryer, tunnel dryer, and heat pump dryer on the composition of volatile compounds of dried nectarine. The result indicated that heat pump dryer is the best system for the preservation of volatile compounds in sliced dried fruits in terms of lactones and terpenoids amongst the three methods of drying. Also, the retention of total chlorophyll content and ascorbic acid content in sweet green pepper was observed by Pal et al. [32] to be more in heat pump-dried samples with higher rehydration ratios and sensory scores than in those hot air-dried.

5.1.4. Aroma and Flavour Loss.

Drying methods that employ low temperature do provide high concentration of key aroma compounds [12]. Ginger dried in a heat pump dryer was found to retain over 26% of gingerol, the principal volatile flavour component responsible for its pungency when compared with the rotary dried commercial samples that have only about 20% [29]. The higher volatile retention in heat pump-dried ginger may be due to reduced degradation of gingerol when low drying temperatures are used instead of high convention dryer temperatures. When HPD is conducted in a closed chamber, any compound that volatilizes will remain within it, and the partial pressure for that compound will gradually build up within the chamber, retarding further volatilization from the product [28]. In addition,

Carrington [29] observed that the color and aroma herbs (e.g., parsley, rosemary, and sweet fennel) can be improved with HPD when compared with other commercial products. Also, the sensory values of heat pump dried herbs will be nearly doubled when compared with commercially dried products.

5.1.5. Viability.

Drying with oxygen-sensitive materials, such as flavor compounds and fatty acids, can undergo oxidation, giving rise to poor flavour, colour, and rehydration properties. Cardona et al. [33] studied the heat pump dehydration of lactic acid bacteria (LAB) to determine the optimum procedure and drying conditions at which LAB can be dehydrated in a heat pump dryer that will not result in unacceptable deterioration of viability and activity. The result of the study indicated that heat pump dehydration of LAB gave favourable results which is comparable with the situation when freeze drying method, which is more costly when used. In addition, use of modified atmospheres obtainable with heat pump drying to replace air would allow new dry products to be developed without oxidative reactions occurring [28], thereby producing seed with high proportion of products with germination potentials.

5.1.6. Rehydration.

During drying, important changes in the structural properties of fruits and vegetables can be observed as water is removed from the moist material. Rehydration is a process of moistening dry food materials. In most cases, dried foods are soaked in water before cooking or consumption, therefore rehydration is a very important criterion. Factors affecting the rehydration process include porosity, capillarity and cavity near the food surface, temperature, trapped air bubbles, amorphous-crystalline state, soluble solids, anion, pH of the soaking water, and dryness level. Faster rehydration had been attributed to apple slices dried with a modified atmosphere heat pump dryer [34]. Also, in another study, heat pump and microwave vacuum-dried tomato slices showed comparatively better rehydration ratios than the hot air- or solar cabinet-dried slices [35].

5.1.7. Shrinkage.

Shrinkage occurs first at the surface and gradually moves to the bottom of drying objects with an increase in the drying time. Also, the cell wall becomes elongated and cracks are formed in the inner structure as drying proceeds at high temperature. From microscopy, it was found that shrinkage of apple slices dried in convection is significantly anisotropic, while less damage to the cell structure during freeze drying leads to more isotropic deformation [36]. Heating produces major changes in structure of products. Shrinkage occurs because polymer food stuffs cannot support their weight and, therefore, collapse under gravitational force in the absence of moisture. Heat pump drying however involves drying at low temperature, making shrinkage less pronounced.

5.2. Drying Efficiency.

The performance of a dryer or drying system is characterized by various indices, including energy efficiency, thermal efficiency, volumetric evaporation rate, specific heat consumption, surface heat losses, and unit steam consumption which were defined to reflect the particularities of various drying technologies [37]. Energy efficiency, the ratio of the energy required (Er) to the energy supplied (Es) in drying, is very important because energy consumption is a very significant factor of drying costs [5]. Due to the complex relationships of the food, the water, and the drying medium, that is, the air, a number of efficiency measures can be worked out, each appropriate to circumstances and therefore selectable to bring out special features important in the particular process.

Efficiency calculations are useful when assessing the performance of a dryer, looking for improvements, and in making comparisons between the various classes of dryers which may be alternatives for a particular drying operation [38]. Energy efficiencies are meant for providing an objective comparison between different dryers and drying processes.

There are three groups of factors affecting drying efficiency [4]. They include

(i) those related to the environment, in particular, ambient air conditions;

(ii) those specific to the crop;

(iii) those specific to the design and operation of the dryer.

The factors relating to the environment are well taken care of by HPD, making it have higher efficiencies than other drying methods [39].

Air-drying efficiency, η, can be defined by

$$\eta = \frac{(T_1 - T_2)}{(T_1 - T_a)}, \tag{1}$$

where T_1 is the inlet air temperature into the dryer, T_2 is the outlet air temperature from the dryer, and T_a is the ambient air temperature. As mentioned earlier, the numerator is a major factor that determines drying efficiency. Also, in order to maximize the efficiency, Strommen and Eikevik [25] suggested that the inlet temperature of the dryer should be maximized in accordance with the product requirement. In addition, energy efficiency is the ratio of the latent heat of evaporation of the moisture removed to the drying air heat input. For HPD systems, drying efficiency is a measure of the quantity of energy used to remove one unit mass of water from the product, normally expressed in kJ kg/water or kWh kg/water.

5.3. Specific Moisture Extraction Ratio (SMER).
An alternative indicator of the energy efficiency for heat pump dryers is the specific moisture extraction ratio which is determined using

$$\text{SMER (kg/kWh)} := \frac{\text{Amount of water evaporated}}{\text{Energy used}}. \tag{2}$$

The SMER can be calculated either as an instantaneous value or as an average value during drying [40]. During the drying process, the SMER value decreases as the removal of moisture becomes more difficult, due to smaller water vapor deficits

on the surface of the product. For heat pump dryers, SMER value can be above the theoretical maximum value. The energy efficiency of HPD can be reflected in the high SMER values and drying efficiency when compared to other drying systems, as shown in Table 1. Consequently, high SMER would then be translated to low operating cost, making the payback period for initial capital considerably short. Other definitions of specific moisture extraction rate with respect to the compressor power are reported by [29]. Also, according to Strommen and Eikevik [25], the refrigeration capacity should not be oversized so as not to reduce the relative humidity and a consequent reduced SMER.

5.4. Coefficient of Performance (COP).
The efficiency of the HPD is indicated by compressor cooling coefficient of performance [12]. COP can be used to evaluate the amount of work converted into heat for two different system operations: cooling and heating. For a heat pump, the heat transfer $\dot{Q}_{out}$ from the system to the hot body is desired, and the coefficient of performance is expressed as

$$\text{COP}_{hp} = \frac{\text{Desired Output}}{\text{Required Input}} = \frac{\text{Heat added}}{\text{Work required}}$$
$$= \frac{\dot{Q}_{out}}{\dot{W}_{cycle}}, \tag{3}$$

where $\dot{W}_{cycle}$ is the electrical power input of the compressor.

As much as possible, the evaporating and condensing temperatures should be selected in order to optimize the product of COP and the thermal efficiency [25].

5.5. Drying Rate.
In air drying, the rate of removal of water depends on the conditions of the air, the properties of the food, and the design of the dryer. Drying rate is expressed as follows:

$$\text{DR} = \frac{m_t - m_{t+\Delta t}}{\Delta t}, \tag{4}$$

where m_t is the mass at time t. Drying rates would decrease as moisture content decreases [38].

Factors affecting the drying rate will vary slightly depending upon the type of drying system used. To attain maximum drying capacity, Strommen and Eikevik [25] enunciated that the air flow should be countercurrent instead of cross flow or cocurrent to the product movement in order to maximize the relative humidity at the dryer's outlet. In addition, Wilhelm et al. [41] suggested the following factors to be considered:

(1) nature of the material: physical and chemical composition, moisture content, and so forth;

(2) size, shape, and arrangement of the pieces to be dried;

(3) wet-bulb depression or relative humidity or partial pressure of water vapor in the air (all are related and indicate the amount of moisture already in the air);

(4) air temperature;

(5) air velocity (drying rate is approximately proportional to $u^{0.8}$).

In general, the drying rate decreases with moisture content, increases with increase in air temperature or decreases with increase in air humidity. At very low air flows, increasing the velocity causes faster drying, but at greater velocities the effect is minimal indicating that moisture diffusion within the grain is the controlling mechanism [38].

5.6. Specific Energy Consumption. Specific energy consumption depends on the dryer efficiency, the product, and the initial moisture content. The moisture near the surface of products needs more energy for its removal than the moisture at the center of the product. This is because the moisture flows from the center to the surface. The specific energy consumption is estimated by considering the drying time involved and energy utilization by the various components of the dryer. It is expressed in terms of MJ/kg of water removed and used as one of the factors in the optimization of process parameters. According to Jokiniemi et al. [14], it can be calculated by integrating the energy used and by calculating the amount of moisture removed as in

$$Qs = \frac{Qh}{\Delta G}, \tag{5}$$

where Qs is the specific energy consumption of drying; Qh is the energy consumption of drying air; and ΔG is the mass of the evaporated water.

6. Conclusion

This paper has reviewed the principles and potentials of heat pump drying of fruits and vegetables for application to sub-Saharan Africa. It has been shown in this paper that heat pump dryers are promising technologies that maintain product quality and reduce energy consumption of drying, particularly for high value products like fruits and vegetables. The application of heat pump drying contributes positively to the following fruit and vegetables quality attributes including improved microbial safety, better colour, vitamin C retention, enhanced volatile compound, aroma and flavor compounds, rehydration, and texture. Finally, some factors that can make heat pump drying cost effective and energy efficient were elucidated. Adoption of heat pump drying technology for drying of fruits and vegetables in sub-Saharan Africa will improve product quality and reduce energy consumed in the process.

Conflict of Interests

The authors declare that there is no conflict of interests regarding the publication of this paper.

References

[1] B. I. Abonyi, H. Feng, J. Tang et al., "Quality retention in strawberry and carrot purees dried with Refractance Window system," *Journal of Food Science*, vol. 67, no. 3, pp. 1051–1056, 2002.

[2] T. Funebo and T. Ohlsson, "Microwave-assisted air dehydration of apple and mushroom," *Journal of Food Engineering*, vol. 38, no. 3, pp. 353–367, 1998.

[3] C. Bonazzi and E. Dumoulin, "Quality changes in food materials as influenced by drying processes," in *Modern Drying Technology, Volume 3: Product Quality and Formulation*, E. Tsotsas and A. S. Mujumdar, Eds., chapter 1, pp. 1–20, Wiley-VCH, 2011.

[4] FAO Agriculture and Consumer Protection, *Grain Storage Techniques—Evolution and Trends in Developing Countries*, FAO Corporate Document Repository, 1994, http://www.fao.org/docrep/t1838e/t1838e0u.htm.

[5] S. V. Jangam and A. S. Mujumdar, "Classification and selection of dryers for foods," in *Drying of Foods, Vegetables and Fruits*, S. V. Jangam, L. C. Lim, and A. S. Mujumdar, Eds., vol. 1, pp. 59–82, Singapore, 2010.

[6] V. Sosle, *A heat pump dehumidifier assisted dryer for agri-foods [Ph.D. dissertation]*, McGill University, Ontario, Canada, 2002, http://www.redalyc.org/pdf/1698/169823914004.pdf.

[7] ESRU, *Heat Pump Background*, A Publication of the Energy Systems Research Unit, University of Strathclyde, Glasgow, UK, 2015, http://www.esru.strath.ac.uk/EandE/Web_sites/01-02/heat_pump/background.html.

[8] V. R. Sagar and P. Suresh Kumar, "Recent advances in drying and dehydration of fruits and vegetables: a review," *Journal of Food Science and Technology*, vol. 47, no. 1, pp. 15–26, 2010.

[9] S. V. Jangam, V. S. Joshi, A. S. Mujumdar, and B. N. Thorat, "Studies on dehydration of sapota (*Achras zapota*)," *Drying Technology*, vol. 26, no. 3, pp. 369–377, 2008.

[10] A. Bailes, "The Shocking Truth about Heat Pumps," 2015, http://www.energyvanguard.com/blog-building-science-HERS-BPI/bid/69996/The-Shocking-Truth-About-Heat-Pumps.

[11] I. C. Kemp, "Fundamentals of energy analysis of dryers," in *Modern Drying Technology*, E. Tsotsas and A. S. Mujumdar, Eds., vol. 4, pp. 1–46, Energy Savings, 2011.

[12] I. Mujić, M. B. Kralj, S. Jokić, K. Jarni, T. Jug, and Ž. Prgomet, "Changes in aromatic profile of fresh and dried fig—the role of pre-treatments in drying process," *International Journal of Food Science and Technology*, vol. 47, no. 11, pp. 2282–2288, 2012.

[13] J. Nipkow and E. Bush, "Promotion of energy-efficient heat pump dryers," in *Proceedings of the 5th International Conference on Energy Efficiency in Domestic Appliances and Lighting (EEDAL '09)*, vol. 9, pp. 16–18, Berlin, Germany, June 2009.

[14] T. Jokiniemi, K. Kautto, E. Kokin, and J. Ahokas, "Energy efficiency measurements in grain drying," *Agronomy Research Biosystem Engineering*, vol. 1, pp. 69–75, 2011.

[15] S. Prasertsan and P. Saen-saby, "Heat pump dryers: research and development needs and opportunities," *Drying Technology*, vol. 16, no. 1-2, pp. 251–270, 1998.

[16] V. Sosle, G. S. V. Raghavan, and R. Kittler, "Experiences with a heat pump dehumidifier assisted drying system," in *Proceedings of the 1st Nordic Drying Conference*, Alves-Filho, Eikevik, and Strømmen, Eds., Paper no. 29, Trondheim, Norway, June 2001.

[17] A. S. Mujumdar, "Drying fundamentals," in *Industrial Drying of Foods*, C. G. J. Baker, Ed., pp. 7–30, Blackie Academic & Professional, London, UK, 1997.

[18] M. R. Islam, J. C. Ho, and A. S. Mujumdar, "Simulation of liquid diffusion-controlled drying of shrinking thin slabs subjected to multiple heat sources," *Drying Technology*, vol. 21, no. 3, pp. 413–438, 2003.

[19] Current Concerns, *Fruit Drying with Heat Pumps*, Current Concerns no. 22, Current Concerns, Zürich, Switzerland, 2009.

[20] A. S. Mujumdar and S. V. Jangam, "Energy issues and use of renewable source of energy for drying of foods," in *Proceedings*

of the International Workshop on Drying of Food and Biomaterials, Bangkok, Thailand, June 2011.

[21] M. R. Islam and A. S. Mujumdar, "Heat pump-assisted drying," in *Drying Technologies in Food Processing*, X. D. Chen and A. S. Mujumdar, Eds., pp. 190–224, Blackwell Publishing, Oxford, UK, 2008.

[22] S. V. Jangam and A. S. Mujumdar, "Heat pump assisted drying technology—overview with focus on energy, environment and product quality," in *Modern Drying Technology: Energy Savings, Volume 4: Energy Savings*, E. Tsotsas and A. S. Mujumdar, Eds., chapter 4, pp. 121–162, Wiley-VCH, 2011.

[23] K. K. Patel and A. Kar, "Heat pump assisted drying of agricultural produce—an overview," *Journal of Food Science and Technology*, vol. 49, no. 2, pp. 142–160, 2012.

[24] T. Kudra and A. S. Mujumdar, *Advanced Drying Technology*, Marcel Dekker, Basel, Switzerland, 2002.

[25] I. Strommen and T. M. Eikevik, "Operational modes for heat pump drying: new technologies and production for a new generation of high quality fish products," in *Presented at 21st International of Congress of Refrigeration (ICR '03)*, Washington, DC, USA, 2003.

[26] A. K. Yagcioglu, V. Demir, and T. Gunhan, "Determination of the drying characteristics of laurel leaves. Ege University Research Fund Project final report," Project 99 ZRF 029, Agricultural Machinery Department, Faculty of Agriculture, 2001.

[27] V. Sosle, "Heat pump drying," *Stewart Postharvest Review*, vol. 2, no. 6, pp. 1–5, 2006.

[28] C. O. Perera and M. S. Rahman, "Heat pump dehumidifier drying of food," *Trends in Food Science and Technology*, vol. 8, no. 3, pp. 75–79, 1997.

[29] C. G. Carrington, "Heat pump and dehumidification drying," in *Food Drying Science and Technology: Microbiology, Chemistry, Applications*, Y. H. Hui, Ed., Technology & Engineering, 2008.

[30] P. Britnell, S. Birchall, S. Fitz-Paine, G. Young, R. Mason, and A. Wood, "The application of heat pump dryer in Australian food industry," *Drying*, vol. 94, pp. 897–903, 1994.

[31] N. Sunthonvit, G. Srzednicki, and J. Craske, "Effects of drying treatments on the composition of volatile compounds in dried nectarines," *Drying Technology*, vol. 25, no. 5, pp. 877–881, 2007.

[32] U. S. Pal, M. K. Khan, and S. N. Mohanty, "Heat pump drying of green sweet pepper," *Drying Technology*, vol. 26, no. 12, pp. 1584–1590, 2008.

[33] T. D. Cardona, R. H. Driscoll, J. L. Paterson, G. S. Srzednicki, and W. S. Kim, "Optimizing conditions for heat pump dehydration of lactic acid bacteria," *Drying Technology*, vol. 20, no. 8, pp. 1611–1632, 2002.

[34] M. N. A. Hawlader, C. O. Perera, and M. Tian, "Properties of modified atmosphere heat pump dried foods," *Journal of Food Engineering*, vol. 74, no. 3, pp. 392–401, 2006.

[35] T. J. Gaware, N. Sutar, and B. N. Thorat, "Drying of tomato using different methods: comparison of dehydration and rehydration kinetics," *Drying Technology*, vol. 28, no. 5, pp. 651–658, 2010.

[36] M. S. Rahman, "Post-drying aspects for meat and horticultural products," in *Drying Technologies in Food Processing*, X. D. Chen and A. S. Mujumdar, Eds., pp. 252–269, John Wiley & Sons, New York, NY, USA, 2008.

[37] T. Kudra, "Energy performance of convective dryers," *Drying Technology*, vol. 30, no. 11-12, pp. 1190–1198, 2012.

[38] R. L. Earle and M. D. Earle, *Unit Operations in Food Processing*, The New Zealand Institute of Food Science & Technology, 2004.

[39] M. R. Islam and A. S. Mujumdar, "Heat pump assisted drying," in *Heat Pump Assisted Drying—in Guide to Industrial Drying*, A. S. Mujumdar, Ed., pp. 187–210, Colour, Mumbai, India, 2004.

[40] G. S. V. Raghavan, T. J. Rennie, P. S. Sunjka, V. Orsat, W. Phaphuangwittayakul, and P. Terdtoon, "Overview of new techniques for drying biological materials with emphasis on energy aspects," *Brazilian Journal of Chemical Engineering*, vol. 22, no. 2, pp. 195–201, 2005.

[41] L. R. Wilhelm, D. A. Suter, and G. H. Brusewitz, "Drying and dehydration," in *Food & Process Engineering Technology*, chapter 10, pp. 259–284, American Society of Agricultural Engineers, 2006.

Permissions

All chapters in this book were first published in IJFS, by Hindawi Publishing Corporation; hereby published with permission under the Creative Commons Attribution License or equivalent. Every chapter published in this book has been scrutinized by our experts. Their significance has been extensively debated. The topics covered herein carry significant findings which will fuel the growth of the discipline. They may even be implemented as practical applications or may be referred to as a beginning point for another development.

The contributors of this book come from diverse backgrounds, making this book a truly international effort. This book will bring forth new frontiers with its revolutionizing research information and detailed analysis of the nascent developments around the world.

We would like to thank all the contributing authors for lending their expertise to make the book truly unique. They have played a crucial role in the development of this book. Without their invaluable contributions this book wouldn't have been possible. They have made vital efforts to compile up to date information on the varied aspects of this subject to make this book a valuable addition to the collection of many professionals and students.

This book was conceptualized with the vision of imparting up-to-date information and advanced data in this field. To ensure the same, a matchless editorial board was set up. Every individual on the board went through rigorous rounds of assessment to prove their worth. After which they invested a large part of their time researching and compiling the most relevant data for our readers.

The editorial board has been involved in producing this book since its inception. They have spent rigorous hours researching and exploring the diverse topics which have resulted in the successful publishing of this book. They have passed on their knowledge of decades through this book. To expedite this challenging task, the publisher supported the team at every step. A small team of assistant editors was also appointed to further simplify the editing procedure and attain best results for the readers.

Apart from the editorial board, the designing team has also invested a significant amount of their time in understanding the subject and creating the most relevant covers. They scrutinized every image to scout for the most suitable representation of the subject and create an appropriate cover for the book.

The publishing team has been an ardent support to the editorial, designing and production team. Their endless efforts to recruit the best for this project, has resulted in the accomplishment of this book. They are a veteran in the field of academics and their pool of knowledge is as vast as their experience in printing. Their expertise and guidance has proved useful at every step. Their uncompromising quality standards have made this book an exceptional effort. Their encouragement from time to time has been an inspiration for everyone.

The publisher and the editorial board hope that this book will prove to be a valuable piece of knowledge for researchers, students, practitioners and scholars across the globe.

List of Contributors

Paul Dawson, Wesam Al-Jeddawi and Nanne Remington
Department of Food, Nutrition and Packaging Sciences, Clemson University, Clemson, SC 29634, USA

Diding Suhandy
Laboratory of Bioprocess and Postharvest Engineering, Department of Agricultural Engineering, The University of Lampung, Jl. Soemantri Brojonegoro No. 1, Gedong Meneng, Bandar Lampung, Lampung 35145, Indonesia

Meinilwita Yulia
Department of Agricultural Technology, Lampung State Polytechnic, Jl. Soekarno Hatta No. 10, Rajabasa, Bandar Lampung, Lampung, Indonesia

Derick Nyabera Malavi
Department of Food Science, Nutrition and Technology, University of Nairobi, Nairobi 00625, Kenya
International Potato Centre (CIP), Sub-Saharan Africa (SSA) Regional Office, Old Naivasha Road, Nairobi 00603, Kenya

Tawanda Muzhingi
International Potato Centre (CIP), Sub-Saharan Africa (SSA) Regional Office, Old Naivasha Road, Nairobi 00603, Kenya

George Ooko Abong'
Department of Food Science, Nutrition and Technology, University of Nairobi, Nairobi 00625, Kenya

Youri Joh
Robert M. Kerr Food & Agricultural Products Center, Oklahoma State University, Stillwater, OK 74078, USA

Niels Maness
Department of Horticulture and Landscape Architecture, Oklahoma State University, Stillwater, OK 74078, USA

William McGlynn
Robert M. Kerr Food & Agricultural Products Center, Oklahoma State University, Stillwater, OK 74078, USA
Department of Horticulture and Landscape Architecture, Oklahoma State University, Stillwater, OK 74078, USA

Carla María Blanco-Lizarazo
Agroindustrial Process Research Group, University of La Sabana, Colombia

Indira Sotelo-Díaz
Titular Professor, EICEA, Agroindustrial Process Research Group, University of La Sabana, Colombia

José Luis Arjona-Roman and Adriana Llorente-Bousquets
Engineering and Technology Department, Faculty of Advanced Studies Cuautitlán, National Autonomous University of Mexico (UNAM), Mexico

René Miranda-Ruvalcaba
Chemistry Department, Faculty of Advanced Studies Cuautitlan, National Autonomous University of Mexico (UNAM), Mexico

Anis Hamizah and Nurrulhidayah binti Ahamad Fadzillah
International Institute for Halal Research and Training (INHART), International Islamic University Malaysia (IIUM), Gombak, Kuala Lumpur, Malaysia

Ademola Monsur Hammed
International Institute for Halal Research and Training (INHART), International Islamic University Malaysia (IIUM), Gombak, Kuala Lumpur, Malaysia
Plant Sciences Department, North Dakota State University, Fargo, ND, USA
Biotechnology Engineering Department, International Islamic University Malaysia (IIUM), Gombak, Kuala Lumpur, Malaysia

Tawakalit Tope Asiyanbi-H
Plant Sciences Department, North Dakota State University, Fargo, ND, USA

Mohamed Elwathig Saeed Mirghani and Irwandi Jaswir
International Institute for Halal Research and Training (INHART), International Islamic University Malaysia (IIUM), Gombak, Kuala Lumpur, Malaysia Biotechnology Engineering Department, International Islamic University Malaysia (IIUM), Gombak, Kuala Lumpur, Malaysia

Arnaud Landry Suffo Kamela
Department of Biochemistry, Faculty of Science, University of Dschang, Dschang, Cameroon
Department of Food Engineering and Design, Indian Institute of Crop and Processing Technology, Thanjavur, India

Raymond Simplice Mouokeu
Institute of Fisheries and Aquatic Sciences, University of Douala, Douala, Cameroon

Rawson Ashish
Department of Food Engineering and Design, Indian Institute of Crop and Processing Technology, Thanjavur, India

Ghislain Maffo Tazoho, Lamye Glory Moh and Jules-Roger Kuiate
Department of Biochemistry, Faculty of Science, University of Dschang, Dschang, Cameroon

Etienne Pamo Tedonkeng
Department of Animal Production, Faculty of Agronomy and Agricultural Science, University of Dschang, Dschang, Cameroon

J. J. Lafont-Mendoza
Universidad de Cordoba, Monteria, Colombia

C. A. Severiche-Sierra
Universidad de Cartagena, Cartagena de Indias, Colombia
Corporacion Universitaria Minuto de Dios (UNIMINUTO), Barranquilla, Colombia

J. Jaimes-Morales
Universidad de Cartagena, Cartagena de Indias, Colombia

Robi Andoyo, Vania Dianti Lestari, Efri Mardawati and Bambang Nurhadi
Department of Food Industrial Technology, Faculty of Agro-Industrial Technology, Universitas Padjadjaran, Jl. Raya Bandung Sumedang Km. 21, Jatinangor, Sumedang 40600, Indonesia

Ahmad Ni'matullah Al-Baarri
Department of Food Technology, Faculty of Animal and Agricultural Sciences, Diponegoro University, Semarang 50275, Indonesia
Laboratory of Food Technology, Integrated Laboratory, Diponegoro University, Semarang 50275, Indonesia

Anang Mohamad Legowo
Department of Food Technology, Faculty of Animal and Agricultural Sciences, Diponegoro University, Semarang 50275, Indonesia

Septinika Kurnia Arum
Department of Animal Science, Faculty of Animal and Agricultural Sciences, Diponegoro University, Semarang 50275, Indonesia

Shigeru Hayakawa
Department of Applied Biological Sciences, Faculty of Agriculture, Kagawa University,Miki-cho 761-0795, Japan

Eisuke Kuraya, Akiko Touyama, Shina Nakada, Osamu Higa and Shigeru Itoh
National Institute of Technology, Okinawa College, 905 Henoko, Nago City, Okinawa 905-2192, Japan

Rahma Belcadi-Haloui and Abdelhakim Hatimi
Laboratory of Plants Biotechnologies, Faculty of Sciences, BP 8016, Agadir 80 000, Morocco

Abderrahmane Zekhnini
Laboratory of Aquatic Systems, Faculty of Sciences, BP 8016, Agadir 80 000, Morocco

Yassine El-Alem
Autonomous Establishment of Control and Coordination of Exports, 23 E, Industrial Zone of Tassila, Agadir 80 000, Morocco

James Owusu-Kwarteng and Fortune Akabanda
Department of Applied Biology, Faculty of Applied Sciences, University for Development Studies, Navrongo, Ghana

Francis K. K. Kori
Department of Applied Physics, Faculty of Applied Sciences, University for Development Studies, Navrongo, Ghana

Ana Gabriela Morachis-Valdez
Departamento de Toxicología Ambiental, Facultad de Química, Universidad Autónoma del Estado de México, Toluca, MEX, Mexico
Departamento de Alimentos, Facultad de Química, Universidad Autónoma del Estado de México, Toluca, MEX, Mexico

Leobardo Manuel Gómez-Oliván and María Dolores Hernández-Navarro
Departamento de Toxicología Ambiental, Facultad de Química, Universidad Autónoma del Estado de México, Toluca, MEX, Mexico

Imelda García-Argueta
Departamento de Nutrición, Facultad de Medicina, Universidad Autónoma del Estado de México, Toluca, MEX, Mexico

Daniel Díaz-Bandera and Octavio Dublán-García
Departamento de Alimentos, Facultad de Química, Universidad Autónoma del Estado de México, Toluca, MEX, Mexico

Alphonse Laya
Department of Life and Earth Sciences, Higher Teachers' Training College of Maroua, University of Maroua, Maroua, Cameroon
Department of Biological Sciences, Faculty of Science, University of Maroua, Maroua, Cameroon

Benoît Bargui Koubala
Department of Life and Earth Sciences, Higher Teachers' Training College of Maroua, University of Maroua, Maroua, Cameroon
Department of Chemistry, Faculty of Science, University of Maroua, Maroua, Cameroon

Habiba Kouninki
Department of Life and Earth Sciences, Higher Teachers' Training College of Maroua, University of Maroua, Maroua, Cameroon

Elias Nchiwan Nukenine
Department of Biological Sciences, Faculty of Science, University of Ngaoundéré, Ngaoundéré, Cameroon

Sally S. Lloyd
CYO'Connor ERADE Village Foundation, Canning Vale South, WA 6155, Australia
CYO'Connor Centre for Innovation in Agriculture, Murdoch University, 5 Del Park Road, North Dandalup, WA 6207, Australia

Jose L. Valenzuela and Roger L. Dawkins
CYO'Connor ERADE Village Foundation, Canning Vale South, WA 6155, Australia
CYO'Connor Centre for Innovation in Agriculture, Murdoch University, 5 Del Park Road, North Dandalup, WA 6207, Australia
Melaleuka Stud, 24 Genomics Rise, Piara Waters, WA 6112, Australia

Edward J. Steele
CYO'Connor ERADE Village Foundation, Canning Vale South, WA 6155, Australia

T. Akonor, H. Ofori and N. T. Dziedzoave
Council for Scientific and Industrial Research-Food Research Institute, Accra, Ghana

N. K. Kortei
Graduate School of Nuclear and Allied Sciences, University of Ghana, Legon, Ghana

Sohana Al Sanjee
Department of Microbiology, Faculty of Biological Sciences, University of Chittagong, Chittagong 4331, Bangladesh

Md. Ekramul Karim
Department of Microbiology, Faculty of Biological Sciences, University of Chittagong, Chittagong 4331, Bangladesh
Environmental Biotechnology Division, National Institute of Biotechnology, Ganakbari, Ashulia, Dhaka 1349, Bangladesh

Joyce Ndunge Musyoka and George Ooko Abong
Department of Food Science, Nutrition and Technology, University of Nairobi, Kangemi, Kenya

Jan Low, Daniel Mahuga Mbogo and Tawanda Muzhingi
International Potato Center (CIP), Sub-Saharan Africa (SSA) Regional Office, Old Naivasha Road, Nairobi, Kenya

Richard Fuchs
Food and Markets Department, Natural Resources Institute of University of Greenwich, Central Avenue, Chatham Maritime, Chatham, KentME4 4TB, UK

Simon Heck
International Potato Center (CIP), Regional Office, Plot 106, Katalima Road, Naguru, Kampala, Uganda

Sarai Villalobos-Chaparro, Erika Salas-Muñóz, Néstor Gutiérrez-Méndez and Guadalupe Virginia Nevárez-Moorillón
Facultad de Ciencias Químicas, Universidad Autónoma de Chihuahua, Chihuahua, 31125, Mexico

Maria Erna Kustyawati
Department of Agriculture Product Technology, University of Lampung, Bandar Lampung 34145, Indonesia

Filli Pratama, Daniel Saputra and Agus Wijaya
Department of Agriculture Technology, University of Sriwijaya, Palembang 30662, Indonesia

Katie R. Kirsch and Jessica C. Hudson
Department of Nutrition and Food Science, Texas A&M University, College Station, TX, USA

Tamra N. Tolen, Alejandro Castillo and T. Matthew Taylor
Department of Animal Science, Texas A&M University, College Station, TX, USA

Davey Griffin
Department of Animal Science, Texas A&M AgriLife Extension, College Station, TX, USA

Tesfemariam Berhe
School of Animal and Range Sciences, Haramaya University, Dire Dawa, Ethiopia
Department of Food Science, University of Copenhagen, Rolighedsvej 30, 1958 Frederiksberg C, Denmark

Eyassu Seifu
Department of Food Science and Technology, Botswana University of Agriculture and Natural Resources, Gaborone, Botswana

Richard Ipsen
Department of Food Science, University of Copenhagen, Rolighedsvej 30, 1958 Frederiksberg C, Denmark

Mohamed Y. Kurtu
School of Animal and Range Sciences, Haramaya University, Dire Dawa, Ethiopia

Egon Bech Hansen
Division for Diet, Disease Prevention and Toxicology, National Food Institute, Technical University of Denmark, 2860 Søborg, Denmark

Belay Tilahun
Department of Biotechnology, College of Natural Sciences, Wolkite University, Wolkite, Ethiopia

Anteneh Tesfaye and Diriba Muleta
Institute of Biotechnology, Addis Ababa University, Addis Ababa, Ethiopia

Andualem Bahiru
Addis Ababa Institute of Technology, Addis Ababa University, Addis Ababa, Ethiopia

Zewdu Terefework
MRC-ET Molecular Diagnostics Laboratory, Addis Ababa, Ethiopia

Gary Wessel
Department of Molecular Biology, Cellular Biology and Biochemistry, Brown University, Providence, RI 02912, USA

Folasayo Fayose
Agricultural and Bio-Resources Engineering, Federal University Oye-Ekiti, PMB 373, Oye-Ekiti 371010, Nigeria

Zhongjie Huan
Tshwane University of Technology, Pretoria, South Africa

Index

www.ingramcontent.com/pod-product-compliance
Lightning Source LLC
Chambersburg PA
CBHW082042190326
41458CB00010B/3432